QUANTUM NANOCHEMISTRY

(A Five-Volume Work)

Volume I:
Quantum Theory and Observability

QUANTUM NANOCHEMISTRY

(A Five-Volume Work)

Volume I:
Quantum Theory and Observability

Mihai V. Putz

Assoc. Prof. Dr. Dr.-Habil. Acad. Math. Chem.
West University of Timişoara,
Laboratory of Structural and Computational Physical-Chemistry
for Nanosciences and QSAR, Department of Biology-Chemistry,
Faculty of Chemistry, Biology, Geography,
Str. Pestalozzi, No. 16, RO-300115, Timişoara, ROMANIA
Tel: +40-256-592638; Fax: +40-256-592620

&

Principal Investigator of First Rank, PI1/CS1
Institute of Research-Development for Electrochemistry
and Condensed Matter (INCEMC) Timisoara,
Str. Aurel Paunescu Podeanu No. 144,
RO-300569 Timişoara, ROMANIA
Tel: +40-256-222-119; Fax: +40-256-201-382

E-mail: mv_putz@yahoo.com
URL: www.mvputz.iqstorm.ro

AAP APPLE ACADEMIC PRESS

Apple Academic Press Inc.
3333 Mistwell Crescent
Oakville, ON L6L 0A2 Canada

Apple Academic Press Inc.
9 Spinnaker Way
Waretown, NJ 08758 USA

Exclusive worldwide distribution by CRC Press, a member of Taylor & Francis Group

Printed in the United States of America on acid-free paper
International Standard Book Number-13: 978-1-77188-133-3 (Hardcover)
International Standard Book Number-13: 978-1-4987-2953-6 (eBook)

Library and Archives Canada Cataloguing in Publication

Putz, Mihai V., author
Quantum nanochemistry / Mihai V. Putz (Assoc. Prof. Dr. Dr. Habil. Acad. Math. Chem.) West University of Timişoara, Laboratory of Structural and Computational Physical-Chemistry for Nanosciences and QSAR, Department of Biology-Chemistry, Faculty of Chemistry, Biology, Geography, Str. Pestalozzi, No. 16, RO-300115, Timişoara, ROMANIA, Tel: +40-256-592638; Fax: +40-256-592620, & Institute of Research-Development for Electrochemistry and Condensed Matter (INCEMC) Timişoara, Str. Aurel Paunescu Podeanu No. 144, RO-300569 Timişoara, ROMANIA Tel: +40-256-222-119; Fax: +40-256-201-382, E-mail: mv_putz@yahoo.com, URL: www.mvputz.iqstorm.ro.

Includes bibliographical references and index.
Contents: Volume I: Quantum theory and observability -- Volume II: Quantum atoms and periodicity -- Volume III: Quantum molecules and reactivity -- Volume IV: Quantum solids and orderability -- Volume V: Quantum structure–activity relationships (Qu-SAR).
Issued in print and electronic formats.
ISBN 978-1-77188-133-3 (volume 1 : hardcover).--ISBN 978-1-77188-134-0 (volume 2: hardcover).--ISBN 978-1-77188-135-7 (volume 3 : hardcover).-- ISBN 978-1-77188-136-4 (volume 4 : hardcover).--ISBN 978-1-77188-137-1 (volume 5 : hardcover).--ISBN 978-1-4987-2953-6 (volume 1 : pdf).--ISBN 978-1-4987-2954-3 (volume 2 : pdf).--ISBN 978-1-4987-2955-0 (volume 3 : pdf).--ISBN 978-1-4987-2956-7 (volume 4 : pdf).--ISBN 978-1-4987-2957-4 (volume 5 : pdf) 1. Quantum chemistry. 2. Nanochemistry. I. Title.
QD462.P88 2016 541'.28 C2015-908030-4 C2015-908031-2

Library of Congress Cataloging-in-Publication Data

Names: Putz, Mihai V., author.
Title: Quantum nanochemistry / Mihai V. Putz.
Description: Oakville, ON, Canada ; Waretown, NJ, USA : Apple Academic Press, [2015-2016] | "2015 | Includes bibliographical references and indexes.
Identifiers: LCCN 2015047099| ISBN 9781771881388 (set) | ISBN 1771881380 (set) | ISBN 9781498729536 (set ; eBook) | ISBN 1498729533 (set ; eBook) | ISBN 9781771881333 (v. 1 ; hardcover) | ISBN 177188133X (v. 1 ; hardcover) | ISBN 9781498729536 (v. 1 ; eBook) | ISBN 1498729533 (v. 1 ; eBook) | ISBN 9781771881340 (v. 2 ; hardcover) | ISBN 1771881348 (v. 2 ; hardcover) | ISBN 9781498729543 (v. 2 ; eBook) | ISBN 1498729541 (v. 2 ; eBook) | ISBN 9781771881357 (v. 3 ; hardcover) | ISBN 1771881356 (v. 3 ; hardcover) | ISBN 9781498729550 (v. 3 ; eBook) | ISBN 149872955X (v. 3 ; eBook) | ISBN 9781771881364 (v. 4 ; hardcover) | ISBN 1771881364 (v. 4 ; hardcover) | ISBN 9781498729567 (v. 4 ; eBook) | ISBN 1498729568 (v. 4 ; eBook) | ISBN 9781771881371 (v. 5 ; hardcover) | ISBN 1771881372 (v. 5 ; hardcover) | ISBN 9781498729574 (v. 5 ; eBook) | ISBN 1498729576 (v. 5 ; eBook) Subjects: LCSH: Quantum chemistry. | Chemistry, Physical and theoretical. | Nanochemistry. | Quantum theory. | QSAR (Biochemistry)
Classification: LCC QD462 .P89 2016 | DDC 541/.28--dc23
LC record available at http://lccn.loc.gov/2015047099

[With the quantum theory of matter…] the underlying physical laws necessary for the Mathematical theory of a large part of Physics and the Whole of Chemistry are thus completely known!
(Dirac, 1929)

To XXI Scholars

CONTENTS

LIST OF ABBREVIATIONS

AM1	Austin Model 1 method
BE	Bose-Einstein distribution
CAS	complete active space
CFD	compact finite difference
CI	configuration interaction approach
COBE	Cosmic Background Explorer Satellite
CP	classical principle
CTAB	cetyltrimethylammonium bromide
DFT	density functional theory
DTAB	n-dodecyl trimethyl ammonium bromide
ELF	electronic localization function
FD	Fermi-Dirac type
FWHM	full width of a half maximum
GCP	gradient corrected Perdew
GS	ground state
GTO	Gaussian type orbitals
HF	Hartree-Fock
HI	iodine-hydrogen system
HK1	first Hohenberg-Kohn theorem
HK2	second Hohenberg-Kohn theorem
HOMO	highest occupied molecular orbital
HUR	Heisenberg uncertainty relationship
INDO	intermediate neglect of differential overlap method
IQ	information-quantum method
IQ	inverse quantum index
IR	infrared spectroscopy
IRS	inertial reference system
KET	key enabling technologies
KS	Kohn-Sham orbitals
LCAO	linear combination of the atomic orbital basis functions

LOA	linear operators' algebra
LOMO	lowest occupied molecular orbital
LUMO	lowest occupied molecular orbital
LYP	Lee, Yang, and Parr
MINDO	modified intermediate neglect of differential overlap method
MNBS	micro-nano-bio-systems
MNDO	modified neglect of diatomic overlap method
MO	molecular orbital
NDDO	neglecting the diatomic differential overlap method
NDO	neglecting the differential overlap method
OECD	Organization for Co-operation and Economical Developing
OS	open-shell
P/W	particle-to-wave ratio
PI	path Integral development
PZ	Perdew and Zunger
QED	quantum electrodynamics
QM	quantum mechanics
QS	quantum statistics
RHF	restricted Hartree-Fock
RJ	Rayleigh-Jeans formula
SCF	self consistent field
SE	semiempirical
SLR	spectral-like resolution
SMILES	simplified molecular-input line-entry system
SP	special postulates
SP	standard Pade scheme
STO	Slater type orbitals
TEOS	Tetraethyl orthosilicate
TO	transversal optical
TOE	theory of everything
UHF	unrestricted Hartree-Fock
UV	ultraviolet catastrophe
VWN	Vosko, Wilk and Nusair
WKB	Wentzel, Kramers and Brillouin framework

PREFACE TO FIVE-VOLUME SET

Dear Scholars (Student, Researcher, Colleague),

I am honored to introduce *Quantum Nanochemistry*, a handbook comprised of the following five volumes:

Volume I: Quantum Theory and Observability
Volume II: Quantum Atoms and Periodicity
Volume III: Quantum Molecules and Reactivity
Volume IV: Quantum Solids and Orderability
Volume V: Quantum Structure–Activity Relationships (Qu-SAR)

This treatise, a compilation of my lecture notes for graduates, postgraduates and doctoral students in physical and chemical sciences as well as my own post-doctoral research, will serve the scientific community seeking information in basic quantum chemistry environments: from the fundamental quantum theories to atoms, molecules, solids and cells (chemical–biological/ligand–substrate/ligand–receptor interactions); and will also creatively explain the quantum level concepts such as observability, periodicity, reactivity, orderability, and activity explicitly.

The book adopts a three-way approach to explain the main principles governing the electronic world:

- firstly, *the introductory principles* of quantumchemistry are stated;
- then, they are analyzed as *primary concepts* employed to understand the microscopic nature of objects;
- finally, they are explained through *basic analytical equations* controlling the observed or measured electronic object.

It explains the first principles of quantum chemistry, which includes quantum mechanics, quantum atom and periodicity, quantum molecule and reactivity, through two levels:

- *fundamental* (or *universal*) character of matter in isolated and interacting states; and
- the primary concepts elaborated for a beginner as well as an advanced researcher in quantum chemistry.

Each volume tells the "story of quantum chemical structures" from different viewpoints offering new insight to some current quantum paradoxes.

- The **first volume** covers the concepts of nuclear, atomic, molecular and solids on the basis of quantum principles—from Planck, Bohr, Einstein, Schrödinger, Hartree–Fock, up to Feynman Path Integral approaches;
- The **second volume** details an atom's quantum structure, its diverse analytical predictions through reviews and an in-depth analysis of atomic periodicities, atomic radii, ionization potential, electron affinity, electronegativity and chemical hardness. Additionally, it also discusses the assessment of electrophilicity and chemical action as the prime global reactivity indices while judging chemical reactivity through associated principles;
- The **third volume** highlights chemical reactivity through molecular structure, chemical bonding (introducing bondons as the quantum bosonic particles of the chemical field), localization from Hückel to Density Functional expositions, especially how chemical principles of electronegativity and chemical hardness decide the global chemical reactivity and interaction;
- The **fourth volume** addresses the electronic order problems in the solid state viewed as a huge molecule in special quantum states; and
- The **fifth volume** reveals the quantum implication to bio-organic and bio-inorganic systems, enzyme kinetics and to pharmacophore binding sites of chemical–biological interaction of molecules through cell membranes in targeting specific bindings modeled by celebrated QSARs (Quantitative Structure–Activity Relationships) renamed here as Qu–SAR (Quantum Structure–Activity Relationships).

Thus, the five-volume set attempts, for the first time ever, to unify the introductory principles, the primary concepts and the basic analytical equations against a background of quantum chemical bonds and interactions (short, medium and long), structures of matter and their properties: periodicity of atoms, reactivity of molecules, orderability of solids, and activity of cells (through an advanced multi-layered quantum structure–activity unifying concepts and algorithms), and observability measured throughout all the introduced and computed quantities (Figure 0.0).

It provides a fresh perspective to the "quantum story" of electronic matter, collecting and collating both research and theoretical exposition the "gold" knowledge of the quantum chemistry principles.

The book serves as an excellent reference to undergraduate, graduate (Masters and PhDs) and post-doctoral students of physical and chemical sciences; for it not only provides basics and essentials of applied quantum theory, but also leads to unexplored areas of quantum science for future research and development. Yet another novelty of the book set is the intelligent unification of the quantum principles of atoms, molecules, solids and cells through the qualitative–quantitative principles underlying the observed quantum phenomena. This is achieved through unitary analytical

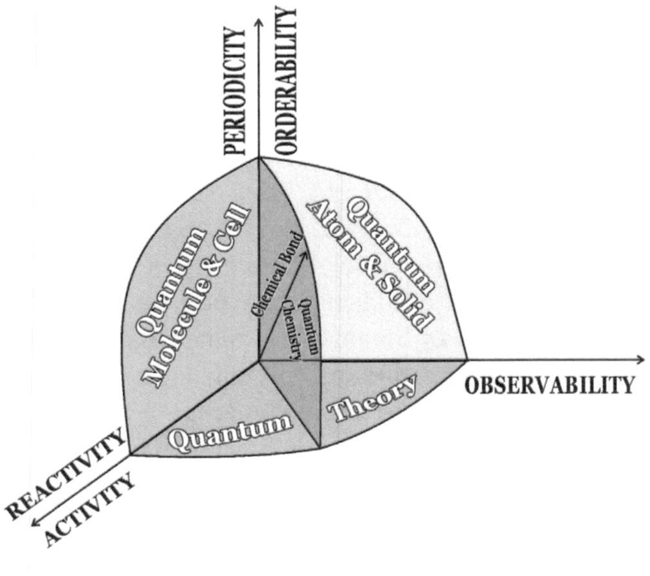

FIGURE 0.0 The featured concepts of the "First Principles of Quantum Chemistry" five-volume handbook as placed in the paradigmatic chemical orthogonal space of atoms and molecules.

exposition of the quantum principles ranging from quanta's nature (either as ondulatory and corpuscular manifestations) to wave function, path integral and electron density tools.

The modern quantum theories are reviewed mindful of their implications to quantum chemistry. Atomic, molecular, solid-state structures along cell/biological activity are analytically characterized. Major quantum aspects of the atomic, molecular, solid and cellular structure, properties/activity features, conceptual and quantitative correlations are unitarily reviewed at basic and advanced physical-chemistry levels of comprehension.

Unlike other available textbooks that are written as monographs displaying the chapters as themes of interests, this book narrates the "story of quantum chemistry" as *an extended review paper*, where theoretical and instructional concepts are appropriately combined with the relevant schemes of quantization of electronic structures, through path integrals, Bohmian, or chemical reactivity indices. The writing style is direct, concise and appealing; wherever appropriate physical, chemical and even philosophical insights are provided to explain quantum chemistry at large.

The author uses his rich university teaching experience of 15 years in physical chemistry at West University of Timisoara, Romania, along with his research expertise in treating chemical bond and bonding through conceptual and analytical quantum mechanical methods to explain the concepts. He has been a regular contributor to many physical-chemical international journals (*Phys Rev, J Phys Chem, Theor Acc Chem, Int J Quantum Chem, J Comp Chem, J Theor Comp Chem, Int J Mol Sci, Molecules, Struct Bond, Struct Chem, J Math Chem, MATCH*, etc.).

In a nutshell, the book amalgamates an analysis of the earlier works of great professors such as Sommerfeld, Slater, Landau and Feynman in a methodological, informative and epistemological way with practical and computational applications. The volumes are layered such that each can be used either individually or in combination with the other volumes. For instance, each volume reviews quantum chemistry from its level: as quantum formalisms in Volume I, as atomic structure and properties in Volume II, as detailed molecular bonding in Volume III, as crystal/solid state (electronic) in Volume IV, and as pharmacophore activity targeting specific bindings in Volume V.

To the best of my knowledge, such a collection does not exist currently in curricula and may not appear soon as many authors prefer to publish well-specialized monographs in their particular field of expertise. This multiple volumes' work, thus, assists academic and research community as a complete basic reference of conceptual and illustrative value.

I wish to acknowledge, with sincerity, the quantum flaws that myself and many researchers and professors make due to stressed delivery of papers using computational programs and software to report and interpret results based on inter-correlation. I feel, therefore, the need of a new comprehensive quantum chemistry reference approach and the present five-volume set fills the gap:

- *Undergraduate students* may use this work as an *introductory and training textbook* in the quantum structure of matter, for basic course(s) in physics and chemistry at college and university;
- *Graduate (Master and Doctoral) students* may use this work as the *recipe book* for analytical research on quantum assessments of electronic properties of matter in the view of chemical reactivity characterization and prediction;
- *University professors and tutors* may use this work as a *reference textbook* to plan their lectures and seminars in quantum chemistry at undergraduate or graduate level;
- *Research (Academic and Institutes) media* may use this work as a *reference monograph* for their results as it contains many tables and original results, published for the first time, on the atomic-molecular quantum energies, atomic radii and reactivity indices (e.g., electronegativity, chemical hardness, ionization and electron affinity results). It also has a collection of original, special and generally recommended literature, integrated results about quantum structure and properties.
- *Industry media* may use this work as a *working tool book* while assessing envisaged theoretical chemical structures or reactions (atoms-in-molecule, atoms-in-nanosystems), including molecular modeling for pharmaceutical purposes, following the presented examples, or simulating the physical–chemical properties before live production;

- *General media* may use this work as an *information book* to get acquainted with the main and actual quantum paradigms of matter's electronic structures and in understanding and predicting the chemical combinations (involving electrons, atoms and molecules) of Nature, because of its educative presentation.

I hope the academia shares the same enthusiasm for my work as the author while writing it and the professionalism and exquisite cooperation of the Apple Academic Press in publishing it.

Yours Sincerely,

Mihai V. Putz,
Assoc. Prof. Dr. Dr.-Habil. Acad. Math. Chem.
West University of Timişoara
& R&D National Institute for Electrochemistry and Condensed Matter Timişoara
(Romania)

ABOUT THE AUTHOR

Mihai V. PUTZ is a laureate in physics (1997), with an MS degree in spectroscopy (1999), and PhD degree in chemistry (2002), with many postdoctorate stages: in chemistry (2002-2003) and in physics (2004, 2010, 2011) at the University of Calabria, Italy, and Free University of Berlin, Germany, respectively. He is currently Associate Professor of theoretical and computational physical chemistry at West University of Timisoara, Romania. He has made valuable contributions in computational, quantum, and physical chemistry through seminal works that appeared in many international journals. He is Editor-in-Chief of the *International Journal of Chemical Modeling* (at NOVA Science Inc.) and the *New Frontiers in Chemistry* (at West University of Timisoara). He is member of many professional societies and has received several national and international awards from the Romanian National Authority of Scientific Research (2008), the German Academic Exchange Service DAAD (2000, 2004, 2011), and the Center of International Cooperation of Free University Berlin (2010). He is the leader of the Laboratory of Computational and Structural Physical Chemistry for Nanosciences and QSAR at Biology-Chemistry Department of West University of Timisoara, Romania, where he conducts research in the fundamental and applicative fields of quantum physical-chemistry and QSAR. In 2010 Mihai V. Putz was declared through a national competition the Best Researcher of Romania, while in 2013 he was recognized among the first Dr.-Habil. in Chemistry in Romania. In 2013 he was appointed Scientific Director of the newly founded Laboratory of Structural and Computational Physical Chemistry for Nanosciences and QSAR in his alma mater of West University of Timisoara, while from 2014, he was recognized by the Romanian Ministry of Research as Principal Investigator of first rank/degree (PI1/CS1) at National Institute for Electrochemistry and Condensed Matter (INCEMC) Timisoara. He is also a full member of International Academy of Mathematical Chemistry.

FOREWORD TO *VOLUME I: QUANTUM THEORY AND OBSERVABILITY*

"The world is not as real as we think.
My personal opinion is that the world is even
weirder than what quantum physics tells us."
– *Anton Zeilinger, The New York Times, December 27, 2005*

During my student days (pre-university, university, PhD), we learned quantum mechanics from the books authored by L. D. Landau and E. M. Lifshitz, A. S. Davydov, D. Bohm, Feynman's course of Lectures on Physics, and from P. A. M. Dirac's "Principles". We were excited with the theories of hidden variables, EPR paradox, decoherence, entanglement, and concerned for a life of 'immortal' Schrödinger's cat – they were in the air at that time! Did I understand it? Yes! – because, due to a conventional wisdom, I used it more than 24 hours a day and every day. I however doubt – doubt together with Feynman who once remarked that "Nobody understands it!" – that I've actually understood it. I touched and used it throughout the molecular world, which is nowadays inhabited by 21 million molecules, and which I studied as a quantum chemist – in fact, by education, I am a theoretical physicist.

Once, after winning the 1908 Nobel Prize in chemistry, Ernest Rutherford, considering himself as a physicist, said, "All science is either physics or stamp collection." (Actually, Feynman was not at all a stamp collector!). True, the name *physics*, $\phi\upsilon\sigma\iota\zeta$ in Greek, means "*nature.*" This is actually a paradox with (of) quantum mechanics 'that everybody uses quantum mechanics and nobody knows how it can be like that', as Yakir Aharonov and Daniel Rohrlich pointed out in their recent book "Quantum

Paradoxes. Quantum Theory for the Perplexed" (Weinheim: Wiley-VCH, 2005. ISBN 3-527-40391-4/pbk): "Our relationship with quantum mechanics recalls a Woody Allen joke: This guy goes to a psychiatrist and says, "Doc, my brother's crazy – he thinks he's a chicken! And, uh, the doctor says, "Well, why don't you turn him in?" And the guy says, "I would, but I need the eggs!" This dialog (not the last one in the book …) precisely mirrors our thoughts: "Quantum mechanics is crazy – but we need the eggs!" I would even add here 'weird', following the above Zeilinger's motto.

In my opinion, this paradox is rooted to the manner of how we were taught quantum mechanics[1]: this manner principally reflects the chronological way it entered the mankind consciousness at the end of the nineteenth century to the beginning of twentieth century: the black-body radiation, the models of atom by Rutherford and Bohr, the slit experiments on a wave–particle duality, and so on. Also taken this road, Mihai V. Putz – everyone agrees with me that writing a book on quantum mechanics is a very hard job – nevertheless tries to offer the presentation of quantum mechanics within the context of nano-dimensional world and "a fresh perspective of the 'quantum story' of electronic matter, collecting in both research and didactical exposition the 'gold' knowledge of the quantum chemistry principles."

True, we all are living in the remarkable time of the *nano-revolution* in science and technology that brings together researchers from many areas: physics, chemistry, material science, electronics, biophysics, biology, and medicine to create and use structures, devices and systems of the extreme tininess, that, of the size of about 0.1–100 nm (nanometer is a billionth of a meter), that is far smaller compared to the world of our everyday objects

[1]In fact, different people differently perceive newness, knowledge. Quite often, a better perception of what is "behind the given" is attained through a practical, concrete, direct "touch," rather than formally. Quantum mechanics is not an exception, rather a rule, mathematics too. I recall the magnificent book "Finite-Dimensional Linear Analysis in Problems" (in Russian) that was published in 1969 and was written by I. M. Glazman and Yu. I. Lyubich (I. M. Glazman, Yu. I. Lyubich, Finite-Dimensional Linear Analysis in Problems; Nauka: Moscow, 1969; The English translation: I. M. Glazman, Yu. I. Lyubich, Finite-Dimensional Linear Analysis: A Systematic Presentation in Problem Form, MIT Press: Cambridge, MA, 1974). This book was definitely a new word in teaching of mathematics, precisely, the functional analysis – it actually preached the form of teaching of the analysis via resolving logically interconnected sequence of 2405 propositions and problems. As Glazman and Lyubich wrote in Preface, their form of teaching rooted to the principles of Violinschule of Louis Spohr, the German composer, violinist, conductor, and teacher of music, who was also known as the inventor of a baton. The analog of such teaching approach to quantum mechanics by S. Flügge (Flügge, S. Practical Quantum Mechanics; Springer: Berlin, 1987) is widely recognized.

which are described by Newton's laws of motion and, on the contrary, bigger than an atom or simple molecules like water for instance – the particles which obey the laws of quantum mechanics. This is the world of nano-sized molecular assemblies, nanostructured materials – the *nano-world* – that has recently become one of the largest areas of chemistry, physics, material science, biology, and medicine with myriad applications in catalysis, biophysics, and the health sciences.

It is true that, on the other hand, many biological molecules, such as the DNA, RNA, proteins, viruses, and biomembranes, are also of a nanometric size – for example, the radius of the DNA double helix is ca. 1 nm, many viruses have dimension of ~10 nm, – and have recently been well recognized for its capability to naturally integrate with nanoparticles in nano-constructions. In the other words, we thus entered the area where, bearing in mind the meaning of '*physics*', we all face the paradigm of whether the *nature* of nature is conceivable or unconceivable?

This is the paradigm, which I do believe is the paradigm for the present generation, the generation, which the present book targets on. It is hard, however, to think that it will be unique on the shelf of the growing series of textbooks on quantum theory and nanotechnology: among them stands the recently appeared "Quantum Mechanics with Applications to Nanotechnology and Information Science" by Yehuda B. Band and Yshai Avishai (Academic Press, 2013. ISBN 978-0-444-53786-7, see also the Book Review by Lee Bassett, Physics Today 50, July 2014). Nevertheless, on the whole, it definitely makes a useful addition to this shelf, and, together with the above book, I indeed recommend the book by Mihai V. Putz to 'quantum engineers', to a broad circle of students and PhD students in chemistry, quantum chemistry, atomic and molecular physics who are expected to perceive the concept of nano-dimension practically, interactively with quanta, and thus to directly apply it in their own research in nanoworld.

Eugene S. Kryachko
Bogolyubov Institute for Theoretical Physics
National Academy of Sciences of Ukraine
Kiev, Ukraine
November 2015

[illegible]

[illegible]

[illegible]

[illegible]

[illegible]

[illegible]

PREFACE TO *VOLUME I: QUANTUM THEORY AND OBSERVABILITY*

THE SCIENTIFIC PREMISES

Life, at any level, involves creation and transformation.

Creation, much discussed and claimed over time by popular beliefs, mystical, theological, philosophical and artistic approaches, still remains a mystery to Science. This is because one cannot make yet a definitive series of cause–effect relationships that reveal both the multitude and multiplicity of the various manifestations of existence into the Unity of Life.

Steps, although important, were made but again Physics – or the *natural philosophy* – has come to formulate a set of principles, connecting one/singularity to the multiple/property concepts, the material point to the statistical ensemble, the electron to the photon, the wave to the corpuscle, the atom particle in the molecule and condensed state, etc. As noted by Romanian Lucian Blaga "*the revealed mysteries open deeper arcanum*"; however, the mystery of primary creation such as causal legitimacy or the "the primal engine" as Aristotle called it, still remains inaccessible despite the necessity for its natural existence. Moreover, Physics itself, as the base of Natural Sciences, established principles of inaccessibility with universal characters such as Heisenberg's uncertainty principle, the inaccessibility of absolute zero temperature, the legacy of adiabatic transformations, and the irreversible growth of entropy of open systems. All of these, among others, contributed to restricting our knowledge beyond any experimental observations or abstract thinking.[1]

And yet, there remains *transformation* by which Physics and Chemistry were formulated and validated through experimentation, modeling and

[1]Putz, M. V. (2006). *The Structure of Quantum Nanosystems* (in Romanian), West University of Timişoara Publishing House, Timişoara.

reasoned implications, a series of principles and theories that allow comprehensive understanding of "mutation" for the manifestation forms of matter, from micro- to macro-universe. Consequently, the matter shaped by natural phenomena involves many developmental sequences and transformation to the otherwise unitary form of life, in a broad sense. Hence, in absence of an immutable definition of creation, life is characterized by a series of changes resulting from natural phenomena (mechanical, thermodynamic, electromagnetic, optical, chemical, biological or in conjunction synergy) which is transposed to the level of principle through transformations and interactions of the basic constituents of matter, the electrons and photons as the basic fermions and bosons, respectively. At this point, one should note that the electrons and particles with electric charge, in general, are responsible for the relative stability of matter, while the photons and matter waves describe and mediate interactions, promoting and catalyzing modification of a substance.[2]

Of all forms of matter, a quantum state represents the complex state of manifestation placed between micro- and macro-cosmos. Therefore, the systematization of the legitimacy governing the structure of matter, from atom to solid crystalline form, based on electronic quantum (orbital/wave/fields) hierarchy stay as the fundamental aim of this volume. How can one model, represent and systematically evaluate a natural transformation? Through symbols and relations between them or more generally by equations; at this point explaining through equations becomes vital. This can be further explained through two steps. The first one relates to the operation of "putting in brackets" to the phenomenon or the system studied—a system supported ideologically through the philosophical *phenomenology* and scientifically through the Cartesian operation of *analysis*. Thus, sequencing in knowledge "freezes" the phenomenon or system being investigated, thus either restricting its interaction with the neighborhood or completely isolating it from the environment. This situation may be defined/described as the *quantum* (proper/eigen) state for which one may formulate the equation associated and then seek solutions. Moving further, relaxation toward observation is equivalent to perturbations encountered during initial analysis, rewriting the phenomenon's equation and seeking

[2]Putz, M. V. (2006). *The Structure of Quantum Nanosystems* (in Romanian), West University of Timişoara Publishing House, Timişoara.

approximate solutions. Thus, the separation, partition, factorization, overlapping, superposition, coupling constitute the fundamental steps and processes for a qualitative–quantitative analysis of a quantum phenomenon or natural quantum state.[3]

Consequently, following the "Greek dream" of gaining knowledge about the basic principle of life or at least the Life's transformation, employment of numbers, symbols and equations become the main tools in the quest and continuous development of scientific knowledge.

VOLUME LAYOUT

The present volume is the *first* in the five-volume set *Quantum Nanochemistry*:

Volume I: Quantum Theory and Observability
Volume II: Quantum Atoms and Periodicity
Volume III: Quantum Molecules and Reactivity
Volume IV: Quantum Solids and Orderability
Volume V: Quantum Structure–Activity Relationships (Qu-SAR)

The volume is a challenge in itself owing to its presentation as an analytical discourse, the only method apt for quantum description of natural phenomena, keeping in mind the subtle treatment of deeper concepts while restraining from lengthy presentation.

All chapters and topics of the book have been previously explained and discussed in various national and international academic presentations, lectures, seminars, scientific sessions and student reunions. However, the contents are far from being just a compilation of themes from various bibliographical sources; it includes visions of teaching and research, rigorously tested and applied. The materials are thus organized so as to cover various quantum paradigms for basic phenomenological stages of organization of matters: from quantification of matter waves including the equivalence of the electromagnetic with photonic radiation, to atomic, molecular, and solid state resumed quantum descriptions.

[3]Putz, M. V. (2006), *The Structure of Quantum Nanosystems* (in Romanian), West University of Timişoara Publishing House, Timişoara.

The book's simple and illustrative presentation of concepts and analyses include both basic physical–chemical quantum principles and observability at each level of matter's organization as well as advanced (more abstract, thus most necessary) formalisms of density matrix, path integrals, Hartree–Fock (as self-consistent quantum methods), original Heisenberg uncertainty by a sub-quantum extension (with more quantum fluctuation insight for the free evolution modeling), eventually leading to a novel undulatory/corpuscular characterization of the Si-based nano/mesosystems.

The contents are, therefore, suitable for students, researchers and teachers either interested as a layman or needing deeper information on quantum description from the angle of physics and chemistry curricula, or a combination of both. To that end, the information in the book is laid out in an equilibrated manner stimulating the creativity of readers beyond just knowing–understanding to predicting the quantum information coined on the nano-scale systems.

Volume I comprises the following chapters:

***Chapter 1* (*Phenomenological Quantification of Matter*)**: Nature, according to Greek belief, is unification between the sea waves and earth dust, whereas the Matter has two sides or manifestations – undulatory and corpuscular. Remarkably, such philosophical view is strongly sustained by the quantum theory of matter as well, as explained through waves and substance quantifications and experimentally through the intrinsic mystery of reciprocal connection at the microscopic level. This chapter reviews the most fundamental Planck (– Einstein) and de Broglie (– Einstein) quantifications of matter while laying the foundation for the second and first quantification of fields and particles, respectively. Thus, the two opposite views of Newton and Huygens, towards the end of seventeenth century, on the corpuscular and undulatory natures of matter seem to be unified at the micro- and macro-scales of the Universe.

***Chapter 2* (*Formalization of Quantum Mechanics*)**: Different approaches to quantum phenomena provide new insights and perspectives on the studied systems either in isolated or interaction states. The chapter reviews the main formalisms and illustrates ground-*nous* of quantum mechanics, i.e., the undulatory quantum mechanics, the quantum propagator (Green function) concept, the semi-classical approach based on Euler–Lagrange and Hamiltonian, and details the *bra-ket* Dirac picture thus opening new perspectives of exploring the quantum theory taking an abstract yet logical and formalized reasoning.

Chapter 3 (Postulates of Quantum Mechanics: Basic Applications): The phenomenological quantum mechanics is practically "re-told" in a more formalized way through the so-called extended quantum postulates. Additionally, meticulous complex analyses of the Nature's phenomena from nuclear to atomic, to molecular, to solid state and scattering observational quantum effects are also provided.

Chapter 4 (Quantum Mechanics for Quantum Chemistry): Based on the first principles of quantum mechanics as explained in the previous chapters and sections, this chapter highlights quantum theory's evolution within the matter systems and with constraints (boundaries). Firstly, the Feynman path integral formalism is discussed and then applied to atomic, quantum barrier and quantum harmonic vibration followed by density-matrix approach. A detailed perspective of the Hartree–Fock and Density Functional pictures of many electronic systems as well as electronic occupancies via the Koopmans theorem finally leads to a further generalization of the Heisenberg observability and its first application to mesosystems.

Thus, this volume covers:

- essential concepts of quantum mechanics: quantification of waves, substance, consequences of matter quantification through quantum statistics, the standard model of fermions–bosons constituents of matter and the Universe' understanding of expansion and anisotropy;
- the Heisenberg principle essential for understanding the observability of quantum phenomena both in formalized and extended (by quantum path integrals) algorithms with focus on its application to the silica sol–gel mesosystems;
- a detailed discussion on how quantum mechanics roots on classical mechanics, including the semi-classical (WKB) approaches;
- the quantum mechanical postulates by wave-function continuity with atomic, molecular and solid state illustrated through eigen-energies and function, semi-classical quantification and variational energy approaches;

- the causal quantum description including Schrodinger to Heisenberg interaction, introducing and applying the Green function to the scattering phenomena;
- the general perturbation algorithm with applications on Zeeman effect, nuclear isotropic corrections, paradigmatic harmonic oscillation, and quasi-free electrons in solids;
- the Feynman path formalism with general and basic information of hydrogen (atomic), oscillatory (molecular) and infinite well (solid state) eigen energies, adopting an integral rather than the differential approach of quantum evolution;
- explanation of the quantum chemical formalism through quantum mechanical algorithms and formalizations taking into account the density matrix, self-consistent Roothaan–Hartree–Fock method, semi-empirical computation schemes, as well as the modern density functional theorems and functionals;
- chemical reactivity indices and chemical hardness explained in the context of the Koopmans theorem and their application on a paradigmatic series of organic molecules, thereby justifying their reliability in assessing freezing orbitals to valence electrons in a molecule;
- extensive appendices containing useful mathematical series expansions, properties and their use to characterizes special functions (Poisson, Euler, etc.), Lagrange interpolation, special relativity concepts, and fundamental constants and equivalents' transformations.

Kind thanks are addressed to individuals, universities, institutions, and publishers that inspired and supported the topics included in the present volume; a few of them include:

- *Supporting individuals*: Prof. Hagen Kleinert (Free University of Berlin); Priv. Doz. Dr. Axel Pelster (Free University of Berlin); Prof. Nino Russo (University of Calabria); Dr. Ottorino Ori (Actinium Chemical Research, Parma); Prof. Eduardo A. Castro (University La Plata, Buenos Aires); Prof. Adrian Chiriac (West University of Timişoara); Dr. Ana-Maria Putz, n. Lacrămă (Chemistry Institute of Romanian Academy, Timişoara);
- *Supporting universities*: West University of Timişoara (Faculty of Chemistry, Biology, Geography/Biology-Chemistry Department/ Laboratory of Computational and Structural Physical Chemistry

for Nanosciences and QSAR); Free University of Berlin (Physics Department/Institute for Theoretical Physics/Research Center for Einstein's Physics, Centre for International Cooperation); University of Calabria (Faculty of Mathematics and Natural Sciences/Chemistry Department);

- *Supporting institutions and grants*: DAAD (German Service for Academic Exchanges) by Grants: 322 A/17690/2004, 322 A/05356/2011; CNCSIS (Romanian National Council for Scientific Research in Higher Education) by Grant: AT54/2006-2007; CNCS-UEFISCDI (Romanian National Council for Scientific Research) by Grant: TE16/2010-2013;
- *Supporting publishers*: Multidisciplinary Digital Publishing Institute – MDPI (Basel); Hindawi Publishers (Cairo Free Zone and New York).

I express special thanks to my family, especially to my lovely daughters *Katy and Ela*, for providing me with a work-and-play atmosphere and always filling me with the necessary energy and drive. Hopefully, the same energy is transmitted to the readers and students too in their quest for scientific knowledge and their approach toward quantum *phenomena* and their *observability* description.

Last but not the least, the author especially thanks the Apple Academic Press (AAP) team and, in particular, Ashish (Ash) Kumar, the AAP President and Publisher, and Sandra (Sandy) Jones Sickels, Vice President, Editorial and Marketing, for professional production of this five-volume set, *Quantum Nanochemistry.*

Quantum chemical theory is an ever-expanding field; thus constructive observations, corrections and suggestions are welcome and peer contribution is appreciated.

Keep close and think high!

Yours Sincerely,

Mihai V. Putz,
Assoc. Prof. Dr. Dr.-Habil. Acad. Math. Chem.
West University of Timişoara
& R&D National Institute for Electrochemistry and Condensed Matter Timişoara
(Romania)

CHAPTER 1

PHENOMENOLOGICAL QUANTIFICATION OF MATTER

CONTENTS

ABSTRACT

Following the "Greek dream" of Nature unification, between the sea waves and earth dust, the Matter is regarded as having two main orders or manifestations, i.e., the undulatory and corpuscular sides. Remarkably, such philosophical view is strongly sustained by the quantum theory of matter, by its prescription through waves and substance quantifications, with experimental counterpart, however, maintaining the intrinsic mystery of reciprocal connection at the microscopic level. This chapter reviews the most fundamental Planck (– Einstein) and de Broglie (– Einstein) quantifications of matter while founding the basis for the second and first quantification of fields and particles, respectively. This way the two opposite visions of Newton and Huygens as advancing on the end of XVII century the corpuscular and undulatory natures of matter seems now be unified either at micro- and macro-scales of Universe.

1.1 INTRODUCTION

The knowledge in general and of Nature's principles in special is grounded on modern understanding of phenomena as based on equations, encrypting by symbols the mutual relationships the various quantities and properties they represent and master. Accordingly, also when is about in elucidating the quantum principles of matter, one should basically be able to:

- assimilate the fundamental theoretical principles and use of mathematics involved;

- use the mathematical tools to understand atomic and molecular structure and properties, as well as chemical structure, properties, reactivity, etc.;
- interpret the experimental results (spectra) to identify the structures and of their properties;
- ability to further model the observed phenomena by assessing specific concepts to various quantities, and inter-correlate them towards seeking the eventual solution by means of the appropriate approximations.
- To this aim, one may usually follow the custom physical-chemistry freshman lectures, at least addressing the following items:
- From quantum mechanics, to quantum physics, to quantum chemistry, to quantum biochemistry. Quantum theory founding fathers and disciples, for example, Einstein, Planck, Bohr, Lewis, de Broglie, Schrödinger, Heisenberg, Fermi, Bose, Feynman, Kohn, Kleinert, etc., struggling to respond to the secular controversy about the answer to the question: "Can the quantum mechanics to afford a complete description for the physics reality?"
- Quantum mechanical principles. Fundamental constants of the universe: the speed of light, the Boltzmann constant, the Planck constant. The wave-particle duality. The link between the Microscopic World of Energetic of Atoms/Molecules and the Macroscopic World: de Broglie relationship, the Heisenberg relationships, and statistical distributions. The Bohr interpretation of the hydrogen atom. The postulates of quantum in the wave function.

These issues are in the following addressed in a didactical yet modern physical-chemistry manner by following the presentation sass quantification of waves-to-quantification of substance-to-consequences of matter's quantification.

1.2 QUANTIFICATION OF WAVES

1.2.1 BLACK-BODY RADIATION

In the end of XIX century one on the most intriguing experiments refers to ability of matter being converted from the thermal to electromagnetic (or spectral) energy through intermediated by "fully" absorbed-emitted object, hereafter called black-body, see Figure 1.1.

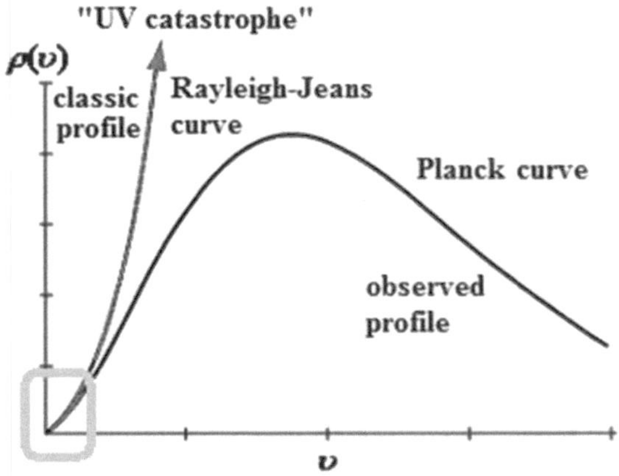

FIGURE 1.1 The dichotomy between classical curve (Rayleigh-Jeans) and the observed one (quantum, Planck) for a typical blackbody radiation (HyperPhysics, 2010; Putz et al., 2010).

In essence, was thought the question of explaining the spectrum of emitted radiation by a "black" body (heated) on a large frequency range, in order to avoid the so-called "ultraviolet catastrophe" (UV), specific of the restrictive approach by classical thermodynamics, see Figure 1.1 and below. Analytically, the problem consists in describing the spectrum obtained through the so-called spectral (energy) density:

$$\rho(\upsilon) = \frac{N(\upsilon)\langle E\rangle}{V} \tag{1.1}$$

calculated as the product between the number of the mode of vibrations and the average energy, reported at the volume of the related blackbody cavity.

As a calculation model, there will be considered a blackbody, like a cube of enclosure L side (later we will see that the result is independent of the geometric shape of the blackbody). Under these conditions, the number of vibration modes, will be calculated evaluating the variation of the number of vibrations by a certain frequency, when the frequency changes, thus

$$N(\upsilon) = \frac{dN_{\upsilon}}{d\upsilon} \tag{1.2}$$

To calculate the number of vibrations, at a given frequency, is considered the abstract construction from Figure 1.2, which represents the blackbody enclosure accommodation, in a eighth from a sphere of frequency, of a given mode of vibration.

Therefore, the number of vibrations for a given frequency, will be written as the ratio between the volume (of frequency) of overlapping geometric bodies, as in Figure 1.2

$$N_{\upsilon} = \frac{\frac{1}{8}\left(\frac{4\pi}{3}\upsilon^{3}\right)}{\frac{1}{2}\left(\frac{c}{2L}\right)^{3}} = \frac{8\pi L^{3}\upsilon^{3}}{3c^{3}} \tag{1.3}$$

where there was considered the radiation from the cavity propagating at the speed of light (as electromagnetic radiation freely propagated inside the cavity of blackbody), with the factor ½ from the denominator, considered in order to avoid the halving of a propagation of electromagnetic radiation, with reverse polarization (the plane of vibration rotated at 180°) for a complete path (2L) between any of the parallel walls of the cavity.

With this expression the number of the mode of vibrations are calculated (based on differential definition above) with the expression

$$N(\upsilon) = \frac{8\pi L^{3}\upsilon^{2}}{c^{3}} \tag{1.4}$$

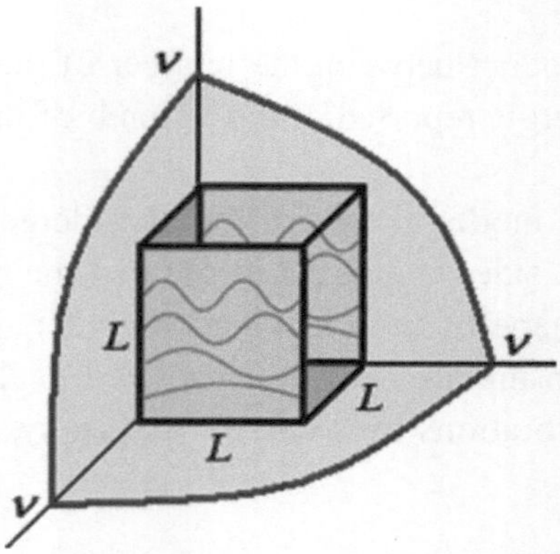

FIGURE 1.2 Insertion of blackbody (considered a cube of side L) in the abstract sphere of radius equal to the frequency v of a mode of vibration (HyperPhysics, 2010; Putz et al., 2010).

It is noted that the number of the mode of vibration in relation to their host cavity volume, does not depend anymore on the geometric structure of the cavity

$$\frac{N(\upsilon)}{V}=\frac{N(\upsilon)}{L^3}=\frac{8\pi\upsilon^2}{c^3} \tag{1.5}$$

certifying the universal value of this expression (with application possibilities to various micro- and macroscopic structures, including planets and stars).

For the average energy on the mode of vibrations, is appealed to the integral formula into possibilities, in order to cover the whole possible spectrum energy.

$$\langle E\rangle=\frac{\int_0^\infty E\wp(E)dE}{\int_0^\infty \wp(E)dE} \tag{1.6}$$

with the expression of probability, given in relation to the thermal energy (k_BT, k_B: Boltzmann's constant)

$$\wp(E)=\frac{1}{k_BT}\exp\left(-\frac{E}{k_BT}\right) \tag{1.7}$$

in order to satisfied the relationship of standardization (in certitude) of the probability (absolute) on the spectrum of all possible energetic events.

$$\int_0^\infty \wp(E)dE=1 \tag{1.8}$$

With these considerations, the average energy on the mode of vibration, is successively written

$$\begin{aligned}
\langle E\rangle_T &= \int_0^\infty E\wp(E)dE \\
&= -\int_0^\infty E\frac{\partial}{\partial E}\left[\exp\left(-\frac{E}{k_BT}\right)\right]dE \\
&= -\int_0^\infty \left\{\frac{\partial}{\partial E}\left[E\exp\left(-\frac{E}{k_bT}\right)\right]-\exp\left(-\frac{E}{k_bT}\right)\right\}dE \\
&= -\left[E\exp\left(-\frac{E}{k_bT}\right)\right]_0^\infty + k_BT = k_BT
\end{aligned} \tag{1.9}$$

again finding the value of termic energy, in agreement with the energy equi-partition theorem from the classical thermodynamic (statistical).

Alternatively, it can be directly calculated the average thermal energy, such as

$$\langle E \rangle_{\beta=1/k_BT} = \frac{\int_0^\infty E\exp(-E\beta)dE}{\int_0^\infty \exp(-E\beta)dE} = -\frac{d}{d\beta}\left[\log\int_0^\infty \exp(-E\beta)dE\right] = \frac{1}{\beta} = k_BT \tag{1.10}$$

Combining the two expressions of the number of mode of vibration with the average energy of vibration and reported to the volume of the enclosure of blackbody, results the expression of the density of spectral energy, as Rayleigh-Jeans formula (RJ)

$$\rho_T(\upsilon) = \frac{8\pi k_B T\upsilon^2}{c^3} \equiv \rho_{R-J}(\upsilon) \tag{1.11}$$

emphasizing through the parabolic dependence of radiation frequency, the crisis of classical physics, applied to radiative phenomena, due to the predicting of UV catastrophe, in contradiction to both phenomena observed in the laboratory and also in the nature as ensemble (according to this prediction, the life on Earth would perish due to the solar radiation!).

1.2.2 PLANCK'S APPROACH

To solve the paradox of blackbody was Planck's hypothesis (inspired) to considered also the radiation's energy as dependent on frequency, with a universal constant of proportionality (h-Planck's constant), as well as the universal Boltzmann constant appears into the thermal energy, In addition quantified in "bundles of energy," which acknowledge as energy quanta; in few words, Planck considered the quantification of electromagnetic radiation as:

$$E_{Planck} = E(\upsilon) = nh\upsilon \tag{1.12}$$

This time, the energy is quantified, or discretized, while its average on mode of vibration will be calculate by series instead of integrals, but keeping the form above, so the average energy becomes

$$\langle E \rangle_{\upsilon} = \frac{\sum_{n=0}^{\infty} E(\upsilon)\wp(E(\upsilon))}{\sum_{n=0}^{\infty} \wp(E(\upsilon))}$$
$$= \frac{\sum_{n=0}^{\infty} \frac{nh\upsilon}{k_B T} \exp\left(-\frac{nh\upsilon}{k_B T}\right)}{\sum_{n=0}^{\infty} \frac{1}{k_B T} \exp\left(-\frac{nh\upsilon}{k_B T}\right)} \tag{1.13}$$

With the replacement

$$x = \frac{h\upsilon}{k_B T} \tag{1.14}$$

the average energy of mode of vibration generically rewrites as the ratio of two mathematical series

$$\langle E \rangle_{\upsilon} = h\upsilon \frac{\sum_{n=0}^{\infty} nx \exp(-nx)}{\sum_{n=0}^{\infty} x \exp(-nx)} \tag{1.15}$$

The series in numerator is calculated with the result

$$\sum_{n=0}^{\infty} nx \exp(-nx) = xe^{-x}\left[1 + 2e^{-x} + 3\left(e^{-x}\right)^2 + \ldots\right] = \frac{xe^{-x}}{\left(1 - e^{-x}\right)^2} \tag{1.16}$$

based on consideration of the function series

$$f(y) = 1 + 2y + 3y^2 + \ldots \tag{1.17}$$

having the argument also as a function

$$y = \exp(-x) \tag{1.18}$$

and for which computation was made by appealing to the integral-differential trick

$$g(y) = \int_0^y f(y)dy \Rightarrow f(y) = \frac{d}{dy} g(y) = \frac{1}{(1-y)^2} \tag{1.19}$$

since recognizing that the integral results in a series under a shape much easily to identify in relation to the geometric series

$$g(y) = y\left(1 + y + y^2 + ...\right) = \frac{y}{1-y} \tag{1.20}$$

Instead, the series of the denominator in the average energy expression above is directly evaluated in relation to the geometric series

$$\sum_{n=0}^{\infty} x\exp(-nx) = x\left[1 + e^{-x} + \left(e^{-x}\right)^2 + ...\right] = \frac{x}{1-e^{-x}} \tag{1.21}$$

Combining the two series, one obtains the Planck expression for average energy, for each mode of vibration of the radiation emitted by a blackbody

$$\langle E \rangle_{\upsilon} = \frac{h\upsilon}{\exp\left(\frac{h\upsilon}{k_B T}\right) - 1} \tag{1.22}$$

Again, alternatively, this energy average can be directly calculated the dependence of thermal frequency regarding the average energy, thereby

$$\begin{aligned} \langle E \rangle_{\upsilon,\beta=1/k_B T} &= \frac{\sum_{n=0}^{\infty} nh\upsilon\exp(-nh\upsilon\beta)}{\sum_{n=0}^{\infty}\exp(-nh\upsilon\beta)} = -\frac{d}{d\beta}\left[\log\sum_{n=0}^{\infty}\exp(-nh\upsilon\beta)\right] \\ &= -\frac{d}{d\beta}\left[\log\frac{1}{1-\exp(-h\upsilon\beta)}\right] = \frac{h\upsilon}{\exp(h\upsilon\beta)-1} \end{aligned} \tag{1.23}$$

Finally, multiplying this average energy with the numbers of modes of vibration from the cavity, and reporting the results to the volume of the cavity, the Planck expression for spectral density is obtained

$$\rho_{Planck}(\upsilon) = \frac{8\pi h}{c^3}\frac{\upsilon^3}{\exp\left(\frac{h\upsilon}{k_B T}\right) - 1} \tag{1.24}$$

in perfect agreement with the entire spectral profile, observed in Figure 1.1.

The extreme limits are checked for confirmation, namely:

- At low frequencies, there can be used the approximation

$$\upsilon, \frac{h\upsilon}{k_B T} << 1 \Rightarrow \exp\left(\frac{h\upsilon}{k_B T}\right) \cong 1 + \frac{h\upsilon}{k_B T} \tag{1.25}$$

which results in finding the Rayleigh-Jeans spectral density

$$\rho_{Planck}(\upsilon) \xrightarrow{\upsilon<<1} \rho_{R-J}(\upsilon) \tag{1.26}$$

- At high frequencies, there can be used the approximation

$$\upsilon, \frac{h\upsilon}{k_B T} >> 1 \Rightarrow \exp\left(\frac{h\upsilon}{k_B T}\right) - 1 \cong \exp\left(\frac{h\upsilon}{k_B T}\right) \tag{1.27}$$

noticing that the "UV catastrophe" from the classic treatment at infinite frequency was this way avoided:

$$\rho_{Planck}(\upsilon) \xrightarrow{\upsilon>>1} 0 \tag{1.28}$$

Moreover, we conclude the presentation with the question: "what would the world look like without Planck's constant?" For this, we will perform the limit for $h \to 0$, in the right spectral density (Planck), obtaining the expression

$$\lim_{h\to 0} \rho_{Planck}(\upsilon) \overset{\frac{0}{0}}{=} \frac{8\pi\upsilon^2}{c^3} \lim_{h\to 0} \frac{\frac{d}{dh}(h\upsilon)}{\frac{d}{dh}\left[\exp\left(\frac{h\upsilon}{k_B T}\right) - 1\right]} = \rho_{R-J}(\upsilon) \tag{1.29}$$

again identical with Rayleigh-Jeans law at low frequencies; this result allows the ascertaining that the world (the Earth, the Universe) without Planck's constant would be catastrophic, i.e., inevitably submissively to ultraviolet catastrophe. From hence the necessity to consider the Planck constant, for the correct approaching (at any frequency) of the phenomena of Nature.

1.2.3 EINSTEIN'S APPROACH

The spontaneous and forced emissions are for the first time solved by Einstein (1917) throughout the introduced coefficients $A_{nn'}$ and $B_{nn'(\text{or } n'n)}$, respectively, relating the probability in unit time once electron going from a state (n) of energy E_n into another one (n') with energy $E_{n'}$ between which the unit photonic energy exists (is postulated):

$$h\upsilon = E_n - E_{n'} \tag{1.30}$$

As such Einstein accepts at once both the wave quantification as well as the Bohr postulate of transitions between stationary states; while his approach will results in the Planck spectral density ρ_υ the Bohr postulate of quantum transitions follows as being with this occasion demonstrated.

Therefore, while assuming N_n and $N_{n'}$ atoms in each of the n and n' states in discussion, with $n>n'$, within the Maxwell-Boltzmann microstates,

$$N_n = \beta \exp(-\beta E_n)$$

$$N_{n'} = \beta \exp(-\beta E_{n'}) \tag{1.31}$$

the spontaneous radiant, emitted and absorbed energies are written accordingly as:

$$W_{emitted-spt.} = N_n A_{nn'}(h\upsilon)$$

$$W_{emitted-forced} = N_n B_{nn'}(h\upsilon)\rho_\upsilon$$

$$W_{absorbed-forced} = N_{n'} B_{n'n}(h\upsilon)\rho_\upsilon \tag{1.32}$$

The energy balance in the closed system of absorbed-emitted-spontaneous radiative phenomena,

$$W_{emitted-spt.} + W_{emitted-forced} = W_{absorbed-forced} \tag{1.33}$$

leads in first instance with

$$N_n A_{nn'} + N_n B_{nn'}\rho_\upsilon = N_{n'} B_{n'n}\rho_\upsilon \tag{1.34}$$

from where one yields:

$$\rho_\upsilon = \frac{N_n A_{nn'}}{N_{n'} B_{n'n} - N_n B_{nn'}} = \frac{\dfrac{A_{nn'}}{B_{nn'}}}{\dfrac{N_{n'} B_{n'n}}{N_n B_{nn'}} - 1} = \frac{\dfrac{A_{nn'}}{B_{nn'}}}{\dfrac{B_{n'n}}{B_{nn'}} \exp\left[\beta\left(E_n - E_{n'}\right)\right] - 1} \tag{1.35}$$

that actually gives:

$$\rho_\upsilon = \frac{\dfrac{A_{nn'}}{B_{nn'}}}{\dfrac{B_{n'n}}{B_{nn'}} \exp[\beta h\upsilon] - 1} \tag{1.36}$$

Now we have to agree this expression with the Rayleigh-Jens one in the classical limit that is performed in two steps. One is to check out the high-temperature limit through first order expanding the denominator exponential in h-Planck constant (i.e., the semi-classical expansion):

$$\infty = \lim_{\beta \to 0} \rho_\upsilon = \frac{\lim\limits_{\beta \to 0} \dfrac{A_{nn'}}{B_{nn'}}}{\lim\limits_{\beta \to 0} \left[\dfrac{B_{n'n}}{B_{nn'}} (1 + \beta h\upsilon) - 1 \right]} \tag{1.37}$$

from where follows that the forced emission and absorption probabilities are intrinsically equal:

$$B_{nn'} = B_{n'n} \tag{1.38}$$

since the nominator is non-infinity expression.

In these conditions the spectral density is reloaded in the first order semi-classical h-expansion and equated with the classical Rayleigh-Jens expression,

$$\frac{8\pi\upsilon^2}{c^3 \beta} = \frac{A_{nn'}}{B_{nn'}} \frac{1}{\beta h\upsilon} \tag{1.39}$$

that further provides:

$$\frac{A_{nn'}}{B_{nn'}} = \frac{8\pi h\upsilon^3}{c^3} \tag{1.40}$$

Finally, the spectral density is obtained as in previously Planck approach.

1.3 QUANTIFICATION OF SUBSTANCE

1.3.1 THE DE BROGLIE FORMULA

Having available two equivalent forms of energy for electromagnetic radiation (respectively, in the relativistic formulation of Einstein and the quantum on of Planck) Louis de Broglie advanced the idea of their unification, corroborated with the generalization for any body in motion (with rest mass m_0)

$$m_0 c^2 = \hbar\omega \tag{1.41}$$

written by pulsation $\omega = \nu / 2\pi$, which involves the introduction of reduced Planck's constant

$$\hbar = \frac{h}{2\pi} \tag{1.42}$$

In fact, the energetic unit is assumed for the own system of a body in motion and it is checked its validity regarding coordinated system-observer (inertial) to which it is moving with a constant velocity v. Moreover, worth taking into consideration how the evolution of a body in motion is equivalent to the coverage (path) of space with an associated wave

$$\psi(x,t) = Ae^{i(k_x x - \omega t)} \tag{1.43}$$

equally representative for the body in the own reference system and also in the observation from an inertial equivalent system (relativistic), see Figure 1.3.

This equivalence would mean (from the wavy perspective of propagation) assuming the present identity (the amplitude does not contain undulatory information related to the movement itself, but rather of the conditions and environment - dispersion, attenuation, etc.).

$$k_x x - \omega t = k_x x' - \omega t' \tag{1.44}$$

depending on the pulsation and wave vector

$$k = \frac{2\pi}{\lambda} \tag{1.45}$$

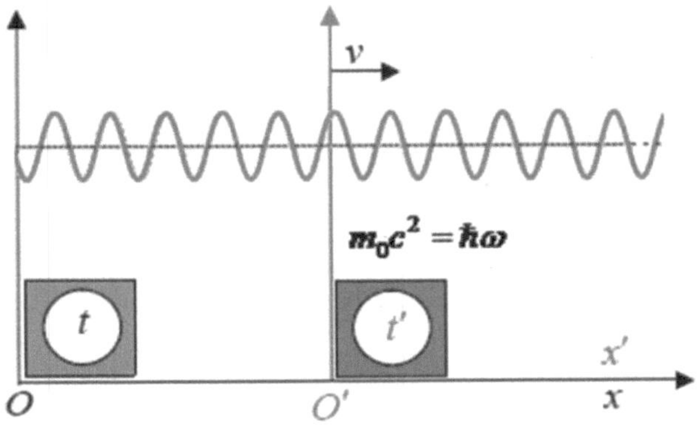

FIGURE 1.3 The relativistic construction for the deduction of Broglie relationship (HyperPhysics, 2010; Putz et al., 2010).

as constants of motion (undulatory). Rewriting the equal equation of phases by relativistic transformations (Lorentz-Einstein) of coordinates, taking into consideration the body (associated) in motion fixed on the reference system O'(x'=0), generates the spatial-temporal relationship (see Appendix A.5–A.7)

$$k_x x - \omega t = -\omega t' = -\frac{m_0 c^2}{\hbar}\frac{t - \frac{v}{c^2}x}{\sqrt{1 - v^2 / c^2}} \tag{1.46}$$

While, recognizing the mass movement (relativistic)

$$m = \frac{m_0}{\sqrt{1 - v^2 / c^2}} \tag{1.47}$$

the last equation generates successive identifications for the wave vector and the pulsation of the associated wavy movement

$$\begin{cases} k_x = \dfrac{1}{\hbar}\dfrac{m_0 v}{\sqrt{1 - v^2 / c^2}} = \dfrac{mv}{\hbar} = \dfrac{p_x}{\hbar} \\ \omega = \dfrac{1}{\hbar}\dfrac{m_0 c^2}{\sqrt{1 - v^2 / c^2}} = \dfrac{mc^2}{\hbar} = \dfrac{E}{\hbar} \end{cases} \tag{1.48}$$

equivalently with the quantification of the substance (de Broglie) and waves (Planck)

$$\begin{cases} \mathrm{p} = \hbar \mathrm{k} \text{ ... substance quantification} \\ E = \hbar\omega \text{ ... } \textit{waves quantification} \end{cases} \tag{1.49}$$

This provides the fundamental and conceptual complete quantification (and mutual) of the matter (field/wave + substance), at the phenomenological level.

1.3.2 THE WAVE PACKET

Once assumed (certified) the undulatory shape of the quantum particle, with simple yet general form

$$\psi(x,t) = Ae^{i(k_x x - \omega t)} = Ae^{i\alpha} \tag{1.50}$$

one may pass to a more in depth characterization of quantum information carried by it. Thus, by combining the quantification of matter (substance/de Broglie and field /wave/Planck) in the expression of the phase velocity

$$v_{phase} = \frac{dx}{dt} \overset{\alpha = ct.}{=} \frac{\omega}{k_x} = \frac{\hbar\omega}{\hbar k_x} = \frac{E}{p_x} = \frac{mc^2}{mv} = \frac{c^2}{v} > c \text{ (!?)} \tag{1.51}$$

there is obtained the paradox (apparently the impossibility) to register a propagation of the wave associated to the quantum particle, with a velocity that exceeds the speed of light!

Resolving this anomaly comes from the removal of the limitation of the undulatory representation by a single wave – considering the representation by a *wave packet* seen as a integral convolution

$$\psi(x,t) = \frac{1}{2\Delta k} \int_{-\Delta k}^{\Delta k} A(k) e^{i(kx - \omega t)} dk \tag{1.52}$$

averaged/normalized on the allowed range of variation of wave vectors included in the wave packet

$$\Delta k = k - k_0, \quad k_0 = 2\pi / \lambda_0 \tag{1.53}$$

where, for simplify the expression, it was noted as $k_x := k$.

In terms of pulsation (the other wavy size but also the quantification size), this is considered as varying slightly (in the first order) comparing to its equilibrium value inside the packet, while the amplitude in mutual space (of wave vector) is considered almost the same for all waves from the packet.

$$\begin{cases} \omega \cong \omega_0 + \left(\dfrac{d\omega}{dk}\right)_0 \zeta, \zeta = k - k_0 \\ A(k) \cong A(k_0) \end{cases} \tag{1.54}$$

Under these conditions the de Broglie wave packet is sequentially given as

$$\begin{aligned} \psi(x,t) &= \frac{A(k_0)e^{i(k_0x-\omega_0t)}}{2\Delta k} \int_{-\Delta k}^{+\Delta k} e^{i\left[x-\left(\frac{d\omega}{dk}\right)_0 t\right]\zeta} d\zeta \\ &= A(k_0) \frac{\sin\left\{\left[x-\left(\dfrac{d\omega}{dk}\right)_0 t\right]\Delta k\right\}}{\left[x-\left(\dfrac{d\omega}{dk}\right)_0 t\right]\Delta k} e^{i(k_0x-\omega_0t)} \\ &\equiv A(x,t)e^{i(k_0x-\omega_0t)} \end{aligned} \tag{1.55}$$

with the representation from Figure 1.4.

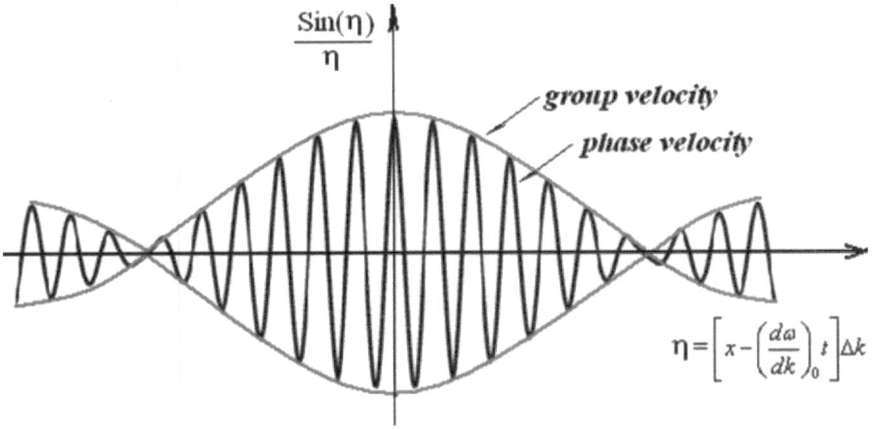

FIGURE 1.4 The de Broglie wave packet structure (HyperPhysics, 2010; Putz et al., 2010).

Since considering the limit

$$\lim_{q \to 0} \frac{\sin q}{q} = 1 \tag{1.56}$$

we have that the maximum of the above wave-function, $\max \psi(x,t)$, is approached under the fulfilling condition

$$x = \left(\frac{d\omega}{dk} \right)_0 t \tag{1.57}$$

leading with the so called *group velocity* of the wave packet:

$$v_{group} = \frac{dx}{dt} = \frac{d\omega}{dk} = \frac{dE}{dp} = \frac{d}{dp}\sqrt{m_0^2 c^4 + c^2 p^2} = \frac{pc^2}{E} = v < c \tag{1.58}$$

achieving the full significance of a physical velocity; however, it is in an interesting relationship with the phase velocity, i.e.

$$v_{phase} v_{group} = c^2 \tag{1.59}$$

Also note that around the point where the group velocity is attached we have the normalization condition preserved at any time:

$$\left| \psi(x,t) \right|_0^2 = A(k_0)^2 = 1 \tag{1.60}$$

giving us the indication that the squared wave-function and its squared amplitude are formally equivalent, although as functions of conjugated or reciprocal variables as space and wave-vector, and should be also normalized. The general case will be treated in what follows.

1.3.3 BORN NORMALIZATION

We are going now to further explore the influence of the non-constant amplitude dependency $A(k)$ and to see its consequences in the de Broglie wave packet evolution. However, worth using the momentum representation from de Broglie quantification in order to better emphasize on the quantum (through Planck constant dependency) influence. That is we consider the general wave function:

$$\psi(x,t) = \int_{-\infty}^{+\infty} A(p) e^{\frac{i}{\hbar}(px - Et)} dp \tag{1.61}$$

and look for an appropriate form to specialize the momentum amplitude *A*(*p*). For doing that we note that from previous section we are leaving with the equivalence

$$|A(x,t)|^2 = \psi^*(x,t)\psi(x,t) = |\psi(x,t)|^2 \equiv \rho_{\wp}(x,t) = \frac{d\wp_x}{dx} \tag{1.62}$$

introducing the so called the probability density $\rho_{\wp}(x,t)$ from where, when the space is extending ad infinitum, one obtains the normalization condition

$$\int |A(x,t)|^2 dx = \int |\psi(x,t)|^2 dx = \int_{-X}^{+X} \rho_{\wp}(x,t) dx = \wp_X \xrightarrow{X \to \infty} 1 \tag{1.63}$$

based of probability $\wp_{\infty}$ certainty condition; such normalization property is associated with Born probabilistic interpretation of the square of the wave-function and may be called as *Born-normalization principle*.

Therefore, we have now the indication that we have to choose the amplitude *A*(*p*) as such to satisfy the similar relation for momentum:

$$\int_{-\infty}^{+\infty} A(p)^2 dp = 1 \tag{1.64}$$

as well as the entirely wave-function normalization condition

$$\int_{-\infty}^{+\infty} |\psi(x,t)|^2 dx = 1 \tag{1.65}$$

at any time of the wave-packet motion.

This can be realized, for instance, since one uses Gaussian momentum function (or spectral function if it viewed in terms of associated wave-vector *k*, throughout the de Broglie relationship) as the driving amplitude for the wave-packet:

$$A(p) = \frac{1}{(2\pi)^{1/4}\sqrt{\sigma_p}} \exp\left[-\frac{(p - p_0)^2}{4\sigma_p^2}\right] \tag{1.66}$$

with the width σ_p (being a spectral characteristic of the non-zero region for the signal) and initial momentum p_0; such function obeys the above momentum amplitude normalization constraint by means of the Poisson integral (see Appendix A.2)

$$\int_{-\infty}^{+\infty} \exp\left(-aq^2\right) = \sqrt{\frac{\pi}{a}} \tag{1.67}$$

The remaining proof concerns the check of the Born normalization condition at the level of the whole wave-function packet. Using the same Poisson integral rule, one computes the wave-function analytical expression in successive steps:

$$\begin{aligned}
\psi(x,t) &= e^{-\frac{i}{\hbar}Et} \int_{-\infty}^{+\infty} A(p) \exp\left(\frac{i}{\hbar} px\right) dp \\
&= \frac{e^{-\frac{i}{\hbar}(p_0 x - Et)}}{(2\pi)^{1/4}\sqrt{\sigma_p}} \int_{-\infty}^{+\infty} \exp\left(-\frac{(p-p_0)^2}{4\sigma_p^2} + \frac{i}{\hbar} px\right) dp \\
&= \frac{e^{-\frac{i}{\hbar}(p_0 x - Et)}}{(2\pi)^{1/4}\sqrt{\sigma_p}} \int_{-\infty}^{+\infty} \exp\left(-\frac{\xi_p^2}{4\sigma_p^2} + \frac{i}{\hbar}\xi_p x\right) d\xi_p \\
&= \frac{e^{-\frac{i}{\hbar}(p_0 x - Et)}}{(2\pi)^{1/4}\sqrt{\sigma_p}} \exp\left(-\frac{\sigma_p^2}{\hbar^2}x^2\right) \int_{-\infty}^{+\infty} \exp\left[\left(\frac{\xi_p}{2\sigma_p}\right) - \left(i\frac{\sigma_p x}{\hbar}\right)\right] d\xi_p \\
&= 2^{3/4}\pi^{1/4}\sqrt{\sigma_p} \exp\left(-\frac{\sigma_p^2}{\hbar^2}x^2\right) e^{-\frac{i}{\hbar}(p_0 x - Et)}
\end{aligned} \tag{1.68}$$

leaving with the squared wave-function expression:

$$|\psi(x,t)|^2 = 2^{3/2}\pi^{1/2}\sigma_p \exp\left(-2\frac{\sigma_p^2}{\hbar^2}x^2\right) \tag{1.69}$$

and finally with the result

$$\int_{-\infty}^{+\infty} |\psi(x,t)|^2 dx = 2\pi\hbar \tag{1.70}$$

From where it follows that in order the correct Born normalization to take place the initial wave-function packet has to be corrected with the quantum factor $1/\sqrt{2\pi\hbar}$ having the normalized general plane-wave expression:

$$\psi(x,t) = \frac{1}{\sqrt{2\pi\hbar}} \int_{-\infty}^{+\infty} A(p) e^{\frac{i}{\hbar}(px - Et)} dp \tag{1.71}$$

Worth noting that this wave-function preserves the Born-normalization all time as far as the $A(p)$ is as well normalized, as previously exposed; for instance, in the case the $A(p) \cong A(p_0)$ but not restricting the integration around the (equilibrium) variables (p_0 or k_0), so that to assure the correct normalization as in previous section employed, the square of the wave-packet function cannot be considered as meaning a probability density since it will be approximated as

$$\psi(x,t)_{p \to p_0} \cong A(p_0) e^{\frac{i}{\hbar}(p_0 x - Et)} \tag{1.72}$$

while the infinite-limit integral of its square would diverge:

$$\int_{-\infty}^{+\infty} \left|\psi(x,t)\right|^2_{p \to p_0} dx = A(p_0)^2 \int_{-\infty}^{+\infty} dx \to \infty \tag{1.73}$$

Therefore caution must be taken when considering which kind of wave-function (normalized or note) to be used to compute probability density on which interval as well; they have to be compatible to assure the correct Born normalization condition at any time.

1.3.4 FORMAL HEISENBERG INDETERMINACY

A directly application and of an extreme importance of the de Broglie wave packet is to consider its normalization by noting that the wave function in the real space and the amplitude in the reciprocal space (or of the impetus by the de Broglie quantification) are conjugate size in the sense of the Fourier mutual transforms.

$$\psi(x,0) = \frac{1}{\sqrt{2\pi\hbar}} \int_{(\infty)} dp A(p) e^{\frac{i}{\hbar}px} \tag{1.74}$$

$$A(p) = \frac{1}{\sqrt{2\pi\hbar}} \int_{(\infty)} dx\psi(x,0) e^{-\frac{i}{\hbar}px} \tag{1.75}$$

Then, by considering their recombination,

$$\psi(x,0) = \frac{1}{2\pi\hbar} \int_{(\infty)} dpdx\psi(x,0) \tag{1.76}$$

in order to keep the formal identity of the wave function from the real space one recognizes the coupling relationship of the variations of transformation in the real space (of the coordinate) and (of the momentum)

$$\Delta p \Delta x \approx 2\pi\hbar = h \tag{1.77}$$

Remarkably, the coupling is of the order of Planck constant - which once again justifies the previous statement, that this constant is universal, is immutable necessary in characterization of the movement, both in the observable space (of the coordinate) or of diffraction (of the momentum).

Note that the last relationship is called the "Heisenberg type", because it only justifies and does not demonstrates - it actually expressing that the impulse and the coordinate are inseparable at the level of Planck constant and can not be distinctly seen at spatiotemporal level, while being driven by it.

Although currently (as a matter of fact even since its publication by Heisenberg in 1927) are heated discussions and attempts to "dismantle" the dogma imposed by limiting/Heisenberg uncertainty in the Planck constant, the utility of this relationship (even borderline) is incontestable, which will be illustrated also by application to the Hydrogen atom (Bohr model), immediately below, and latter in a more elaborate framework.

1.3.5 BOHR HYDROGENIC QUANTIFICATION

Considering the hydrogen atom as the model (simple) of the circular motion of the electron around the nucleus, the coordinate variation (on circular direction) and momentum (on radial direction) recorded by the electron are expressed such as

$$\begin{cases} \Delta x = \Delta r = 2\pi r \\ \Delta p = p / n = \begin{cases} 0 \ \ldots n \to \infty \ldots \textit{atomic} \text{ frontier} \\ \infty \ \ldots n \to 0 \ \ldots \textit{atomic} \text{ kern} \end{cases} \end{cases} \tag{1.78}$$

which, through combining in it the Heisenberg-type relationship (from previous lesson) the quantification condition is generated

$$rp = \hbar n \ , \ n = 1, 2, \ldots \tag{1.79}$$

where it was considered (*ab initio*) a meshing over the set of natural numbers, as in the quantification of Planck, excluding the zero value (atomic collapse). Under these conditions, for the electron motion in a central field (Coulomb potential) the total energy is formulated as

$$\begin{aligned} E &= T(p) + V(r) \\ &= \frac{p^2}{2m_0} - \frac{Ze_0^2}{4\pi\varepsilon_0}\frac{1}{r} = \frac{p^2}{2m_0} - \frac{Ze_0^2}{4\pi\varepsilon_0\hbar}\frac{p}{n} \end{aligned} \tag{1.80}$$

This achieve to itself the state of dynamic equilibrium (optimal energy charge balance) satisfying the variational condition

$$\left(\frac{\partial E}{\partial p}\right)_{optimum} = 0 \tag{1.81}$$

which generates the quantification of the momentum variables, coordinate and optimum total energy in a consistent manner

$$\begin{cases} p_{opt,n} = \dfrac{e_0^2 m_0}{4\pi\varepsilon_0\hbar}\left(\dfrac{Z}{n}\right) \\ r_{opt,n} = \dfrac{4\pi\varepsilon_0\hbar^2}{e_0^2 m_0}\left(\dfrac{n^2}{Z}\right) \\ E_{opt,n} = -\dfrac{1}{2}\dfrac{e_0^4 m_0}{(4\pi\varepsilon_0)^2\hbar^2}\left(\dfrac{Z}{n}\right)^2 = -\dfrac{1}{2}\dfrac{Z^2}{n^2}[a.u.] \end{cases} \tag{1.82}$$

Worth introducing a fundamental chemical-physical energy constant, namely the atomic unit [a.u.] or the hartree (Hartree) unit as:

$$1a.u. \equiv 1E_h(hartree) = \frac{e_0^4 m_0}{(4\pi\varepsilon_0)^2 \hbar^2} \equiv 2R_\infty hc \equiv m_0 c^2 \alpha^2 \cong 27.2eV \quad (1.83)$$

with R_∞ the Rydberg constant (see Appendix A.8 for numerical value):

$$R_\infty = \frac{m_0 e_0^4}{8c\varepsilon_0^2 h^3} = \frac{m_0 c\alpha^2}{2h}\left[m^{-1}\right] \quad (1.84)$$

and were

$$\alpha = \frac{e_0^2}{4\pi\varepsilon_0}\frac{1}{\hbar c} \cong \frac{1}{137} \quad (1.85)$$

is introduced as the fine-structure (universal) constant (see Appendix A.8 for numerical value). Detailed energy conversion rules are given in Appendix A.8.1.

The energetic terms, i.e., kinetic and potential energy provided by the Bohr hydrogenic quantification, looks like:

$$T_{opt,n} = \frac{p_{opt,n}^2}{2m_0} = \frac{1}{2}\frac{e_0^4 m_0}{(4\pi\varepsilon_0)^2 \hbar^2}\left(\frac{Z}{n}\right)^2 = \frac{e_0^2}{4\pi\varepsilon_0}\frac{Z}{2r_{opt,n}} = -\frac{1}{2}V_{opt,n} = -E_{opt,n} \quad (1.86)$$

while fulfilling the virial relationships at the atomic level.

For the shake of completeness, the *virial theorem* can be rather elementary be proved by assuming the total energy composed by kinetic and potential terms depending on the second and first power of the space displacement about the equilibrium position, respectively, i.e.,

$$E_{tot} = E_{kin} + E_{pot}, \quad E_{kin} \propto x^2 \; ; E_{pot} \propto x \quad (1.87)$$

Then, if the space is parametrically inflated (rescaled) as

$$x \to \lambda x, \quad \lambda \in \Re \quad (1.88)$$

one has now from the energy

$$E_{tot}^{\lambda} = \lambda^2 E_{kin} + \lambda E_{pot} \quad (1.89)$$

whereas the equilibrium condition now demands:

$$\left.\frac{\partial E_{tot}^{\lambda}}{\partial \lambda}\right|_{\lambda=1} = 0 \tag{1.90}$$

leading with the general relationship:

$$E_{kin} = -0.5E_{pot} = -E_{tot} \tag{1.91}$$

that is the virial theorem affirms on.

Note that also for the central potential, as for any other, the potential energy variation is correctly considered as being proportional with the space displacement from the equilibrium position, as above, since it may always be written as related with the associated work through the consecrated relationship:

$$\Delta E_{pot} = \mathrm{F} \cdot \Delta \mathrm{x} = -\partial_x V(x) \cdot \Delta \mathrm{x} \tag{1.92}$$

Thus, the above demonstration of the virial theorem holds in general (for conservative systems).

1.3.6 BOHR'S CORRESPONDENCE PRINCIPLE

Turning back to Bohr atomic description it provides the frequency of emitted waves (photons) when the transition between two states in an atom takes place:

$$\upsilon_{n_1 \to n_2} = \frac{E_2 - E_1}{h} = \frac{m_0 e_0^4}{8\varepsilon_0^2 h^3} Z^2 \left(\frac{1}{n_1^2} - \frac{1}{n_2^2}\right) = R_\infty c Z^2 \left(\frac{1}{n_1^2} - \frac{1}{n_2^2}\right)$$

$$= R_\infty c Z^2 \frac{(n_2 - n_1)(n_2 + n_1)}{n_1^2 n_2^2},\ n_2 > n_1 \tag{1.93}$$

Now, the main question is when this transition frequency is becoming eventually equal with the frequencies associated with the orbital circular motion on states "1" and "2", individually. For responding in that, one may notice that the Bohr quantification supports the classical counterpart picture of electronic circular movement at optimum distance around the nucleus, with the angular velocity $\dot{\varphi}$ and the revolution frequency *f*, linked by the equation:

$$T_{opt,n} = \frac{m_0 v_{opt,n}^2}{2} = \frac{1}{2} m_0 \left(r_{opt,n} \dot{\varphi} \right)^2 = \frac{1}{2} m_0 r_{opt,n}^2 \left(2\pi f_n \right)^2 \tag{1.94}$$

from where we have:

$$f_n = \frac{1}{2\pi\hbar n} \frac{e_0^4 m_0}{\left(4\pi\varepsilon_0\right)^2 \hbar^2} \left(\frac{Z}{n} \right)^2 = \frac{2T_{opt,n}}{nh} = \frac{2R_\infty c Z^2}{n^3} \tag{1.95}$$

Now, the relation between the quantum transition frequency $\upsilon_{n_1 n_2}$ and the "classical" ones associate to the quantum states n_1 and n_2, f_{n_1} and f_{n_2}, can be clarified in two limiting cases.

- When is about first neighbor high levels, i.e., $\Delta n = n_2 - n_1 = 1$ and $n_1, n_2 >> 1$, we have in asymptotical sense that $n_1 \cong n_2$ leading with classical-quantum equivalency

$$\upsilon_{n_1 \cong n_2} = \frac{2R_\infty c Z^2}{n_1^3} = f_{n_1} = \frac{2R_\infty c Z^2}{n_2^3} = f_{n_2} \tag{1.96}$$

- When is about non-first neighbor but still high levels, i.e., $\Delta n = n_2 - n_1 > 1$ and $n_1, n_2 >> 1$, so that $n_1 \cong n_2 >> \Delta n$ we get the classical-quantum connection as:

$$\upsilon_{n_1 \cong n_2} = \frac{2R_\infty c Z^2}{n_1^3} \Delta n = \frac{2R_\infty c Z^2}{n_2^3} \Delta n = f_{n_1 \cong n_2} \Delta n \tag{1.97}$$

Therefore, the rule is that as much the quantum levels are higher as the quantum and classical frequencies approaches each other, establishing the so called Bohr correspondence principle between the quantum and classical "worlds".

An even more striking and practical form of Bohr correspondence principle may be unfold since we introduce the counter of the quantum transition states as:

$$\Delta I = h\Delta n \tag{1.98}$$

that combined with Bohr quantum transition principle

$$\Delta E = h\upsilon_{\Delta n} \tag{1.99}$$

provides the quantum frequency under the form

$$\upsilon_{\Delta n} = \frac{\Delta E}{\Delta I}\Delta n \tag{1.100}$$

to be compared with the classical frequency of the state with E_n, recognized to can be written as

$$f_n = \frac{1}{h}\frac{dE_n}{dn} = \frac{dE_n}{dI} = \frac{\upsilon_{\Delta n}}{\Delta n} \tag{1.101}$$

Leading with the idea that both frequencies approach each other when the slopes of the secant of spectrum lines equals the slopes of the tangents on the initial and final points on the graph

$$E_n = E_n(I_{n,\varphi}) \tag{1.102}$$

being here $I_{n,\varphi}$ recognized as the phase integral

$$I_{n,\varphi} = \oint p_\varphi d\varphi = nh \tag{1.103}$$

since providing the angular momentum quantification

$$p_\varphi = n\hbar = m_0 r_{opt,n}^2 \dot{\varphi} \tag{1.104}$$

in accord with above kinetic energy quantification – where the classical frequency was rooted. Note that the accompanying radial integral

$$I_{n,r} = \oint p_r dr = nh \tag{1.105}$$

gives nothing less than the starting de Broglie-Heisenberg-Bohr quantification. Also note that while Bohr model considers angular and radial integrals being quantified by the same quantum number n, further discrimination between them opens the way to treat 2D-eliptic Sommerfeld orbital description. However, while the last approach is still not the general one we prefer to directly treat the 3D-case, however in detail in the next volume of the set.

The Table 1.1 further illustrates the Bohr quantum-classical correspondence principle.

TABLE 1.1 Check of the Correspondence Principle for Asymptotic Bohr's Hydrogen Atom Levels (White, 1934; Putz et al., 2010)

Quantum states		Orbit frequency (s–1)		Transition Frequency (s–1)
Initial	*Final*	*Initial*	*Final*	
2	1	$0.82 \cdot 10^{15}$	$2.47 \cdot 10^{15}$	$6.58 \cdot 10^{15}$
6	5	$3.04 \cdot 10^{13}$	$4.02 \cdot 10^{13}$	$5.26 \cdot 10^{13}$
10	9	$6.58 \cdot 10^{12}$	$7.71 \cdot 10^{12}$	$9.02 \cdot 10^{12}$
25	24	$4.21 \cdot 10^{11}$	$4.48 \cdot 10^{11}$	$4.76 \cdot 10^{11}$
101	100	$6.383 \cdot 10^{9}$	$6.479 \cdot 10^{9}$	$6.576 \cdot 10^{9}$
501	500	$5.229 \cdot 10^{7}$	$5.245 \cdot 10^{7}$	$5.261 \cdot 10^{7}$

1.4 CONSEQUENCES OF MATTER QUANTIFICATION

1.4.1 *MOSELEY LAW AND SPECTRAL ATOMIC PERIODICITY*

Atomic Bohr's spectra are based from the transition energies given before; however, we may identify the so called spectral term

$$\mathrm{T}_n = \frac{R_\infty Z^2}{n^2} \tag{1.106}$$

allowing to rewrite the spectral transition in terms of the wave-number (the so called *Rayleight-Ritz principle*):

$$\tilde{\upsilon}_{n_1 n_2} = \frac{1}{\lambda_{n_1 n_2}} = \frac{\Delta E_{n_1 n_2}}{hc} = \mathrm{T}_{n_1} - \mathrm{T}_{n_2} \tag{1.107}$$

Still, for a many-electronic atom, the hydrogenic Bohr treatment can be still preserved with the price of introducing the so-called shielding constant σ

$$\mathrm{T}_n^* = \frac{R_\infty (Z-\sigma)^2}{n^2} \tag{1.108}$$

that eventually depends on the shell's quantum number n (and of its further sub-shell's refinement). The last relationship may be further transformed to get the linear Z-dependency:

$$\sqrt{\frac{\mathrm{T}_n^*}{R_\infty}} = \frac{1}{n}(Z-\sigma) \tag{1.109}$$

telling that the spectral terms for a given shell are proportional with the inverse of the quantum number of that shell. This has fundamental phenomenological interpretation for periods and groups of periodic Table: practically, for the different periods down groups it displays the increasing n so the diminishing angle of the fitted lines among the K, L, M, etc. transitions; whereas within periods, as the Z increases the squared of the spectral term to the Rydberg constant contribution increases.

Yet, for practical use the frequency is to be employed, rather than individual spectral terms; in this case one firstly has in generally that:

$$\upsilon_{n_1n_2} = \frac{c}{\lambda_{n_1n_2}} = c\left(\mathrm{T}_{n_1} - \mathrm{T}_{n_2}\right) = cR_{\infty}\left[\frac{(Z-\sigma_1)^2}{n_1^2} - \frac{(Z-\sigma_2)^2}{n_2^2}\right] \tag{1.110}$$

while if assuming the same initial and final shielding constant, $\sigma_1 \cong \sigma_2 = \sigma$, the simplified squared root transition frequency is obtained as the generalized Moseley law:

$$\sqrt{\upsilon_{n_1n_2}} = (Z-\sigma)\sqrt{cR_{\infty}\left[\frac{1}{n_1^2} - \frac{1}{n_2^2}\right]} \tag{1.111}$$

As a specialization, for instance, for K lines (n_1=1, n_2=2) the shielding constants can be further assumed as $\sigma \cong 1$ so that the Moseley working formula is obtained:

$$\sqrt{\upsilon_{K\alpha}} = \frac{(Z-1)}{2}\sqrt{3cR_{\infty}} \cong (Z-1)\sqrt{2.48\times10^{15}(\mathrm{Hz})} \tag{1.112}$$

where the energy-frequency conversion follows from the rules of Appendix A.8.2.

Note that the Moseley rule is less exact comparing with the spectral terms' interpretation of periodic spectra of elements through assuming the same shielding constants between the paired levels considered. Still, for Kα lines it behaves in fair agreement with experiment see Table 1.2 though comparing with theoretical yield:

$$h\upsilon_{K\alpha} \cong 13.6[eV]\frac{3}{4}(Z-1)^2 = 10.2(Z-1)^2[\mathrm{eV}] \tag{1.113}$$

TABLE 1.2 Experimental Kα2 (or KL2: transition from the level L2 with n=2, l=1 and j=1/2 to the level K with n=1 and l=0) X-Ray Energies, in eV, with Experimental Uncertainty in Parenthesis, as Compared with Computed Ones from the Moseley Law, for the Third and Fourth Periodic Groups of Elements (NIST 2009; Putz et al., 2010)

Z	*Element*	*Ka2 (eV)*	
		Experimental (unc.)	*Computed*
11	Na	1040.98 (12)	1020
12	Mg	1253.437 (13)	1234.2
13	Al	1486.295 (10)	1468.8
14	Si	1739.394 (34)	1723.8
15	P	2012.70 (48)*	1999.2
16	S	2306.700 (38)	2295
17	Cl	2620.846 (39)	2611.2
18	Ar	2955.566 (16)	2947.8
19	K	3311.1956 (60)	3304.8
20	Ca	3688.128 (49)	3682.2
21	Sc	4085.9526 (85)	4080.
22	Ti	4504.9201 (94)	4498.2
23	V	4944.671 (59)	4936.8
24	Cr	5405.5384 (71)	5395.8
25	Mn	5887.6859 (84)	5875.2
26	Fe	6391.0264 (99)	6375.
27	Co	6915.5380 (39)	6895.2
28	Ni	7461.0343 (45)	7435.8
29	Cu	8027.8416 (26)	7996.8
30	Zn	8615.823 (73)	8578.2
31	Ga	9224.835 (27)	9180.
32	Ge	9855.42 (10)	9802.2
33	As	10507.50 (15)	10444.8
34	Se	11181.53 (31)	11107.8
35	Br	11877.75 (34)	11791.2
36	Kr	12595.424 (56)	12495.

* Interpolated from nearby elements.

Finally, note that the difference between the spectral term and frequency pictures of the Moseley law is the same as that between the orbital motion frequencies and the transition frequencies between two (Bohr) levels.

Moreover, the Moseley law may be regarded also as provided the atomic number Z in terms of structural quantum information including spectral terms, energies, shielding constants; in other words, atomic Z may be seen as a measure of such inner quantum structure merely as the given constant for an atom.

1.4.2 SYSTEMS WITH IDENTICAL PARTICLES

The natural modeling of macro- and microscopy systems is made through the reduction of the components at structural units: *the elementary particles* (or more precisely – fundamental), essentially the same (electrons, photons, protons, neutrons, nuclei, atoms, etc.). These are also known as *ideal*, if from the macroscopic point of view the systems that belong are isolated, while from the microscopic point of view they do not interact between them.

The quality of *isolated system* ensures *the conservation of particles numbers from the system.*

$$N = \sum_i N_i = ct \tag{1.114}$$

The quality of isolated system is accomplished if the particles do not interact: it is considered that each particle is associated (quantified) with a given energy $\varepsilon_i \neq \varepsilon_j$ and around each thus energy there is a microscopic range of relocation, of volume $a = h^f$, with h Planck's constant and f degree of freedom allowed to each particle, is identical to the dimension of the so called *small space of the phases*

$$d\gamma = \underbrace{dp_1 dq_1}_{\approx h} \underbrace{dp_2 dq_2}_{\approx h} \dots \underbrace{dp_f dq_f}_{\approx h} \cong h^f \tag{1.115}$$

results from the cumulated application of the Heisenberg uncertainty, which actually drives the possible *action,* which is associated to each thus particles

$$S \cong \Delta E \Delta t \cong \Delta p \Delta q \tag{1.116}$$

Under these conditions, the system of particles is called ideal, being isolated and with particles considered as independent; while the particles or the individual states of energy are considered in very ample numbers, then can be considered as forming a *set of statistics,* with a certain arrangement of the particles on the available energy levels, called *microstate*. Note that there may be several iso-states, which correspond at the same macroscopic energy of the system. Then, it define the so-called *thermodynamic probability W* (reverse toward the mathematical one) as the number of distinct microstates corresponding to a macrostate given by the energy

$$E = \sum_i N_i \varepsilon_i = ct \tag{1.117}$$

constant in the context of the isolated systems.

The calculation of the number of microstates is made following a rigorous algorithm:

1. it is considered the dependence of thermodynamic probability (the number of microstates) on the number of existing particles N_i with a certain energy ε_i and on the number g_i which express the degeneration of the "i" energetic level, i.e., the number of sub-energetic levels (all with the same energy ε_i existing in the volume h^f associated with this energy); in a few words *"as in the large - so in the small"*, or analytical

$$W_E = W_E\left(N_i, g_i\right) \tag{1.118}$$

2. the implementation of the constraints regarding the conservation of numbers of particles and of the total energy for the isolated system of ideal particles (statistical) through *Lagrange multipliers*, α & β, respectively, toward the macroscopic balance registered by the *entropy,* in relation to the thermodynamic possibility, thus

$$S = k_B \ln W \tag{1.119}$$

with

$$k_B = = 1.3806503 \times 10^{-23} \text{ (m}^2 \text{ kg s}^{-2} \text{ K}^{-1}) \tag{1.120}$$

the Boltzmann constant (universal). Then, generically, *the thermodynamic macrostate function* will be expressed as

$$\underbrace{\underbrace{\Psi = \ln W_E\left(N_i, g_i\right)}_{DISORDER} + \underbrace{\alpha\left(N - \sum_i N_i\right) + \beta\left(E - \sum_i N_i \varepsilon_i\right)}_{ORDER}}_{MACROSTATE\ \ FUNCTION} \tag{1.121}$$

3. the equilibrium macrostate is achieved by highly condition required of the function macrostate

$$\frac{\partial \Psi}{\partial N_i} = 0 \tag{1.122}$$

wherefrom results the expression of the *distribution* of particles on the energetic states from each configuration of microstates compatible with the macrostate

$$N_i = N_i(\varepsilon_i, g_i) \tag{1.123}$$

Note that, in addition to those discussed, the statistical effect of large numbers, of particles N_i or of available energetic (sub) levels g_i – it manifested, beyond the thermodynamic considerations, also analytically by the systematic application of Stirling's approximation (see Appendix A.2)

$$\ln(n!) \cong n \ln n - n \tag{1.124}$$

And also of its form equivalent for permutations

$$n! \cong e^{n \ln n} e^{-n} = \frac{n^n}{e^n} = \left(\frac{n}{e}\right)^n \tag{1.125}$$

Further, we will present the three types of fundamental distributions in Nature, with related applications:

- The (classical) distribution of Maxwell-Boltzmann, applied to the physics of the atmosphere through the barometric pressure formula;
- The (quantum) distributions Fermi-Dirac and Bose-Einstein, with application to the classification of fundamental particles and forces.

1.4.3 MAXWELL-BOLTZMANN STATISTICS: THE PARTITION FUNCTION

For the classical particles, or associable to the classical description, one recognizes the number of microstates possible configurations for a given macrostate as given by the product (as mathematical probabilities, indicating the simultaneity of the events) between the particle permutation, $N!$ (since the "name/label" of the particles can be inter-changed at any time, based on their identity) with the number of the available energetic levels of the number of possible configurations for each of the levels (with its sub-levels) $g_i^{N_i}$; this is illustrated in Table 1.3 for the i=1, N_i=2, g_i=2, which is standardized at the number of possible permutations between particles in the sub-energetic levels (for not counting the identical configurations with multiple particles on the same sub-level, the particles being identical).

Thus, the probability of thermodynamic work for this general case is expressed as

$$W^B = N!\prod_i \frac{g_i^{N_i}}{N_i!} \tag{1.126}$$

being known as the *Maxwell-Boltzmann probability*. Using the second identity Stirling (see Appendix A.2), it can be also write as

$$W^B = N!\prod_i \left(\frac{g_i}{N_i}\right)^{N_i} e^{N_i} \tag{1.127}$$

Further, applying the above algorithm, the thermodynamic function is expressed and converted with the first type of Stirling identity

$$\begin{aligned}\Psi^B &= \ln W^{MB}(N_i, g_i) + \alpha\left(N - \sum_i N_i\right) + \beta\left(E - \sum_i N_i\varepsilon_i\right)\\ &= N\ln N - N + \sum_i (N_i \ln g_i - N_i \ln N_i + N_i)\\ &\quad + \alpha\left(N - \sum_i N_i\right) + \beta\left(E - \sum_i N_i\varepsilon_i\right)\end{aligned} \tag{1.128}$$

TABLE 1.3 The Illustrations of the Distribution of Two Classical Particles in Microstates Covering a Single Energy Level (i=1) with Two Sub-Levels (g=2) (Putz, 2010)

i=1	*g*=2	
ε	X1 X2	
	or	
ε		X1 X2
	or	
ε	X1	X2
	or	
ε	X2	X1

whose extreme condition

$$0 = \frac{\Psi^B}{\partial N_i} = \ln g_i - \ln N_i - \alpha - \beta\varepsilon_i \tag{1.129}$$

generates the *Boltzmann distribution*

$$N_i^B = g_i \exp(-\alpha - \beta\varepsilon_i) \tag{1.130}$$

Worthy of note Eq. (1.30) may be re-writing with the so-called *sum statistic* or *partition function*

$$Z = \sum_i g_i \exp(-\beta\varepsilon_i) \tag{1.131}$$

which intervenes in the evaluation of the total number of particles of the system

$$N = \sum_i N_i^B = \sum_i g_i \exp(-\alpha - \beta\varepsilon_i) = e^{-\alpha} \sum_i g_i \exp(-\beta\varepsilon_i) = Ze^{-\alpha} \tag{1.132}$$

wherefrom there is immediately obtained the shape (and implicitly the value) of the α-parameter

$$e^{-\alpha} = \frac{N}{Z} \tag{1.133}$$

which allows the final result

$$N_i^B = \frac{N}{Z} \exp(-\beta\varepsilon_i) \tag{1.134}$$

wherefrom there is emphasized the partition function normalizing role of the Boltzmann distribution.

Next, to determine the parameter β in Eq. (1.134), one appeals to the physical situation which is characteristic for constancy of the total energy which modulates in the thermodynamic function of the macrostate (formulated as a Lagrange expansion); in this respect we consider that the system exchanges the thermal energy with the environment after which is again isolated, so that the total number of particles of the system do not ever change. Thus we have

$$dE = \sum_i \varepsilon_i dN_i + \sum_i N_i d\varepsilon_i \tag{1.135}$$

where the first term represents the heat changed by the global system

$$\bar{d}Q = \sum_i \varepsilon_i dN_i \tag{1.136}$$

when, as a result of the transfer of heat, the transfer (jump) of particles N_i between the energetic levels ε_i from one level to another there is recorded, but without affecting the total number of particles from the system. Therefore, the condition that the *total number of particles* remains constant

$$0 = dN = d\left(\sum_i g_i e^{-\alpha-\beta\varepsilon_i}\right) = \sum_i g_i e^{-\alpha-\beta\varepsilon_i}\left(-d\alpha - \beta d\varepsilon_i - \varepsilon_i d\beta\right) \tag{1.137}$$

Equation (1.137) leads, by successively equivalent writings, to the useful expression of the r.h.s. term (which perhaps records a displacement of the spectrum, i.e., of all the energetic levels of the system), including the total energy above

$$-d\alpha - \beta d\varepsilon_i - \varepsilon_i d\beta = 0$$

$$\Leftrightarrow -\sum_i N_i d\alpha - \beta\sum_i N_i d\varepsilon_i - \underbrace{\sum_i N_i \varepsilon_i}_{E} d\beta = 0$$

$$\Leftrightarrow \beta\sum_i N_i d\varepsilon_i = -Ed\beta - Nd\alpha \tag{1.138}$$

In these circumstances, we can write for the entropy variation of the system the successive formulas by using the second kind of Stirling identity [the Eq. (1.127)] with the conservation of particles as in Eq. (1.132) along the terms of variation in Eq. (1.138)

$$\begin{aligned}\frac{1}{T}\bar{d}Q &= dS^B \\ &= d\left(k_B \ln W^B\right) = k_B d\left\{\ln\left[N!\prod_i e^{N_i(\alpha+\beta\varepsilon_i)} e^{N_i}\right]\right\} \\ &= k_B d\left[N\ln N - N + \sum_i N_i(\alpha + \beta\varepsilon_i) + \sum_i N_i\right]\end{aligned}$$

$$= k_B(d\alpha \underbrace{\sum_i N_i}_{N} + \alpha \underbrace{dN}_{0} + \beta \underbrace{\sum_i \varepsilon_i dN_i}_{\not{d}Q} + \beta \underbrace{\sum_i N_i d\varepsilon_i}_{-Ed\beta - Nd\alpha} + d\beta \underbrace{\sum_i N_i \varepsilon_i}_{E})$$

$$= k_B \beta \not{d}Q \tag{1.139}$$

wherefrom immediately results the envisaged parameter

$$\beta = \frac{1}{k_B T} \tag{1.140a}$$

to produce the final form of Boltzmann distribution

$$N_i^B = \frac{N}{Z} \exp\left(-\frac{\varepsilon_i}{k_B T}\right) \tag{1.141}$$

Beyond the basic applications of the Boltzmann distribution, particularly in statistical thermodynamics or even in quantum applications such as modeling populations of atoms in laser media, they are fundamental also in Environmental Physics by the famous modeling of the pressure with the atmosphere known as the barometric formula.

1.4.4 FERMI-DIRAC STATISTICS

For the statistics of semi-integer spin particles the rank of occupancy with particles is maximum one particle, through the Pauli principle, for each sub-level of the potential (of degeneracy g_i) for an energetic level ε_i – in this case, or in other words, it can take two values: zero (vacant sub-level) and one (occupied sub-level), *this being the quantum effect of fermions in thermal equilibrium state*. Under these conditions, the number of possible configurations (arrangements) of N_i particles on g_i sub-state for the energetic level ε_i is calculated by normalizing (removing from) the total number of possible permutations of sub-state g_i to the number of permutations of particles (single occupied cells) and of possible holes (unfilled/unoccupied cells), in order to not over-count their identity (e.g., the boxes/cells available in Table 1.4).

Under these conditions, the thermodynamic probability is said of a Fermi-Dirac (FD) type, or in a few words, of fermionic type (particle with half-integer spin) and is genuinely written as

TABLE 1.4 The Illustration of a Mode (from the possible ones) for the Quantum Distribution for Fermionic Type Particles with Half-Integer Spin on an Energetic Level with Sub-Levels g_i (Putz, 2010)

Level i	g_i						
ε_i	X1		X2		X3	X4

$$W^{FD} = \prod_i \frac{g_i!}{N_i!(g_i - N_i)!} \tag{1.142}$$

It is nevertheless equivalently written through the application of Stirling transformation of permutation statistics (see Appendix A.2)

$$\begin{aligned} W^{FD} &= \prod_i \frac{\left(\frac{g_i}{e}\right)^{g_i}}{\left(\frac{N_i}{e}\right)^{N_i}\left(\frac{g_i - N_i}{e}\right)^{g_i - N_i}} = \prod_i \frac{g_i^{\,g_i}}{N_i^{\,N_i}(g_i - N_i)^{g_i - N_i}} \\ &= \prod_i \frac{\left(\frac{g_i}{N_i}\right)^{g_i}}{\left(\frac{g_i}{N_i} - 1\right)^{g_i - N_i}} \end{aligned} \tag{1.143}$$

Furthermore, the thermodynamic function of the macrostate is formed, using the last form of Fermi-Dirac probability to look like

$$\begin{aligned} \Psi^{FD} &= \sum_i \left[g_i \ln g_i - N_i \ln N_i - (g_i - N_i)\ln(g_i - N_i) \right] \\ &+ \alpha\left(N - \sum_i N_i\right) + \beta\left(E - \sum_i N_i \varepsilon_i\right) \end{aligned} \tag{1.144}$$

for which the extreme (stable) condition of equilibrium

$$0 = \frac{\Psi^{FB}}{\partial N_i} = \ln N_i - \ln(g_i - N_i) - \alpha - \beta\varepsilon_i \tag{1.145}$$

generates the *Fermi-Dirac distribution*

$$N_i^{FD} = \frac{g_i}{\exp(\alpha + \beta\varepsilon_i) + 1} \tag{1.146}$$

with Lagrange parameters α and β to be (re)determined in terms of the current thermodynamic probability.

In order to determine the parameter β the procedure of the Boltzmann distribution will be repeated, while adapted to the Fermi-Dirac probability for the equivalently writing of the entropy's variation to the opening and then re-isolation of the system at an (infinitesimal) energy variation, including the energy exchange, but not the particles' exchange. This way, considering the Fermi-Dirac distribution [the Eq. (1.146)] in order to express the N_i / g_i size in the FD probability [the Eq. (1.143)] we successively have

$$
\begin{aligned}
\frac{1}{T} đQ &= dS^{FD} \\
&= d\left(k_B \ln W^{FD}\right) = k_B d\left\{\ln\left[\prod_i \frac{\left(e^{\alpha+\beta\varepsilon_i}+1\right)^{g_i}}{\left(e^{\alpha+\beta\varepsilon_i}\right)^{g_i-N_i}}\right]\right\} \\
&= k_B d\left\{\ln\left[\prod_i \left(e^{\alpha+\beta\varepsilon_i}\right)^{N_i}\left(1+e^{-\alpha-\beta\varepsilon_i}\right)^{g_i}\right]\right\} \\
&= k_B d\left\{\sum_i N_i(\alpha+\beta\varepsilon_i)+\sum_i g_i \ln\left(1+e^{-\alpha-\beta\varepsilon_i}\right)\right\} \\
&= k_B d\left\{\alpha N+\beta E+\sum_i g_i \ln\left(1+e^{-\alpha-\beta\varepsilon_i}\right)\right\} \\
&= k_B[N d\alpha+\beta dE+E d\beta+\sum_i \underbrace{\frac{g_i e^{-\alpha-\beta\varepsilon_i}}{1+e^{-\alpha-\beta\varepsilon_i}}}_{N_i}(-d\alpha-\beta d\varepsilon_i-\varepsilon_i d\beta)] \\
&= k_B[N d\alpha+\beta dE+E d\beta-N d\alpha-\beta\sum_i N_i d\varepsilon_i-E d\beta] \\
&= k_B\beta[\underbrace{dE}_{\sum_i \varepsilon_i dN_i+\sum_i N_i d\varepsilon_i}-\sum_i N_i d\varepsilon_i] \\
&= k_B\beta\underbrace{\sum_i \varepsilon_i dN_i}_{đQ} = k_B\beta đQ
\end{aligned}
\tag{1.147}
$$

wherefrom results that, as before in the Boltzmann's case, the identification of the thermal parameter

$$\beta = \frac{1}{k_B T} \tag{1.140b}$$

The α parameter determination is easily made, through the recognition of other identity of the entropy variation, this time when the total number of particles in the system fluctuates, specific of Lagrange's constraint, so that it modulates the thermodynamic macrostate function. This way, one appeals to the classical thermodynamic expansion

$$\begin{aligned} dS &= \frac{1}{T} đQ \\ &= \frac{1}{T}(dE + pdV - \mu dN) \\ &= \left(\frac{\partial S}{\partial E}\right)_{V,N} dE + \left(\frac{\partial S}{\partial V}\right)_{E,N} dV + \left(\frac{\partial S}{\partial N}\right)_{E,V} dN \end{aligned} \tag{1.148}$$

wherefrom the equation that connects the entropy variation to the total number of particles with chemical potential μ of the system results from the state of thermodynamic equilibrium at temperature T

$$\frac{\partial S}{\partial N} = -\frac{\mu}{T} \tag{1.149}$$

The relationship (1.149) customized with the actual expression of FD entropy, see inside of Eq. (1.147), provides equivalent identities

$$\frac{\partial}{\partial N}\left\{k_B\left[\alpha N + \beta E + \sum_i g_i \ln\left(1 + e^{-\alpha-\beta\varepsilon_i}\right)\right]\right\} = -\frac{\mu}{T} \tag{1.150a}$$

$$\alpha k_B = -\frac{\mu}{T} \tag{1.150b}$$

wherefrom the searched expression results

$$\alpha = -\frac{\mu}{k_B T} \tag{1.151a}$$

Note that from the thermodynamic equivalence (1.148) for the total entropy variation, one can write also the companion equation for the entropy variation with the energy

$$\frac{\partial S}{\partial E}=\frac{1}{T} \tag{1.152}$$

which provides a direct way also find to the β energy parameter from the FD entropy expression

$$S^{FD}=k_B\left\{\alpha N+\beta E+\sum_i g_i \ln\left(1+e^{-\alpha-\beta\varepsilon_i}\right)\right\} \tag{1.153}$$

through the immediate derivative

$$\frac{1}{T}=\frac{\partial S^{FD}}{\partial E}=k_B\beta \Rightarrow \quad \beta=\frac{1}{k_B T} \tag{1.140c}$$

With these two thermodynamic parameters, Eqs. (1.151a) and (1.140c), the Fermi-Dirac distribution is finally written under the explicit form

$$N_i^{FD}=\frac{g_i}{\exp\left(\dfrac{\varepsilon_i-\mu}{k_B T}\right)+1} \tag{1.154}$$

This distribution is successfully applied, for instance, to explain the electronic conduction in metals, and also to modeling the equilibrium and the physical parameters of white dwarfs, before entering the supernovae state!

1.4.5 BOSE-EINSTEIN STATISTICS

In agreement with Pauli principle, for the particles with integer spin, there is no restriction to the number of particles that can be placed in an available sub-level of an energy level. However, the fundamental difference toward the Boltzmann distribution, somehow similar in occupancy, is that the quantum one (the Bose-Einstein spin based ne) it can be obtained by arranging particles-understate also through the permutation of the "walls" between sub-states in the same way in which are permuted the particles; *id est*, the particles and the "walls" that are separating them, can be identical treated, having the same effect, which is purely quantum effect of the bosons in thermal equilibrium states: the permutation of two walls is equivalent to "the passage through the walls" (the wall's tunneling, aka of the potential barriers) of the particles in the system – generating the Bose-Einstein (BE) distribution, see Table 1.5.

Under these conditions, the thermodynamic Bose-Einstein probability that is by cumulating the entire permutation particles + walls $(N_i + g_i - 1)!$ excluding the permutation of two particles $N_i!$ and two walls $(g_i - 1)!$, based on their identity; so it can be written in an elementary manner

$$W^{BE} = \prod_i \frac{(N_i + g_i - 1)!}{N_i!(g_i - 1)!} \tag{1.155}$$

or, by applying the statistical considerations of the large numbers (of particles and walls), $N_i >> 1$, $g_i >> 1$, it firstly provides the operational expression

$$W^{BE} \cong \prod_i \frac{(N_i + g_i)!}{N_i! g_i!} \tag{1.156}$$

which can be further transformed, as in Fermi-Dirac case, with the aid of Stirling relationship for permutations (see Appendix A.2)

$$W^{BE} \cong \prod_i \frac{(N_i + g_i)^{N_i + g_i}}{N_i^{N_i} g_i^{g_i}} = \prod_i \left(\frac{N_i}{g_i}\right)^{g_i} \left(1 + \frac{g_i}{N_i}\right)^{N_i + g_i} \tag{1.157}$$

The standard construction requires that the thermodynamic function of the macrostate it will be written by combining the statistical information contained within the thermodynamic probability (1.157) with the Langrange constraints of particle and energy conservation

TABLE 1.5 The Illustration of a Possible Mode (Arrangement) for the Quantum Distribution of Bosonic-Type Particles (with Integer Spin) on an Energetic Level with g_i Sub-Levels (Putz, 2010)

Level i	g_i					
ε_i	X1		→...		...←	X2X3X4

$$\Psi^{BE} = \sum_i \left[(N_i + g_i)\ln(N_i + g_i) - N_i \ln N_i - g_i \ln g_i \right]$$
$$+\alpha\left(N - \sum_i N_i \right) + \beta\left(E - \sum_i N_i \varepsilon_i \right) \tag{1.158}$$

to form the equilibrium condition

$$0 = \frac{\Psi^{BE}}{\partial N_i} = \ln(N_i + g_i) - \ln N_i - \alpha - \beta\varepsilon_i \tag{1.159}$$

wherefrom the famous Bose-Einstein statistics is obtained

$$N_i^{BE} = \frac{g_i}{\exp(\alpha + \beta\varepsilon_i) - 1} \tag{1.160}$$

Mathematically, the BE distribution differs from the FD one just by a sign in the denominator, yet with deep consequences in the physical interpretation, as it will be seen below. We know that the determination of Lagrange parameters is based on the evaluation of entropy variation at the energy variation, in the absence of the variation of the total numbers of particles – for β determination, respectively when the total number of particles slightly varies – for α determination.

Thus, for the first instance of entropy variation one will use the form with the Stirling approximation for the probability (1.157) combined with the BE statistics (1.160) to yield

$$S^{BE} = k_B \ln W^{FD} = k_B \ln\left[\prod_i \left(\frac{1}{e^{\alpha+\beta\varepsilon_i} - 1} \right)^{g_i} \left(e^{\alpha+\beta\varepsilon_i} \right)^{N_i + g_i} \right]$$
$$= k_B \ln\left[\prod_i \left(\frac{e^{\alpha+\beta\varepsilon_i}}{e^{\alpha+\beta\varepsilon_i} - 1} \right)^{g_i} \left(e^{\alpha+\beta\varepsilon_i} \right)^{N_i} \right]$$
$$= k_B \ln\left[\prod_i \left(\frac{1}{1 - e^{-\alpha-\beta\varepsilon_i}} \right)^{g_i} \left(e^{\alpha+\beta\varepsilon_i} \right)^{N_i} \right]$$
$$= k_B \left[-\sum_i g_i \ln\left(1 - e^{-\alpha-\beta\varepsilon_i}\right) + \sum_i N_i (\alpha + \beta\varepsilon_i) \right]$$
$$= k_B \left[\alpha N + \beta E - \sum_i g_i \ln\left(1 - e^{-\alpha-\beta\varepsilon_i}\right) \right] \tag{1.161}$$

From Eq. (1.161) the Lagrange particle parameter immediately results by applying the already customary thermodynamic equation

$$-\frac{\mu}{T}=\frac{\partial S^{BE}}{\partial N}=\alpha k_B \Rightarrow \qquad \alpha=-\frac{\mu}{k_B T} \tag{1.151b}$$

Instead, the Lagrange's energy parameter is find out through the companion equation (1.152)

$$\frac{1}{T}=\frac{\partial S^{BE}}{\partial E}=k_B\beta \Rightarrow \qquad \beta=\frac{1}{k_B T} \tag{1.140d}$$

Finally, the explicit Bose-Einstein distribution is expressed as

$$N_i^{BE}=\frac{g_i}{\exp\left(\frac{\varepsilon_i-\mu}{k_B T}\right)-1} \tag{1.162}$$

and will represent the basis for modeling the photon radiation, and implicitly of blackbody radiation too, with implications in Environmental Physics, for example for the stellar classification and the establishment of the thermal age of the Earth.

1.4.6 FUNDAMENTAL FORCES AND ELEMENTARY PARTICLES: THE THEORY OF EVERYTHING (TOE)

1. The first level of unification of distributions consists in reducing the FD and BE quantum distribution to the Boltzmann one's, for system achieving large mass for particles and in any case much larger than the one of the electron/fermions and of the photon/bosons in moving at high temperatures (classical mode). Analytically, the first condition is rendered into

$$\varepsilon_i \cong nk_B T,\ n>>1 \tag{1.163}$$

which corroborated with the second one (T>>) gives the limit

$$\lim_{T\to\infty} N_i^{FD/BE}=\lim_{T\to\infty}\frac{g_i}{\exp\left(\frac{\varepsilon_i-\mu}{k_B T}\right)\pm 1}$$

$$\cong g_i \exp\left(-\frac{\varepsilon_i - \mu}{k_B T}\right) = g_i \exp(-\alpha - \beta\varepsilon_i) = N_i^B \quad (1.164)$$

2. The second level of unification of quantum distributions, is more subtle and relates to: (i) the quality of fermions (half-integer spin) to characterize the substance elementary particles of matter; (ii) the bosons (integer spin) as particles associated to the fundamental fields (to the forces implicitly) of matter that intermediate the interactions between the elementary particles. For clarity, we present in Table 1.6, face-to-face, the elementary particles for substance and the characteristic particle-carriers to the fundamental forces of Nature.

The unification of fundamental forces, respectively of the first three types of the interactions from the Nature generates the so-called GUT (Grand Unification Theory), which when extended to the fourth interaction - the gravitational one (see Figure 1.5) constitutes the TOE (Theory of Everything) table, with the schematic representation in Figure 1.6.

Phenomenological, the final unification takes place at the so-called Planck era or scale, as in Figure 1.6, which can be analytically explicated in a direct manner through considering the unification of macro-micro cosmos, equating the gravitational and quantum levels, yet using basic formulas of gravitation, of special relativity theory, and of the Planck's quantification

$$E = h\upsilon = \frac{h}{t} \quad (1.165)$$

From the Newtonian equality of the inertial force to the gravitational one (the principle of the equivalence between inertia and gravity is by this way "activated")

$$ma = G\frac{mM}{r^2} \quad (1.166)$$

one obtains the relation for the universal constant of gravitation

$$G = \frac{ar^2}{M} \quad (1.167)$$

TABLE 1.6 The Short Description of the Elementary Particles Which Quantify the Substance (elementary fermions) and the Interaction Fields/Forces (elementary bosons) (Putz, 2010)

Elementary Fermions

***A.* The lighter fermions: leptons**

- three fundamental families/generations of leptons

$$\begin{matrix} Q=-1 \\ Q=0 \end{matrix} : \begin{pmatrix} e^- \\ \upsilon_e \end{pmatrix}, \begin{pmatrix} \mu^- \\ \upsilon_\mu \end{pmatrix}, \begin{pmatrix} \tau^- \\ \upsilon_\tau \end{pmatrix}$$

- each generation being a combination of charged lepton tied to the correspondent neutral lepton (neutrinos)
- e^-, the electron, the oldest known lepton, was discovered by JJ Thomson, in 1897, while the electronic neutrino υ_e was anticipated by Ettore Majorana (student of E. Fermi, disappeared in 1938, still in unknown circumstances) and theorized by Pauli in 1930 was experimentally discovered in 1956, by Reines and Peierls
- μ^-, the muon, was discovered by Anderson in 1937 (the discoverer of the positron too) and further, in 1962, was highlighted the neutrino muons υ_μ
- τ^-, the triton (from tritos, literal Greek, meaning "third" in English), or tauonic lepton, is a lepton with a high mass (of baryonic order – see below) considered as heavy lepton, discovered in 1977 by the Peierls with tau neutrino (tauon) experimentally identified in 1990.

***B.* The heavy fermions: the quarks**

- three fundamental families/generations

$$\begin{matrix} Q=+2/3 \\ Q=-1/3 \end{matrix} : \underbrace{\begin{pmatrix} u \\ d \end{pmatrix}}_{\begin{pmatrix} up \\ down \end{pmatrix}}, \underbrace{\begin{pmatrix} c \\ s \end{pmatrix}}_{\begin{pmatrix} charm \\ strange \end{pmatrix}}, \underbrace{\begin{pmatrix} t \\ b \end{pmatrix}}_{\begin{pmatrix} top/truth \\ botton/beauty \end{pmatrix}}$$

Fundamental Forces

***A.* The strong force**

- Occurs between baryonic quarks;
- It is intermediated by particle g-GLUON (from glue = which binds);
- The spin s(G) = 1
- With zero mass m (G) = 0
- Interaction by gluons is the central subject of quantum chromodynamics (Q.C.D); the feature *cromo* cames from the fact gluons are further quantified by the so called "color" property: red (r), green (g), blue (b), turquoise (anti red $\overline{r}$), lilac (anti-green $\overline{g}$), yellow (anti-blue $\overline{b}$), with associated combinations;
- Specific potential form

$$V_t = -k_1\frac{1}{r} + k_2 r = \begin{cases} \infty ... r \to 0 \\ \infty ... r \to \infty \end{cases}$$

***B.* The Electro-Magnetic Interaction**

- Occurs between electrically charged bodies, being intermediated by the PHOTON-γ
- The spin s(γ) = 1
- With zero mass $m_0(\gamma)=0$
- Is the subject of quantum electrodynamics (QED)
- Specific potential form

$$V_{EM} = -\frac{1}{4\pi\varepsilon_0}\frac{Q_1 Q_2}{r}$$

***C.* The Weak Interaction**

- Models the transformations of nucleons, being intermediated by WEAKONS (from the meaning of weak), both charged $W^{\pm}$ or neutral Z^0

TABLE 1.6 Continued

Elementary Fermions	Fundamental Forces
• the quark concept was introduced by M. Gell-Mann in 1964–1965, at Caltech (although there was a preprint with similar ideas, at CERN, from Georg Zweig, he called them ace), assuming the expression from the opera Finnegans Wake, by James Joyce, "Three quarks for Muster Mark!", at page 383. • In 1964 there were about 200 particles called elementary particles, without being systematically explained, until the naive theory (with 3 quarks q = u, d, s) of Gell-Mann; for example, the proton was a combination of 3q: p=uud, while the neutron has the structure n=3q=udd; • The strangeness of name come to underline the strangeness of these particles with fractional charge, in fact sub-electronic! • The combination of three quarks qqq composes the bound state called baryons (nucleons, hiperons, etc.). • The combination of two quarks (one quark and one antiquark) $q\tilde{q}$ generates the bounded states of mesons (i.e., π, k mesons or charmonium particle $J/\psi = c\tilde{c}$) • From the leptons and quarks it can be formed the substance's Matter, however complex • Might as also the quarks themselves have internal structure!	• The specific potential is unspecified • Charged weakens intervene in β reactions of transformations occurring in proton-neutron interchanging inside the nucleus, so $\beta^-: n \rightarrow p + W^-$, $W^- \downarrow e^- + \tilde{\upsilon}_e$ $\beta^+: p \rightarrow n + W^+$, $W^+ \downarrow e^+ + \upsilon_e$ • with the spin $s(W^{\pm}, Z^0)$=1 • With the masses $m_0(W^{\pm})$~84 GeV (or proton mass m_p), $m_0(Z^0)$~94 GeV • Have been discovered at CERN in 1983 (in January $W^{\pm}$, in June Z^0) by the group of Carlo Rubia and Simon Van der Meer (rewarded with a Nobel Prize in 1984) ***D.* The gravitational interaction** • Occurs between bodies with mass, being mediated by the GRAVITON-Γ • The spin s (Γ) = 2 • With zero rest mass m_0(Γ)=0 • It moves at the speed of light! • The graviton is carried by gravitational waves (still undetected) • Specific potential form $V_{EM} = -G\frac{M_1 M_2}{r}$

which it can eventually written as equivalent, using "the Planck's manner" involving only universal constants (since assumed as unified or at least interchangeable at the level of the grand unification of forces, particles and energies, immediately after the Big Bang singularity).

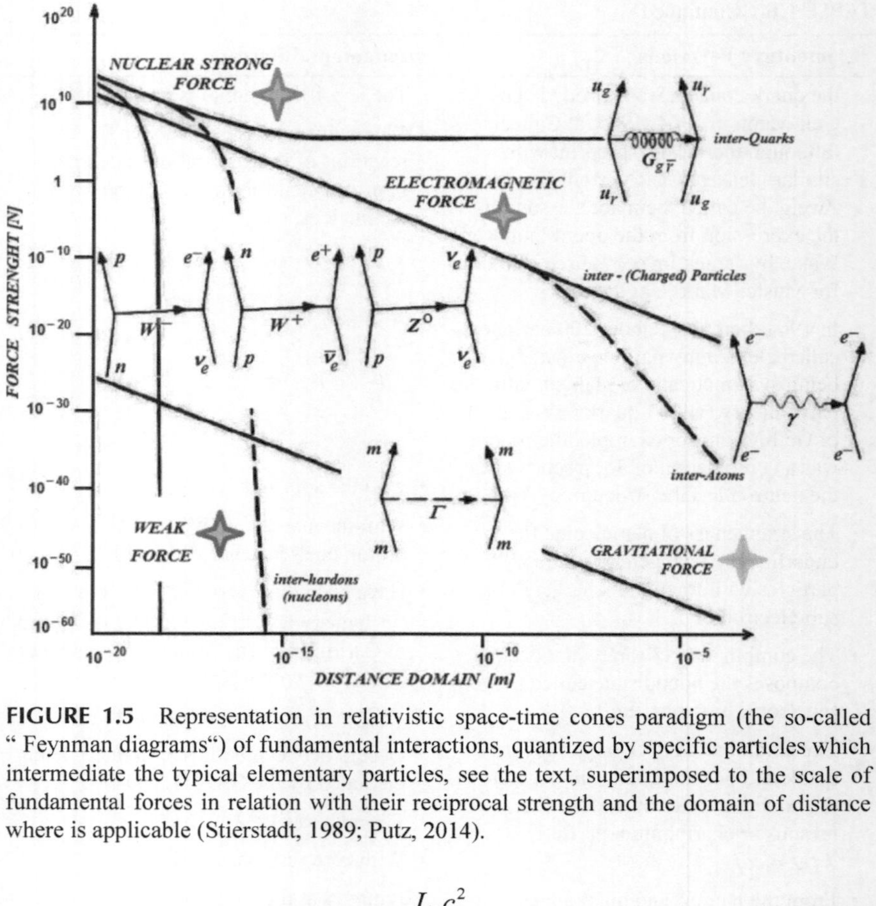

FIGURE 1.5 Representation in relativistic space-time cones paradigm (the so-called " Feynman diagrams") of fundamental interactions, quantized by specific particles which intermediate the typical elementary particles, see the text, superimposed to the scale of fundamental forces in relation with their reciprocal strength and the domain of distance where is applicable (Stierstadt, 1989; Putz, 2014).

$$G = \frac{L_P c^2}{M_P} \tag{1.168}$$

Note that, for consistency, everything that is not universal constant in Eq. (1.168) becomes of Planck (P) size, specific for the birth of the Universe! On the other hand, the quantum de Broglie relationship

$$p\lambda = h \tag{1.169}$$

is rewritten at the level of Planck's Universe, to obtain a second working relation

$$(M_P c)L_P = h \tag{1.170}$$

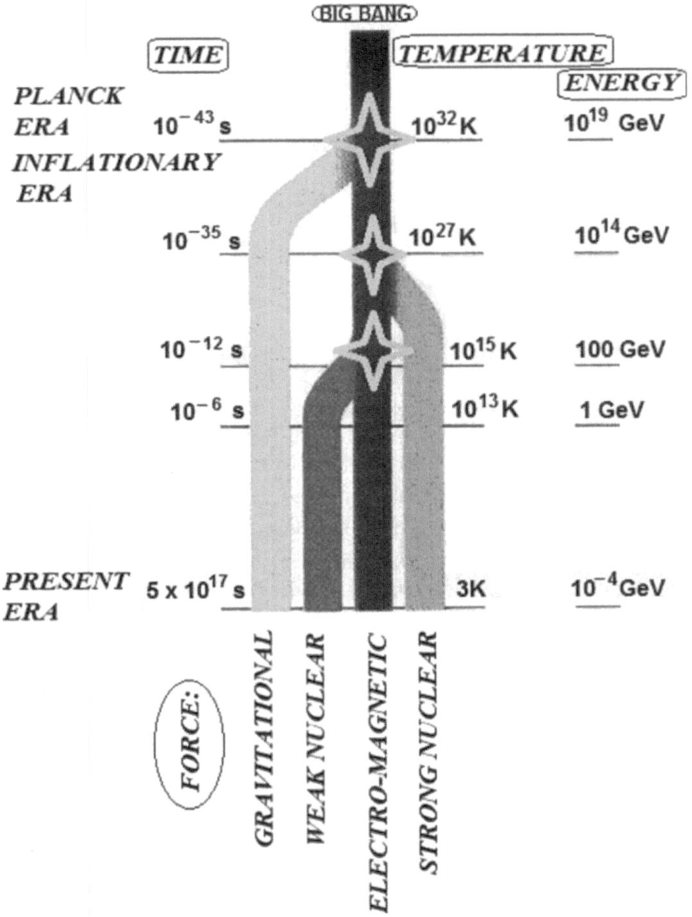

FIGURE 1.6 Unification paradigm of fundamental forces of Nature, in relation with time, temperature and energy of dominant particles in the age's specific to the evolution of Universe (Evans, 2010; Putz, 2010).

which combined with Eq. (1.168), allows the solution for the Planck mass

$$M_P = \sqrt{\frac{ch}{G}} \cong 5 \cdot 10^{-8} [kg] \tag{1.171}$$

which allows the finding of early (Planck) Universe's length

$$L_P = \sqrt{\frac{Gh}{c^3}} \cong 4 \cdot 10^{-35} [m] \tag{1.172}$$

They generate a plethora of further determinations, as for instance the Planck's time

$$t_P = \frac{L_P}{c} = \sqrt{\frac{Gh}{c^5}} \cong 10^{-43}[s] \tag{1.173}$$

closely related to the Planck's energy calculation

$$E_P = \frac{h}{t_P} = \sqrt{\frac{hc^5}{G}} \cong 5 \cdot 10^9 [J] \cong 3 \cdot 10^{19} [GeV] \tag{1.174}$$

and further to the Planck's temperature

$$T_P = \frac{E_P}{k_B} = \sqrt{\frac{hc^5}{Gk_b^2}} \cong 3 \cdot 10^{32} [K] \tag{1.175}$$

being accompanied by a colossal Planck density of the early Universe

$$\rho_P = \frac{M_P}{L_P^3} = \frac{c^5}{hG^2} \cong 10^{96} [kg / m^3] \tag{1.176}$$

This is an elementary picture for the unification of quantum, gravitational and relativist concepts actually characterizing the Big Bang moment until the Planck's horizon, through the spatial-temporal parameters as mass, thermal and energetic, expressed only in terms universal constants (Planck's constant, Boltzmann's constant, speed of light in vacuum, universal gravitation constant) as all these would be merged into one single entity!

1.4.7 STEFAN-BOLTZMANN LAW OF RADIATION

The Bose-Einstein distribution (1.162) may be considered to recover the Planck law of black body radiation, i.e., the photon radiation modeling, by considering the following peculiarities:

- One considers the fact that the photons, at equilibrium, do not operate under the condition of constancy for the total number of particles $N \neq ct$, wherefrom there results that one can not applied the appropriate Lagrange multiplier, setting therefore the associated chemical potential to zero

$$\alpha = 0 \Rightarrow \mu = 0 \tag{1.177}$$

- The degeneration of multiplicities does not apply regarding the spin, but at the fact that there are two types of electromagnetic polarization ($g = 2$), given that the transition from the *discrete* statistics to the continuous one is done through the small space of the phases, quantum normalized in the sense of Heisenberg localization/ delocalization through the analytical transformation

$$N_i \rightarrow dN = \frac{1}{\exp\left(\dfrac{\varepsilon}{k_B T}\right) - 1}\left(\frac{g}{a} d\gamma\right) \tag{1.178}$$

Note that in Eq. (1.178), for the photon particles (with three degrees of freedom) we have

$$a = \left(2\pi\hbar\right)^3 = h^3 \tag{1.179}$$

Equally, the infinitesimal volume of the phases

$$d\gamma = \left(dxdp_x\right)\left(dydp_y\right)\left(dzdp_z\right) = d\mathbf{r}d\mathbf{p} = dV\left(4\pi p^2 dp\right) \tag{1.180}$$

is rewritten by adapting to the photon the quantum energy-momentum particle relation

$$\varepsilon = c \cdot p \tag{1.181}$$

to become

$$d\gamma_p = 4\pi V \frac{\varepsilon^2}{c^2} \frac{d\varepsilon}{c} \tag{1.182}$$

With Eqs. (1.179) and (1.182) the photon statistics (1.178) becomes

$$dN = \frac{8\pi}{ac^3} V \frac{\varepsilon^2 d\varepsilon}{\exp\left(\dfrac{\varepsilon}{k_B T}\right) - 1} \tag{1.183}$$

Based on such statistics, the total energy of photon radiation is obtained by passing from the discreet definition to the continuous one by the sum-to-integral (statistical) conversion

$$U = \sum_i \varepsilon_i N_i \to \int \varepsilon dN \tag{1.184}$$

allowing the successive calculations

$$U = \frac{8\pi}{ac^3} V \int_0^\infty \frac{\varepsilon^3 d\varepsilon}{\exp\left(\frac{\varepsilon}{k_B T}\right) - 1}$$

$$\overset{\frac{\varepsilon}{k_B T}=x}{=} \frac{8\pi V}{ac^3} (k_B T)^4 \underbrace{\int_0^\infty \frac{x^3 dx}{\exp(x) - 1}}_{\pi^4/15}$$

$$= \frac{8\pi V}{ac^3} (k_B T)^4 \frac{\pi^4}{15} \tag{1.185}$$

where the value of Riemann-zeta function of the third order it had been used (see Appendix A.4)

$$\int_0^\infty \frac{x^3 dx}{\exp(x) - 1} = \frac{\pi^4}{15} \tag{1.186}$$

Next, when one considers the volume density of energy

$$u = \frac{U}{V} = \frac{8\pi^5 k_B^4}{15 h^3 c^3} T^4 \tag{1.187}$$

one can define the so-called radiance or energy flux density or radiant flux (energy emitted by a blackbody per unit area per unit time, or the radiant power per unit area)

$$L = \int dL = \frac{P}{A} = \frac{U}{A \cdot t} = \frac{x}{t} \frac{U}{A \cdot x} = \frac{c}{4} u \tag{1.188}$$

simply reshaped under the so-called Stefan-Boltzmann law

$$L = \sigma T^4 \tag{1.189}$$

when collecting the Stefan's constant from Eq. (1.187)

$$\sigma = \frac{2\pi^5 k_B^4}{15 h^3 c^3} = 5.670400 \times 10^{-8} [J \cdot s^{-1} \cdot m^{-2} \cdot K^{-4}] \tag{1.190}$$

Note that the factor ¼, from the definition of radiation in Eq. (1.188), comes from the geometrically condition for integration after the values of possible radiation angles, toward a given source, as shown in Figure 1.7, with the associate short proof:

$$\begin{aligned}\Delta U &= L\cdot t\cdot \Delta A\\ &= \underbrace{2}_{\pm x}\frac{\Delta L}{\cos\theta}\cdot\frac{\Delta t}{\cos\theta}\cdot\Delta A\\ &= 2\frac{1}{\cos^2\theta}\cdot\frac{\Delta x}{c}\cdot\Delta L\Delta A\end{aligned} \tag{1.191}$$

wherefrom results the customary expression

$$L=\int_{\theta}\Delta L d\theta=\frac{\Delta U}{\Delta V}\frac{c}{2}\frac{1}{\pi}\overbrace{\underbrace{\int_{-\pi/2}^{\pi/2}\cos^2\theta d\theta}_{1/2}}^{\langle\cos^2\theta\rangle}=\frac{c}{4}u \tag{1.192}$$

Further, if we rewrite in Eq. (1.185) the density of total radiant energy by employing the Planck quantification of radiation, $\varepsilon = h\upsilon$, the newly integral is formed

$$u=\frac{U}{V}=\frac{1}{h^3}\int_0^\infty\frac{8\pi}{c^3}\frac{(h\upsilon)^3 d(h\upsilon)}{\exp\left(\frac{h\upsilon}{k_B T}\right)-1}$$

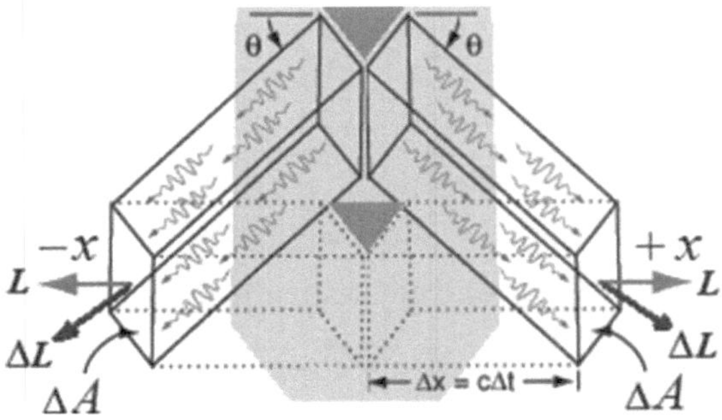

FIGURE 1.7 The geometrical configuration, which express the photonic radiance (HyperPhysics, 2010; Putz, 2010).

$$= \int_0^\infty \frac{8\pi h}{c^3} \frac{\upsilon^3 d\upsilon}{\exp\left(\frac{h\upsilon}{k_B T}\right) - 1}$$

$$\equiv \int_0^\infty \rho(\upsilon, T) d\upsilon \tag{1.193}$$

wherefrom the energy spectral density (in frequency) is identified

$$\rho(\upsilon, T) = \frac{8\pi h}{c^3} \frac{\upsilon^3}{\exp\left(\frac{h\upsilon}{k_B T}\right) - 1} \tag{1.24b}$$

which corresponds to the Planck formula (distribution) of the blackbody radiation as earlier given in Eq. (1.24a)! Note that actually, Planck had used for spectral density distribution the formula under the form of the spectral intensity obtained as the energy emitted by a blackbody at temperature T per unit area per unit time per unit solid angle, in the $(\upsilon, \upsilon + d\upsilon)$ interval that is:

$$I(\upsilon, T) = \frac{1}{\Omega} \frac{1}{t} \frac{1}{A} \frac{dU}{d\upsilon}$$

$$= \frac{1}{4\pi} \underbrace{\frac{1}{t} \underbrace{\frac{V}{A}}_{x}}_{c} \frac{8\pi h}{c^3} \frac{\upsilon^3}{\exp\left(\frac{h\upsilon}{k_B T}\right) - 1}$$

$$= \frac{2h\upsilon^3}{c^2} \frac{1}{\exp\left(\frac{h\upsilon}{k_B T}\right) - 1}$$

$$= \frac{c}{4\pi} \rho(\upsilon, T) \tag{1.194}$$

However, another important quantity related to the energy density refers to the photon radiation pressure, phenomenological deduced from the general relations

$$P_\gamma = \frac{Force}{Area} = \frac{1}{A} \frac{dp}{dt} \overset{\varepsilon = c \cdot p}{=} \frac{1}{c} \underbrace{\frac{1}{A} \frac{d\varepsilon}{dt}}_{\sim c \cdot u} \propto \frac{1}{c}(c \cdot u) = u \tag{1.195}$$

and which is specialized by taking into consideration the statistical geometric conditions, i.e., considering the equipartition on the 3D coordinates, casting as

$$P_{\gamma} = \frac{1}{3}u = \frac{1}{3}\sigma^{*}T^{4} \tag{1.196}$$

It is worth noting that in astrophysics, is often used the changed form of the Stefan-Boltzmann law, in practice written at the energy density level

$$u = \sigma^{*}T^{4} \tag{1.197}$$

with the new Stefan (Stefan-star) constant

$$\sigma^{*} = \frac{4\sigma}{c} = \frac{8\pi^{5}k_{B}^{4}}{15h^{3}c^{3}} = 7.565767\times10^{-16}[J\cdot m^{-3}\cdot K^{-4}] \tag{1.198}$$

These are the theoretical premises that will allow in characterizing various bodies of the universe (e.g., the Sun, the Earth, the Stars and even the Cosmos as a statistical ensemble) based on the electromagnetic radiation that they emit or store.

1.4.8 THE WIEN LAW: THE UNIVERSE'S TEMPERATURE AND ANISOTROPY

In the relations of the previous section, see Eq. (1.193), it appears that the energy spectral density per unit frequency can be written as

$$\rho(\upsilon,T) = \frac{1}{V}\frac{dU}{d\upsilon} \tag{1.199}$$

which be generalized also to the case of energy spectral density per unit pulse

$$\rho(\omega,T) = \frac{1}{V}\frac{dU}{d\omega} \tag{1.200}$$

and respectively, for the case of energy spectral density per unit wavelength

$$\rho(\lambda,T) = \frac{1}{V}\frac{dU}{d\lambda} \tag{1.201}$$

However, the connection between these three quantities is immediately (taking into consideration that the frequency and pulsation increase when the relative wavelength decreases ($d\upsilon \,\&\, d\omega > 0 \Rightarrow d\lambda < 0$) thus yielding

$$dU = \rho(\upsilon,T)d\upsilon = \rho(\omega,T)d\omega = -\rho(\lambda,T)d\lambda \tag{1.202a}$$

Such analytical variable interchanging is of great importance because it allows the transcription from one spectral quantity to another, depending on the conceptual/computational needs and issues involved. For example, the spectral energy density per unit wavelength, is successively obtained from Eq. (1.24a/b)

$$\begin{aligned}\rho(\lambda,T) &= -\rho\left(\upsilon = \frac{c}{\lambda},T\right)\frac{d\upsilon}{d\lambda}\\ &= -\frac{8\pi h}{c^3}\frac{\frac{c^3}{\lambda^3}}{\exp\left(\frac{hc}{k_B T\lambda}\right)-1}\left(-\frac{c}{\lambda^2}\right)\\ &= \frac{8\pi hc}{\lambda^5}\frac{1}{\exp\left(\frac{hc}{k_B T\lambda}\right)-1}\end{aligned} \tag{1.202b}$$

with representations in Figure 1.8. From the Figure 1.8, there is apparent that the maximum of the spectral energy density lays on a curve, on a geometrical locus of $\lambda = \lambda(T)$ type, with universal value – defined by what had been established as the Wien's law. Analytically, one searched for the solution of the maximum condition

$$\frac{d\rho(\lambda,T)}{d\lambda} = 0 \tag{1.203}$$

with the specific form for photonic radiation

$$\frac{d}{d\lambda}\frac{\lambda^{-5}}{\exp\left(\frac{\xi}{T\lambda}\right)-1} = 0 \tag{1.204}$$

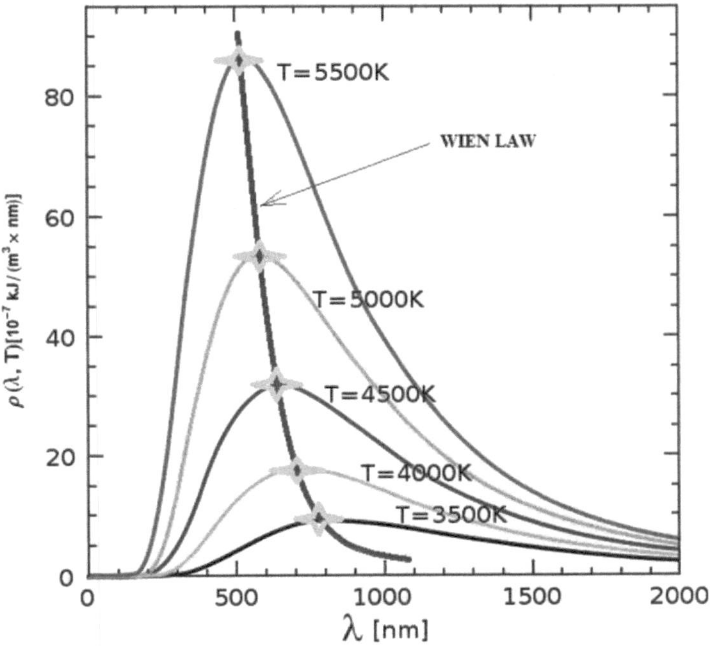

FIGURE 1.8 The representation of energy spectral density per unit wavelength, for the "black" body, with high temperatures (such as for the Sun roughly about 6000 K); note that the curves at lower temperature have the same allure (at T = 300 K for Earth surface type, or at T = 2.7 K for the whole Cosmos) - with the modification of the wavelength domain; the region of the visible spectrum lies between 380 and 750 nm (HyperPhysics, 2010; Putz, 2010).

where it had been introduced the constant factor

$$\xi \cong \frac{hc}{k_B} = 1.43878 \cdot 10^{-2}[m \cdot K] \tag{1.205}$$

From Eq. (1.204) there follows that the equation to be solved takes the shape

$$\lambda T = \frac{\xi}{5\left[1 - \exp\left(-\frac{\xi}{\lambda T}\right)\right]} = f(\lambda T) \tag{1.206}$$

which can be numerically solved by self-iterations until it reaches the equality between the argument and function, starting by the order of size of the λT product similar with those of the ξ size. Thus, we successively have

$$f(\lambda T = 10^{-2}) = 3.77245 \cdot 10^{-3}[m \cdot K] \tag{1.207a}$$

$$f(\lambda T = 3.77245 \cdot 10^{-3}) = 2.94248 \cdot 10^{-3}[m \cdot K] \tag{1.207b}$$

$$f(\lambda T = 2.94248 \cdot 10^{-3}) = 2.89937 \cdot 10^{-3}[m \cdot K] \tag{1.207c}$$

$$f(\lambda T = 2.89937 \cdot 10^{-3}) = 2.89783 \cdot 10^{-3}[m \cdot K] \tag{1.207d}$$

$$f(\lambda T = 2.89783 \cdot 10^{-3}) = 2.89778 \cdot 10^{-3}[m \cdot K] \tag{1.207e}$$

and finally

$$f(\lambda T = 2.89778 \cdot 10^{-3}) = 2.89778 \cdot 10^{-3}[m \cdot K] \tag{1.207f}$$

which generates the analytical form of Wien's law

$$\lambda T = 2.89778 \cdot 10^{-3}[m \cdot K] \tag{1.208}$$

Remarkably, for the temperature values around $T \cong 3K$ there are obtained wavelength in order size of micro-waves (millimeters) $\lambda \cong [10^{-3} m] = [mm]$, which corresponds to the remnant background radiation of the Universe, systematically detected by the research satellite COBE (The Cosmic Background Explorer), launched in 1989, Figure 1.9.

Since the interpolation with an error less than 1%, was achieved for an exact temperature $T \cong 2.74[K]$ it is therefore assumed as the current temperature of the Universe, in excellent agreement with the physical-mathematical modeling of blackbody radiation, according to Planck photon statistics (Figure 1.9). The fundamental consequence of the data about the microwave background radiation around 3K on knowledge of the Universe is extremely important. For example, from the discovery of background radiation (in 1965 by Arno A. Penzias and Robert W. Wilson of Bell Laboratories, winners of the Nobel Prize for this discovery in 1978) the launching of the COBE satellite was followed, in 1989, which confirmed the distribution of blackbody radiation in the Universe, extended backwards in time from its current form, as in Figure 1.11 until its decoupling from the substance (400,000 years after the Big Bang).

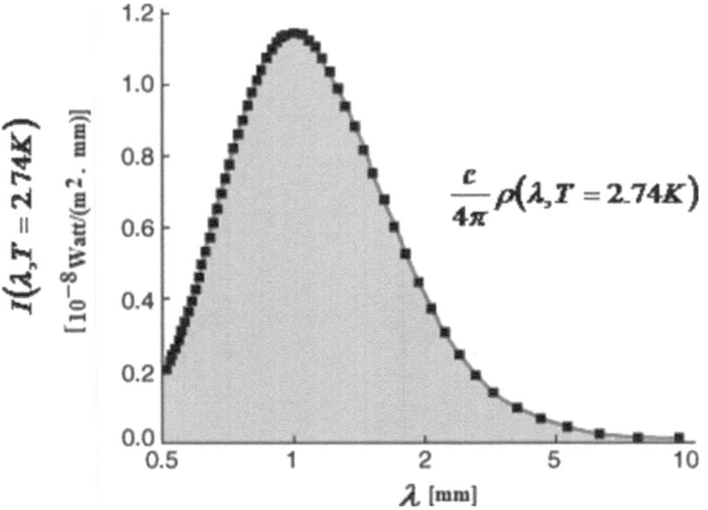

FIGURE 1.9 The representation of intensity of spectral energy density per unit wavelength in the micro-waves domain, for the residual temperature of the Universe, about 2.74K, as had been recorded by the COBE satellite in 1989; data assumed and processed from Mather et al. (1990) and further adapted from HyperPhysics (2010) and Putz (2010).

More recently, were found temperature fluctuations, in the structure itself of the background radiation; this was possible with the aid of the WMAP (Wilkinson Microwave Anisotropy Probe), launched in one of the Lagrangian points of the Earth-Moon-Sun system (L2 more accurately, to ensure a minimum energy consumption), see Figure 1.10, from where the temperature fluctuations around the background one of 2.74 K was observed with an error about $\Delta T/T = 6\text{x}10^{-6}$ which, in the reversed ratio framework represent a higher value (or sensible close) to the temporal horizon of radiation-substance decoupling ($t=10^5$ years $\leftrightarrow$ $T=10^5$ K). From these results that the (quantum) fluctuations have a temporal scale higher than the time wherefrom the background radiation comes, Figure 1.11, from where the quantum indeterminacy at the universe scales, after Big-Bang. The discovery made by the WMAP satellite is particularly exciting since it seems that exactly these quantum fluctuations generate the so-called Background Radiation *anisotropy* of the Universe (after Big Bang), visible in the polarization of the cosmic radiation, indicated at the top of Figure 1.11, which may be the quantum cause of the major global effects such as:

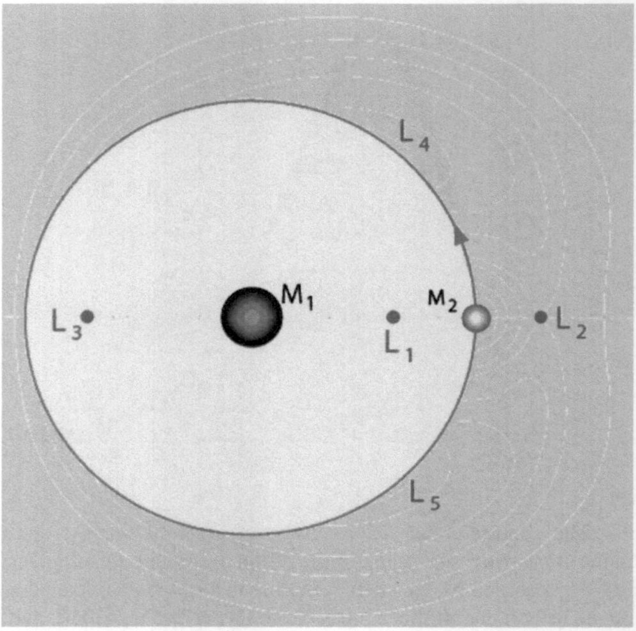

FIGURE 1.10 The illustration of Lagrange points (L1, L2, L3 – instable, L4 and L5 – stable) for the Earth-Moon binary system, representing the points of binary minimum potential corresponding to the minimum attraction of both bodies in a reciprocal motion (HyperPhysics, 2010; Putz, 2010).

- the separation of matter from the anti-matter (under the so-called violation of C-charge symmetry related to the invariance of physical laws at the reverse of electric charges);
- the preferential formation of the substance in certain areas of the Universe, with the presence of the dark energy, probably also responsible of the accelerated expansion of the Universe too (the so-called violation of P-parity symmetry related to the invariance of physical laws for the inversion of spatial coordinates);
- the establishment of an arrow of time leading to a sense of evolution for the Universe (the so-called violation of T-time symmetry related to the invariance of physical laws for the inversion of the temporal coordinates).

Yet these last important issues are only quoted here being left for a well deserved detailed presentation and discussion for other occasion and textbook.

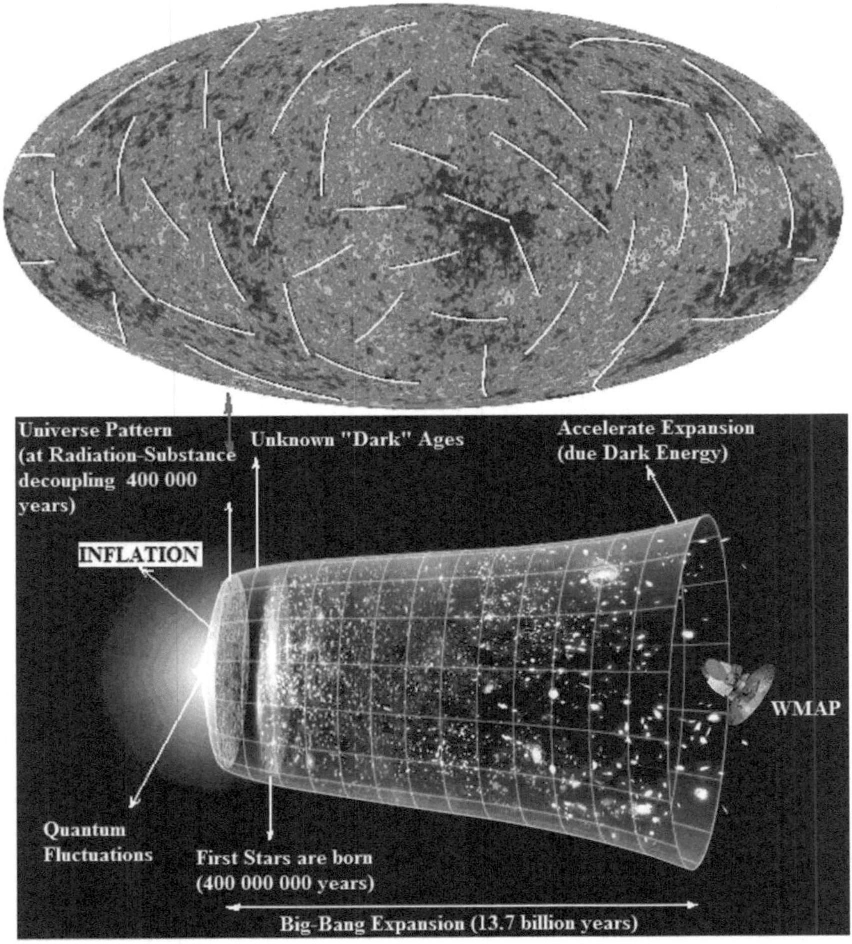

FIGURE 1.11 The map of microwave background radiation, observed by the WMAP satellite (top) and its causal carryover to the evolution of the Universe; adapted from NASA (2010) and Putz (2010).

1.5 CONCLUSION

The main lessons to be kept for the further theoretical and applicative investigations of the phenomenological matter's quantification that are approached in the present chapter pertain to the following:

- identifying the necessity for wave quantification as arisen from the problem/paradox of the black-body radiation;

- employing the Einstein assumptions of "quantum light" by absorption and stimulated and spontaneous emission in recovering Planck' distribution of photons;
- writing the de Broglie formula for matter-to-wave quantification for a given system with rest mass;
- dealing with wave packet form and properties in jointly recovering the Planck and de Broglie quantification formula for waves and substance, respectively;
- characterizing any objective entity by wave function with normalization property over its existence space;
- understanding the stable systems quantification through the complementarity between corpuscular and undulatory nature of the structural constituents;
- describing the quantum world by classical correspondence for higher quantum numbers (e.g., valence shells of atoms);
- learning the quantum causes of electronic configuration in atoms, including the shielding (Moseley) effects;
- treating systems of identical particles as statistical samples;
- solving the statistical paradox even for small particles or single particles: due to the quantum (wave) nature the quantum statistics prescribes a practically infinite number of states to be occupied by low number or even single particle with its existence probability spread over the entire spectrum of states;
- formulating the microscopic statistics for fermions (Fermi-Dirac) and bosons (Bose-Einstein);
- interpreting the matter structure by fundamental particles and forces, associated with fermions and bosons and of their statistics, respectively;
- connecting the fundamental fermions with the particles constitutive of matter and the bosons with the particles (usually with zero of very low rest mass) carrying the interactions/forces between the first ones;
- developing the thermal radiative phenomenology by quantum nature of the light/electromagnetic field, with universal value, i.e., with astrophysical relevance as well as for substance heating;
- finding applications for quantum-thermal correlation via the Wien law connecting temperature with the wavelength of the emitter electromagnetic radiation, with projection from micro- to macro- scale;
- modeling the Universe's "evolution arrow" and the noted experimental anisotropy by symmetry breakings in fundamental particles' charge, time and space.

KEYWORDS

- **atomic periodicity**
- **black-body radiation**
- **Bohr correspondence principle**
- **Bohr quantification**
- **Born normalization**
- **Bose-Einstein statistics**
- **de Broglie formula**
- **Einstein quantification**
- **elementary particles**
- **Fermi-Einstein statistics**
- **Fundamental forces**
- **Heisenberg indeterminacy**
- **Maxwell-Boltzmann statistics**
- **Moseley Law**
- **partition function**
- **Planck quantification**
- **Stefan-Boltzmann law**
- **the wave-packet**
- **the Wien law**

REFERENCES

AUTHOR'S MAIN REFERENCES

Putz, M. V. (2010). *Environmental Physics and the Universe* (in Romanian), The West University of Timişoara Publishing House, Timişoara.

Putz, M. V. (2014). Nanouniverse expanding macrouniverse: from elementary particles to dark matter and energy. In: *Research Horizons of Nanosystems Structure, Properties and Interactions*, Putz, M. V. (Ed.), Apple Academics, Toronto, Canada.

Putz, M. V., Lazea, M., Chiriac, A. (2010). *Introduction in Physical Chemistry. The Structure and Properties of Atoms and Molecules* (in Romanian), Mirton Publishing House, Timişoara.

SPECIFIC REFERENCES

Evans, C. R. (at University of North Carolina): http://user.physics.unc.edu/~evans/pub/A31/Lecture24-25-Big-Bang/

HyperPhysics (2010). http://hyperphysics.phy-astr.gsu.edu/Hbase/hframe.html

Mather, J. C., Cheng, E. S., Eplee, R. E. Jr., Isaacman, R. B., Meyer, S. S., Shafer, R. A., Weiss, R., Wright, E. L., Bennett, C. L., Boggess, N. W., Dwek, E., Gulkis, S., Hauser, M. G., Janssen, M., Kelsall, T., Lubin, P. M., Moseley S H. Jr., Murdock, T. L., Silverberg, R. F., Smoot, G. F., Wilkinson, D. T. (1990). A Preliminary Measurement of the Cosmic Microwave Background Spectrum by the Cosmic Background Explorer (COBE) Satellite, *Astrophysical Journal, Part 2 - Letters*, 354(May 10), L37–L40.

NASA (National Agency for Space Administration): http://map.gsfc.nasa.gov.

NIST (National Institute for Standards and Technology): http://physics.nist.gov/PhysRef-Data

Stierstadt, K. (1989). Physik der Materie, VCH-Wiley, Weinheim.

White, H. E. (1934). *Introduction to Atomic Spectra*, McGraw-Hill, Inc., New York, pp. 40.

FURTHER READINGS

Bohm, D. (1980). *Wholeness and the Implicate Order*, Routledge, London.

Bohr, N. (1921). *Abhandlungen* über *Atombau au den Jahren 1913–1916*, Vieweg & Sohn, Braunschweig.

De Broglie, L. (1966). *Certitudes et Incertitudes de la Science*, Editions Albin Michel, Paris.

Einstein, A. (1955, 2005). *Mein Weltbild*, Ullstein, Frankfurt am Main.

Hawking, S. (1994). *Black Holes and Baby Universes*, Bantam Books, New York.

Heisenberg, W. (1955). *Das Naturbild der Heutigen Physik*, Rowohlt Taschenbuch Verlag GmbH, Hamburg.

Heisenberg, W. (1969, 1981). *Der Tail und das Ganze*, Piper Verlag GmbH, München.

Kastler, A. (1976). *Cette* Étrange *Matière*, Stock, Paris.

Lightman, A. P. (1993). *Einstein's dreams*, Pantheon Books of Random House, Inc., New York.

Prigogine, I., Stengers, I. (1988). *Entre le Temps et L'Éternité*, Librairie Arthème Fayard, Paris.

Rees, M. J. (1999). *Just Six Numbers. The Deep Forces that Shape the Universe*, Phoenix of Orion Publishing Group Ltd., London.

Weinberg, S. (1977). *The First Three Minutes*, Bantam Books, New York.

CHAPTER 2

FORMALIZATION OF QUANTUM MECHANICS

CONTENTS

ABSTRACT

Although equivalent to certain points, various approaches of quantum phenomena may add new insights and perspective on studied systems either in isolate or interaction states. Here, the main formalisms are reviewed and illustrated, especially those making the ground-*nous* of the quantum mechanics itself, i.e., the undulatory quantum mechanics, while introducing the quantum propagator (Green function) concept, the semi-classical approach based on Euler-Lagrange and Hamiltonian, while the *bra-ket* Dirac picture opens the new perspectives of further developing quantum theory by abstract yet logical and formalized reasoning.

2.1 INTRODUCTION

One attempts to formalize the quantum observed phenomena of Chapter 1 of the present volume. In this respect, while following the didactic line for a modern physical-chemical course, one should expect to get acquaintance with the main formal concepts of quantum theory in general, namely the basic curricula:

- *Schrödingerequation*.Energyeigenstates.Freepropagator.Variat-ional principle. Hamilton's equation. Potential and ground state. Temporal Schrödinger equation. Klein-Gordon equation. Electronic spin.
- *The principles of action in classical and quantum mechanics' perspective* – short history. Basic concepts of force, motion, mass and units of physical quantities used in laws of motion. Quick survey of laws of motion. The Lagrangian function and its main role in the principle of least action. The motion by Euler-Lagrange equation. Newton equation and the second principle of classical mechanics. Correspondence with quantum mechanics.
- *Equation of quantum state*. The Dirac bra-ket formalism of quantum mechanics. Representation of the wave-momentum and coordinates. The adjunct operators. Hermiticity. Normal and adjunct operators. Scalar multiplication. Hilbert space. Dirac function. Orthogonality and orthonormality. Commutators. The completely set of commuting operators.

This way the quantum concepts are therefore introduced in three steps' phenomenology: formulating the quantum mechanics main equation-looking for classical roots of quantum mechanics-exploring the general formalization of quantum equation, states, and solutions. The corresponding issues are in the following unfolded, in grand detail in the main text, with the formal results summarized at the end of the chapter.

2.2 WAVE FUNCTION PICTURE

2.2.1 GREEN AND DIRAC FUNCTIONS

Consider 3D representation of coordinate space

$$\mathbf{r} = x\mathbf{i} + y\mathbf{j} + z\mathbf{k} \tag{2.1}$$

in terms of orthogonal vectors (unity vectors)

$$|\mathbf{i}| = |\mathbf{j}| = |\mathbf{k}| = 1 \quad \mathbf{i}\cdot\mathbf{j} = \mathbf{i}\cdot\mathbf{k} = \mathbf{j}\cdot\mathbf{k} = 0 \tag{2.2}$$

with the help of which one may use the generalized forms of Eqs. (1.74)–(1.76), employing the Fourier transformation of coordinate to momentum wave function dependency

$$\psi(\mathbf{r},t)=\frac{1}{(2\pi\hbar)^{3/2}}\int_{(\infty)} d\mathbf{p}\psi(\mathbf{p})e^{\frac{i}{\hbar}(\mathbf{p}\cdot\mathbf{r}-Et)} \tag{2.3}$$

There will be interesting to re-express the wave function (2.3) as a coordinate dependency only; to this aim one reconsiders eq. (2.3) for initial time

$$\psi(\mathbf{r},0)=\frac{1}{(2\pi\hbar)^{3/2}}\int_{(\infty)} d\mathbf{p}\psi(\mathbf{p})e^{\frac{i}{\hbar}\mathbf{p}\cdot\mathbf{r}} \tag{2.4}$$

From where having the inverse Fourier specialization of momentum wave function in terms of initial coordinate wave function (distribution)

$$\psi(\mathbf{p})=\frac{1}{(2\pi\hbar)^{3/2}}\int_{(\infty)} d\mathbf{r}\psi(\mathbf{r},0)e^{-\frac{i}{\hbar}\mathbf{p}\cdot\mathbf{r}} \tag{2.5}$$

With the form (2.5) back in time dependent wave function (2.3) one gets its re-shape

$$\psi(\mathbf{r},t)=\frac{1}{(2\pi\hbar)^{3}}\int_{(\infty)} d\mathbf{p}d\mathbf{r}'\psi(\mathbf{r}',0)e^{\frac{i}{\hbar}(\mathbf{p}\cdot\mathbf{r}-Et)}e^{-\frac{i}{\hbar}\mathbf{p}\cdot\mathbf{r}'} \tag{2.6}$$

in terms of the so called Green-function,

$$G(\mathbf{r}\text{-}\mathbf{r}',t)=\frac{1}{(2\pi\hbar)^{3}}\int_{(\infty)} d\mathbf{p}e^{\frac{i}{\hbar}[\mathbf{p}\cdot(\mathbf{r}-\mathbf{r}')-Et]} \tag{2.7}$$

a major concept in quantum theory, which practically links the initial wave function (quantum state) with any further evolved quantum state,

$$\psi(\mathbf{r},t)=\int_{(\infty)} d\mathbf{r}'G(\mathbf{r}\text{-}\mathbf{r}',t)\psi(\mathbf{r}',0) \tag{2.8}$$

containing in this regard a sort of determinism inside the quantum free evolution (due to integral representation which practically covers all coordinate space and possible evolution histories – a matter on which we will deeply return with the occasion presenting the path integral approach of quantum mechanics in the third part of the present volume).

Nevertheless, observe the bi-local nature of the Green function which allows its interpretation as a distribution function, being just equivalent with eth celebrated Dirac delta-function

$$G(\mathbf{r}\text{-}\mathbf{r}',0)=\frac{1}{(2\pi\hbar)^3}\int_{(\infty)} d\mathbf{p}e^{\frac{i}{\hbar}\mathbf{p}\cdot(\mathbf{r}\text{-}\mathbf{r}')}=\delta(\mathbf{r}\text{-}\mathbf{r}')=\begin{cases}0, & \mathbf{r}\neq\mathbf{r}'\\ +\infty, & \mathbf{r}=\mathbf{r}'\end{cases} \tag{2.9}$$

Fulfilling the basic conditions of normalization

$$\int_{-\infty}^{+\infty}\delta(\mathbf{r}\text{-}\mathbf{r}')d\mathbf{r}=1 \tag{2.10}$$

And filtration

$$\int_{-\infty}^{+\infty}f(\mathbf{r})\delta(\mathbf{r}\text{-}\mathbf{r}')d\mathbf{r}=f(\mathbf{r}') \tag{2.11}$$

One note however that for initial times the working coordinate function can be identified with the initial coordinate wave function ($t=0: f\to\psi$). These physical-mathematical tools will be further employed in formalizing one of the subtlest theories of matter, the quantum theory of Nature.

2.2.2 MOMENTUM AND ENERGY OPERATORS

From above de Broglie wave-function packet one can identify at $t=0$ the functions

$$u_{\mathbf{p}}(\mathbf{r})=\frac{1}{(2\pi\hbar)^{3/2}}e^{\frac{i}{\hbar}\mathbf{p}\cdot\mathbf{r}},\quad \overline{u}_{\mathbf{p}}(\mathbf{r})=\frac{1}{(2\pi\hbar)^{3/2}}e^{-\frac{i}{\hbar}\mathbf{p}\cdot\mathbf{r}} \tag{2.12}$$

with the coordinate-momentum scalar products unfolded as

$$\mathbf{p}\cdot\mathbf{r}=p_x x+p_y y+p_z z \tag{2.13}$$

However, note the very intriguing fact that, while the wave function in *momentum representation* takes the form associable with a (generalized) *functional scalar product*

$$\psi_0(\mathbf{p})=\int_{(\infty)}\overline{u}_{\mathbf{p}}(\mathbf{r})\psi_0(\mathbf{r})d\mathbf{r}:=\left(u_{\mathbf{p}}(\mathbf{r}),\psi_0(\mathbf{r})\right) \tag{2.13}$$

this is not the case of the wave-function in the *coordinate representation* since the integral performed on convoluted function belonging to different spaces (and representations):

$$\psi_0(\mathbf{r}) = \int_{(\infty)} u_{\mathbf{p}}(\mathbf{r})\psi_0(\mathbf{p})d\mathbf{p} \tag{2.14}$$

Such striking situation is solved (explained) by the fact the function $u_{\mathbf{p}}(\mathbf{r})$ is not assimilated with ordinary wave-function because not obeys the normalization condition through providing a divergent square integral:

$$\int_{-\infty}^{+\infty} \left|u_{\mathbf{p}}(\mathbf{r})\right|^2 d\mathbf{r} = \int_{-\infty}^{+\infty} \bar{u}_{\mathbf{p}}(\mathbf{r})u_{\mathbf{p}}(\mathbf{r})d\mathbf{r} = \frac{1}{(2\pi\hbar)^3}\int_{-\infty}^{+\infty} d\mathbf{r} \to \infty \tag{2.15}$$

although being indefinite derivable respecting the space coordinates:

$$\frac{d^n}{dx^n}u_{\mathbf{p}}(\mathbf{r}) = \left(-\frac{i}{\hbar}\mathbf{p}_x\right)^n u_{\mathbf{p}}(\mathbf{r}) \tag{2.16}$$

and with finite norm in both zero- and n-th order of such derivation:

$$\left|u_{\mathbf{p}}(\mathbf{r})\right| = \frac{1}{(2\pi\hbar)^{3/2}}, \quad \left|\frac{d^n}{dx^n}u_{\mathbf{p}}(\mathbf{r})\right| = \left(\frac{p_x}{\hbar}\right)^n \frac{1}{(2\pi\hbar)^{3/2}} \tag{2.17}$$

Such functions are called distributions, so being more general than the ordinary wave functions since the chain spaces inclusion:

$$\underbrace{\mathscr{L}(\Re_{\mathbf{r}})}_{\int_{(\infty)} f d\mathbf{r}<\infty} \subset \underbrace{\mathscr{L}^2(\Re_{\mathbf{r}})}_{\int_{(\infty)} f^2 d\mathbf{r}<\infty} \subset \underbrace{\mathscr{D}(\Re_{\mathbf{r}})}_{\int_{(\infty)} f^2 d\mathbf{r}\to\infty} \tag{2.18}$$

among the simple integrable (trial wave functions), square integrable (true wave functions) and square non-integrable (temperate distributions) functions to be used as describing quantum particles and events.

The fact the function $u_{\mathbf{p}}(\mathbf{r})$ is in fact a temperate distribution can be seen also from its inner scalar product

$$\left(u_{\mathbf{p}'}, u_{\mathbf{p}}\right) = \int_{(\infty)} \bar{u}_{\mathbf{p}'}(\mathbf{r})u_{\mathbf{p}}(\mathbf{r})d\mathbf{r} = \delta(\mathbf{p}-\mathbf{p}') \tag{2.19}$$

leading with another distribution function, the Dirac (delta) function, previously introduced.

Nevertheless, one fundamental consequence of dealing with $u_{\mathbf{p}}(\mathbf{r})$ as a generalized function is that it can be considering as patterning a special (generalized in the sense of resumed) wave function abstracted from the *de Broglie wave-packet*, namely

$$\vartheta_t(\mathbf{r}) = u_{\mathbf{p}}(\mathbf{r})e^{-\frac{i}{\hbar}Et} = \frac{1}{(2\pi\hbar)^{3/2}}e^{\frac{i}{\hbar}(\mathbf{p}\cdot\mathbf{r}-Et)} \tag{2.20}$$

leading with the basic identities

$$\frac{\partial}{\partial t}\vartheta_t(\mathbf{r}) = -\frac{i}{\hbar}E\vartheta_t(\mathbf{r}) \tag{2.21}$$

$$\frac{\partial}{\partial x}\vartheta_t(\mathbf{r}) = \frac{i}{\hbar}p_x\vartheta_t(\mathbf{r}) \tag{2.22}$$

producing the *quantum operatorial* definitions for energy and momentum:

$$\hat{E}\bullet = i\hbar\frac{\partial}{\partial t}\bullet \tag{2.23}$$

$$\hat{p}_x\bullet = -i\hbar\frac{\partial}{\partial x}\bullet \tag{2.24}$$

while the 3D-momentum operatorial definition will look like

$$\hat{\mathbf{p}}\bullet = -i\hbar\nabla\bullet \tag{2.25}$$

in terms of nabla-differential operator over the space coordinates:

$$\nabla = \partial_i\partial^i \quad \partial^i := \partial/\partial x_i, \quad i = 1(for\, x), 2(for\, y), 3(for\, z) \tag{2.26}$$

These operators, along the multiplicative space-coordinate rules (equally for coordinates and vectors):

$$\hat{x}\bullet = x\bullet, \quad \hat{V}(x,t)\bullet = V(\hat{x},t)\bullet = V(x,t)\bullet \tag{2.27}$$

are of prime importance in developing the forthcoming quantum equations and modeling the driven (or measurable or observable) events.

2.2.3 KLEIN-GORDON AND SCHRLIDINGER EQUATIONS

Combining relativistic energy and momentum through velocity elimination (fro their derivations see Appendices A.5–A.7):

$$E = mc^2 = \frac{m_0 c^2}{\sqrt{1-(v/c)^2}} \tag{2.28}$$

$$p = mv = \frac{m_0 v}{\sqrt{1-(v/c)^2}} \tag{2.29}$$

one gets the general momentum energy relationship:

$$E^2 = m_0^2 c^4 + p^2 c^2 \tag{2.30}$$

from where the two possible energy solutions are obtained as

$$E = \pm\sqrt{m_0^2 c^4 + p^2 c^2} \tag{2.31}$$

with the Dirac celebrated energy solutions for the trapped particles ($\mathbf{p} = 0$)

$$E = \pm m_0 c^2 \tag{2.32}$$

while fixing the intriguing Dirac seas of positive and negative energy limits within which the any given substantial particle ($m_0 \neq 0$) evolves, see Figure 2.1; eventually, it collapses into the referential zero level for the photonic case ($m_0 \to 0$).

However, the quantum correspondence principle applied upon the square of the above energy with energetic and momentum considered as operators and applied on the working wave-function $\psi_t(\mathbf{r})$,

$$\hat{E}^2 \psi_t(\mathbf{r}) = c^2 \hat{\mathbf{p}}^2 \psi_t(\mathbf{r}) + m_0^2 c^4 \hat{1} \psi_t(\mathbf{r}) \tag{2.33}$$

leading with the so called Klein-Gordon equation:

$$\frac{1}{c^2}\frac{\partial^2}{\partial t^2}\psi_t(\mathbf{r}) = \left[\nabla^2 - \left(\frac{m_0 c}{\hbar}\right)^2\right]\psi_t(\mathbf{r}) \tag{2.34}$$

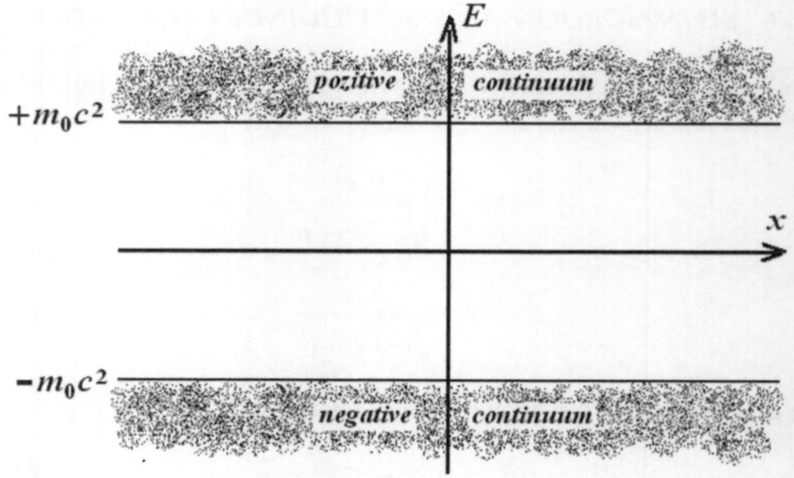

FIGURE 2.1 The separation ($2m_0c^2$) gap between the positive (E>+ m_0c^2) and negative (E<–m_0c^2) continuum ("Dirac seas") for the energy of a relativistic particle (Putz et al., 2010).

or written in the (1+3)D form

$$\left[\nu + \left(\frac{m_0 c}{\hbar} \right)^2 \right] \psi_t(\mathbf{r}) = 0 \tag{2.35}$$

where the D'Alambertian was defined in the (x_0=ct, x_1=x, x_2=y, x_3=z) ≈ (+, –, –, –) space-time relativistic metric

$$\nu = \partial_\mu \partial^\mu = \partial_0^2 - \partial_i \partial^i, \quad \partial^{\mu/i} = \partial / \partial x_{\mu/i}, \quad \mu = 0,1,2,3; i = \overline{1,3} \tag{2.36}$$

Turning now to the non-relativistic motion, one expands the positive (or *electronic*) energy-momentum relativistic above solution in term of (v/c) *yielding* in the first order expansion,

$$\left(1 + a\right)^{1/2}\Big|_{a \to 0} \cong 1 + a/2 \tag{2.37}$$

the actual relationship:

$$E \cong m_0 c^2 + \frac{p^2}{2m_0} \tag{2.38}$$

However, when the electronic motion is driven by a potential too, say $V(x)$, the last expression can be modified as such the energy spectrum be shifted with origin in the positive ($+m_0c^2$) rest energy so that the working energy expression become

$$E \cong V(\mathbf{r}) + \frac{p^2}{2m_0} \tag{2.39}$$

Now, considering once more the operatorial version of this equation with energy and momentum operatorial rules applied on the wave-function, while noting the space operator as well as space dependent function(s) do not modify the space dependence, see above, the final result unfold as the famous Schrödinger temporal equation

$$i\hbar\frac{\partial}{\partial t}\psi_t(\mathbf{r}) = \left[-\frac{\hbar^2}{2m_0}\nabla^2 + V(\mathbf{r})\right]\psi_t(\mathbf{r}) \tag{2.40}$$

in terms of Laplacian only this time:

$$\nabla^2 = \partial_i\partial^i,\ \partial^i = \partial/\partial x_i,\ i = \overline{1,3} \tag{2.41}$$

2.2.4 ELECTRONIC AND PHOTONIC SPINS

The major consequence of having both Klein-Gordon and Schrödinger equation springs out through looking on the symmetry and dissymmetry with which these equations transforms space in time, respectively. However, in order to achieve the symmetry of space-time transformation in both equations one can rewrite the time-power as being quantified by the s number (hereafter called as *spin*). This way, one can unfold the photonic and electronic wave equation in unitary manner since specializing the Klein-Gordon and Schrödinger equation with common features through imposing the shift $m_0c^2 \to 0$ on the first (regaining the ordinary wave-equation for electromagnetic fields, quantified by photons) and free motion $V(\mathbf{r}) \to 0$ for the second; the resulted photonic and electronic wave equations are:

$$\frac{\partial^{2s(\gamma)}}{\partial t^{2s(\gamma)}}\psi_t(\mathbf{r}) = c^2\nabla^2\psi_t(\mathbf{r}) \tag{2.42}$$

$$\frac{\partial^{2s(e)}}{\partial t^{2s(e)}}\psi_t(\mathbf{r}) = \frac{i\hbar}{2m_0}\nabla^2\psi_t(\mathbf{r}) \tag{2.43}$$

from where there follows the *photonic and electronic spin quantization* as:

$$s(\gamma) = 1,\ s(e) = \frac{1}{2} \tag{2.44}$$

in order to be in accordance with the operatorial deduced equations, respectively. With this we can affirm that:

- the spin concept is an intrinsic space-time transformation effect;
- the spin quantification depends on the relativistic or non-relativistic level of wave function expression whom the particle evolution is attributed;
- the spin quantification contains, at either relativistic or non-relativistic levels, the quantum influence of motion.

2.2.5 EIGEN-FUNCTIONS AND EIGEN-VALUES

If once consider appropriate factorized time-coordinate resumed wave function from above patterned distribution based solution, and then the working wave function will look like

$$\psi_t(\mathbf{r}) = \psi_0(\mathbf{r})e^{-\frac{i}{\hbar}Et} \tag{2.45}$$

while through considering it into the temporal Schrödinger equation it leaves in the first instance with

$$\left[-\frac{\hbar^2}{2m_0}\nabla^2 + V(\mathbf{r})\right]\psi_0(\mathbf{r}) = E\psi_0(\mathbf{r}) \tag{2.46}$$

thus providing the so called stationary Schrödinger equation that in operatorial form simply casts as an eigen-value problem:

$$\widehat{H}\psi = E\psi \tag{2.47}$$

being ψ the eigen-function and E the eigen-value to be determined once the Hamiltonian

$$\widehat{H} = -\frac{\hbar^2}{2m_0}\nabla^2 + V(\mathbf{r}) \tag{2.48}$$

is specialized in terms of applied potential $V(\mathbf{r})$.

In terms of Hamiltonian the corresponding temporal Schrödinger equation would be displayed as generalization of that stationary:

$$\widehat{H}\psi = i\hbar\partial_t\psi \tag{2.49}$$

where we have introduced the short-notation of time-derivation

$$\partial_t := \frac{\partial}{\partial t} \tag{2.50}$$

2.2.6 HERMITIC OPERATORS

Giving a general operator $\widehat{A}$ it is said to be *hermitic* or *self-adjoin* and is written as

$$\widehat{A}^{+} = \widehat{A} \tag{2.51}$$

if fulfills the general identity:

$$\int \psi^*\left(\widehat{A}\psi\right)d\text{“} = \int \psi\left(\widehat{A}\psi\right)^* d\text{“} = \int\left(\widehat{A}\psi\right)^*\psi d\text{“} \tag{2.52}$$

with $d\Gamma$ formally denoting the elementary (space-time) volume of integration.

As a useful illustration let's check the coordinate, momentum and Hamiltonian hermiticity. The position operator is hermitic

$$\widehat{x}^{+} = \widehat{x} \tag{2.53}$$

because fulfills the successive identities:

$$\int \psi^*\widehat{x}\psi dx = \int \psi^* x\psi dx = \int \psi x\psi^* dx = \int \psi\left(x\psi\right)^* dx = \int \psi\left(\widehat{x}\psi\right)^* dx \tag{2.54}$$

since $x=x^*$ due to its real nature; the momentum operator is as well hermitic

$$\widehat{p}_x^{+} = \widehat{p}_x \tag{2.55}$$

throughout the identities:

$$\int \psi^* \widehat{p}_x \psi \, dx = \int \psi^* \left(-i\hbar \frac{\partial}{\partial x} \right) \psi \, dx = -i\hbar \int \left[\frac{\partial}{\partial x} \left(\psi^* \psi \right) - \psi \frac{\partial}{\partial x} \psi^* \right] dx$$

$$= -i\hbar \int \frac{\partial}{\partial x} \left(\psi^* \psi \right) dx + \int \psi \left(i\hbar \frac{\partial}{\partial x} \right) \psi^* dx$$

$$= -i\hbar \underbrace{\left(\psi^* \psi \right)_{-\infty}^{+\infty}}_{0} + \int \psi \left(-i\hbar \frac{\partial}{\partial x} \psi \right)^* dx = \int \psi \left(\widehat{p}_x \psi \right)^* dx \quad (2.56)$$

since either direct or conjugate wave-function cancels at infinity; for the squared of momentum operator the hermitic property is even more direct proofed because it is a real operator

$$\widehat{p}_x^2 \bullet = \widehat{p}_x \left(\widehat{p}_x \right) \bullet = i\hbar \frac{\partial}{\partial x} \left(i\hbar \frac{\partial}{\partial x} \right) \bullet = -\hbar^2 \frac{\partial^2}{\partial x^2} \quad (2.57)$$

that automatically fulfills this condition; and the same for all other coordinates; all in all there is clear that the Hamiltonian operator

$$\widehat{H} = \frac{\widehat{p}^2}{2m} + V(\hat{x}, \hat{y}, \hat{z}) \quad (2.58)$$

as a sum of hermitic operators is an operator as well:

$$\widehat{H}^+ = \widehat{H} \quad (2.59)$$

The hermitic property of Hamiltonian may also directly be checked (or cross-checked) through the following construction; since assuming a prepared normalized state one successively has:

$$\partial_t \int \psi^* \psi \, d\Gamma = 0$$

$$\Leftrightarrow \int \left(\partial_t \psi^* \right) \psi \, d\Gamma + \int \psi^* \left(\partial_t \psi \right) d\Gamma = 0 \quad (2.60)$$

However, when considering in the last equality the direct and conjugate variants of temporal Schrödinger equation,

$$\widehat{H}\psi = i\hbar \partial_t \psi \quad \left(\widehat{H}\psi \right)^* = -i\hbar \partial_t \psi^* \quad (2.61)$$

one gets the hermiticity condition for Hamiltonian fulfilled:

$$\int \left(\widehat{H}\psi\right)^{*} \psi d\Gamma = \int \psi^{*} \left(\widehat{H}\psi\right) d\Gamma \tag{2.62}$$

In general, the hermitic property is usually associated with operators that correspond with observables, i.e., providing a unique measured value when applied (operates) on a given (prepared) state characterized by an eigen-function ψ. This can be easily check out by defining the observed value of an operator as its average measure

$$\left\langle \widehat{A} \right\rangle = \int \psi^{*} \widehat{A}\psi d\Gamma \tag{2.63}$$

for a (Born) normalized (prepared) eigen-state:

$$1 = \int \psi^{*} \psi d\Gamma \tag{2.64}$$

and then by applying it to calculate its the zero observed dispersion (square of standard deviation). Actually, for observed (average) dispersion one has the value:

$$\begin{aligned} \left\langle \left(\Delta\widehat{A}\right)^{2} \right\rangle &= \int \psi^{*} \left(\widehat{A} - \left\langle \widehat{A} \right\rangle\right)^{2} \psi d\Gamma = \int \grave{E}^{*} \left(\widehat{A} - \left\langle \widehat{A} \right\rangle\right)\left(\widehat{A} - \left\langle \widehat{A} \right\rangle\right) \grave{E} d\Gamma \\ &= \int \left[\left(\widehat{A} - \left\langle \widehat{A} \right\rangle\right)\grave{E}\right]^{*} \left(\widehat{A} - \left\langle \widehat{A} \right\rangle\right)\grave{E} d\Gamma = \int \left|\left(\widehat{A} - \left\langle \widehat{A} \right\rangle\right)\grave{E}\right|^{2} d\Gamma \end{aligned} \tag{2.65}$$

that when goes to zero leaves with the eigen-value problem from the hermitic operator:

$$\widehat{A}\psi = a_{\grave{E}}\Psi, \quad a_{\psi} = \left\langle \widehat{A} \right\rangle_{\psi} \tag{2.66}$$

assuring therefore its fully observable character.

Next, worth treating the superposition problem: how eigen-functions of the same operator, having the same eigen-value, i.e., being called *degenerate functions* for degenerate eigen-states, behave if they are considered composed in a linear manner? For better illustration of the answer let's take the two eigen-function case

$$\widehat{A}\varphi_{1,2} = a\varphi_{1,2} \tag{2.67}$$

producing the superposition wave-function:

$$\psi = c_1\varphi_1 + c_2\varphi_2,\ c_{1,2} \in \Im \tag{2.68}$$

Then, one can check directly that the eigen-problem is conserved at the superimposed level by the successive determinations:

$$\hat{A}\psi = \hat{A}(c_1\varphi_1 + c_2\varphi_2) = c_1\hat{A}\varphi_1 + c_2\hat{A}\varphi_2 = a(c_1\varphi_1 + c_2\varphi_2) = a\psi \tag{2.69}$$

supporting the appropriate generalization:

$$\exists\hat{A}\varphi_n = a\varphi_n,\ n = \overline{1,g} \Rightarrow \forall\psi_g = \sum_{n=1}^{g} c_n\varphi_n,\ c_n \in \Im \left| \hat{A}\psi_g = a\psi_g \right. \tag{2.70}$$

In the same general context, one can say that the eigen-value a is g-fold degenerate if there exist exactly g – linear independent eigen-functions, i.e., having the closure property

$$\sum_{n=1}^{g} c_n\varphi_n = 0 \Leftrightarrow c_1 = c_2 = \ldots = c_g = 0 \tag{2.71}$$

carrying the same eigen-value problem. Then their superposition gives another eigen-function of the same operator, with the same eigen-value; this is the consecration of the so-called *superposition principle* in quantum theory.

There eventually remains to unfold the meaning of hermiticity condition for the superposition eigen-function in terms of its g-eigen-components. To this aim, one rewrites the hermiticity expression

$$\int \psi_g^*\left(\hat{A}\psi_g\right)d\Gamma = \int\left(\hat{A}\psi_g\right)^*\psi_g d\Gamma \tag{2.72}$$

under its generalized superposition form:

$$\int \sum_{m=1}^{g} c_m^*\varphi_m^*\left(\hat{A}\sum_{n=1}^{g} c_n\varphi_n\right)d\Gamma = \int\left(\hat{A}\sum_{m=1}^{g} c_m\varphi_m\right)^*\sum_{n=1}^{g} c_n\varphi_n d\Gamma \tag{2.73}$$

and equivalently as:

$$\sum_{m,n=1}^{g} c_m^* c_n\left[\int \varphi_m^*\left(\hat{A}\varphi_n\right)d\Gamma - \int\left(\hat{A}\varphi_m\right)^*\varphi_n d\Gamma\right] = 0 \tag{2.74}$$

from where there follows the generalized hermiticity condition for an operator in terms of its degenerate eigen-functions:

$$\int \varphi_m^* \left(\hat{A}\varphi_n\right) d\Gamma = \int \left(\hat{A}\varphi_m\right)^* \varphi_n d\Gamma \tag{2.75}$$

With these, one may consider the further case in which there are two wave functions, both as eigen-values of the same hermitic operator, yet producing two different eigen-values:

$$\hat{A}\psi_n = a_n\psi_n, \quad \hat{A}\psi_m = a_m\psi_m, \quad a_n \uparrow a_m \tag{2.76}$$

In these conditions, how we should regard the eigen-functions ψ_n, ψ_m? The answer is that they have to be *orthogonal*; the proof starts from considering one the above eigen-value problem, say that of ψ_n, multiplied on left by the conjugated of the remaining eigen-function, here ψ_m, integrating the resulted equation over the space-time volume:

$$\int \psi_m^* \left(\hat{A}\psi_n\right) d\Gamma = a_n \int \psi_m^* \psi_n d\Gamma \tag{2.77}$$

taking its conjugate (not forget that a_n is a real value):

$$\int \psi_m \left(\hat{A}\psi_n\right)^* d\Gamma = a_n \int \psi_m \psi_n^* d\Gamma \tag{2.78}$$

performing the index inversion $m \leftrightarrow n$:

$$\int \psi_n \left(\hat{A}\psi_m\right)^* d\Gamma = a_m \int \psi_m^* \psi_n d\Gamma \tag{2.79}$$

using the hermiticity property of the involved operator

$$\int \psi_m^* \hat{A}\psi_n d\Gamma = a_m \int \psi_m^* \psi_n d\Gamma \tag{2.80}$$

and being finally subtracted from the initial one to give:

$$0 = (a_n - a_m) \int \psi_m^* \psi_n d\Gamma \tag{2.81}$$

leaving with the general ortho-normal condition for the eigen-functions belonging to the same hermitic operator:

$$\int \psi_m^* \psi_n d\Gamma = \delta_{mn} \tag{2.82}$$

in terms of the delta-Kronecker tensor

$$\delta_{mn} = \begin{cases} 1, m = n \\ 0, m \neq n \end{cases} \tag{2.83}$$

Further use of hermiticity and eigen-value properties are in next employed in obtaining specific quantum theorems.

2.2.7 HEISENBERG UNCERTAINTY THEOREM

THEOREM: Two non-commutative hermitic operators cannot provide simultaneous observable measurements with the same precision on a given (prepared) eigen-state.

To proceed with the proof, note that two operators are said to be non-commutative if their commutator

$$\left[\hat{A}, \hat{B}\right] = \hat{A}\hat{B} - \hat{B}\hat{A} \tag{2.84}$$

is non-zero, and commutative otherwise.

Now, considering two hermitic non-commutative operators one would be interested in behavior their standard deviation forms, namely the behavior of the new operators

$$\hat{\alpha} = \hat{A} - \left\langle \hat{A} \right\rangle, \ \hat{\beta} = \hat{B} - \left\langle \hat{B} \right\rangle \tag{2.85}$$

towards measurements. Then the simultaneous measurement of both operators on an eigen-state for both operators ϕ, i.e., prepared as $1 = \int \varphi^* \varphi d\Gamma$, would mean that the above standard deviation operators be applied simultaneously on that state in accordance with the superposition principle; say that the resulting state has the general (imaginary) form:

$$\psi_M = \left(\hat{\alpha} + i\lambda\hat{\beta}\right)\phi, \ \lambda \in \Re \tag{2.86}$$

However, this resulted state has to be characterized at least by the positive probability of existence, that is:

$$\begin{aligned}0 \le \int \psi_M^* \psi_M d\Gamma &= \int \left[\left(\hat{\alpha} + i\lambda\hat{\beta}\right)\varphi\right]^* \left(\hat{\alpha} + i\lambda\hat{\beta}\right)\phi d\Gamma \\ &= \int \varphi^* \left(\hat{\alpha} + i\lambda\hat{\beta}\right)\left(\hat{\alpha} + i\lambda\hat{\beta}\right)\phi d\Gamma \\ &= \int \left\{\phi^* \hat{\alpha}^2 \varphi + i\lambda\lambda^* \left(\hat{\alpha}\hat{\beta} - \hat{\beta}\hat{\alpha}\right)\varphi + \lambda^2 \varphi^* \hat{\beta}^2 \phi\right\} d\Gamma \\ &= \lambda^2 \left\langle \hat{\beta}^2 \right\rangle - \lambda \left\langle \hat{\gamma} \right\rangle + \left\langle \hat{\alpha}^2 \right\rangle \\ &= \underbrace{\left\langle \hat{\beta}^2 \right\rangle \left[\lambda - \frac{\left\langle \hat{\gamma} \right\rangle}{2\left\langle \hat{\beta}^2 \right\rangle}\right]^2}_{\ge 0} + \frac{1}{4\left\langle \hat{\beta}^2 \right\rangle}\left[4\left\langle \hat{\alpha}^2 \right\rangle \left\langle \hat{\beta}^2 \right\rangle - \left\langle \hat{\gamma} \right\rangle^2\right]\end{aligned} \tag{2.87}$$

yielding with the operatorial identity:

$$\sqrt{\left\langle \hat{\alpha}^2 \right\rangle}\sqrt{\left\langle \hat{\beta}^2 \right\rangle} \ge \frac{1}{2}\left|\left\langle \hat{\gamma} \right\rangle\right| \tag{2.88}$$

where we have introduced the new operator as the commutator:

$$\hat{\gamma} = -i\left[\hat{\alpha}, \hat{\beta}\right] = -i\left[\hat{A} - \left\langle \hat{A} \right\rangle, \hat{B} - \left\langle \hat{B} \right\rangle\right] = -i\left[\hat{A}, \hat{B}\right] \tag{2.89}$$

Now, noting that, in fact, we are dealing with the dispersion statistical definition, see for instance the averages' equivalencies for the first operator:

$$\sqrt{\left\langle \hat{\alpha}^2 \right\rangle} = \sqrt{\left\langle \left(\hat{A} - \left\langle \hat{A} \right\rangle\right)\left(\hat{A} - \left\langle \hat{A} \right\rangle\right) \right\rangle} = \sqrt{\left\langle \hat{A}^2 \right\rangle - \left\langle \hat{A} \right\rangle^2} = \Delta\hat{A} \tag{2.90}$$

and for the other operator as well:

$$\sqrt{\left\langle \hat{\beta}^2 \right\rangle} = \Delta\hat{B} \tag{2.91}$$

one yields the general operatorial proof for the Uncertainty Theorem through the basic inequality:

$$\Delta\hat{A}\Delta\hat{B} \ge \frac{1}{2}\left|\left\langle i\left[\hat{A}, \hat{B}\right]\right\rangle_\phi\right| \tag{2.92}$$

There is clear now that if the commutator of the two (observed) operators is eventually zero,

$$\left[\hat{A},\hat{B}\right]=0 \tag{2.93}$$

the uncertainty relation becomes

$$\Delta\hat{A}\Delta\hat{B}\geq 0 \tag{2.94}$$

so that allowing the limiting situation in which

$$\Delta\hat{A}=0\Leftrightarrow\Delta\hat{B}=0 \tag{2.95}$$

that gives the same precision in measuring both operators/observables on the prepared eigen-state.

As a natural consequence when the commutator of two operators/observables is non-zero, then the high precision in measuring one of it implies the infinite error in observing the eigen-values of the other. This is the case for the space and momentum operators on a certain direction that, while obeying the formal eigen-value problems

$$\hat{x}\phi=x\phi\ ,\ \hat{p}_x\phi=-i\hbar\partial_x\phi \tag{2.96}$$

are linked by the non-zero effect of the commutator:

$$\left[\hat{x},\hat{p}_x\right]\phi=\left(\hat{x}\hat{p}_x-\hat{p}_x\hat{x}\right)\phi=-i\hbar x\partial_x\phi+i\hbar\partial_x\left(x\phi\right)=i\hbar\phi \tag{2.97}$$

this way consecrating the relationship

$$\left[\hat{x},\hat{p}_x\right]=i\hbar\hat{1} \tag{2.98}$$

to provide uncertainty Heisenberg space-momentum relationship:

$$\Delta\hat{x}\Delta\hat{p}_x\geq\frac{1}{2}\hbar\hat{1} \tag{2.99}$$

calling that the average of the unity operator is unity due to normalized prepared ϕ state.

However, a special discussion holds when the energy and time variables are to be linked, eventually through a spontaneous emission from a certain eigen-energy level where the electrons are inherently (spectroscopic) associated with a certain lifetime as well. Still, since the energy has an operator while the time has not, they cannot be combined in the above prescribed uncertainty way. In this case another construction should be considered as follows.

Let a system being characterized by a Hamiltonian with an energy spectrum consisting by two eigen-energies $E_n \neq E_m$ and two eigen-functions$\phi_n \neq \phi_m$, while the generalization to a generalized spectrum is straightforward. According with the above exposed eigen-values and eigen-function properties it follows the eigen-functions as orthogonal, while their superposition will generate another eigen-function of the Hamiltonian:

$$\psi(\mathbf{r},t) = c_n\phi_n(\mathbf{r})e^{-i\frac{E_n t}{\hbar}} + c_m\phi_m(\mathbf{r})e^{-i\frac{E_m t}{\hbar}} \tag{2.100}$$

with the normalized (preparation) condition leading with condition

$$1 = \int \psi(\mathbf{r},t)^* \psi(\mathbf{r},t) d\mathbf{r} \Rightarrow 1 = |c_n|^2 + |c_m|^2 \tag{2.101}$$

since the ortho-normalized functions ϕ_n, ϕ_m.

However, measuring the energy in the above superposition state will give out either E_n or E_m energies, being therefore the energy uncertainty in the system given as

$$\delta E_{nm} = |E_n - E_m| \tag{2.102}$$

i.e., associated with their reciprocally distance.

On the other side, while measuring other hermitic observable property of the system in the superposition state, one has its expectation value successively written as:

$$\begin{aligned}
\left\langle \hat{O} \right\rangle_\psi &= \int \psi(\mathbf{r},t)^* \hat{O} \psi(\mathbf{r},t) d\mathbf{r} \\
&= |c_n|^2 O_{nn} + |c_m|^2 O_{mm} + c_n^* c_m O_{nm} e^{i\frac{E_n - E_m}{\hbar}t} + c_m^* c_n O_{mn} e^{-i\frac{E_n - E_m}{\hbar}t} \\
&= |c_n|^2 O_{nn} + |c_m|^2 O_{mm} + 2\Re e\left[c_n^* c_m O_{nm} e^{i\frac{E_n - E_m}{\hbar}t} \right]
\end{aligned} \tag{2.103}$$

where was considered the notation and its hermiticity property

$$O_{nm} = \int \phi_n^*(\mathbf{r})\widehat{\mathbf{O}}\phi_m(\mathbf{r})d\mathbf{r} = O_{mn}^* \tag{2.104}$$

to resume the oscillatory terms above. The time-dependency of the measured property may induce the time indeterminacy only if it is driven by lower transition frequency (i.e., by appreciable *nm* transition time) respecting that expected between the considered eigen-states at the instantaneous emission-absorption process:

$$\upsilon_{nm} = \frac{1}{\delta t_{nm}} \le \frac{|E_n - E_m|}{\hbar} = \frac{\delta E_{nm}}{\hbar} \tag{2.105}$$

There is obvious now that the spectroscopic (or time-energy) Heisenberg uncertainty relation

$$\delta E_{nm}\delta t_{nm} \ge \hbar \tag{2.106}$$

stands as a special realization or as a generalization of Planck-Bohr transition frequency postulate, namely:

$$\delta E_{nm} \ge \hbar\upsilon_{nm} \tag{2.107}$$

2.2.8 EHRENFEST THEOREM

THEOREM: Quantum hermitic operators of coordinate and momentum fulfill the Newtonian laws of motion

$$\dot{x} = p_x / m_0 \text{ ... kinetic equation of motion} \tag{2.108}$$

$$\dot{p}_x = F_x = -\partial_x V(x) \text{ ... dynamic equation of motion} \tag{2.109}$$

in terms of their expectation values.

The proof is based on evaluation of the time-evolution for the expectation value for a given hermitic operator

$$
\begin{aligned}
\frac{d}{dt}\left\langle \widehat{A} \right\rangle &= \partial_t \int \psi^* \widehat{A} \psi \, d\Gamma \\
&= \int \left(\partial_t \psi^*\right) \widehat{A} \psi \, d\Gamma + \int \psi^* \left(\partial_t \widehat{A}\right) \psi \, d\Gamma + \int \psi^* \widehat{A} \left(\partial_t \psi\right) d\Gamma \\
&= \int \left(\frac{1}{i\hbar} \widehat{H} \psi \right)^* \widehat{A} \psi \, d\Gamma + \left\langle \partial_t \widehat{A} \right\rangle + \int \psi^* \widehat{\mathrm{A}} \left(\frac{1}{i\hbar} \widehat{H} \psi \right) d\Gamma \\
&= -\frac{1}{i\hbar} \int \psi^* \widehat{H} \widehat{A} \psi \, d\Gamma + \frac{1}{i\hbar} \int \psi^* \widehat{A} \widehat{H} \psi \, d\Gamma + \left\langle \partial_t \widehat{A} \right\rangle \\
&= \frac{1}{i\hbar} \int \psi^* \left[\widehat{A}, \widehat{H} \right] \psi \, d\Gamma + \left\langle \partial_t \widehat{A} \right\rangle = \frac{1}{i\hbar} \left\langle \left[\widehat{A}, \widehat{H} \right] \right\rangle + \left\langle \partial_t \widehat{A} \right\rangle
\end{aligned}
\tag{2.110}
$$

from where there follows the time-dependent operator expectation equation:

$$
i\hbar \frac{d}{dt}\left\langle \widehat{A} \right\rangle = \left\langle \left[\widehat{A}, \widehat{H} \right] \right\rangle + i\hbar \left\langle \partial_t \widehat{A} \right\rangle \tag{2.111}
$$

We are going now to apply this equation to space and momentum operators. However, since the commutator to evaluate involves Hamiltonian that contains the squared of momentum operator one needs to employ the distribution of multiplication commutator rule (easy to be checked out):

$$
\left[\widehat{A}, \widehat{B}\widehat{C} \right] = \left[\widehat{A}, \widehat{B} \right] \widehat{C} + \widehat{B} \left[\widehat{A}, \widehat{C} \right] \tag{2.112}
$$

Therefore, for the space operator we have:

$$
\begin{aligned}
\left[\hat{x}, \widehat{H} \right] &= \left[\hat{x}, \frac{1}{2m_0} \hat{p}_x^2 \right] + \underbrace{\left[\hat{x}, V(\hat{x}) \right]}_{0} \\
&= \frac{1}{2m_0} \underbrace{\left[\hat{x}, \hat{p}_x \right]}_{i\hbar} \hat{p}_x + \frac{1}{2m_0} \hat{p}_x \underbrace{\left[\hat{x}, \hat{p}_x \right]}_{i\hbar} = \frac{i\hbar}{m_0} \hat{p}_x
\end{aligned}
\tag{2.113}
$$

and assuming the Schrödinger picture of non-temporal operators (see also Section 2.2.3)

$$
\left\langle \partial_t \hat{x} \right\rangle = 0 \tag{2.114}
$$

the space operatorial time-dependent equation takes the form

$$\frac{d}{dt}\left\langle \hat{x} \right\rangle = \frac{1}{m_0}\left\langle \hat{p}_x \right\rangle \tag{2.115}$$

that consecrates the expectation value counterpart of the classical momentum definition.

In the same manner for the momentum operator we have:

$$\begin{aligned}\left[\hat{p}_x, \hat{H}\right] &= \left[\hat{p}_x, \frac{1}{2m_0}\hat{p}_x^2\right] + \left[\hat{p}_x, V(\hat{x})\right] \\ &= \frac{1}{2m_0}\underbrace{\left[\hat{p}_x, \hat{p}_x\right]}_{0}\hat{p}_x + \frac{1}{2m_0}\hat{p}_x\underbrace{\left[\hat{p}_x, \hat{p}_x\right]}_{0} + \left[\hat{p}_x, V(\hat{x})\right] \\ &= \left[\hat{p}_x, V(\hat{x})\right]\end{aligned} \tag{2.116}$$

The evaluation of the last commutator we unfold as successively acting on a trial wave-function to get:

$$\begin{aligned}\underbrace{\left[\hat{p}_x, f(\hat{x})\right]}_{\mapsto}\varphi &= \left[-i\hbar\partial_x, f(x)\right]\varphi = -i\hbar\partial_x\left[f(x)\varphi\right] + i\hbar f(x)\partial_x\varphi \\ &= \underbrace{-i\hbar\left[\partial_x f(x)\right]}_{\uparrow}\varphi\end{aligned} \tag{2.117}$$

With this the momentum-Hamiltonian commutator becomes:

$$\left[\hat{p}_x, \hat{H}\right] = -i\hbar\partial_x V(x) \tag{2.118}$$

that together with the expectation of the time change in momentum,

$$\left\langle \partial_t \hat{p}_x \right\rangle = -i\hbar\left\langle \partial_t\left(\partial_x\right)\right\rangle = 0 \tag{2.119}$$

the associate time evolution of the momentum expectation yields

$$\partial_t\left\langle \hat{p}_x \right\rangle = -\left\langle \partial_x V(x)\right\rangle \tag{2.120}$$

proofing the second part or the Ehrenfest theorem.

Overall, the Ehrenfest theorem shows that quantum description is compatible with classical mechanics, under the expectation values of its main operators, i.e., the space, momentum and energy (Hamiltonian). Moreover, it says us that what we can know from quantum mechanical description of

Nature is not the detailed evolution but its average; nevertheless, it seems that *quantum mechanically we can not know everything in causes but the remaining knowledge is not absolutely necessary to be revealed for explaining the observed world*! With this answer we opened the direction in which the first Kantian universal interrogative may be approached with the quantum theory.

However, worth remarking that the time-dependent operator expectation equation simply becomes:

$$\frac{d}{dt}\langle\hat{A}\rangle = \langle\partial_t\hat{A}\rangle \tag{2.121}$$

i.e. the expectation value of the given observable is computed over stationary eigen-functions, if the concerned observable commutes with Hamiltonian:

$$\left[\hat{A},\hat{H}\right] = 0 \tag{2.122}$$

The fundamental consequence of this assertion is that *the stationary eigen-functions of a quantum system may be found from the eigen-problems of the operators that commute with the Hamiltonian of the system*. Moreover, if two operators give eigen-values on the same eigen-function of a quantum system,

$$\hat{A}\psi = a\psi; \quad \hat{B}\psi = b\psi; \quad a,b \in \Re e \tag{2.123}$$

they necessary commute:

$$\left[\hat{A},\hat{B}\right]\psi = \left(\hat{A}\hat{B} - \hat{B}\hat{A}\right)\psi = (ba - ab)\psi = 0 \tag{2.124}$$

Combining the last two ideas, one may conclude that all operators that commute among them and commute with Hamiltonian build up the so-called *complete set of commutative operators*:

$$\text{CoSCOpe: } \left\{\hat{A},\hat{B},...,\hat{H}\right\} \mid \left[\hat{A},\hat{B}\right] = 0,...,\left[\hat{A},\hat{H}\right] = 0,\left[\hat{B},\hat{H}\right] = 0 \tag{2.125}$$

with the help of which all stationary eigen-values and eigen-functions of a system may be determined throughout the associate eigen-problems:

$$\begin{aligned}\hat{A}\psi &= a\psi \\ \hat{B}\psi &= b\psi \\ &\ldots \\ \hat{H}\psi &= E\psi\end{aligned} \tag{2.126}$$

This is a fundamental property of operators and will be of the first importance in furnishing the complete solution of the Hydrogen atomic problem, in the next volume of this five-volume set.

2.2.9 CURRENT DENSITY PROBABILITY CONSERVATION THEOREM

THEOREM: Schrödinger equation is compatible with charge conservation at the probability density level.

Firstly, let's unfold the meaning of "conservation at the probability density level" in an analytical manner; that is considering a certain volume region of the space, say Γ_Σ, that is characterized by the localization density probability

$$\wp_t(\Gamma_\Sigma) = \int_{\Gamma_\Sigma} |\psi_t(\mathbf{r})|^2\, d\mathbf{r} \tag{2.127}$$

for an electronic containing system with a normalized wave-function:

$$1 = \int_{(\infty)} |\psi_t(\mathbf{r})|^2\, d\mathbf{r} \tag{2.128}$$

remembering that the normalization condition is not depending on time when all space involved.

Therefore, the time-variation of the probability $\wp(\Gamma_\Sigma)$ produces the appearance of the correspondent probability density current according with the basic charge variation-current generation principle

$$\frac{d}{dt}\wp_t(\Gamma_\Sigma) = -\oint_\Sigma \mathbf{j}(\mathbf{r},t)\,d\sigma_\Sigma \tag{2.129}$$

where the minus sign means that the diminishing the charge localization in the region Γ_Σ, i.e., the increase of charge probability density outside of this region, is associated with increasing of the appeared current probability density in the complementing region $(\infty \setminus \Gamma_\Sigma)$. Next, the above equation may be rewritten in its equivalent form:

$$\frac{d}{dt}\int_{\Gamma_\Sigma} |\psi_t(\mathbf{r})|^2 \, d\mathbf{r} = -\oint_\Sigma \mathbf{j}(\mathbf{r},t) d\sigma_\Sigma \tag{2.130}$$

end even more as

$$\int_{\Gamma_\Sigma} \frac{d}{dt} |\psi_t(\mathbf{r})|^2 \, d\mathbf{r} = -\int_{\Gamma_\Sigma} div\mathbf{j}(\mathbf{r},t) d\mathbf{r} \tag{2.131}$$

through the Gauss surface-to-volume integral transformation in the r.h.s. of the last two equations. Thus the pattern charge-current probability density conservation equation casts as:

$$\frac{d}{dt}\rho_t(\mathbf{r}) + div\mathbf{j}(\mathbf{r},t) = 0 \tag{2.132}$$

since recognizing that:

$$\rho_t(\mathbf{r}) = |\psi_t(\mathbf{r})|^2 = \psi_t(\mathbf{r})^* \psi_t(\mathbf{r}) \tag{2.133}$$

in any region of the space.

With this, the remaining proof of the theorem regards the possibility to regain the above charge-current probability density conservation from the Schrödinger equation. This may be achieved quite straightforward throughout considering both the direct and conjugated temporal Schrödinger equations multiplied by wave-function and conjugated wave-function, in reciprocal manner:

$$-\frac{\hbar^2}{2m_0}\psi^* \nabla^2 \psi + V\psi^* \psi = i\hbar \psi^* \partial_t \psi \tag{2.134a}$$

$$-\frac{\hbar^2}{2m_0}\psi \nabla^2 \psi^* + V\psi \psi^* = -i\hbar \psi \partial_t \psi^* \tag{2.134b}$$

Then, their subtraction firstly leads with expression:

$$-\frac{\hbar^2}{2m_0}\left(\psi^* \nabla^2 \psi - \psi \nabla^2 \psi^*\right) = i\hbar\left(\psi^* \partial_t \psi + \psi \partial_t \psi^*\right) \tag{2.135}$$

that can be rearranges as:

$$\frac{i\hbar}{2m_0}\nabla\left(\psi^* \nabla \psi - \psi \nabla \psi^*\right) = \partial_t\left(\psi^* \psi\right) \tag{2.136}$$

Now, recognizing the electronic density probability,

$$\rho_e = (-e)\psi^*\psi \tag{2.137}$$

and introducing the current density probability as:

$$\mathbf{j}_e = (-e)\mathbf{j} = (-e)\frac{i\hbar}{2m_0}\left(\psi\nabla\psi^* - \psi^*\nabla\psi\right) \tag{2.138}$$

the wave-function charge conservation law is obtained under the form:

$$\partial_t\rho_e + div\mathbf{j} = 0 \tag{2.139}$$

standing as another quantum counter-part for the corresponding electromagnetic law of charge-current conservation; moreover this equation will be the main checking stage when developing the quantification of the chemical bond through charge circulation in bonding, either within the de Broglie-Schrödinger-Bohm or Dirac treatment of quantification of the chemical bonding field throughout the associate quantum particles bondons, see the Volume III of the present five-volume work (Putz, 2016).

Finally, there is remarkable that the current density probability current has as one of its major consequences the property of wave functions to be square integrable, i.e., with finite constant value for the integral

$$\int_{-\infty}^{+\infty} \psi_t(\mathbf{r})^*\psi_t(\mathbf{r})d\mathbf{r} = \int_{-\infty}^{+\infty} \left|\psi_t(\mathbf{r})\right|^2 d\mathbf{r} = ct, \forall t \tag{2.140}$$

under the (natural) assumptions that wave-function asymptotically vanishes at the infinite frontier of the domain of integration

$$\lim_{\mathbf{r}\to\pm\infty}\psi_t(\mathbf{r}) = 0 \tag{2.141}$$

and the applied potential is of real nature:

$$V(\mathbf{r}) \in \Re e \tag{2.142}$$

This can be immediately see when integrating over the infinite domain the above theorem

$$\frac{d}{dt}\int_{-\infty}^{+\infty}|\psi_t(\mathbf{r})|^2\,d\mathbf{r} = -\int_{-\infty}^{+\infty}\nabla\mathbf{j}(\mathbf{r},t)d\mathbf{r} = -\left[\mathbf{j}(\mathbf{r},t)\right]_{-\infty}^{+\infty}$$
$$= \frac{ei\hbar}{2m_0}\left[\psi\nabla\psi^* - \psi^*\nabla\psi\right]_{-\infty}^{+\infty}$$
$$= \frac{ei\hbar}{2m_0}\left(\underbrace{\left[\psi\right]_{-\infty}^{+\infty}}_{0}\left[\nabla\psi^*\right]_{-\infty}^{+\infty} - \underbrace{\left[\psi^*\right]_{-\infty}^{+\infty}}_{0}\left[\nabla\psi\right]_{-\infty}^{+\infty}\right) = 0 \quad (2.143)$$

from where appears that:

$$\int_{-\infty}^{+\infty}|\psi_t(\mathbf{r})|^2\,d\mathbf{r} = ct < \infty \quad (2.144)$$

This special feature of wave-function assures the background on which the postulates of quantum mechanics (especially the integrability of the wave-function, its continuity, asymptotic continuity and representation by means of the scalar product in Hilbert spaces of vectors and operators) may be appropriately formulated and applied.

2.3 CLASSICAL TO QUANTUM MECHANICS' CORRESPONDENCE

2.3.1 *CLASSICAL EULER-LAGRANGE, HAMILTON, AND HAMILTON-JACOBI EQUATIONS*

While modeling the time-space evolution of the systems the so called generalized coordinate of motion may be introduction through considering the set of triplets containing time, space coordinate q and its time derivative us the velocity or as the generalized momentum p:

$$t, q(t), p(t) \text{ or } t, q(t), \dot{q}(t)$$

as the sufficient and necessary variables for construction of the driving equations. When entering the energy functionals they may be usually combined in two different, however equivalent, forms, namely under the (classical) Hamiltonian

$$H = H(t, q(t), p(t)) = T(p(t)) + V(t, q(t)) \quad (2.145)$$

or Lagrangian

$$L = L(t, q(t), \dot{q}(t)) = T(\dot{q}(t)) - V(t, q(t)) \tag{2.146}$$

of the concerned systems, accounting for the total and "effective free" energies, respecting the adding or subtracting of the external potential, respectively.

The relationships between the Hamiltonian and Lagrangian of a system may be better visualized as the Laguerre transformation involving the conjugate variables concerned:

$$H(t, q, p) = p\dot{q} - L(t, q, \dot{q}) \tag{2.147}$$

with

$$p = \frac{\partial L}{\partial \dot{q}} \tag{2.148}$$

introduced as the *conjugated momentum*, while the equation of motion is fixed throughout the variational principle, see Figure 2.2,

$$\delta S = 0 \tag{2.149}$$

for the so-called action functional

$$S = \int_{t_a}^{t_b} L(t, q, \dot{q}) dt \tag{2.150}$$

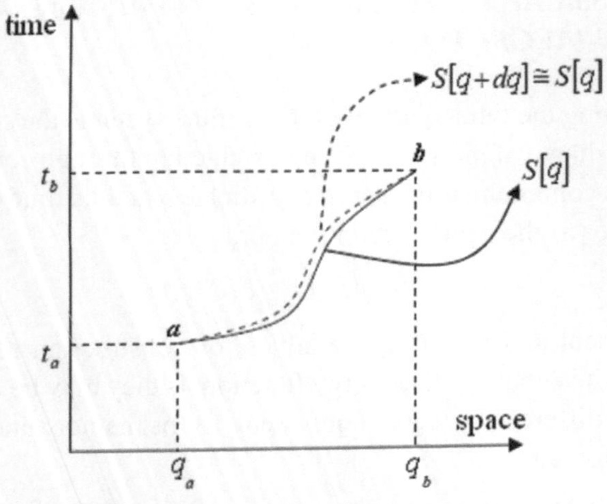

FIGURE 2.2 The graphical illustration of the variational action principle.

One may note that this equation is primarily written in terms of Lagrangian since it is intimately related with the "effective free" energy to be engaged in motion, i.e., obtained as the rest of the kinetic energy upon the external energy influence (and constraint) is subtracted. As well, worth noted that the present approach is a phenomenological one and can be at any moment extended to a complete set of conjugate variables (q, p) in what they build up as the so-called phase space. However, going on with a single pair of coordinate-momentum variables one firstly gets for the action variational expression

$$\delta S = \frac{\partial S}{\partial q}\delta q + \frac{\partial S}{\partial \dot{q}}\delta \dot{q} = \frac{\partial}{\partial q}\left[\int_a^b L dt\right]\delta q + \frac{\partial}{\partial \dot{q}}\left[\int_a^b L dt\right]\delta \dot{q}$$
$$= \int_a^b \left(\frac{\partial L}{\partial q}\delta q\right)dt + \int_a^b \left(\frac{\partial L}{\partial \dot{q}}\delta \dot{q}\right)dt \tag{2.151}$$

Worth noting that the variation in time is not explicitly taken in functional derivative of action functional since wile intervening in its integral definition respecting the Lagrangian is to be integrated out before explicit derivation; however, using the derivative equivalence

$$\delta \dot{q} = \delta\left(\frac{dq}{dt}\right) = \frac{d}{dt}(\delta q) \tag{2.152}$$

the variation of action functional further writes employing also the paths' integration rule

$$\delta S = \int_a^b \left(\frac{\partial L}{\partial q}\delta q\right)dt + \int_a^b \left[\frac{\partial L}{\partial \dot{q}}\frac{d}{dt}(\delta q)\right]dt$$
$$= \int_a^b \left(\frac{\partial L}{\partial q}\delta q\right)dt + \int_a^b \frac{d}{dt}\left(\frac{\partial L}{\partial \dot{q}}\delta q\right)dt - \int_a^b \left[\frac{d}{dt}\left(\frac{\partial L}{\partial \dot{q}}\right)\right]\delta q dt$$
$$= \underbrace{\left(\frac{\partial L}{\partial \dot{q}}\delta q\right)_a^b}_{\substack{=0 \\ (\delta q_a = \delta q_b = 0)}} + \int_a^b \left[\frac{\partial L}{\partial q} - \frac{d}{dt}\left(\frac{\partial L}{\partial \dot{q}}\right)\right]\delta q dt \tag{2.153}$$

from where the variational principle leads with the famous Euler-Lagrange equation:

$$\frac{\partial L}{\partial q}=\frac{d}{dt}\left(\frac{\partial L}{\partial \dot{q}}\right) \tag{2.154}$$

There is immediate that the Euler-Lagrange equation regain the celebrated Newton dynamical equation

$$\vec{F}(orce)=m\vec{a}(cceleration) \tag{2.155}$$

when employed upon a conservative potential Lagrangian

$$L_{classic}=\frac{m\dot{q}}{2}-V(q) \tag{2.156}$$

by means of associate derivatives:

$$\frac{\partial L}{\partial q}=-\nabla V\equiv\vec{F} \tag{2.157}$$

$$\frac{d}{dt}\left(\frac{\partial L}{\partial \dot{q}}\right)=\frac{d}{dt}(m\dot{q})=m\ddot{q}\equiv m\vec{a} \tag{2.158}$$

Alternatively, through repeating the same variational algorithm of action functional in terms of Hamiltonian one firstly has

$$\begin{aligned}
\delta S&=\delta\int_{t_a}^{t_b}L(t,q,\dot{q})dt=\int_{t_a}^{t_b}\delta\left[p\dot{q}-H(q,p)\right]dt\\
&=\int_a^b\left[p\delta\dot{q}+\dot{q}\delta p-\frac{\partial H}{\partial q}\delta q-\frac{\partial H}{\partial p}\delta p\right]dt\\
&=\int_a^b\left[p\frac{d}{dt}(\delta q)+\dot{q}\delta p-\frac{\partial H}{\partial q}\delta q-\frac{\partial H}{\partial p}\delta p\right]dt\\
&=\int_a^b\left[\frac{d}{dt}(p\delta q)-\dot{p}\delta q+\dot{q}\delta p-\frac{\partial H}{\partial q}\delta q-\frac{\partial H}{\partial p}\delta p\right]dt\\
&=\int_a^b\left[\frac{d}{dt}(p\delta q)\right]dt+\int_a^b\left[\left(-\dot{p}-\frac{\partial H}{\partial q}\right)\delta q+\left(\dot{q}-\frac{\partial H}{\partial p}\right)\delta p\right]dt\\
&=\underbrace{(p\delta q)_a^b}_{\substack{=0\\(\delta q_a=\delta q_b=0)}}+\int_a^b\left[\left(-\dot{p}-\frac{\partial H}{\partial q}\right)\delta q+\left(\dot{q}-\frac{\partial H}{\partial p}\right)\delta p\right]dt
\end{aligned} \tag{2.159}$$

releasing upon the application of the variational principle of this action variation with the so called *Hamilton canonical equations* for the conjugate variables:

$$\dot{p} = -\frac{\partial H}{\partial q} \tag{2.160}$$

$$\dot{q} = \frac{\partial H}{\partial p} \tag{2.161}$$

in a clear phenomenological equivalents with the above Euler-Lagrange equation; however, the main difference consist in that they represent a coupled first order equations of motion instead of single second order differential equation as Euler-Lagrange equation is unfolded.

Still, a third equation of motion may be found from the action variation when the time is eventually explicitly considered through writing its definition as:

$$S = \int_a^b dS = \int_a^b [p\dot{q} - H(q,p,t)]dt \tag{2.162}$$

equivalently with

$$dS = pdq - H(q,p,t)dt \tag{2.163}$$

By comparing this expression with its total differential expansion

$$dS = \frac{\partial S}{\partial q}dq + \frac{\partial S}{\partial t}dt \tag{2.164}$$

there result the new expressions:

$$p = \frac{\partial S}{\partial q} \tag{2.165}$$

$$H(q,p,t) = -\frac{\partial S}{\partial t} \tag{2.166}$$

releasing through their combination with the so called *Hamilton-Jacobi equation*:

$$H\left(q, \frac{\partial S}{\partial q}, t\right) + \frac{\partial S}{\partial t} = 0 \tag{2.167}$$

linking the Hamiltonian with the action derivatives either in explicit (in time) and implicit (in space) forms.

All these equations will be in next transposed on quantum level for the wave function.

2.3.2 WAVE-FUNCTION QUANTUM FIELD

The passage from classical analytical mechanics to quantum mechanics may be realized through assuming the basic *coordinate-to-field correspondence principle*

$$q \to \phi(x_i),\ i = 1,2,3 \tag{2.168}$$

along replacing the Lagrangian and Hamiltonian functionals with their space-densities

$$\mathcal{L} = \mathcal{L}\left(\phi, \dot{\phi}, \nabla\phi\right) = \mathcal{L}\left(\phi, \phi_{,\mu}\right),\quad L = \int \mathcal{L} d^3x \tag{2.169}$$

$$\mathcal{H} = \mathcal{H}\left(\phi, \dot{\phi}, \nabla\phi\right) = \mathcal{H}\left(\phi, \phi_{,\mu}\right),\quad H = \int \mathcal{H} d^3x \tag{2.170}$$

where there the Einstein's derivative notation was used:

$$\phi_{,\mu} = \partial_\mu \phi = \frac{\partial \phi}{\partial x_\mu},\ \mu = 0,1,2,3 \tag{2.171}$$

in terms of 4D derivative operator

$$\partial_\mu \equiv \frac{\partial}{\partial x_\mu} \equiv \left(\partial_0 = \frac{\partial}{\partial t}, \partial_1 = \frac{\partial}{\partial x_1}, \partial_2 = \frac{\partial}{\partial x_2}, \partial_3 = \frac{\partial}{\partial x_3}\right) \tag{2.172}$$

In these conditions, the working related concepts are introduced as:

- canonic conjugated moment field:

$$\pi(\mathbf{r},t) = \frac{\delta \mathcal{L}}{\delta \dot{\phi}} \tag{2.173}$$

- Hamiltonian-Lagrange field transformation:

$$\mathcal{H} = \pi\dot{\phi} - \mathcal{L} \tag{2.174}$$

- 4D Euler-Lagrange field equation is resumed as:

$$\frac{\delta \mathcal{L}}{\delta \dot{\phi}} = \partial_{\mu} \left(\frac{\delta \mathcal{L}}{\delta \dot{\phi}_{,\mu}} \right) \tag{2.175}$$

with the unfolded version extended as

$$\frac{\delta \mathcal{L}}{\delta \phi} = \frac{d}{dt} \left(\frac{\delta \mathcal{L}}{\delta \dot{\phi}} \right) + \nabla \left(\frac{\delta \mathcal{L}}{\delta \nabla \phi} \right) \tag{2.176}$$

- Hamiltonian canonic field equations:

$$\dot{\pi} = -\frac{\delta \mathcal{H}}{\delta \phi} = \frac{\delta \mathcal{L}}{\delta \phi} \tag{2.177}$$

$$\dot{\phi} = \frac{\delta \mathcal{H}}{\delta \pi} \tag{2.178}$$

Going now to check whether the Schrödinger equation may be recovered from these Lagrangian or Hamiltonian field formalisms, one may employ the so-called Schrödinger Lagrangian

$$\mathcal{L}_{Sch} = i\hbar \phi^{*} \dot{\phi} - \frac{\hbar^{2}}{2m_{0}} \left(\nabla \phi^{*} \right) \left(\nabla \phi \right) - V \phi^{*} \phi \tag{2.179}$$

constructed in pairs of direct-complex conjugated products of fields since the Lagrangian should be a real quantity (the complex factor "*i*" assure on the other side the complex nature of the Schrödinger equation). Application of the Euler-Lagrange field equation is done respecting the conjugated field to obtain:

$$\frac{\delta \mathcal{L}_{Sch}}{\delta \phi^{*}} = i\hbar \dot{\phi} - V \phi \tag{2.180}$$

$$\frac{d}{dt} \left(\frac{\delta \mathcal{L}_{Sch}}{\delta \dot{\phi}^{*}} \right) = 0 \tag{2.181}$$

$$\nabla \left(\frac{\delta \mathcal{L}_{Sch}}{\delta \nabla \phi^{*}} \right) = -\frac{\hbar^{2}}{2m_{0}} \nabla \left(\nabla \phi \right) = -\frac{\hbar^{2}}{2m_{0}} \nabla^{2} \phi \tag{2.182}$$

aggregating on the Schrödinger equation for the direct field:

$$-i\hbar\dot{\phi} = \left(-\frac{\hbar^2}{2m_0}\nabla^2 + V\right)\phi \tag{2.183}$$

However, if one wants to apply the Euler-Lagrange field equation on the direct wave function field the above Lagrangian should be firstly equivalently transformed through the derivative replacement:

$$\nabla\phi^* \cdot \nabla\phi = \nabla\left[\left(\nabla\phi^*\right)\phi\right] - \phi\nabla^2\phi^* \tag{2.184}$$

to became

$$\mathcal{L}'_{Sch} = i\hbar\phi^*\dot{\phi} - \frac{\hbar^2}{2m_0}\nabla\left[\left(\nabla\phi^*\right)\phi\right] + \frac{\hbar^2}{2m_0}\phi\nabla^2\phi^* - V\phi^*\phi \tag{2.185}$$

Now, performing the Euler-Lagrange derivatives

$$\frac{\delta\mathcal{L}'_{Sch}}{\delta\phi} = \frac{\hbar^2}{2m_0}\nabla^2\phi^* - V\phi^* \tag{2.186}$$

$$\frac{d}{dt}\left(\frac{\delta\mathcal{L}'_{Sch}}{\delta\dot{\phi}}\right) = \frac{d}{dt}\left(i\hbar\phi^*\right) = i\hbar\dot{\phi}^* \tag{2.187}$$

$$\nabla\left(\frac{\delta\mathcal{L}'_{Sch}}{\delta\nabla\phi}\right) = 0 \tag{2.188}$$

the Schrödinger equation of the conjugated field is this time revealed:

$$-i\hbar\dot{\phi}^* = \left(-\frac{\hbar^2}{2m_0}\nabla^2 + V\right)\phi^* \tag{2.189}$$

that nevertheless confirm the correctness of the Lagrange formalism and of the classical-to-quantum above enounced correspondence.

Then, we may check whether the Hamiltonian formalism agrees with Schrödinger picture as well; for that we may build up the field Hamiltonian from the field Lagrangian

$$\begin{aligned}\mathcal{H}_{Sch} &= \pi\dot{\phi} - \mathcal{L}_{Sch} = \frac{\delta\mathcal{L}_{Sch}}{\delta\dot{\phi}}\dot{\phi} - \mathcal{L}_{Sch} = i\hbar\phi^*\dot{\phi} - \mathcal{L}_{Sch} \\ &= \frac{\hbar^2}{2m_0}\left(\nabla\phi^*\right)\left(\nabla\phi\right) + V\phi^*\phi = -i\frac{\hbar}{2m_0}\left(\nabla\pi_{Sch}\right)\left(\nabla\phi\right) - \frac{i}{\hbar}V\pi_{Sch}\phi\end{aligned} \tag{2.190}$$

when employing the conjugated momentum field

$$\pi_{Sch} = \frac{\delta \mathcal{L}_{Sch}}{\delta \dot{\phi}} = i\hbar\phi^* \tag{2.191}$$

to replace with it the conjugated field and its gradient

$$\phi^* = -\frac{i}{\hbar}\pi_{Sch}, \ \nabla\phi^* = -\frac{i}{\hbar}\nabla\pi_{Sch} \tag{2.192}$$

As before, since we consider the Hamiltonian's product of gradients re-expressed as

$$\nabla\pi \cdot \nabla\phi = \nabla\left[(\nabla\pi)\phi\right] - \phi\nabla^2\pi \tag{2.193}$$

or as

$$\nabla\pi \cdot \nabla\phi = \nabla\left[(\nabla\phi)\pi\right] - \pi\nabla^2\phi \tag{2.194}$$

the alternative Hamiltonians will cast accordingly

$$\begin{aligned}\mathcal{H}_{Sch} &= -i\frac{\hbar}{2m_0}\nabla\left[(\nabla\pi_{Sch})\phi\right] + i\frac{\hbar}{2m_0}\phi\nabla^2\pi_{Sch} - \frac{i}{\hbar}V\pi_{Sch}\phi \\ &= -i\frac{\hbar}{2m_0}\nabla\left[(\nabla\phi)\pi_{Sch}\right] + i\frac{\hbar}{2m_0}\pi_{Sch}\nabla^2\phi - \frac{i}{\hbar}V\pi_{Sch}\phi\end{aligned} \tag{2.195}$$

Now, the Hamiltonian canonic field equations provide the direct Schrödinger equation when the field evolution equation is considered,

$$\dot{\phi} = \frac{\delta \mathcal{H}_{Sch}}{\delta \pi_{Sch}} = i\frac{\hbar}{2m_0}\nabla^2\phi - \frac{i}{\hbar}V\phi \tag{2.196}$$

whereas, when the momentum field equation is derived,

$$\dot{\pi}_{Sch} = -\frac{\delta \mathcal{H}_{Sch}}{\delta \phi} = -i\frac{\hbar}{2m_0}\nabla^2\pi_{Sch} + \frac{i}{\hbar}V\pi_{Sch} \tag{2.197}$$

the Schrödinger equation for the conjugated field springs out trough replacing the abode derived Schrödinger momentum field and of its temporal derivative.

Finally, there may be straightforwardly proven that the Hamilton-Jacobi classical equation may be recovered within *classical limit*

$$m \to \infty; \hbar \to 0 \tag{2.198}$$

when quantum fields of type:

$$\phi = \phi_0 \exp\left(\frac{i}{\hbar} S(\mathbf{r}, t)\right) \tag{2.199}$$

is assumed as the natural solution for the Schrödinger equation, where, for the de Broglie wave function there appears that the action functional may be identified as

$$S(\mathbf{r}, t) = \mathbf{p} \cdot \mathbf{r} - Et \tag{2.200}$$

thus combining the space-time coordinates with energy-momentum information. This way, once all involved derivatives are performed,

$$\phi = \frac{i}{\hbar} \dot{S} \phi \tag{2.201}$$

$$\nabla^2 \phi = \phi_0 \nabla \left(\frac{i}{\hbar} e^{\frac{i}{\hbar} S} \nabla S \right) = \frac{i}{\hbar} \phi \nabla^2 S - \frac{1}{\hbar^2} \phi \nabla S \cdot \nabla S \tag{2.202}$$

their replacement in general Schrödinger equation

$$i\hbar \dot{\phi} = \frac{\hbar^2}{2m} \nabla^2 \phi + V\phi \tag{2.203}$$

leaves with the *quantum Hamilton-Jacobi equation*

$$-\dot{S} = \frac{\nabla S \cdot \nabla S}{2m} - \frac{i\hbar}{2m} \nabla^2 S + V \tag{2.204}$$

Now, there is clear that this equation may collapses to the classical Hamilton-Jacobi one

$$-\dot{S} = \frac{\nabla S \cdot \nabla S}{2m} + V \tag{2.205}$$

recognizing the Hamiltonian

$$H = \frac{\nabla S \cdot \nabla S}{2m} + V = \frac{p^2}{2m} + V \tag{2.206}$$

since recalling the momentum-action relationship

$$p = \nabla S \tag{2.207}$$

when the above asserted classical limit is fulfilled in the resumed limit

$$\frac{\hbar}{m} \to 0 \tag{2.208}$$

Concluding, we may affirm that *any physical system which in no experiment is able to produce values for $\hbar / m$ larger than the experimental errors is essentially manifested as a classical system.*

2.3.3 SEMI-CLASSICAL EXPANSION AND THE WKB APPROXIMATION

One may observe that the Hamilton-Jacobi equation was recovered by assuming the resumed wave function in terms of action:

$$S(\mathbf{r},t) = \tilde{S}(\mathbf{r},E) - tE \tag{2.209}$$

while leaving open the question: "where the wave-packet is ?". The response is that it may be formed from summing up waves of action with different energies resulting eventually in the integral

$$\psi(\mathbf{r},t) = \int dE \exp\left[\frac{i}{\hbar}\left(\tilde{S}(\mathbf{r},E) - tE\right)\right] \tag{2.210}$$

from where its localization (towards measuring of observing it) is realized from the variational constraint:

$$0 = \frac{\partial}{\partial E}\psi(\mathbf{r},t) \Rightarrow \frac{\partial}{\partial E}\left(\tilde{S}(\mathbf{r},E) - tE\right) = 0 \tag{2.211}$$

leaving with the stationary condition

$$t = \frac{\partial}{\partial E}\tilde{S}(\mathbf{r}, E) \tag{2.212}$$

This result, above of its inverse Legendre transformation form, stands as an implicit equation for the particle's position; further identification as

$$\tilde{S}(\mathbf{r}, E) = xp_x = \pm x\sqrt{2m(E - V(x))} \tag{2.213}$$

helps in recognizing the above stationary condition as the classical equation of motion giving the actual picture of the quantum action evolution: the wave packet it exists as an infinite superposition of action waves, however, due to the small Planck constant, pose a phase that oscillate very fast producing a strong cancellation of the containing waves except that releasing with the stationary condition equivalent with the classical equation of motion; even shortly, the quantum motion of the wave packet, once measured, is observed as the classical trajectories; or even more: the quantum evolution of a system remains hidden except its classical manifestation (or observation).

Nevertheless, the bridge between the classical and quantum regimes may be smoothly accounted from the so-called action *semiclassical expansion* in powers of $\hbar$ or the *eikonal* of action:

$$S(\mathbf{r}, E) = S_0 - i\hbar S_1 + (-i\hbar)^2 S_2 + (-i\hbar)^2 S_3 ... \tag{2.214}$$

The replacement of this expansion back into the quantum Hamilton-Jacobi equation provides the successive equations for the various orders of $\hbar$:

$$-\frac{\partial S_0}{\partial t} = \frac{1}{2m}(\nabla S_0)^2 + V \tag{2.215a}$$

$$-\frac{\partial S_1}{\partial t} = \frac{1}{2m}\left[\nabla^2 S_0 + 2(\nabla S_0)(\nabla S_1)\right] \tag{2.215b}$$

$$-\frac{\partial S_2}{\partial t} = \frac{1}{2m}\left[\nabla^2 S_1 + (\nabla S_1)^2 + 2(\nabla S_0)(\nabla S_2)\right] \tag{2.215c}$$

$$-\frac{\partial S_3}{\partial t} = \frac{1}{2m}\left[\nabla^2 S_2 + 2(\nabla S_1)(\nabla S_2) + 2(\nabla S_0)(\nabla S_3)\right] \tag{2.215d}$$

...

$$-\frac{\partial S_{n+1}}{\partial t}=\frac{1}{2m}\left[\nabla^2 S_n+\sum_{m=0}^{n+1}(\nabla S_m)(\nabla S_{n+1-m})\right] \tag{2.215e}$$

...

However, one may immediately observe that the entire action eikonal may be explicitly written once the successive iterative equations are solved; moreover there is noted that the zero-th order in action corresponds entirely to the classical Hamilton-Jacobi equation; the restriction to the first order in $\hbar$ makes nonetheless the Wentzel, Kramers and Brillouin (WKB) framework for semiclassical wave-function approximation:

$$\psi(\mathbf{r},t)=e^{\frac{i}{\hbar}(S_0-i\hbar S_1)}e^{-\frac{i}{\hbar}Et} \tag{2.216}$$

with S_0 and S_1 to be determined, while assuming that only S_0 has the time dependence and that is of the form

$$S_0(\mathbf{r},t)=S_0(\mathbf{r})-tE \tag{2.217}$$

with the higher terms carrying only the spatial dependence:

$$S_i(\mathbf{r},t)=S_i(\mathbf{r})\ ,i>0 \tag{2.218}$$

Let's consider the uni-dimensional problems, restricted on axis $0x$, for better emphasis of the method's principle. In this context, the S_0 above equation simply becomes:

$$E=\frac{1}{2m}(\partial_x S_0)^2+V(x) \tag{2.219}$$

with the integration solution

$$S_0(x)=\pm\int^x\sqrt{2m(E-V(x'))}dx'=\int^x p(x')dx' \tag{2.220}$$

Going to the next eikonal level, the equation for S_1 under its stationary condition leaves with

$$\partial_x^2 S_0=-2(\partial_x S_0)(\partial_x S_1) \tag{2.221}$$

from where the solution

$$S_1(x) = -\frac{1}{2}\int^x \frac{\partial_{x'}^2 S_0(x')}{\partial_{x'} S_0(x')} dx' = -\frac{1}{2}\int^x \frac{\partial_{x'} p(x')}{p(x')} dx'$$
$$= -\frac{1}{2}\int^x \frac{dp(x')}{p(x')} = -\frac{1}{2}\log p(x) + ct. \tag{2.222}$$

All in all the 1D-WKB wave-function turns out to be:

$$\psi(x,t) = \frac{ct}{\sqrt{p(x)}} e^{\frac{i}{\hbar}\int^x p(x')dx'} e^{-\frac{i}{\hbar}Et}$$
$$= \frac{ct}{\left[2m(E - V(x))\right]^{1/4}} e^{\pm\frac{i}{\hbar}\int^x \sqrt{2m(E-V(x'))}dx'} e^{-\frac{i}{\hbar}Et} \tag{2.223}$$

with two warnings: one is regarding the fact that is expression is yet not-normalized; and the second is that it is related with the particle (wave-packet) moving in the classically allowed (accessible) region (regime) with $E > V(x)$ that prescribe the evolution is described by an oscillating wave function. Instead, in the classical inaccessible regime $E < V(x)$ it decreased or increased exponentially, i.e., it should look like

$$\psi(x,t) = \frac{ct}{\left[2m(V(x) - E)\right]^{1/4}} e^{\pm\frac{1}{\hbar}\int^x \sqrt{2m(V(x')-E)}dx'} e^{-\frac{i}{\hbar}Et} \tag{2.224}$$

However, this approximation clearly breaks down at the *classical turning points* $E = V(x)$, i.e., at the points were the classical particle (or the manifestation of the wave packet) stops, since $p(x) = 0$ there, and turns due to the potential equalizes its (eigen-value) energy. This is the WKB approximation holds for the cases where S_1 is much smaller than S_0 equivalently with the requirement that quantum Hamilton-Jacobi equation to collapse on its classical variant through the action condition:

$$\left|(\nabla S)^2\right| >> \hbar\left|\nabla^2 S\right| \tag{2.225}$$

that in the light of above S_0 form of solution may be rewritten in terms of momentum as:

$$p(x)^2 >> \hbar\left|\partial_x p(x)\right| \tag{2.226}$$

or when recalling the present momentum-energy relationship it is even more transformed as

$$\left|\frac{\hbar \partial_x V(x)}{[E-V(x)]p(x)}\right| << 1 \tag{2.227}$$

from where there is one more clear appearance of the classical turning points $E = V(x)$ that limits the WKB viability approximation.

Lastly, but not with less importance, the above WKB momentum condition may be related with the de Broglie wavelength

$$\lambda_B = \frac{2\pi\hbar}{p(x)} \tag{2.228}$$

in the form

$$\frac{\lambda_B}{2\pi}\left|\frac{\partial_x p(x)}{p(x)}\right| << 1 \tag{2.229}$$

telling that the WKB framework is limited to momentum functions encountering little relative change over the de Broglie wavelength. This condition is also useful in closely defining (or postulating) the wave function structure when an ansatz form is to be tested in solving specific quantum systems (see the corresponding section of QM postulates).

Yet, the learned lesson of WKB approximation, other – more powerful – forms of semiclassical eikonal expansion for the action will be considered in the forthcoming volume of this five-volume set with the occasion it will be also applied on the valence states (i.e., treated as semiclassical states) of atomic systems within the path integral formalism that will be soon in next exposed.

2.3.4 FROM FIELD INTERNAL SYMMETRY TO CURRENT CONSERVATION

One important property of the Schrödinger Lagrangian is that it is invariant for the phase transformations of the fields

$$\phi' = \phi \exp(i\alpha); \quad \phi'^{*} = \phi^{*} \exp(-i\alpha), \quad \alpha \in \Re \tag{2.230}$$

as it may immediately be observed since the specific pairs of direct-conjugated products of wave functions. There is said that the Schrödinger Lagrangian is *symmetric* for the phase transitions of the containing wave fields; moreover this symmetry is also called *internal*.

Yet, assuming the infinitesimal field transformations

$$\phi' = \phi + i\alpha\phi; \quad \phi'^{*} = \phi^{*} - i\alpha\phi^{*} \tag{2.231}$$

we deal in fact the field variations

$$\delta\phi = i\alpha\phi; \quad \delta\phi^{*} = -i\alpha\phi^{*} \tag{2.232}$$

and with associate 4D field derivatives

$$\delta\phi_{,\mu} = i\alpha\phi_{,\mu}; \quad \delta\phi^{*}_{,\mu} = -i\alpha\phi^{*}_{,\mu}; \quad \mu = 0,1,2,3 \tag{2.233}$$

In these conditions, the field Lagrangian variation will be successively written

$$\begin{aligned}
\delta\mathscr{L} &= \frac{\partial\mathscr{L}}{\partial\phi}\delta\phi + \frac{\partial\mathscr{L}}{\partial\phi_{,\mu}}\delta\phi_{,\mu} + \frac{\partial\mathscr{L}}{\partial\phi^{*}}\delta\phi^{*} + \frac{\partial\mathscr{L}}{\partial\phi^{*}_{,\mu}}\delta\phi^{*}_{,\mu} \\
&= i\alpha\left(\frac{\partial\mathscr{L}}{\partial\phi}\phi + \frac{\partial\mathscr{L}}{\partial\phi_{,\mu}}\partial_{\mu}\phi - \frac{\partial\mathscr{L}}{\partial\varphi^{*}}\phi^{*} - \frac{\partial\mathscr{L}}{\partial\phi^{*}_{,\mu}}\partial_{\mu}\phi^{*}\right) \\
&= i\alpha\left\{\begin{array}{l}
\phi\underbrace{\left[\frac{\partial\mathscr{L}}{\partial\phi} - \partial_{\mu}\left(\frac{\partial\mathscr{L}}{\partial\phi_{,\mu}}\right)\right]}_{\substack{=0 \\ Euler-Lagrange\ \ equation}} + \partial_{\mu}\left(\frac{\partial\mathscr{L}}{\partial\phi_{,\mu}}\phi\right) \\
-\phi^{*}\underbrace{\left[\frac{\partial\mathscr{L}}{\partial\phi^{*}} - \partial_{\mu}\left(\frac{\partial\mathscr{L}}{\partial\phi^{*}_{,\mu}}\right)\right]}_{\substack{=0 \\ Euler-Lagrange\ \ equation}} - \partial_{\mu}\left(\frac{\partial\mathscr{L}}{\partial\phi^{*}_{,\mu}}\phi^{*}\right)
\end{array}\right\} \\
&= i\alpha\partial_{\mu}\left(\frac{\partial\mathscr{L}}{\partial\phi_{,\mu}}\phi - \frac{\partial\mathscr{L}}{\partial\phi^{*}_{,\mu}}\phi^{*}\right)
\end{aligned} \tag{2.234}$$

Now, under the condition of invariant Lagrangian for the fields' phase transformation, implicitly also for their infinitesimal variation,

$$\delta\mathscr{L} = 0 \tag{2.235}$$

there results the 4D-current conservation law

$$\partial_{\mu} j_{\mu} = 0 \tag{2.236}$$

with the 4D-current defined as

$$j_{\mu} = \frac{\partial \mathcal{L}}{\partial \phi_{,\mu}} \phi - \frac{\partial \mathcal{L}}{\partial \phi^{*}_{,\mu}} \phi^{*} \tag{2.237}$$

The application of this conservation law to the Schrödinger Lagrangian will firstly provide the expanded 4D-current on the time and space derivatives

$$\begin{aligned} j_{Sch} &= \underbrace{\frac{\partial \mathcal{L}_{Sch}}{\partial \dot{\phi}} \phi - \frac{\partial \mathcal{L}_{Sch}}{\partial \dot{\phi}^{*}} \phi^{*}}_{temporal\ part} + \underbrace{\frac{\partial \mathcal{L}_{Sch}}{\partial \nabla \phi} \phi - \frac{\partial \mathcal{L}_{Sch}}{\partial \nabla \phi^{*}} \phi^{*}}_{spatial\ part} \\ &= \underbrace{i\hbar \phi^{*} \phi - 0}_{\substack{temporal \\ contribution}} \underbrace{- \frac{\hbar^{2}}{2m_{0}} \phi \nabla \phi^{*} + \frac{\hbar^{2}}{2m_{0}} \phi^{*} \nabla \phi}_{spatial\ contribution} \end{aligned} \tag{2.238}$$

that then produces the associated density-current conservation law

$$\partial_{0} j_{0} + \nabla \mathbf{j} = 0 \tag{2.239}$$

with temporal and spatial wave field contributions, respectively

$$\partial_{0} j_{0} = \frac{\partial}{\partial t} \left(\phi^{*} \phi \right) \tag{2.240}$$

$$\nabla \mathbf{j} = \frac{i\hbar}{2m_{0}} \nabla \left(\phi \nabla \phi^{*} - \phi^{*} \nabla \phi \right) \tag{2.241}$$

this way regaining the previously deduced density-current probability density conservation within the phenomenological wave-function picture.

2.3.5 FROM POISSON PARENTHESES TO QUANTUM COMMUTATORS AND HEISENBERG PICTURE

In classical mechanics, if there is about a function that depends on the conjugate canonical variables and explicitly on time, $F = F(q, p, t)$, its total variation in time (both by explicit and implicit dependence) may be expressed using the Hamilton canonical equations successively as:

$$\frac{dF}{dt}=\frac{\partial F}{\partial t}+\frac{\partial F}{\partial q}\dot{q}+\frac{\partial F}{\partial p}\dot{p}=\frac{\partial F}{\partial t}+\left(\frac{\partial F}{\partial q}\frac{\partial H}{\partial p}-\frac{\partial F}{\partial p}\frac{\partial H}{\partial q}\right) \tag{2.242}$$

allowing the rewriting of equation of motion for function *F*:

$$\dot{F}=\frac{\partial F}{\partial t}+\{F,H\} \tag{2.243}$$

in terms of the introduced classical Poisson parenthesis

$$\{F,H\}=\frac{\partial F}{\partial q}\frac{\partial H}{\partial p}-\frac{\partial F}{\partial p}\frac{\partial H}{\partial q} \tag{2.244}$$

At this moment there appears as striking the analogy between this classical equation of motion and the previously deduced equation of an operator average (observed value)

$$\frac{d}{dt}\left\langle\widehat{F}\right\rangle=\left\langle\partial_t\widehat{F}\right\rangle+\frac{1}{i\hbar}\left\langle\left[\widehat{F},\widehat{H}\right]\right\rangle \tag{2.245}$$

that, apart of the involved average operation, allows for the direct classical-to-quantum correspondence principle turning Poisson parentheses to commutators and functions to operators while considering the multiplicative factor *iħ*:

$$i\hbar\{F,H\}\leftrightarrow\left[\widehat{F},\widehat{H}\right] \tag{2.246}$$

However, worth mentioning that the above transfer from classical to quantum picture has to follow also the rule according which the result of the Poison equation to be written in such manner so that when quantized to provide a Hermitic operator. An example is here given for particular functions q^2 and p^2 of conjugated variables q and p; their commutator looks like:

$$\begin{aligned}\left[\hat{q}^2,\hat{p}^2\right]&=\hat{q}\left[\hat{q},\hat{p}^2\right]+\left[\hat{q},\hat{p}^2\right]\hat{q}\\&=\hat{q}\left(\underbrace{\left[\hat{q},\hat{p}\right]}_{i\hbar}\hat{p}+\hat{p}\underbrace{\left[\hat{q},\hat{p}\right]}_{i\hbar}\right)+\left(\underbrace{\left[\hat{q},\hat{p}\right]}_{i\hbar}\hat{p}+\hat{p}\underbrace{\left[\hat{q},\hat{p}\right]}_{i\hbar}\right)\hat{q}=2i\hbar\left(\hat{q}\hat{p}+\hat{p}\hat{q}\right)\end{aligned} \tag{2.247}$$

while the classical Poisson bracket yields:

$$\left\{q^2, p^2\right\} = \frac{\partial q^2}{\partial q}\frac{\partial p^2}{\partial p} - \underbrace{\frac{\partial q^2}{\partial p}}_{0}\underbrace{\frac{\partial p^2}{\partial q}}_{0} = 4qp = 2(qp + pq) \tag{2.248}$$

There is clear now if restricting only to the result

$$\left\{q^2, p^2\right\} = 4qp \tag{2.249}$$

since this result may not be transformed in an hermitic operator,

$$\left(\hat{q}\hat{p}\right)^+ = \hat{p}\hat{q} \neq \hat{q}\hat{p} \tag{2.250}$$

it cannot be considered for proper quantification; instead if employing the other equivalent classical result

$$\left\{q^2, p^2\right\} = 2(qp + pq) \tag{2.251}$$

it is true it may be quantified because the associate operatorial expression is indeed hermitic:

$$\left(\hat{q}\hat{p} + \hat{p}\hat{q}\right)^+ = \left(\hat{q}\hat{p}\right)^+ + \left(\hat{p}\hat{q}\right)^+ = \hat{p}^+\hat{q}^+ + \hat{q}^+\hat{p}^+ = \hat{p}\hat{q} + \hat{q}\hat{p} = \hat{q}\hat{p} + \hat{p}\hat{q} \tag{2.252}$$

since was previously proved that momentum and coordinate are hermitic operators.

Finally, a very interesting quantum feature is obtained from considering the classical canonical Hamilton equations rewritten with Poisson parentheses

$$\frac{\partial H}{\partial p} = \dot{q} = \frac{dq}{dt} = \frac{\partial q}{\partial q}\frac{\partial H}{\partial p} - \underbrace{\frac{\partial q}{\partial p}}_{0}\frac{\partial H}{\partial q} = \{q, H\} \tag{2.253}$$

$$-\frac{\partial H}{\partial q} = \dot{p} = \frac{dp}{dt} = \underbrace{\frac{\partial p}{\partial q}}_{0}\frac{\partial H}{\partial p} - \frac{\partial p}{\partial p}\frac{\partial H}{\partial q} = \{p, H\} \tag{2.254}$$

with the quantum counterpart:

$$\frac{d\hat{q}}{dt} = \frac{1}{i\hbar}\left[\hat{q}, \widehat{H}\right] = \frac{\partial \widehat{H}}{\partial \hat{p}} \tag{2.255}$$

$$\frac{d\hat{p}}{dt} = \frac{1}{i\hbar}\left[\hat{p}, \widehat{H}\right] = -\frac{\partial \widehat{H}}{\partial \hat{q}} \tag{2.256}$$

Displaying the dynamical (with energy thus with force involvement) commutator expressions with the first equals in each equation, while providing with the second equal the pattern for the general commutation rules for functions of conjugated variables of coordinate and momentum:

$$\left[\hat{q}(t), \widehat{F}(\hat{q}(t), \hat{p}(t), t)\right] = i\hbar \frac{\partial \widehat{F}}{\partial \hat{p}} \tag{2.257}$$

$$\left[\hat{p}(t), \widehat{F}(\hat{q}(t), \hat{p}(t), t)\right] = -i\hbar \frac{\partial \widehat{F}}{\partial \hat{q}} \tag{2.258}$$

leaving with the so-called the kinematical commutator relationships:

$$\left[\hat{q}(t), \hat{q}(t)\right] = 0;\ \left[\hat{p}(t), \hat{p}(t)\right] = 0\ \left[\hat{q}(t), \hat{p}(t)\right] = i\hbar \tag{2.259}$$

However, another subtle consequence of these relations is that from dynamical quantum equation may be abstracted (beyond the coordinate and momentum operators) the general equation of motion for an arbitrary operator:

$$\frac{d\widehat{A}}{dt} = \frac{1}{i\hbar}\left[\widehat{A}, \widehat{H}\right] + \frac{\partial \widehat{A}}{\partial t} \tag{2.260}$$

where the last term was added for generality. This equation is the instantaneous version of the above identical one in terms of observed quantum averages; it is called as *Heisenberg type equation* and this is motivated by the *Heisenberg quantum picture*, equivalent with Schrödinger one, that we are going to describe in next.

Actually, when considering the general Schrödinger equation

$$i\hbar \frac{d\psi_{Sch}(t)}{dt} = \hat{H}_{Sch}\psi_{Sch}(t) \tag{2.261}$$

it can be formally solved for the Schrödinger wave function to be

$$\psi_{Sch}(t) = CT(t_0)\exp\left(-\frac{i}{\hbar}\hat{H}_{Sch}t\right) \tag{2.262}$$

from where one may introduce the Heisenberg wave-function

$$\psi_{Hei} = \psi_{Sch}(t)\exp\left(+\frac{i}{\hbar}\hat{H}_{Sch}t\right) \tag{2.263}$$

as a non-time dependent wave-function. More generally, the reciprocal connection between the Schrödinger and Heisenberg pictures may be established as

$$\psi_{Sch}(t) = \hat{U}\psi_{Hei} \tag{2.264}$$

$$\psi_{Hei} = \hat{U}^{+}\psi_{Sch}(t) \tag{2.265}$$

throughout the introducing of the operator

$$\hat{U} = \exp\left(-\frac{i}{\hbar}\hat{H}_{Sch}t\right);\ \hat{U}^{+} = \exp\left(\frac{i}{\hbar}\hat{H}_{Sch}t\right);\ \left(\hat{U}^{+}\right)^{+} = \hat{U} \tag{2.266}$$

that is recognized as *unitary operator* since fulfilling the condition

$$\hat{U}\hat{U}^{+} = 1 \tag{2.267}$$

With this, the relation between an operator written in Schrödinger and Heisenberg pictures is achieved under the constraint that the observed average be the same in either picture:

$$\left\langle \hat{O}_{Hei} \right\rangle_{\psi_{Hei}} = \left\langle \hat{O}_{Sch} \right\rangle_{\psi_{Sch}} \tag{2.268}$$

that can be unfolded as

$$\left\langle \hat{O}_{Hei} \right\rangle_{\psi_{Hei}} = \int \psi^{*}_{Hei}\hat{O}_{Hei}\psi_{Hei}d\Gamma = \int \psi^{*}_{Sch}\hat{U}\hat{O}_{Hei}\hat{U}^{+}\psi_{Sch}d\Gamma$$

$$=\int \psi^{*}_{Sch}\widehat{O}_{Sch}\psi_{Sch}d\Gamma=\left\langle \widehat{O}_{Sch}\right\rangle_{\psi_{Sch}} \tag{2.269}$$

so releasing with the operatorial quantum interchange picture

$$\widehat{O}_{Sch}=\widehat{U}(t)\widehat{O}_{Hei}(t)\widehat{U}^{+}(t) \tag{2.270}$$

$$\widehat{O}_{Hei}(t)=\widehat{U}^{+}(t)\widehat{O}_{Sch}\widehat{U}(t) \tag{2.271}$$

Having the wave-function and operatorial expressions in Heisenberg picture one may try to establish their equation of motion; for the wave function we have:

$$\begin{aligned}
i\hbar\frac{d}{dt}\psi_{Hei} &= i\hbar\left[\frac{d\widehat{U}^{+}}{dt}\psi_{Sch}+\widehat{U}^{+}\frac{d\psi_{Sch}}{dt}\right]\\
&= i\hbar\left[\frac{d\widehat{U}^{+}}{dt}\widehat{U}\psi_{Hei}-\frac{i}{\hbar}\widehat{U}^{+}\left(\widehat{H}_{Sch}\psi_{Sch}\right)\right]\\
&= i\hbar\left[\frac{d\widehat{U}^{+}}{dt}-\frac{i}{\hbar}\widehat{U}^{+}\widehat{H}_{Sch}\right]\widehat{U}\psi_{Hei}
\end{aligned} \tag{2.272}$$

Since the Heisenberg wave-function was assumed as non-depending on time the last equality should go to zero, that is equivalently with fulfilling the operatorial equation

$$-i\hbar\frac{d\widehat{U}^{+}}{dt}=\widehat{U}^{+}\widehat{H}_{Sch} \tag{2.273}$$

leading by conjugation with the operatorial Heisenberg equation:

$$i\hbar\frac{d\widehat{U}(t)}{dt}=\widehat{H}_{Sch}\widehat{U}(t) \tag{2.274}$$

with identical structure as that of Schrödinger picture, having only the unitary operator instead of the wave-function.

As a general vision, within Schrödinger picture the operators are not depending of time but the wave-function, while in the Heisenberg picture the wave-function is fixed and the operators are variable in time; physically, the two pictures are equivalent as much as the *inertial* and *non-inertial reference* systems are considered for Schrödinger and Heisenberg quantum description, respectively.

Finally, the correctness of all this equivalent construction and commutators may be proved by considering the dynamical equation of a given operator in the Heisenberg picture; it becomes successively:

$$\begin{aligned}
\frac{d}{dt}\hat{O}_{Hei}(t) &= \frac{d}{dt}\left[\hat{U}^{+}(t)\hat{O}_{Sch}\hat{U}(t)\right] \\
&= \frac{\partial \hat{U}^{+}(t)}{\partial t}\hat{O}_{Sch}\hat{U}(t) + \hat{U}^{+}(t)\frac{\partial \hat{O}_{Sch}}{\partial t}\hat{U}(t) + \hat{U}^{+}(t)\hat{O}_{Sch}\frac{\partial \hat{U}(t)}{\partial t} \\
&= \frac{i}{\hbar}\hat{U}^{+}(t)\hat{H}_{Sch}\hat{O}_{Sch}\hat{U}(t) + \hat{U}^{+}(t)\frac{\partial \hat{O}_{Sch}}{\partial t}\hat{U}(t) \\
&\quad - \frac{i}{\hbar}\hat{U}^{+}(t)\hat{O}_{Sch}\hat{H}_{Sch}\hat{U}(t)
\end{aligned} \tag{2.275}$$

or equivalently

$$i\hbar\frac{d}{dt}\hat{O}_{Hei}(t) = \underbrace{\hat{U}^{+}(t)\left[\hat{O}_{Sch},\hat{H}_{Sch}\right]\hat{U}(t)}_{\left[\hat{O}_{Hei},\hat{H}_{Hei}\right]} + \underbrace{i\hbar\hat{U}^{+}(t)\frac{\partial \hat{O}_{Sch}}{\partial t}\hat{U}(t}_{\frac{\partial \hat{O}_{Hei}}{\partial t}} \tag{2.276}$$

producing *the Heisenberg equation for a general operator*:

$$i\hbar\frac{d}{dt}\hat{O}_{Hei}(t) = \left[\hat{O}_{Hei},\hat{H}_{Hei}\right] + i\hbar\frac{\partial \hat{O}_{Hei}}{\partial t} \tag{2.277}$$

in close agreement with the earlier result based on inspection of Poisson parenthesis, Hamilton canonic equations and their quantization. Further insight in the Schrödinger and Heisenberg quantum pictures will be providing in the forthcoming sections.

2.4 BRA-KET (DIRAC) FORMALISM

2.4.1 VECTORS IN HILBERT SPACE

Any quantum dynamical state of a physical system may be represented by a vector (*bra* or *ket*) with a unitary norm (see below) within the so-called *space of the quantum states* or the *Hilbert space*:

$$\begin{array}{ccc} & \textit{QUANTUM} & \\ & \textit{DYNAMICAL} & \\ & \textit{STATE} & \\ \swarrow & & \searrow \\ \langle\psi| \in \bar{\mathcal{H}} & \overset{DUAL}{\underset{SPACES}{\leftrightarrow}} & |\psi\rangle \in \mathcal{H} \\ \textbf{bra}-\textbf{vectors} & & \textbf{ket}-\textbf{vectors} \end{array}$$

The Hilbert space is a vectorial space with *scalar product*, which is *complete*. A metrical space is called *complete* (or Banach space) if any convergent sequence of space elements has its limits within the space. On the other side, the scalar product is defined as the functional constructed on an abelian (commutative) vectorial space $\mathcal{V}$

$$\langle\ |\ \rangle : \mathcal{V} \times \mathcal{V} \to \mathbf{C} \tag{2.278}$$

with properties

I. *The norm; the positive definite self scalar product*:

$$\left\| \langle\psi|\psi\rangle \right\|^2 = \langle\psi|\psi\rangle \geq 0\,;\ \langle\psi|\psi\rangle = 0 \Leftrightarrow |\psi\rangle = 0 \tag{2.279}$$

II. *Ket-Distributivity*:

$$\langle\psi|a_1\alpha_1 + a_2\alpha_2\rangle = a_1\langle\psi|\alpha_1\rangle + a_2\langle\psi|\alpha_2\rangle \tag{2.280}$$

III. *Hermiticity*:

$$\langle\psi|\alpha\rangle = \overline{\langle\alpha|\psi\rangle} \tag{2.281}$$

These properties allow the following consequences:

Ibis. *The null vector property*:

$$|0\rangle = 0|\psi\rangle, \forall|\psi\rangle \in \mathcal{H} \tag{2.282}$$

IIbis. *Superposition property*:

$$\langle\psi|a_1\alpha_1 + a_2\alpha_2\rangle = \langle\psi|\left(a_1|\alpha_1\rangle + a_2|\alpha_2\rangle\right), \forall|\psi\rangle,|\alpha_1\rangle,|\alpha_2\rangle \in \mathcal{H} \tag{2.283}$$

IIIbis. *Bra-Distributivity*:

$$\begin{aligned}\langle a_1\alpha_1 + a_2\alpha_2|\psi\rangle &= \overline{\langle\psi|a_1\alpha_1 + a_2\alpha_2\rangle} = \overline{a_1\langle\psi|\alpha_1\rangle + a_2\langle\psi|\alpha_2\rangle} \\ &= a_1^*\overline{\langle\psi|\alpha_1\rangle} + a_2^*\overline{\langle\psi|\alpha_2\rangle} = a_1^*\langle\alpha_1|\psi\rangle + a_2^*\langle\alpha_2|\psi\rangle\end{aligned} \tag{2.284}$$

I2bis. *Schwartz inequality*:

$$\left\|\langle\alpha|\beta\rangle\right\|^2 \le \langle\alpha|\alpha\rangle\langle\beta|\beta\rangle, \ \forall|\alpha\rangle,|\beta\rangle \in \mathcal{H} \tag{2.285}$$

Proof: using the above rules one can consider the successive equivalences:

$$\begin{aligned}&\langle\alpha + a\beta|\alpha + a\beta\rangle \ge 0, \ \forall a \in \mathbf{R} \\ &\Leftrightarrow \left(\langle\alpha| + a\langle\beta|\right)\left(|\alpha\rangle + a|\beta\rangle\right) \ge 0 \\ &\Leftrightarrow \langle\alpha|\alpha\rangle + a[\langle\alpha|\beta\rangle + \underbrace{\langle\beta|\alpha\rangle}_{\overline{\langle\alpha|\beta\rangle}}] + a^2\langle\beta|\beta\rangle \ge 0 \\ &\Leftrightarrow a^2\langle\beta|\beta\rangle + 2a\,\mathrm{Re}\left(\langle\alpha|\beta\rangle\right) + \langle\alpha|\alpha\rangle \ge 0 \\ &\Leftrightarrow \Delta \le 0 : \mathrm{Re}^2\left(\langle\alpha|\beta\rangle\right) - \langle\alpha|\alpha\rangle\langle\beta|\beta\rangle \le 0 \\ &\Leftrightarrow \mathrm{Re}^2\left(\langle\alpha|\beta\rangle\right) \le \langle\alpha|\alpha\rangle\langle\beta|\beta\rangle, \ \forall|\beta\rangle \in \mathcal{H}\end{aligned} \tag{2.286}$$

therefore true also for the specialization:

$$|\beta'\rangle = e^{i\theta}|\beta\rangle; \ \langle\beta'| = \langle\beta|e^{-i\theta} \tag{2.287}$$

providing the further inequality:

$$\mathrm{Re}\left(e^{i\theta}\langle\alpha|\beta\rangle\right)\leq\sqrt{\langle\alpha|\alpha\rangle\langle\beta|\beta\rangle} \tag{2.288}$$

Employing the left side expression only, one gets:

$$\begin{aligned}\mathrm{Re}\left(e^{i\theta}\langle\alpha|\beta\rangle\right)&=\mathrm{Re}\left((\cos\theta+i\sin\theta)\langle\alpha|\beta\rangle\right)\\&=\mathrm{Re}\left[(\cos\theta+i\sin\theta)\left(\mathrm{Re}\langle\alpha|\beta\rangle+i\,\mathrm{Im}\langle\alpha|\beta\rangle\right)\right]\\&=\mathrm{Re}\left[\begin{array}{l}\cos\theta\,\mathrm{Re}\langle\alpha|\beta\rangle-\sin\theta\,\mathrm{Im}\langle\alpha|\beta\rangle\\+i\left(\cos\theta\,\mathrm{Im}\langle\alpha|\beta\rangle+\sin\theta\,\mathrm{Re}\langle\alpha|\beta\rangle\right)\end{array}\right]\\&=\cos\theta\,\mathrm{Re}\langle\alpha|\beta\rangle-\sin\theta\,\mathrm{Im}\langle\alpha|\beta\rangle\ ,\ \forall\theta\end{aligned} \tag{2.289}$$

While choosing θ so that

$$\tan\theta=-\frac{\mathrm{Im}\langle\alpha|\beta\rangle}{\mathrm{Re}\langle\alpha|\beta\rangle} \tag{2.290}$$

we firstly have:

$$\cos\theta=\frac{1}{\sqrt{1+\tan^2\theta}}=\sqrt{\frac{\mathrm{Re}^2\langle\alpha|\beta\rangle}{\left\|\langle\alpha|\beta\rangle\right\|^2}}=\frac{\mathrm{Re}\langle\alpha|\beta\rangle}{\left\|\langle\alpha|\beta\rangle\right\|} \tag{2.291}$$

$$\sin\theta=\frac{\tan\theta}{\sqrt{1+\tan^2\theta}}=-\frac{\mathrm{Im}\langle\alpha|\beta\rangle}{\left\|\langle\alpha|\beta\rangle\right\|} \tag{2.292}$$

and then

$$\mathrm{Re}\left(e^{i\theta}\langle\alpha|\beta\rangle\right)=\frac{\mathrm{Re}^2\langle\alpha|\beta\rangle+\mathrm{Im}^2\langle\alpha|\beta\rangle}{\left\|\langle\alpha|\beta\rangle\right\|}=\left\|\langle\alpha|\beta\rangle\right\| \tag{2.293}$$

that replaced in the last inequality proofs the Schwartz theorem.

I3bis. *The triangle inequality theorem*:

$$\left\||\alpha\rangle+|\beta\rangle\right\|\leq\left\||\alpha\rangle\right\|+\left\||\beta\rangle\right\| \tag{2.294}$$

may be immediately proofed with the help of Schwartz's one equivalent forms by the chain of relations:

$$\left\| |\alpha\rangle + |\beta\rangle \right\|^2 \le \left(\left\| |\alpha\rangle \right\| + \left\| |\beta\rangle \right\| \right)^2$$

$$\Leftrightarrow \left(\langle\alpha| + \langle\beta| \right)\left(|\alpha\rangle + |\beta\rangle \right) \le \langle\alpha|\alpha\rangle + \langle\beta|\beta\rangle + 2\sqrt{\langle\alpha|\alpha\rangle\langle\beta|\beta\rangle}$$

$$\mathrm{Re}\left(\langle\alpha|\beta\rangle \right) \le \sqrt{\langle\alpha|\alpha\rangle\langle\beta|\beta\rangle} \tag{2.295}$$

that recovers one of the above Schwartz proof's inequality.

I4bis. *Cosines definition between states' vectors*

$$\cos\left(|\alpha\rangle, |\beta\rangle \right) \equiv \frac{\langle\alpha|\beta\rangle}{\sqrt{\langle\alpha|\alpha\rangle}\sqrt{\langle\beta|\beta\rangle}} \le 1,\ \forall |\alpha\rangle, |\beta\rangle \in \mathcal{H} \tag{2.296}$$

appears as a natural consequence of the Schwartz theorem. However, the validity of this definition may be seen also by considering the metric distance between two vectors in Hilbert space releasing with the generalized cosines theorem:

$$\begin{aligned} d\left(|\alpha\rangle, |\beta\rangle \right) &= \left\| |\alpha\rangle - |\beta\rangle \right\| = \left\| \alpha - \beta\rangle \right\| = \sqrt{\langle\alpha - \beta|\alpha - \beta\rangle} \\ &= \sqrt{\langle\alpha|\alpha\rangle + \langle\beta|\beta\rangle - \left(\langle\alpha|\beta\rangle + \langle\beta|\alpha\rangle \right)} \\ &= \sqrt{\langle\alpha|\alpha\rangle + \langle\beta|\beta\rangle - 2\mathrm{Re}\langle\alpha|\beta\rangle} \end{aligned} \tag{2.297}$$

where one can recognize the classical scalar product definition generalized through the present Hilbert states' vectors:

$$\langle\alpha|\beta\rangle = \left\| \alpha\rangle \right\| \left\| \beta\rangle \right\| \cos\left(|\alpha\rangle, |\beta\rangle \right) = \sqrt{\langle\alpha|\alpha\rangle}\sqrt{\langle\beta|\beta\rangle} \frac{\langle\alpha|\beta\rangle}{\sqrt{\langle\alpha|\alpha\rangle}\sqrt{\langle\beta|\beta\rangle}} \tag{2.298}$$

In next, the vector states are to be combined with operators to provide further transformations on Hilbert space of quantum reality.

2.4.2 LINEAR OPERATORS IN HILBERT SPACE

The basic definition of the linear operators acting on Hilbert space reads as:

$$\hat{A}:\mathcal{H}\rightarrow\mathcal{H}_1\subset\mathcal{H} \tag{2.299}$$

They have the *linearity* role in quantum (eigen) equations,

$$\hat{A}\left(a_1|\alpha_1\rangle+a_2|\alpha_2\rangle\right)=a_1\hat{A}|\alpha_1\rangle+a_2\hat{A}|\alpha_2\rangle \tag{2.300}$$

and in *observability* through transforming the observables' averages or the transition probabilities by the conjugate property:

$$\overline{\langle\alpha|\hat{A}|\beta\rangle}=\langle\beta|\hat{A}^{+}|\alpha\rangle \tag{2.301}$$

Such feature implies important consequences in bra-ket formalism. For instance, if one has the operatorial-ket equation

$$\hat{A}|\alpha\rangle=|\psi\rangle \tag{2.302}$$

the corresponding bra-equation is found through equivalences

$$\overline{\langle\beta|\hat{A}|\alpha\rangle}=\overline{\langle\beta|\psi\rangle}\Leftrightarrow\langle\alpha|\hat{A}^{+}|\beta\rangle=\langle\psi|\beta\rangle$$

$$\Rightarrow\langle\psi|=\langle\alpha|\hat{A}^{+} \tag{2.303}$$

In the same manner further operatorial properties may be unfolded, as follows.

I. *double conjugation*

$$\left(\hat{A}^{+}\right)^{+}=\hat{A} \tag{2.304}$$

$$\langle\alpha|\left(\hat{A}^{+}\right)^{+}|\beta\rangle=\overline{\langle\beta|\hat{A}^{+}|\alpha\rangle}=\overline{\overline{\langle\alpha|\hat{A}|\beta\rangle}}=\langle\alpha|\hat{A}|\beta\rangle \tag{2.305}$$

II. *observables' product conjugation*

$$\left(\hat{A}\hat{B}\right)^{+}=\hat{B}^{+}\hat{A}^{+} \tag{2.306}$$

$$\langle\alpha|\left(\hat{A}\hat{B}\right)^{+}|\beta\rangle=\overline{\underbrace{\langle\beta|\hat{A}}_{\langle\phi|}\underbrace{\hat{B}|\alpha\rangle}_{|\psi\rangle}}=\overline{\langle\phi|\psi\rangle}=\langle\psi|\phi\rangle=\underbrace{\langle\alpha|\hat{B}^{+}}_{\langle\psi|}\underbrace{\hat{A}^{+}|\beta\rangle}_{|\phi\rangle} \tag{2.307}$$

that can be by induction generalized to the rule:

$$\left(\hat{A}_1 \ldots \hat{A}_n\right)^+ = \hat{A}_n^+ \ldots \hat{A}_1^+ \tag{2.308}$$

III. *distributivity conjugation*

$$\left(a\hat{A} + b\hat{B}\right)^+ = a^*\hat{A}^+ + b^*\hat{B}^+ \tag{2.309}$$

$$\begin{aligned}\langle\alpha|\left(a\hat{A} + b\hat{B}\right)^+|\beta\rangle &= \overline{\langle\beta|\left(a\hat{A} + b\hat{B}\right)|\alpha\rangle} = \overline{a\langle\beta|\hat{A}|\alpha\rangle + b\langle\beta|\hat{B}|\alpha\rangle} \\ &= a^*\overline{\langle\beta|\hat{A}|\alpha\rangle} + b^*\overline{\langle\beta|\hat{B}|\alpha\rangle} = a^*\langle\alpha|\hat{A}^+|\beta\rangle + b^*\langle\alpha|\hat{B}^+|\beta\rangle \\ &= \langle\alpha|a^*\hat{A}^+ + b^*\hat{B}^+|\beta\rangle\end{aligned} \tag{2.310}$$

Next, the hermiticity property of operators is formalized in direct way through fulfilling the auto-adjunct condition:

$$\hat{A}^+ = \hat{A} \tag{2.311}$$

while *anti-hermitic* operators behave like:

$$\hat{A}^+ = -\hat{A} \tag{2.312}$$

The direct consequences regard the hermiticity and anti-hermiticity of the next combinations:

IV. $\hat{A} + \hat{A}^+$ – *hermitic*:

$$\left(\hat{A} + \hat{A}^+\right)^+ = \hat{A}^+ + \left(\hat{A}^+\right)^+ = \hat{A} + \hat{A}^+ \tag{2.313}$$

V. $\hat{A} - \hat{A}^+$ – *anti-hermitic*:

$$\left(\hat{A} + (-1)\hat{A}^+\right)^+ = \hat{A}^+ + (-1)\left(\hat{A}^+\right)^+ = -\left(\hat{A} - \hat{A}^+\right) \tag{2.314}$$

VI. $\pm i\left(\hat{A}-\hat{A}^{+}\right)$ – *hermitic*:

$$\left[\pm i\left(\hat{A}+(-1)\hat{A}^{+}\right)\right]^{+}=\mp i\left[\hat{A}^{+}+(-1)\left(\hat{A}^{+}\right)^{+}\right]=\pm i\left(\hat{A}-\hat{A}^{+}\right) \tag{2.315}$$

VII. Any linear operator $\hat{F}$ may be decomposed on *one hermitic and one anti-hermitic* or on *only hermitic operators* as the context demands:

$$\begin{aligned}\hat{F}&=\frac{1}{2}\left(\hat{F}+\hat{F}^{+}+\hat{F}-\hat{F}^{+}\right)\\&=\underbrace{\frac{1}{2}\left(\hat{F}+\hat{F}^{+}\right)}_{hermitic}+\underbrace{\frac{1}{2}\left(\hat{F}-\hat{F}^{+}\right)}_{anti-hermitic}=\underbrace{\frac{1}{2}\left(\hat{F}+\hat{F}^{+}\right)}_{hermitic}+i\underbrace{\left[-\frac{1}{2}i\left(\hat{F}-\hat{F}^{+}\right)\right]}_{hermitic}\end{aligned} \tag{2.316}$$

With these there can be constructed the so called *linear operators' algebra* (LOA): $\{+_{\text{operators,}}\bullet_{\text{with scalars,}}\bullet_{\text{with operators}}\}$ on the Hilbert space of state vectors respecting the following operations:

LOA-I. *the sum of operators*:

$$\left(\hat{A}+\hat{B}\right)|\alpha\rangle=\hat{A}|\alpha\rangle+\hat{B}|\alpha\rangle \tag{2.317}$$

LOA-II. *the product of operators with scalars*:

$$\left(a\hat{A}\right)|\alpha\rangle=a\left(\hat{A}|\alpha\rangle\right) \tag{2.318}$$

LOA-III. *the product among operators*:

$$\left(\hat{A}\hat{B}\right)|\alpha\rangle=\hat{A}\left(\hat{B}|\alpha\rangle\right) \tag{2.319}$$

However, note that this algebra is non-commutative, the measure of this non-commutativity being the introduced *commutator*

$$\left[\hat{A},\hat{B}\right]=\hat{A}\hat{B}-\hat{B}\hat{A} \tag{2.320}$$

and the anti-commutator

$$\left\{\hat{A},\hat{B}\right\}=\hat{A}\hat{B}+\hat{B}\hat{A} \tag{2.321}$$

of two operators, respectively.

If the operators $\hat{A}\ \&\ \hat{B}$ are hermitic then the bellow assertions holds:

$$\left\{\hat{A},\hat{B}\right\}\text{-}hermitic;\ \left[\hat{A},\hat{B}\right]\text{-}anti\text{-}hermitic;\ i\left[\hat{A},\hat{B}\right]\text{-}hermitic \qquad (2.322)$$

as may be immediately proofed. Moreover, with the same assumption there is true that the product $\hat{A}\hat{B}$ is hermitic if the operators $\hat{A}\ \&\ \hat{B}$ commute:

$$\underbrace{\left(\hat{A}\hat{B}\right)^{*}}_{\mapsto\cdots}=\hat{B}^{*}\hat{A}^{*}=\hat{B}\hat{A}=\underbrace{\hat{A}\hat{B}}_{\cdots\uparrow}+\underbrace{\left[\hat{B},\hat{A}\right]}_{0} \qquad (2.323)$$

while $\hat{A}^{+}\hat{A}\ \&\ \hat{A}\hat{A}^{+}$ are hermitic combination for any operators (non-necessary hermitic) $\hat{A}$. As well, there is interesting to note that if an operator $\hat{C}$ is hermitic, then any combination $\hat{G}=\hat{A}\hat{C}\hat{A}^{+}$, for any arbitrary operator $\hat{A}$, is as well hermitic:

$$\hat{G}^{+}=\left(\hat{A}\hat{C}\hat{A}^{+}\right)^{+}=\hat{A}\hat{C}^{+}\hat{A}^{+}=\hat{A}\hat{C}\hat{A}^{+}=\hat{G} \qquad (2.324)$$

Back to commutators and anti-commutators, there is useful quoting their main properties

$$\left[\hat{A},\hat{B}\right]=-\left[\hat{B},\hat{A}\right] \qquad (2.325)$$

$$\left[\hat{A},\left[\hat{B},\hat{C}\right]\right]+\left[\hat{C},\left[\hat{A},\hat{B}\right]\right]+\left[\hat{B},\left[\hat{C},\hat{A}\right]\right]=0 \text{ (the Jacobi identity)} \qquad (2.326)$$

$$\left[\hat{A},\hat{B}+\hat{C}\right]=\left[\hat{A},\hat{B}\right]+\left[\hat{A},\hat{C}\right] \qquad (2.327)$$

$$\left[\hat{A},a\hat{B}\right]=\left[a\hat{A},\hat{B}\right]=a\left[\hat{A},\hat{B}\right] \qquad (2.328)$$

$$\left[\hat{A},\hat{B}_1\hat{B}_2\right]=\left[\hat{A},\hat{B}_1\right]\hat{B}_2+\hat{B}_1\left[\hat{A},\hat{B}_2\right] \qquad (2.329)$$

$$\left[\hat{A},\hat{B}_1\hat{B}_2...\hat{B}_n\right]=\left[\hat{A},\hat{B}_1\right]\hat{B}_2...\hat{B}_n+\hat{B}_1\left[\hat{A},\hat{B}_2\right]\hat{B}_3...\hat{B}_n+\hat{B}_1...\hat{B}_{n-1}\left[\hat{A},\hat{B}_n\right] \qquad (2.330)$$

$$\left[\hat{A},\hat{B}_1\hat{B}_2...\hat{B}_n\right]=\left\{\hat{A},\hat{B}_1\right\}\hat{B}_2...\hat{B}_n-\hat{B}_1\left\{\hat{A},\hat{B}_2\right\}\hat{B}_3...\hat{B}_n$$
$$+(-1)^{n-1}\hat{B}_1...\hat{B}_{n-1}\left\{\hat{A},\hat{B}_n\right\} \tag{2.331}$$

as one may show directly or by induction based on the above definitions and properties.

Finally, we are introducing the so-called *functions of operators*, based on the analyticity of the complex functions that are expanded in the Taylor series:

$$f:\mathbf{C}\to\mathbf{C}\left|\ f(z)=f(0)+\sum_{n=1}^{\infty}\frac{f^{(n)}(0)}{n!}z^n\right. \tag{2.332}$$

which allows in the base of the formal correspondence:

$$z^n\to\hat{A}^n \tag{2.333}$$

the operatorial function expansion:

$$f(\hat{A})=f(0)+\sum_{n=1}^{\infty}\frac{f^{(n)}(0)}{n!}\hat{A}^n \tag{2.334}$$

Some examples for operatorial function expansion are:

$$\frac{1}{\hat{1}-\hat{A}}=1+\hat{A}+\hat{A}^2+...+\hat{A}^n+... \tag{2.335}$$

$$\exp\left(\hat{A}\right)=\sum_{n=0}^{\infty}\frac{\hat{A}^n}{n!}=1+\hat{A}+\frac{\hat{A}^2}{2!}+...+\frac{\hat{A}^n}{n!}+... \tag{2.336}$$

The last expansion helps in proving the important identity:

$$e^{\hat{A}}\hat{B}e^{-\hat{A}}=\hat{B}+\sum_{n=1}^{\infty}\frac{1}{n!}\underbrace{\left[\hat{A},\left[\hat{A},...\left[\hat{A},\hat{B}\right]\right]\right]}_{n\ parentheses}=\hat{B}+\left[\hat{A},\hat{B}\right]+\frac{1}{2}\left[\hat{A},\left[\hat{A},\hat{B}\right]\right]+... \tag{2.337}$$

Restricting to the second order only, one can arrange successively that

$$\begin{aligned}
e^{\hat{A}}\hat{B}e^{-\hat{A}} &= \left(\hat{1}+\hat{A}+\frac{\hat{A}^{2}}{2!}+\frac{\hat{A}^{3}}{3!}+\ldots\right)\hat{B}\left(\hat{1}-\hat{A}+\frac{\hat{A}^{2}}{2!}-\frac{\hat{A}^{3}}{3!}+\ldots\right)\\
&= \hat{B}+\left(\hat{A}\hat{B}-\hat{B}\hat{A}\right)+\left(\frac{\hat{A}^{2}}{2!}\hat{B}+\hat{B}\frac{\hat{A}^{2}}{2!}-\hat{A}\hat{B}\hat{A}\right)+\ldots\\
&= \hat{B}+\left[\hat{A},\hat{B}\right]+\frac{1}{2}\left(\hat{A}\hat{A}\hat{B}+\hat{B}\hat{A}\hat{A}-2\hat{A}\hat{B}\hat{A}\right)+\ldots\\
&= \hat{B}+\left[\hat{A},\hat{A}\right]+\frac{1}{2}\left(\hat{A}\left[\hat{A},\hat{B}\right]+\left[\hat{B},\hat{A}\right]\hat{A}\right)+\ldots\\
&= \hat{B}+\left[\hat{A},\hat{B}\right]+\frac{1}{2}\left(\hat{A}\left[\hat{A},\hat{B}\right]-\left[\hat{A},\hat{B}\right]\hat{A}\right)+\ldots\\
&= \hat{B}+\left[\hat{A},\hat{B}\right]+\frac{1}{2}\left[\hat{A},\left[\hat{A},\hat{B}\right]\right]+\ldots
\end{aligned} \tag{2.338}$$

while for the higher terms the induction method can be applied.

Such expansion is extremely useful in showing that for operators, in general we have

$$\exp\left(\hat{A}\right)\exp\left(\hat{B}\right)\neq\exp\left(\hat{A}+\hat{B}\right) \tag{2.339}$$

In fact, the left hand side product can be evaluated through considering the more general expression, say:

$$f(\lambda)=\exp\left(\lambda\hat{A}\right)\exp\left(\lambda\hat{B}\right),\ \lambda\in\mathbf{R} \tag{2.340}$$

that through derivation provides expression:

$$\frac{df(\lambda)}{d\lambda}=\hat{A}\exp\left(\lambda\hat{A}\right)\exp\left(\lambda\hat{B}\right)+\underbrace{\exp\left(\lambda\hat{A}\right)\hat{B}}_{?}\exp\left(\lambda\hat{B}\right) \tag{2.341}$$

to be then integrated. However, one recognizes that through reconsidering the above formula within the actual parameter involvement, namely

$$\begin{aligned}
f(\lambda)=e^{\lambda\hat{A}}\hat{B}e^{-\lambda\hat{A}} &= \hat{B}+\lambda\left[\hat{A},\hat{B}\right]+\frac{\lambda^{2}}{2!}\left[\hat{A},\left[\hat{A},\hat{B}\right]\right]\\
&+\ldots\frac{\lambda^{n}}{n!}\underbrace{\left[\hat{A},\left[\hat{A},\ldots\left[\hat{A},\hat{B}\right]\right]\right]}_{n\ parenthesis}+\ldots
\end{aligned} \tag{2.342}$$

is obtained that

$$\exp\left(\lambda\hat{A}\right)\hat{B}=\left\{\begin{array}{l}\hat{B}+\lambda\left[\hat{A},\hat{B}\right]+\frac{\lambda^2}{2!}\left[\hat{A},\left[\hat{A},\hat{B}\right]\right]\\+\ldots\frac{\lambda^n}{n!}\left[\hat{A},\left[\hat{A},\ldots\left[\hat{A},\hat{B}\right]\right]\right]+\ldots\end{array}\right\}\exp\left(\lambda\hat{A}\right) \tag{2.343}$$

helping in rewriting the above derivative as:

$$\frac{df(\lambda)}{d\lambda}=\left\{\begin{array}{l}\left(\hat{A}+\hat{B}\right)+\lambda\left[\hat{A},\hat{B}\right]+\frac{\lambda^2}{2!}\left[\hat{A},\left[\hat{A},\hat{B}\right]\right]\\+\ldots\frac{\lambda^n}{n!}\left[\hat{A},\left[\hat{A},\ldots\left[\hat{A},\hat{B}\right]\right]\right]+\ldots\end{array}\right\}e^{\lambda\hat{A}}e^{\lambda\hat{B}} \tag{2.344}$$

Integration of this expression, respecting the parameter λ, while keeping in mind that $f(0)=1$, provides the searched result:

$$\begin{aligned}f(\lambda)&=\exp\left(\lambda\hat{A}\right)\exp\left(\lambda\hat{B}\right)\\&=\exp\left\{\begin{array}{l}\left(\hat{A}+\hat{B}\right)\lambda+\frac{\lambda^2}{2!}\left[\hat{A},\hat{B}\right]+\frac{\lambda^3}{3!}\left[\hat{A},\left[\hat{A},\hat{B}\right]\right]\\+\ldots\frac{\lambda^{n+1}}{(n+1)!}\left[\hat{A},\left[\hat{A},\ldots\left[\hat{A},\hat{B}\right]\right]\right]+\ldots\end{array}\right\}\end{aligned} \tag{2.345}$$

producing the particularization for $\lambda=1$:

$$\exp\left(\hat{A}\right)\exp\left(\hat{B}\right)=\exp\left\{\begin{array}{l}\left(\hat{A}+\hat{B}\right)+\frac{1}{2}\left[\hat{A},\hat{B}\right]+\frac{1}{6}\left[\hat{A},\left[\hat{A},\hat{B}\right]\right]\\+\ldots\frac{1}{(n+1)!}\underbrace{\left[\hat{A},\left[\hat{A},\ldots\left[\hat{A},\hat{B}\right]\right]\right]}_{n\ parenthesis}+\ldots\end{array}\right\} \tag{2.346}$$

or, for the particular case in which $\hat{A}$ commutes with $\left[\hat{A},\hat{B}\right]$ the further simplification is obtained as (the so called Baker-Hausdorff formula):

$$\exp\left(\hat{A}\right)\exp\left(\hat{B}\right)=\exp\left\{\left(\hat{A}+\hat{B}\right)+\frac{1}{2}\left[\hat{A},\hat{B}\right]\right\} \tag{2.347}$$

that still do not allow the direct summation of operators under exponential function unless they commutes as well.

Both vectors and operators on Hilbert space of quantum states admit also various representations with which occasion additional properties should be revealed, as will be in next section exposed.

2.4.3 SPECTRAL REPRESENTATIONS OF VECTORS AND OPERATORS

Let be a vectorial (linear) finite space with scalar (dot) product operation included, $\mathcal{V}/\mathrm{C}$, with the (ortho-normal) basis or vectors:

$$\left\{|e_i\rangle\right\}_{1,\dots,n=\dim\mathcal{V}}\left|\langle e_i|e_j\rangle=\delta_{ij}=\begin{cases}1, & i=j\\0, & i\neq j\end{cases}\right. \tag{2.348}$$

with the column matrix representation:

$$|e_1\rangle=\begin{pmatrix}1\\0\\\vdots\\0\end{pmatrix},\ |e_2\rangle=\begin{pmatrix}0\\1\\\vdots\\0\end{pmatrix},\dots,\ |e_n\rangle=\begin{pmatrix}0\\0\\\vdots\\1\end{pmatrix} \tag{2.349}$$

such that any other vector $|\alpha\rangle$ of the space will be represented on this basis by the linear decomposition:

$$|\alpha\rangle=\alpha_1|e_1\rangle+\alpha_2|e_2\rangle+\dots+\alpha_n|e_n\rangle=\begin{pmatrix}\alpha_1\\\alpha_2\\\vdots\\\alpha_n\end{pmatrix}=\sum_{k=1}^{n}\alpha_k|e_k\rangle \tag{2.350}$$

while its conjugation looks like

$$\begin{aligned}\langle\alpha|&=\overline{|\alpha\rangle}=\overline{\alpha_1|e_1\rangle+\alpha_2|e_2\rangle+\dots+\alpha_n|e_n\rangle}\\&=\alpha_1^*\langle e_1|+\alpha_2^*\langle e_2|+\dots+\alpha_n^*\langle e_n|=\begin{pmatrix}\alpha_1^* & \alpha_2^* & \cdots & \alpha_n^*\end{pmatrix}\\&=\sum_{k=1}^{n}\alpha_k^*\langle e_k|\end{aligned} \tag{2.351}$$

Within this representation the scalar product may be written as:

$$\langle \alpha | \beta \rangle = \begin{pmatrix} \alpha_1^* & \alpha_2^* & \cdots & \alpha_n^* \end{pmatrix} \begin{pmatrix} \beta_1 \\ \beta_2 \\ \vdots \\ \beta_n \end{pmatrix} \equiv [\pm]^{+}[\,^{2}\,] = \sum_{k=1}^{n} \alpha_k^* \beta_k \tag{2.352}$$

with the specializations:

$$\langle \alpha | e_i \rangle = \sum_{k=1}^{n} \alpha_k^* \langle e_k | e_i \rangle = \sum_{k=1}^{n} \alpha_k^* \delta_{ki} = \alpha_i^* = \overline{\langle e_i | \alpha \rangle} \tag{2.353}$$

$$\langle e_i | \alpha \rangle = \alpha_i = \overline{\langle \alpha | e_i \rangle} \tag{2.354}$$

leading with the general vectorial rewriting:

$$|\alpha\rangle = \sum_{k=1}^{n} |e_k\rangle \underbrace{\langle e_k | \alpha \rangle}_{\substack{Fourier \\ coefficients}} \tag{2.355}$$

as well as the scalar product equivalences:

$$\langle \alpha | \beta \rangle = \sum_{k=1}^{n} \alpha_k^* \beta_k = \sum_{k=1}^{n} \overline{\langle e_k | \alpha \rangle} \langle e_k | \beta \rangle = \sum_{k=1}^{n} \langle \alpha | e_k \rangle \langle e_k | \beta \rangle \tag{2.356}$$

from where there follows the existence of the so called *unity operator*:

$$\hat{1} = \sum_{k=1}^{n} |e_k\rangle\langle e_k| \tag{2.357}$$

that assures the so called spectral decomposition of the states' product on the ortho-normal basis of the Hilbert space.

Moreover, the last operator constitutes the link between the vector and operatorial representation, i.e., representing an operator (the unity one) written in terms of bra- and ket-vectors or the Hilbert space.

Before proceeding further worth noting that from the inner structure of the Hilbert space it can be completed in the sense of *norm convergence* when its basis is extended ad infinitum into a sequence of vectors:

$$|e_n\rangle \to |e\rangle \Big| \; \| \, |e_n\rangle - |e\rangle \, \| \xrightarrow[n\to\infty]{} 0 \tag{2.358}$$

allowing the general state (vectorial) expansion:

$$|\alpha\rangle = \hat{1}_\infty |\alpha\rangle = \sum_{k=1}^{\infty} |k\rangle\langle k|\alpha\rangle, \quad \forall|\alpha\rangle \in \mathcal{H} \tag{2.359}$$

and its associate norm or self-scalar product casting:

$$\langle\alpha|\alpha\rangle = \langle\alpha|\hat{1}_\infty|\alpha\rangle = \sum_{k=1}^{\infty}\langle\alpha|k\rangle\langle k|\alpha\rangle = \sum_{k=1}^{\infty}\overline{\langle k|\alpha\rangle}\langle k|\alpha\rangle = \sum_{k=1}^{\infty}|\langle k|\alpha\rangle|^2 = \| \, |\alpha\rangle \, \|^2 \tag{2.360}$$

in terms of the infinite (discrete) unity (or projector) operator:

$$\hat{1}_\infty = \sum_{k=1}^{\infty} |k\rangle\langle k| \tag{2.361}$$

Now, going to better understand the properties of the projector operators, let's take firstly the 3D-case where the unitary projector has the representation:

$$\hat{1}_3 = \begin{pmatrix} 1 & 0 & 0 \\ 0 & 1 & 0 \\ 0 & 0 & 1 \end{pmatrix} = \hat{\Lambda}_X + \hat{\Lambda}_Y + \hat{\Lambda}_Z \tag{2.362}$$

where the sub-spaces' projectors are defined as:

$$\hat{\Lambda}_X = \begin{pmatrix} 1 & 0 & 0 \\ 0 & 0 & 0 \\ 0 & 0 & 0 \end{pmatrix}; \; \hat{\Lambda}_Y = \begin{pmatrix} 0 & 0 & 0 \\ 0 & 1 & 0 \\ 0 & 0 & 0 \end{pmatrix}; \; \hat{\Lambda}_Z = \begin{pmatrix} 0 & 0 & 0 \\ 0 & 0 & 0 \\ 0 & 0 & 1 \end{pmatrix} \tag{2.363}$$

with the obvious features:

$$\hat{\Lambda}_X|\alpha\rangle = \begin{pmatrix} 1 & 0 & 0 \\ 0 & 0 & 0 \\ 0 & 0 & 0 \end{pmatrix}\begin{pmatrix} \alpha_1 \\ \alpha_2 \\ \alpha_3 \end{pmatrix} = \begin{pmatrix} \alpha_1 \\ 0 \\ 0 \end{pmatrix} = \alpha_1\begin{pmatrix} 1 \\ 0 \\ 0 \end{pmatrix} = \alpha_1|e_X\rangle, \ldots \tag{2.364}$$

$$\hat{\Lambda}_X \hat{\Lambda}_Y = \hat{\Lambda}_X \hat{\Lambda}_Z = \hat{\Lambda}_Y \hat{\Lambda}_Z = 0 \tag{2.365}$$

thus being confirmed as true projectors in the sense that they clearly discriminate between the orthogonal directions (and of associate space's basis vectors) while preserving the orthogonalization property at the operatorial level (abstracted from orthogonal condition of the space's basis vectors).

As such, in general, the *k*-subspace of the Hilbert space can be selected by the projection operator

$$\hat{\Lambda}_k = |k\rangle\langle k| \tag{2.366}$$

fulfilling the following rules:

$$\hat{\Lambda}_k^2 = \hat{\Lambda}_k = \hat{\Lambda}_k^+ \tag{2.367}$$

as one can immediately checking out. Therefore, a complete (with finite norm) Hilbert space admits the operatorial decomposition:

$$\hat{1} = \sum_k \hat{\Lambda}_k = \sum_k |k\rangle\langle k| \tag{2.368}$$

with orthogonalization condition:

$$\hat{\Lambda}_i \hat{\Lambda}_j = \begin{cases} \hat{\Lambda}_i, & i = j \\ 0 \ , & i \neq j \end{cases} \tag{2.369}$$

Let's remark that through the projection operators the orthogonalization and spectral decomposition that are specific to vectors was transferred or reformulate in terms of operators; nevertheless, this gives the *bra-ket formalism* the consistency in respect with states and operators that are quantum mechanically linked by eigen-equation in general, and here by insertion of the unity operator providing the spectral orthogonal decomposition in term of projection operators selecting the orthogonal vectors (or states). Moreover, the power of the formalism resides in this possibility that the unity (operator) may be at any time considered as a multiplication operation and it unfolds or resumes the entire Hilbert space information of the quantum evolution and existence of the states; in other words, the quantum nature is hidden or compressed in unity projector operators written with the help of $|ket\rangle\langle bra|$ vectors; with these the formalism is complete since the operators may be written with a special combination of the vectors as well.

Let's see some formal consequences of this formalism of projection operators.

Firstly, let's note that the complement of a projection operator $\left(\hat{1}-\hat{\Lambda}\right)$ is orthogonal on it $\left(\hat{\Lambda}\right)$

$$\left(\hat{1}-\hat{\Lambda}\right)\hat{\Lambda}=\hat{\Lambda}-\underbrace{\hat{\Lambda}^{2}}_{\hat{\Lambda}}=0 \tag{2.370}$$

stands itself as a valid projection operator through fulfilling the basic requirement:

$$\left(\hat{1}-\hat{\Lambda}\right)^{2}=\hat{1}-2\hat{\Lambda}+\underbrace{\hat{\Lambda}^{2}}_{\hat{\Lambda}}=\hat{1}-\hat{\Lambda} \tag{2.371}$$

while completing the space when added with the direct one:

$$\left(\hat{1}-\hat{\Lambda}\right)+\hat{\Lambda}=\hat{1} \tag{2.372}$$

Let's see in next the relation of an arbitrary operator with the projector operators; it may be at once re-expressed as:

$$\hat{A}=\hat{1}\hat{A}\hat{1}=\sum_{i,j}\hat{\Lambda}_{i}\hat{A}\hat{\Lambda}_{j}=\sum_{i,j}|i\rangle\underbrace{\langle i|\hat{A}|j\rangle}_{\substack{matrix\\ elements}}\langle j| \tag{2.373}$$

Now the discussion regards the types of terms the projector operators provide for operatorial decomposition on a general Hilbert space (i.e., with an arbitrary vectorial basis); for clearly specifying the cases one has to introduce the notion of *invariant subspace* $\hat{\Lambda}_k\mathcal{H}$: it exists whenever an associate projector operator $\hat{\Lambda}_k$ exists with the property that transforms into itself under the action of an arbitrary operator:

$$\hat{A}:\hat{\Lambda}_k\mathcal{H}\rightarrow\hat{\Lambda}_k\mathcal{H} \tag{2.374}$$

while producing the diagonalization of the terms of the form:

$$\hat{\Lambda}_{i}\hat{A}\hat{\Lambda}_{k}=\hat{\Lambda}_{i}\hat{\Lambda}_{k}\hat{A}=\begin{cases}\hat{\Lambda}_{k}^{2}\hat{A}=\hat{\Lambda}_{k}\hat{A}\,, & i=k\\ 0 & ,\, i\neq k\end{cases} \tag{2.375}$$

An operator that admits one or many non-trivial (different than zero and than the whole space itself) invariant subspaces it is called *reducible operator* and is recognized from the upper branch of the last equation by fulfilling the condition

$$\widehat{\Lambda}_k\left[\widehat{A},\widehat{\Lambda}_k\right]=0 \tag{2.376}$$

If the action of the operator on the complement $\left(\hat{1}-\widehat{\Lambda}_k\right)\mathcal{H}$ of the invariant subspace $\widehat{\Lambda}_k\mathcal{H}$ is as well invariant then the operator is called *completely reducible operator*, whose analytical condition simply looks like:

$$\left[\widehat{A},\widehat{\Lambda}_k\right]=0 \tag{2.377}$$

since it implies both the above reducible condition as well as the invariant condition for the complement operator

$$\left[\widehat{A},\left(\hat{1}-\widehat{\Lambda}_k\right)\right]=0 \tag{2.378}$$

The link between two cases is made by the operators that are self-adjunct $\widehat{A}=\widehat{A}^+$ and reducible – then they are automatically complete reducible:

$$\widehat{\Lambda}_k\left[\widehat{A},\widehat{\Lambda}_k\right]=0 \Leftrightarrow \underbrace{\widehat{\Lambda}_k\widehat{A}=\widehat{\Lambda}_k\widehat{A}\widehat{\Lambda}_k \overset{conjugation}{\Leftrightarrow} \widehat{A}\widehat{\Lambda}_k=\widehat{\Lambda}_k\widehat{A}\widehat{\Lambda}_k}_{\widehat{A}\widehat{\Lambda}_k-\widehat{\Lambda}_k\widehat{A}=\left[\widehat{A},\widehat{\Lambda}_k\right]=0} \tag{2.379}$$

All in all, restricting to one invariant subspace ($k=1$), a complete reducible operator will have the diagonalized projection on the Hilbert space as:

$$\widehat{A}/\mathcal{H}\rightarrow\begin{pmatrix}\widehat{\Lambda}\widehat{A}\widehat{\Lambda} & 0\\ 0 & \left(\hat{1}-\widehat{\Lambda}\right)\widehat{A}\left(\hat{1}-\widehat{\Lambda}\right)\end{pmatrix} \tag{2.380}$$

throughout the (formal) decomposition

$$\begin{aligned}\widehat{A}&=\left(\widehat{\Lambda}+\hat{1}-\widehat{\Lambda}\right)\widehat{A}\left(\widehat{\Lambda}+\hat{1}-\widehat{\Lambda}\right)\\&=\widehat{\Lambda}\widehat{A}\widehat{\Lambda}+\left(\hat{1}-\widehat{\Lambda}\right)\widehat{A}\left(\hat{1}-\widehat{\Lambda}\right)+\underbrace{\widehat{\Lambda}\widehat{A}\left(\hat{1}-\widehat{\Lambda}\right)}_{0}+\underbrace{\left(\hat{1}-\widehat{\Lambda}\right)\widehat{A}\widehat{\Lambda}}_{0}\end{aligned} \tag{2.381}$$

where the last two terms vanishes on the basis of the simple invariant subspace conditions.

Following the same line of properties, there is interesting to remark that one general vector (state) may be identically reshaped in terms of its projector as:

$$|a_k\rangle = |a_k\rangle\langle a_k|a_k\rangle = \hat{\Lambda}_k|a_k\rangle \tag{2.382}$$

being this another fundamental passage from the vectors to operators by means of the projectors. The fundamental consequence regards the rewriting of the eigen-value problem

$$\hat{A}|a_k\rangle = a_k|a_k\rangle \tag{2.383}$$

with the help of projector representation

$$\hat{A}\hat{\Lambda}_k|a_k\rangle = a_k\hat{\Lambda}_k|a_k\rangle, \quad \forall|a_k\rangle \in \mathcal{H} \tag{2.384}$$

leading with the operatorial equivalence

$$\hat{A}\hat{\Lambda}_k = a_k\hat{\Lambda}_k \tag{2.385}$$

involving only the eigen-values of a given operator acting on associate projector. This result allows the alternative of an arbitrary operator expansion on all its eigen-values and associate projectors, on discrete Hilbert space, as:

$$\hat{A} = \hat{1}\hat{A}\hat{1} = \sum_k \hat{\Lambda}_k \hat{A} \sum_{k'} \hat{\Lambda}_{k'} = \sum_k \hat{\Lambda}_k \sum_{k'} \underbrace{\hat{A}\hat{\Lambda}_{k'}}_{a'_k\hat{\Lambda}_{k'}} = \sum_{k,k'} a'_k \underbrace{\hat{\Lambda}_k\hat{\Lambda}_{k'}}_{\delta_{kk'}} = \sum_k a_k\hat{\Lambda}_k \tag{2.386}$$

This result is of outmost importance since establishes the direct way of generalizing the operators' representation in terms of their eigen-values coupled with their projection directions in Hilbert space; for instance, through induction, the previous result may be extended to the operatorial powers:

$$\widehat{A}^2 = \widehat{A}\widehat{A} = \left(\sum_n a_n \widehat{\Lambda}_n\right)\left(\sum_m a_m \widehat{\Lambda}_m\right) = \left(\sum_n a_n |a_n\rangle\langle a_n|\right)\left(\sum_m a_m |a_m\rangle\langle a_m|\right) \tag{2.387}$$

$$= \sum_{n,m} a_n a_m |a_n\rangle \underbrace{\langle a_n||a_m\rangle}_{\delta_{nm}} \langle a_m| = \sum_n a_n^2 |a_n\rangle\langle a_n| = \sum_n a_n^2 \widehat{\Lambda}_n \tag{2.388}$$

...

$$\widehat{A}^p = \sum_n a_n^p |a_n\rangle\langle a_n| = \sum_n a_n^p \widehat{\Lambda}_n \tag{2.389}$$

or, even more under the general form for the functions of operators:

$$f\left(\widehat{A}\right) = \sum_n f(a_n)|a_n\rangle\langle a_n| = \sum_n f(a_n)\widehat{\Lambda}_n \tag{2.390}$$

The next level of generalization is to consider both the discrete and continuum spectra of the eigen-values of an operator that is to work with the unity operators on the entirely Hilbert space whose basis is constructed from the reunion of the discrete and continuum eigen-vectors for the given operator:

$$\left\{|a_n\rangle \middle| a_n \in S_{Discrete}\left(\widehat{A}\right)\right\} \cup \left\{|a\rangle \middle| a \in S_{Continuum}\left(\widehat{A}\right)\right\} = \mathcal{H}_{\mathbf{BASIS}} \tag{2.391}$$

with the discrete and continuum contribution represented by the sum and integrals respectively:

$$\hat{1}_{\mathcal{H}} = \sum_{S_D(\widehat{A})} |a_n\rangle\langle a_n| + \int_{S_C(\widehat{A})} |a\rangle\langle a| da \tag{2.392}$$

with the direct consequence in representing the operator itself as:

$$\widehat{A}_{\mathcal{H}} = \sum_{S_D(\widehat{A})} a_n |a_n\rangle\langle a_n| + \int_{S_C(\widehat{A})} |a\rangle\langle a| a da \tag{2.393}$$

and of the vectors as:

$$|\phi\rangle_{\mathcal{H}} = \sum_{S_D(\hat{A})} |a_n\rangle \underbrace{\langle a_n|\phi\rangle}_{\substack{Fourier\\coefficients}} + \int_{S_C(\hat{A})} |a\rangle \underbrace{\langle a|\phi\rangle}_{\substack{Wave\\function}} da \tag{2.394}$$

while the (square of the) norm obeys the form:

$$\langle\phi|\phi\rangle_{\mathcal{H}} = \| \, |\phi\rangle \, \|^2 = \sum_{S_D(\hat{A})} |\langle a_n|\phi\rangle|^2 + \int_{S_C(\hat{A})} |\langle a|\phi\rangle|^2 \, da \tag{2.395}$$

that when considered as normalized condition recovers the probability conservation (or the Parseval relationship):

$$1 = \sum_{S_D(\hat{A})} \mathsf{p}_a + \int_{S_C(\hat{A})} \wp(a) da \tag{2.396}$$

Very interesting, in order the above relations to hold one notes that the discrete and continuous projection operators are to be orthogonal, therefore also in a complementary relationship:

$$\hat{\Lambda}_D = \sum_{S_D(\hat{A})} |a_n\rangle\langle a_n| \tag{2.397}$$

$$\hat{1} - \hat{\Lambda}_D = \hat{\Lambda}_C = \int_{S_C(\hat{A})} |a\rangle\langle a| da \tag{2.398}$$

with the basic ortho-normal relationships at the level of eigen-vectors themselves from both discrete and continuum $\hat{A}$- spectrum sectors:

$$\langle a_n | a_m \rangle = \delta_{nm} \tag{2.399}$$

$$\langle a|a'\rangle = \delta(a-a') = \begin{cases} \infty \, , \; a = a' \\ 0 \, , \; a \neq a' \end{cases} \tag{2.400}$$

$$\langle a_n|a\rangle = 0, \; a_n \in S_D(\hat{A}), \; a \in S_C(\hat{A}) \tag{2.401}$$

In the light of these properties one could immediately observe that the object:

$$\left(|a\rangle\langle a|\right)^2 = |a\rangle\underbrace{\langle a|a\rangle}_{\delta(a-a)}\langle a| = \delta(0)|a\rangle\langle a| \tag{2.402}$$

is not a projector but a *projector's density.*

With the same recipe, when two operators are involved and their discrete and continuous spectra of eigen-values

$$a_n \in S_D\left(\hat{A}\right), \quad a \in S_C\left(\hat{A}\right), \quad b_k \in S_D\left(\hat{B}\right), \quad b \in S_C\left(\hat{B}\right) \tag{2.403}$$

fulfill the eigen-equations

$$\hat{A}|a_n,b_k\rangle = a_n|a_n,b_k\rangle, \quad \hat{B}|a_n,b_k\rangle = b_k|a_n,b_k\rangle \tag{2.404a}$$

$$\hat{A}|a_n,b\rangle = a_n|a_n,b\rangle, \quad \hat{A}|a,b_k\rangle = a|a,b_k\rangle \tag{2.404b}$$

$$\hat{B}|a,b_k\rangle = b_k|a,b_k\rangle, \quad \hat{B}|a_n,b\rangle = b|a_n,b\rangle \tag{2.404c}$$

$$\hat{A}|a,b\rangle = a|a,b\rangle, \quad \hat{B}|a,b\rangle = b|a,b\rangle \tag{2.404d}$$

are used as *basis* for representations in Hilbert space of the of vectorial states, therefore satisfying the ortho-normal constraints

$$\langle a_n,b_k|a_m,b_l\rangle = \delta_{nm}\delta_{kl} \tag{2.405a}$$

$$\langle a_n,b_k|a',b_l\rangle = \delta(a-a')\delta_{kl} \tag{2.405b}$$

$$\langle a_n,b|a_m,b'\rangle = \delta_{nm}\delta(b-b') \tag{2.405c}$$

$$\langle a,b|a',b'\rangle = \delta(a-a')\delta(b-b') \tag{2.405d}$$

one has therefore to use the extended (mixed) unity projector:

$$\hat{1}_{\mathcal{H}} = \sum_{\substack{S_D(\hat{A}),\\ S_D(\hat{B})}} |a_n,b_k\rangle\langle a_n,b_k| + \sum_{S_D(\hat{A})}\int_{S_C(\hat{B})} |a_n,b\rangle\langle a_n,b|db$$

$$+\sum_{S_D(\hat{B})}\int_{S_C(\hat{A})}|a,b_k\rangle\langle a,b_k|da+\int_{\substack{S_C(\hat{A}),\\ S_C(\hat{B})}}|a,b\rangle\langle a,b|dadb \tag{2.406}$$

with the direct generalization for the number of operators considered as composed the so called *complete set of commutative operators* (CoSCOpe), see the final discussion of the above section on Ehrenfest theorems, furnishing the minimal set of operators whose discrete and continuum spectra provide the unity projection decomposition that may help to represent *any* eigen-state (vector) of the concerned quantum system.

2.4.4 COORDINATE AND MOMENTUM REPRESENTATIONS

For 1D motion, the eigen-value problem of the coordinate operator $\hat{x}$

$$\hat{x}|a\rangle = a|a\rangle \tag{2.407}$$

may be rearranged as

$$(x-a)\langle x|a\rangle = 0 \tag{2.408}$$

whose solution is the Dirac function (distribution):

$$\langle x|a\rangle = \delta(x-a) = \begin{cases}\infty\ ,\ x=a\\ 0\ ,\ x\neq a\end{cases} \tag{2.409}$$

Therefore, since the eigen-value "a" may take any all values in the range $(-\infty,+\infty)$ in a continuous way, one concludes the coordinate spectrum is continuous, thus carrying the basic continuous closure and scalar product relationships

$$\hat{1}_x = \int_{-\infty}^{+\infty}|x\rangle\langle x|dx \tag{2.410}$$

$$\langle x|x'\rangle = \delta(x-x') \tag{2.411}$$

leaving with the so called coordinate representation with the continuous base $\{|x\rangle\}$ of position vectors. The direct consequence is that the space operator itself has the coordinate representation as a diagonal continuous matrix:

$$\langle x'|\underbrace{\hat{x}|x\rangle}_{x|x\rangle} = x\delta(x'-x) \tag{2.412}$$

while an arbitrary state (vector) $|\psi\rangle$ in coordinate representation will provide the traditional wave-function:

$$\psi(x) \equiv \langle x|\psi\rangle \tag{2.413}$$

with its conjugate

$$\psi^*(x) \equiv \overline{\langle x|\psi\rangle} = \langle\psi|x\rangle \tag{2.414}$$

recovering the normalization rule in a (not only) formal manner as:

$$1 = \langle\psi|\psi\rangle = \langle\psi|\hat{1}_x|\psi\rangle = \int_{-\infty}^{+\infty}\langle\psi|x\rangle\langle x|\psi\rangle dx = \int_{-\infty}^{+\infty}\psi^*(x)\psi(x)dx \tag{2.415}$$

Moreover, when considering the Fourier expansion of an arbitrary state vector in other discrete ortho-normalized base $\{|n\rangle\}$,

$$|\varphi\rangle = \hat{1}_n|\varphi\rangle = \sum_n |n\rangle\underbrace{\langle n|\varphi\rangle}_{C_n} = \sum_n C_n|n\rangle \tag{2.416}$$

it may be equivalently written as:

$$\int\varphi^*(x)\varphi(x)dx = \langle\varphi|\varphi\rangle = \sum_{m,n} C_m^* C_n \underbrace{\langle m|n\rangle}_{\delta_{mn}} = \sum_n |C_n|^2 \tag{2.417}$$

resulting in the celebrated *Parseval relationship*, linking the probability of localization of a given wave-function in terms of its Fourier coefficients.

Other interesting consequence is the coordinate representation of a function of coordinate operator $f(\hat{x})$ acting on a state vector $|u\rangle$:

$$\begin{aligned}\langle x|f(\hat{x})|u\rangle &= \langle x|f(\hat{x})\hat{1}_{x'}|u\rangle = \int\langle x|f(\hat{x})|x'\rangle\langle x'|u\rangle dx' \\ &= \int f(x)\underbrace{\langle x|x'\rangle}_{\delta(x-x')}\underbrace{\langle x'|u\rangle}_{u(x')}dx' = f(x)\underbrace{\int\delta(x-x')u(x')dx'}_{u(x)} \\ &= f(x)u(x)\end{aligned} \tag{2.418}$$

where the filtration property of the Dirac distribution was employed.

At this moment worth presenting a very interesting feature of the projector operators: let it be the eigen-problem

$$\hat{A}|\varphi_n\rangle = a_n|\varphi_n\rangle \tag{2.419}$$

Once the eigen-values and eigen-vectors are determined, the original operator may be written in terms of them with the help of operator projector (above proofed) property as:

$$\hat{A} = \sum_n a_n \hat{\Lambda}_n = \sum_n a_n |\varphi_n\rangle\langle\varphi_n| \tag{2.420}$$

Now, interpreting the last expression as a *Lagrange interpolation polynomial* the individual projectors can be approximated as (see Appendix A.4):

$$\hat{\Lambda}_n = \frac{\ldots\left(\hat{A}-a_{n-k}\right)\ldots\left(\hat{A}-a_{n-1}\right)\left(\hat{A}-a_{n+1}\right)\ldots\left(\hat{A}-a_{n+k}\right)\ldots}{\ldots\left(a_n-a_{n-k}\right)\ldots\left(a_n-a_{n-1}\right)\left(a_n-a_{n+1}\right)\ldots\left(a_n-a_{n+k}\right)\ldots} = |\varphi_n\rangle\langle\varphi_n| \tag{2.421}$$

A relevant exemplification of this operatorial approximation may be unfolded for the momentum rooting operator

$$\hat{A} = \frac{1}{i}\frac{d}{dx} \tag{2.422}$$

whose eigen-equation

$$\hat{A}|\varphi_n\rangle = n|\varphi_n\rangle \Leftrightarrow \frac{1}{i}\frac{d}{dx}\langle x|\varphi_n\rangle = n\langle x|\varphi_n\rangle \tag{2.423}$$

has the formal solution

$$\langle x|\varphi_n\rangle = \varphi_n(x) = \frac{1}{\sqrt{2\pi}}\exp(inx) \tag{2.424}$$

whose eigen-values take the discrete values $n = 0, \pm 1, \pm 2, \ldots$

There is immediate that it fulfills the normalization condition through the Dirac distribution function normalization integral:

$$\langle\varphi_n|\varphi_m\rangle = \langle\varphi_n|\hat{1}_x|\varphi_m\rangle = \frac{1}{2\pi}\int_{-\infty}^{+\infty}\underbrace{\langle\varphi_n|x\rangle}_{\overline{\langle x|\phi_n\rangle}}\langle x|\varphi_m\rangle dx$$

$$= \frac{1}{2\pi}\underbrace{\int_{-\infty}^{+\infty}\exp\left[-i(n-m)x\right]dx}_{2n\delta(n-m)} = \delta(n-m) \tag{2.425}$$

However, one retains the idea of discrete eigen-values to construct the associate Lagrange projector operator $\widehat{\Lambda}_n$ through the successive transformations:

$$\widehat{\Lambda}_n = \frac{...\left(\widehat{A}-n+k\right)...\left(\widehat{A}-n+1\right)\left(\widehat{A}-n-1\right)...\left(\widehat{A}-n-k\right)...}{...k...1(-1)...(-k)...}\langle\varphi_n|\varphi_m\rangle$$

$$= \langle\varphi_n|\hat{1}_x|\varphi_m\rangle = \frac{1}{2\pi}\int_{-\infty}^{+\infty}\underbrace{\langle\varphi_n|x\rangle}_{\overline{\langle x|\phi_n\rangle}}\langle x|\varphi_m\rangle dx \cong \prod_{k=1}^{\infty}\exp\left[-\left(\frac{\widehat{A}-n}{k}\right)^2\right]$$

$$= \exp\left[-\left(\widehat{A}-n\right)^2\sum_{k=1}^{\infty}\frac{1}{k^2}\right] = \exp\left[-\left(\widehat{A}-n\right)^2\frac{\pi^2}{6}\right]$$

$$\cong 1-\frac{\left[\left(\widehat{A}-n\right)\pi\right]^2}{6} = \frac{1}{\left(\widehat{A}-n\right)\pi}\left\{\left(\widehat{A}-n\right)\pi - \frac{\left[\left(\widehat{A}-n\right)\pi\right]^3}{6}\right\} \tag{2.426}$$

leading with the final result

$$\widehat{\Lambda}_n = |\varphi_n\rangle\langle\varphi_n| \cong \frac{\sin\left[\left(\widehat{A}-n\right)\pi\right]}{\left(\widehat{A}-n\right)\pi} \tag{2.427}$$

where, beside the exponential and sinus functions approximations used, also the celebrated Riemann series limit was involved (see Appendix A.3):

$$\sum_{k=1}^{\infty}\frac{1}{k^2} = \frac{\pi^2}{6} \tag{2.428}$$

Yet, although of symbolical value, the above result reflects the quantum (wave) nature of the operatorial formalism, especially in this case where the working operator was that roots the momentum, i.e., the free quantum motion. For that, the associate projector operator remembers of the de Broglie wave-packet, however one wave-packet for each orthogonal direction of motion, being this result a more formal and generalized result. Nevertheless, with all approximations included the obtained projector expression fulfills the basic constraints for a well behaving projection operator, namely:

$$\hat{\Lambda}_n|\varphi_m\rangle \cong \left.\frac{\sin\left[\left(\hat{A}-n\right)\pi\right]}{\left(\hat{A}-n\right)\pi}\right||\varphi_m\rangle = \left.\frac{\sin\left[(m-n)\pi\right]}{(m-n)\pi}\right||\varphi_m\rangle = \delta_{mn}|\varphi_m\rangle$$

$$= \begin{cases} |\varphi_n\rangle, & n=m \\ 0, & n\neq m \end{cases} \tag{2.429}$$

in totally accordance with the formal development:

$$\hat{\Lambda}_n|\varphi_m\rangle = |\varphi_n\rangle\underbrace{\langle\varphi_n|\varphi_m\rangle}_{\delta_{nm}} = \begin{cases} |\varphi_n\rangle, & n=m \\ 0, & n\neq m \end{cases} \tag{2.430}$$

since the special behavior of the sin-function normalized on its argument:

$$\frac{\sin\left[(m-n)\pi\right]}{(m-n)\pi} = \begin{cases} 1=\lim\limits_{\xi\to 0}\dfrac{\sin\xi}{\xi}, & n=m \\ 0, & n\neq m \end{cases} \tag{2.431}$$

when taking into account the eigen-values of the actual problem as being integer numbers. Last note here regards the fact the actual problem can be immediately generalized to the momentum problem with $\hat{A}=\hat{p}_x=-i\hbar\partial_x$ with the only changes relating the changing of the normalization constant, $\sqrt{2\pi}\to\sqrt{2\pi\hbar}$ (see also bellow) and with that according which the actual eigen-values are replaced by the momentum eigen-values $n\to p_{n/x}$; the only "philosophical" problems remains the fact that the free momentum spectra is continuous while the above deductions were made under the assumption of discrete spectra; yet, the actual formalism may be identically transposed fro the momentum spectra assuming its spectra as being composed by "cuts" on the real axis of the momentum eigen-values, or, otherwise, being considered those eigen-momentum that corresponds to the observed coordinates: the spectra is infinite but in a continuous-cutting fashion; this is possible since the closure (normalization) condition is still preserved in terms of delta-Dirac distribution.

Going now to explicitly consider the momentum operator as a function of the coordinate, $\widehat{p}_x = -i\hbar\partial_x$, one has immediately the formal relationships:

$$\langle x|\widehat{p}_x|u\rangle = -i\hbar\partial_x u(x) \tag{2.432}$$

$$\langle x|f(\widehat{p}_x)|u\rangle = f(-i\hbar\partial_x)u(x) \tag{2.433}$$

written on the same grounds as previously done for the coordinate operators and functions.

As well, similar expression involving the space-momentum commutator is drawn as:

$$\langle x|\left[\hat{x},\widehat{p}_x\right]|u\rangle = \langle x|i\hbar\hat{1}|u\rangle = i\hbar\langle x|u\rangle = i\hbar u(x) \tag{2.434}$$

At this moment there is clear that the set of operators $\left\{\hat{1},\hat{x},\widehat{p}_x\right\}$ may not constitute a *complete set of commutative operators* (CoSCOpe) since while the commutations with unity operator is fulfilled

$$\left[\hat{x},\hat{1}\right] = \left[\widehat{p}_x,\hat{1}\right] = 0 \tag{2.435}$$

this is not the true with the coordinate and momentum operators. Yet, starting from their fundamental commutation

$$\left[\hat{x},\widehat{p}_x\right] = i\hbar\hat{1} \tag{2.436}$$

one may successively derive the following identities:

$$\left[\hat{x}^2,\widehat{p}_x\right] = \left[\hat{x}\hat{x},\widehat{p}_x\right] = \hat{x}\left[\hat{x},\widehat{p}_x\right] + \left[\hat{x},\widehat{p}_x\right]\hat{x} = \left(2\hat{x}\right)i\hbar \tag{2.437}$$

$$\left[\hat{x},\hat{p}_x^2\right]=\left[\hat{x},\hat{p}_x\hat{p}_x\right]=\hat{p}_x\left[\hat{x},\hat{p}_x\right]+\left[\hat{x},\hat{p}_x\right]\hat{p}_x=\left(2\hat{p}_x\right)i\hbar \tag{2.438}$$

and by the induction the general ones:

$$\left[\hat{x}^k,\hat{p}_x\right]=\left(k\hat{x}^{k-1}\right)i\hbar \tag{2.439}$$

$$\left[\hat{x},\hat{p}_x^k\right]=\left(k\hat{p}_x^{k-1}\right)i\hbar \tag{2.440}$$

The generalization to functions of these operators may be as well proofed for analytical functions, i.e., for convergent functions though the series expansion. For instance, for coordinate operator analytical functions one may write that:

$$f\left(\hat{x}\right)=\sum_{k=0}^{\infty}\alpha_k\hat{x}^k \tag{2.441}$$

yielding for the associate commutator:

$$\left[f\left(\hat{x}\right),\hat{p}\right]=\left[\sum_{k=0}^{\infty}\alpha_k\hat{x}^k,\hat{p}_x\right]=i\hbar\sum_{k=0}^{\infty}\alpha_k k\hat{x}^{k-1}=i\hbar\partial_{\hat{x}}f\left(\hat{x}\right) \tag{2.442}$$

and analogous for the momentum operator functions:

$$\left[\hat{x},g\left(\hat{p}_x\right)\right]=i\hbar\partial_{\hat{p}_x}g\left(\hat{p}_x\right) \tag{2.443}$$

Now there is clear that in order that the coordinate operator to commute with some other operatorial object we should have that

$$\left[\hat{x},g\left(\hat{p}_x\right)\right]=0\Rightarrow\partial_{\hat{p}_x}g\left(\hat{p}_x\right)=0\Rightarrow g\left(\hat{p}_x\right)=const \tag{2.444}$$

a condition that restricts the operators' class to unity or to those that depend only by coordinate; the same rationale holds for momentum operator as well; therefore, the operators $\left\{\hat{x}\right\}$ and $\left\{\hat{p}_x\right\}$ may constitute themselves a CoSCOpe, meaning that their representations may be used independently; even more, this implies that the momentum's representation is based on similar relationships as was the coordinate representation case, namely:

$$\hat{p}\left|p\right\rangle=p\left|p\right\rangle \tag{2.445}$$

$$\hat{1}_p = \int_{-\infty}^{+\infty} |p\rangle\langle p| dp \tag{2.446}$$

$$\langle p|p'\rangle = \delta(p - p') \tag{2.447}$$

as the momentum eigen-problem, closure formula and the dot product rule, respectively.

However, there still remains the problem of reciprocal representation, i.e., to determine the quantum objects $\langle x|p\rangle, \langle p|x\rangle$, understanding from now on that the momentum p is associated with the motion on the direction x, for simplifying the writings as far as no confusion may arise. To this aim worth introducing an interesting unitary operator

$$\widehat{U}(a) = \exp\left(-\frac{i}{\hbar} a \widehat{p}\right) \tag{2.448}$$

called as *the translation operator* since it at once fulfills the transformations:

$$\widehat{U}(a)\widehat{U}(a') = \widehat{U}(a + a') \tag{2.449a}$$

$$\widehat{U}(0) = \hat{1} \tag{2.449b}$$

$$\widehat{U}^{+}(a) = \widehat{U}^{-1}(a') = \widehat{U}(-a) \tag{2.449c}$$

corresponding to the translational summation, identity for the absence of movement, and to inversion, respectively. Nevertheless, when applied on a given space eigen-problem

$$\hat{x}|x\rangle = x|x\rangle \tag{2.450}$$

produces the successive transformations:

$$\begin{aligned}\hat{x}\widehat{U}(a)|x\rangle &= \underbrace{\left[\hat{x}, \widehat{U}(a)\right]}_{i\hbar\partial_a \widehat{U}(a)}|x\rangle + \widehat{U}(a)\underbrace{\hat{x}|x\rangle}_{x|x\rangle} \\ &= i\hbar\left(-\frac{i}{\hbar}a\right)\widehat{U}(a)|x\rangle + x\widehat{U}(a)|x\rangle = (x+a)\widehat{U}(a)|x\rangle \end{aligned} \tag{2.451}$$

from where one recognizes the possibility of introducing the *translation eigen-vector*

$$\hat{U}(a)|x\rangle \equiv |x+a\rangle \tag{2.452}$$

associated with the eigen-problem:

$$\hat{x}|x+a\rangle = (x+a)|x+a\rangle \tag{2.453}$$

Yet, the translation vector equivalently becomes by operatorial and vectorial series expansion in left and right hand sides, respectively:

$$\begin{aligned}\exp\left(-\frac{i}{\hbar}a\hat{p}\right)|x\rangle \equiv |x+a\rangle \Leftrightarrow \left[\hat{1}-\frac{i}{\hbar}a\hat{p}+\frac{(-i)}{2!\hbar^2}a^2\hat{p}^2+\ldots\right]|x\rangle \\ \equiv |x\rangle + \frac{\partial|x\rangle}{\partial x}a+\ldots \Leftrightarrow -\frac{i}{\hbar}\hat{p}|x\rangle + a\left(-\frac{\hat{p}^2}{2!\hbar^2}+\ldots\right)|x\rangle \\ \equiv \frac{\partial|x\rangle}{\partial x}+a(\ldots)\end{aligned} \tag{2.454}$$

which for the limit $a \to 0$ provides the mixed eigen-equations:

$$\hat{p}_x|x\rangle = i\hbar\partial_x|x\rangle \tag{2.455}$$

$$\langle x|\hat{p}_x = -i\hbar\partial_x\langle x| \tag{2.456}$$

Very interesting, with the last relation one may get back the momentum operatorial definition when a vector state $|\phi\rangle$:

$$\langle x|\hat{p}_x|\phi\rangle = -i\hbar\partial_x\langle x|\phi\rangle \tag{2.457}$$

through being transcribed in the light of above coordinate representation consideration of functions as:

$$\hat{p}_x\phi(x) = -i\hbar\partial_x\phi(x) \tag{2.458}$$

On the other side, when identifying $\phi = x'$ the matrix elements of the momentum in the coordinate representation are obtained in relation with Dirac distribution

$$\langle x|\hat{p}_x|x'\rangle = -i\hbar\partial_x\langle x|x'\rangle = -i\hbar\partial_x\delta(x-x') \quad (2.459)$$

displaying the form of a non-diagonal continuum matrix.

Going further, the above coordinate-momentum eigen-problem may be employed by applying the momentum ket-vectorial action at left to get the equation:

$$\langle x|\underbrace{\hat{p}_x|p_x\rangle}_{p_x|p_x\rangle} = -i\hbar\partial_x\langle x|p_x\rangle \Leftrightarrow p_x\langle x|p_x\rangle = -i\hbar\partial_x\langle x|p_x\rangle \quad (2.460)$$

with the immediate solution

$$\langle x|p_x\rangle = Ct\exp\left(\frac{i}{\hbar}p_x x\right) \quad (2.461)$$

whose constant is determined by applying the (continuous) normalization condition in momentum representation:

$$\begin{aligned}\delta(p-p') &= \langle p|p'\rangle = \langle p|\hat{1}_x|p'\rangle \\ &= \int_{-\infty}^{+\infty}\langle p|x\rangle\langle x|p'\rangle dx = |Ct|^2\underbrace{\int_{-\infty}^{+\infty}\exp\left(-\frac{i}{\hbar}x(p-p')\right)dx}_{2\pi\hbar\delta(p-p')}\end{aligned} \quad (2.462)$$

resulting in the value:

$$Ct = \frac{1}{\sqrt{2\pi\hbar}} \quad (2.463)$$

and in the final 1D-distribution for the quantum free motion:

$$\langle x|p_x\rangle = \frac{1}{\sqrt{2\pi\hbar}}\exp\left(\frac{i}{\hbar}p_x x\right) \equiv u_p(x) \quad (2.464)$$

With this one may formulate also the action of the coordinate operator in the momentum basis (representation) throughout the identities:

$$\langle p_x| \rightarrow |\ \hat{x}|x\rangle = x|x\rangle \Rightarrow \langle p_x|\hat{x}|x\rangle = x\langle p_x|x\rangle$$
$$= x\overline{\langle x|p_x\rangle} = x\underbrace{\frac{1}{\sqrt{2\pi\hbar}}\exp\left(-\frac{i}{\hbar}p_x x\right)}_{\langle p_x|x\rangle} = i\hbar\frac{d}{dp_x}\langle p_x|x\rangle \tag{2.465}$$

so that the following operatorial action rules are derived:

$$\langle p_x|\hat{x} = i\hbar\frac{d}{dp_x}\langle p_x| \text{ ... for } bra\text{-momentum vectors} \tag{2.466}$$

$$\hat{x}|p_x\rangle = -i\hbar\frac{d}{dp_x}|p_x\rangle \text{ ... for } ket\text{-momentum vectors} \tag{2.467}$$

The reciprocal coordinate-momentum *bra-ket* representation transformation provides the formalism for Fourier transformation between the wave-function $\psi(x)$ and momentum function amplitude $\psi(p)$:

$$\psi(x) = \langle x|\psi\rangle = \langle x|\hat{1}_p|\psi\rangle = \int_{-\infty}^{+\infty}\underbrace{\langle x|p_x\rangle}_{u_p(x)}\underbrace{\langle p_x|\psi\rangle}_{\psi(p)}dp_x$$
$$= \frac{1}{\sqrt{2\pi\hbar}}\int_{-\infty}^{+\infty}\psi(p_x)\exp\left(\frac{i}{\hbar}p_x x\right)dp_x \tag{2.468}$$

$$\psi(p_x) = \langle p_x|\psi\rangle = \langle p_x|\hat{1}_x|\psi\rangle = \int_{-\infty}^{+\infty}\underbrace{\langle p_x|x\rangle}_{\overline{u_p(x)}}\underbrace{\langle x|\psi\rangle}_{\psi(x)}dx$$
$$= \frac{1}{\sqrt{2\pi\hbar}}\int_{-\infty}^{+\infty}\psi(x)\exp\left(-\frac{i}{\hbar}p_x x\right)dx \tag{2.469}$$

Finally, worth showing that the 1D analysis may be easily generalized to *N*-dimensionally space with associate coordinate and momentum representations. The starring point is the generating of the *N*-unity operator as the direct product of the N-1D-unity operators from the direct product of the N-1D Hilbert spaces:

$$\hat{1}_{\mathcal{H}_1 \otimes \mathcal{H}_2 \otimes \ldots \otimes \mathcal{H}_N} = \hat{1}_1 \otimes \hat{1}_2 \otimes \ldots \otimes \hat{1}_N \tag{2.470}$$

with the help of which the individual and products of operators are constructed as:

$$\hat{x}_1 = \hat{x}_1 \otimes \hat{1}_2 \otimes \ldots \otimes \hat{1}_N \tag{2.471a}$$

…

$$\hat{x}_N = \hat{1}_1 \otimes \hat{1}_2 \otimes \ldots \otimes \hat{x}_N \tag{2.471b}$$

$$\hat{x}_1 \hat{x}_2 = \hat{x}_1 \otimes \hat{x}_2 \otimes \ldots \otimes \hat{1}_N \tag{2.471c}$$

$$\hat{x}_2 \hat{x}_1 = \hat{x}_2 \otimes \hat{x}_1 \otimes \ldots \otimes \hat{1}_N \tag{2.471d}$$

…

$$\hat{x}_1 \hat{x}_2 \ldots \hat{x}_N = \hat{x}_1 \otimes \hat{x}_2 \otimes \ldots \otimes \hat{x}_N \tag{2.471e}$$

from where the generalized N-dimensional vectors are constructed

$$|x_1, x_2, \ldots, x_N\rangle = |x_1\rangle \otimes |x_2\rangle \otimes \ldots \otimes |x_N\rangle \equiv |\vec{x}_N\rangle \tag{2.472}$$

whose particular eigen-value problems are selected on individual directions:

$$\hat{x}_k |x_1, x_2, \ldots, x_N\rangle = x_k |x_1, x_2, \ldots, x_N\rangle, \; k = \overline{1, N} \tag{2.473}$$

providing the ortho-normalization condition

$$\begin{aligned} \langle x_1, x_2, \ldots, x_N | x'_1, x'_2, \ldots, x'_N\rangle &= \delta(x_1 - x'_1)\delta(x_2 - x'_2)\ldots\delta(x_N - x'_N) \\ &\equiv \delta^N(x - x') \end{aligned} \tag{2.474}$$

and the closure relation

$$\hat{1}_{\mathcal{H}_N \equiv \mathcal{H}_1 \otimes \mathcal{H}_2 \otimes ... \otimes \mathcal{H}_N} = \int_{S(\hat{x}_1)\otimes ... \otimes S(\hat{x}_N)} |x_1, x_2, ..., x_N\rangle \langle x_1, x_2, ..., x_N| dx_1 dx_2 ... dx_N \tag{2.475}$$

The same expressions may be re-written for the momentum representation, while for the mixed products one has the uni-directional distribution:

$$\langle x_k | p_k \rangle = \frac{1}{\sqrt{2\pi\hbar}} \exp\left(\frac{i}{\hbar} p_k x_k\right) \equiv u_{p_k}(x_k) \tag{2.476}$$

that enters N-times the N-dimensional distribution:

$$\begin{aligned} u_{p_{1...N}}(x_{1...N}) &= \langle x_{1...N} | p_{1...N} \rangle = \left(\langle x_1 | \otimes ... \otimes \langle x_N | \right) \left(| p_1 \rangle \otimes ... \otimes | p_N \rangle \right) \\ &\equiv \langle \vec{x}_N | \vec{p}_N \rangle = \langle x_1 | p_1 \rangle ... \langle x_k | p_k \rangle ... \langle x_N | p_N \rangle \\ &= \frac{1}{(2\pi\hbar)^{N/2}} \exp\left(\frac{i}{\hbar} (p_1 x_1 + ... + p_k x_k + ... + p_N x_N) \right) \\ &= \frac{1}{(2\pi\hbar)^{N/2}} \exp\left(\frac{i}{\hbar} \vec{p}_N \vec{x}_N \right) \end{aligned} \tag{2.477}$$

entering at its time on the generalized N-dimensional wave-function:

$$\langle \vec{x}_N | \psi \rangle = \langle \vec{x}_N | \hat{1}_{\vec{p}_N} | \psi \rangle = \int_{\vec{\mathcal{H}}_N} \langle \vec{x}_N | \vec{p}_N \rangle \langle \vec{p}_N | \psi \rangle d\vec{p}_N \tag{2.478}$$

These considerations will be most helpful when introducing the N-body systems, specific to chemical samples, and most useful for the second quantification techniques.

2.4.5 ENERGY REPRESENTATION

When about the energy representation, one should restrict to the stationary states (eigen-energies),

$$\widehat{H}|E\rangle = E|E\rangle \tag{2.479}$$

while the general quantum evolution fulfills the temporal Schrödinger equation:

$$i\hbar\partial_t|t\rangle = \widehat{H}(t)|t\rangle \tag{2.480}$$

Within this context, the dynamical state $|t\rangle$ fulfills the conservation of its norm (scalar product in fact):

$$\begin{aligned}\partial_t\langle t|t\rangle &= \left(\partial_t\langle t|\right)|t\rangle + \langle t|\left(\partial_t|t\rangle\right)\\ &= \frac{1}{i\hbar}\left(-\langle t|\underbrace{\widehat{H}^{+}}_{\widehat{H}}|t\rangle + \langle t|\widehat{H}|t\rangle\right) = 0\end{aligned} \tag{2.481}$$

employing the hermiticity property of the Hamiltonian; the conservation norm actually looks like:

$$\langle t|t\rangle = \langle t_0|t_0\rangle = 1 \tag{2.482}$$

The second application of the temporal vectorial state regards the temporal variation of the average of an operator $\widehat{A}$, which in Schrödinger equation is constant in time $\partial_t\widehat{A} = 0$, leading with:

$$\begin{aligned}i\hbar\partial_t\langle t|\widehat{A}|t\rangle &= i\hbar\left[\left(\partial_t\langle t|\right)\widehat{A}|t\rangle + \langle t|\widehat{A}\left(\partial_t|t\rangle\right)\right]\\ &= -\langle t|\widehat{H}\widehat{A}|t\rangle + \langle t|\widehat{A}\widehat{H}|t\rangle = \langle t|\left[\widehat{A},\widehat{H}\right]|t\rangle\end{aligned} \tag{2.483}$$

providing the basic relation that fundaments the Ehrenfest theorems.

The third consequence appears through the temporal-stationary factorization of the dynamical states:

$$|t\rangle = |E\rangle f(t) \tag{2.484}$$

that together with above (stationary) energy eigen-equation reduces the temporal Schrödinger equation to the ordinary differential equation

$$i\hbar\partial_t f(t) = Ef(t) \tag{2.485}$$

from where the particular solution:

$$f(t) = \exp\left(-\frac{i}{\hbar}Et\right) \quad (2.486)$$

preserves the previous norm conservation for the dynamical $|t\rangle$ state, while through evolution it may be seen as the general superposition of the discrete and continuum energy states representations multiplied by the evolution factor above:

$$|t\rangle = \sum_{S_D(\widehat{H})} C_n |E_n\rangle e^{-\frac{i}{\hbar}E_n(t-t_0)} + \int_{S_C(\widehat{H})} C(E)|E\rangle e^{-\frac{i}{\hbar}E(t-t_0)} dE \quad (2.487)$$

The immediate specializations of this instantaneous state can be achieved through setting $t = t_0$ for initial (prepared or measured) state

$$|t_0\rangle = \sum_{S_D(\widehat{H})} C_n |E_n\rangle + \int_{S_C(\widehat{H})} C(E)|E\rangle dE \quad (2.488)$$

from where, with the help of ortho-normal relationships between the states belonging to the discrete, continuum and mixed spectra, respectively,

$$\langle E_n | E_m \rangle = \delta_{nm} \quad (2.489a)$$

$$\langle E | E' \rangle = \delta(E - E') \quad (2.489b)$$

$$\langle E | E_m \rangle = 0 \quad (2.489c)$$

the individual constants are determined as:

$$\langle E_k | t_0 \rangle = \delta_{k,n} C_n + 0 = C_k \quad (2.490a)$$

$$\langle E' | t_0 \rangle = 0 + \int_{S_C(\widehat{H})} C(E)\delta(E - E') dE = C(E') \quad (2.490b)$$

with the signification of being the Fourier coefficients for the initial state.

With these the general dynamical state is re-written as:

$$|t\rangle = \sum_{S_D(\widehat{H})} |E_n\rangle\langle E_n|t_0\rangle e^{-\frac{i}{\hbar}E_n(t-t_0)} + \int_{S_C(\widehat{H})} |E\rangle\langle E|t_0\rangle e^{-\frac{i}{\hbar}E(t-t_0)} dE \quad (2.491)$$

from where there is recognized the closure projector relationship for the energy representation:

$$\hat{1}_E = \sum_{S_D(\widehat{H})} |E_n\rangle\langle E_n| + \int_{S_C(\widehat{H})} |E\rangle\langle E| dE \quad (2.492)$$

The energy representation is one of the most important in treating the stationary state analysis and will be used in characterizing the density of states in the section dedicated to density matrix formalism.

2.4.6 HEISENBERG MATRIX QUANTUM MECHANICS: THE HARMONIC OSCILLATOR

Recognizing the fact that representation of observable quantum operators, in various bases, is made by (hermitic) matrices, Heisenberg had generalized the commutation rules to operators and thus to matrix level, while this way constructing the so-called quantum matrix mechanics. It is basically founded by the commutation rules among the coordinate $\{[Q]_n\}$ and momentum $\{[P]_n\}$ matrices,

$$[Q]_m[Q]_n - [Q]_n[Q]_m = 0 \quad (2.493)$$

$$[P]_m[P]_n - [P]_n[P]_m = 0 \quad (2.494)$$

$$[P]_m[Q]_n - [Q]_n[P]_m = -i\hbar\delta_{mn}[1] \quad (2.495)$$

with $[1]$ the unity matrix, while the Hamiltonian one, $[H](\{[Q]_m\},\{[P]_m\})$, constructed from the classical-to-matrix rewriting, will be rendered diagonal, with the diagonal elements being the measurable eigenvalues concerned.

Yet, the internal consistency of the matrix approach is based on the important *Lemma* according which the relationships that hold between two operators, say $\hat{A}, \hat{B}$, also hold between the matrices and their elements formed in one given basis (or representation):

$$[A]_{ij} = \langle \phi_i | \hat{A} | \phi_j \rangle, [B]_{ij} = \langle \phi_i | \hat{B} | \phi_j \rangle \tag{2.496}$$

The true of this fact may be efficiently exemplified in the case of summation and multiplication operations:

$$[A+B]_{ij} = [A]_{ij} + [B]_{ij} \tag{2.497}$$

$$[AB]_{ij} = \sum_k [A]_{ik} [B]_{kj} \tag{2.498}$$

While the first operation is obvious from the distributivity of the operatorial action on the vectorial space (see earlier sections), the second one may be easily proofed noting that, for instance

$$\hat{B}|\phi_j\rangle = \hat{1}_\phi \hat{B}|\phi_j\rangle = \sum_k |\phi_k\rangle \underbrace{\langle \phi_k | \hat{B} | \phi_j \rangle}_{[B]_{kj}} = \sum_k [B]_{kj} |\phi_k\rangle \tag{2.499}$$

With this there is immediate to form the succession of operations:

$$\begin{aligned} [AB]_{ij} &= \langle \phi_i | \hat{A}\hat{B} | \phi_j \rangle = \langle \phi_i | \hat{A} \sum_k [B]_{kj} |\phi_k\rangle \\ &= \sum_k [B]_{kj} \underbrace{\langle \phi_i | \hat{A} | \phi_k \rangle}_{[A]_{ik}} = \sum_k [A]_{ik} [B]_{kj} \end{aligned} \tag{2.500}$$

that proofs the multiplication specialization of the operatorial-to-matrix Lemma.

With these the formalism is open of being applied of whatever quantum systems providing the chosen coordinate and momentum matrices provide through their combination in the Hamiltonian a diagonal matrix that can be "read" at once for the stationary eigen-values. Moreover, those operators have to fulfill the basic commutation rules above; from this reason the choice of coordinate an momentum operators may eventually not being unique, although there is presumed only one combination that provides also the diagonalization of the Hamiltonian matrix. For instance, the operators

$$\widehat{Q} = \exp\left(i\hat{x}\right), \quad \widehat{P} = -\hbar\exp\left(-i\hat{x}\right)\frac{d}{d\hat{x}} \tag{2.501}$$

may be formally seen as associate with matrices that satisfy the above commutation rule:

$$\widehat{P}\widehat{Q} - \widehat{Q}\widehat{P} = -i\hbar\hat{1} \tag{2.502}$$

as one may check immediately:

$$\left[\widehat{P}\widehat{Q} - \widehat{Q}\widehat{P}\right]f\left(\hat{x}\right) = -\hbar\exp\left(-i\hat{x}\right)\begin{bmatrix} if\left(\hat{x}\right)\exp\left(i\hat{x}\right) + \\ \exp\left(i\hat{x}\right)\partial_{\hat{x}}f\left(\hat{x}\right)\end{bmatrix} + \hbar\partial_{\hat{x}}f\left(\hat{x}\right) = -i\hbar f\left(\hat{x}\right) \tag{2.503}$$

Yet, the same is the case also for the operators

$$\widehat{Q} = \hat{x}, \ \widehat{P} = -i\hbar\frac{d}{d\hat{x}} \tag{2.504}$$

that happened to build the Schrödinger equation. Therefore, the associate matrices constructed according with the rules

$$[Q] = q, \ [P] = -i\hbar\frac{d}{dq} \tag{2.505}$$

provide the Hamiltonian matrix:

$$[H] = -\frac{\hbar^2}{2m}[P]^2 + V([Q]) \tag{2.506}$$

with the diagonal elements (according with the above matrix elements' definition):

$$[H]_{ii} = \left[-\frac{\hbar^2}{2m}[P]^2 + V([Q])\right]_{ii} = \underbrace{\langle\phi_i|\widehat{H}|\phi_i\rangle}_{E_i} = E_i \tag{2.507}$$

recovering the basic Schrödinger eigen-values. This way, the Schrödinger's wave-function (or eigen-state) representation is recovering the Heisenberg matrix (operatorial) formulation, both describing the same (quantum) reality! Nevertheless, the cumbersome with Heisenberg approach is the guessing stage in advancing the coordinate matrix $[Q]$ such that in combination with the associate momentum one $[P]$ to diagonalize the Hamiltonian of the problem. As with Schrödinger equation that cannot be exactly solved in general, also for matrix approach cannot be formulated a general recipe. However, in next, the case of the harmonic oscillator is unfolded to show how the formalism eventually works.

The starting point in treating the harmonic oscillator by Heisenberg quantum matrix approach resides in employing its classical Hamiltonian

$$H_{\omega} = \frac{p^2}{2m} + \frac{1}{2} m\omega_0^2 q^2 \tag{2.508}$$

though the Hamilton equations of motion, that gives:

$$\dot{p} = -\frac{\partial H_{\omega}}{\partial q} = -m\omega_0^2 q \tag{2.509}$$

$$\dot{q} = \frac{\partial H_{\omega}}{\partial p} = \frac{p}{m} \tag{2.510}$$

The matrix analysis starts from the coordinate equation, however once more derived so that to become integrable:

$$\ddot{q} = \frac{\dot{p}}{m} = -\omega_0^2 q \tag{2.511}$$

that has to be satisfied by each element of the matrix $[Q]$, independently:

$$\ddot{q}_{nm} + \omega_0^2 q_{nm} = 0 \tag{2.512}$$

Once the solution of this equation is considered under the (natural) form:

$$q_{nm}(t) = q_{nm}(0)\exp\left(i\omega_{nm}t\right) \tag{2.513}$$

its replacement in the original equation leads with the constraint:

$$\left(\omega_0^2-\omega_{nm}^2\right)q_{nm}=0 \tag{2.514}$$

This new equation has, obviously, two ways for being fulfilled: one regarding the ω's and one respecting q's; the resumation of the frequencies related analytical solutions looks like:

$$q_{nm}\neq 0\Rightarrow\begin{cases}\omega_{nm}=+\omega_0 ...emission...|n\rangle\rightarrow|m=n-1\rangle\\ \omega_{nm}=-\omega_0 ...absorption...|n\rangle\rightarrow|m=n+1\rangle\end{cases} \tag{2.515}$$

while otherwise:

$$q_{nm}=0\Rightarrow m\neq n\pm 1 \tag{2.516}$$

With these, the coordinate matrix displays as

$$[Q]=\begin{bmatrix}0 & q_{01} & 0 & \cdots\\ q_{10} & 0 & q_{12} & \cdots\\ 0 & q_{21} & 0 & \cdots\\ \vdots & \vdots & \vdots & \ddots\end{bmatrix} \tag{2.517}$$

and from it the rest of involved matrices as:

$$[P]=m\left[\dot{Q}\right]=im\left[\omega_{nm}Q\right]=im\begin{bmatrix}0 & \omega_{01}q_{01} & 0 & \cdots\\ \omega_{10}q_{10} & 0 & \omega_{12}q_{12} & \cdots\\ 0 & \omega_{21}q_{21} & 0 & \cdots\\ \vdots & \vdots & \vdots & \ddots\end{bmatrix}$$

$$\overset{\omega_{n,n+1}=-\omega_{n,n-1}=-\omega_0}{=} im\omega_0\begin{bmatrix}0 & -q_{01} & 0 & \cdots\\ q_{10} & 0 & -q_{12} & \cdots\\ 0 & q_{21} & 0 & \cdots\\ \vdots & \vdots & \vdots & \ddots\end{bmatrix} \tag{2.518}$$

$$[Q]^2 = [Q][Q] = \begin{bmatrix} q_{01}q_{10} & 0 & q_{01}q_{12} & \cdots \\ 0 & q_{10}q_{01} + q_{12}q_{21} & 0 & \cdots \\ q_{21}q_{10} & 0 & q_{21}q_{12} + q_{23}q_{32} & \cdots \\ \vdots & \vdots & \vdots & \ddots \end{bmatrix} \quad (2.519)$$

$$[P]^2 = [P][P] = m^2\omega_0^2 \begin{bmatrix} q_{01}q_{10} & 0 & -q_{01}q_{12} & \cdots \\ 0 & q_{10}q_{01} + q_{12}q_{21} & 0 & \cdots \\ -q_{21}q_{10} & 0 & q_{21}q_{12} + q_{23}q_{32} & \cdots \\ \vdots & \vdots & \vdots & \ddots \end{bmatrix}$$
$$(2.520)$$

$$[H_\omega] = \frac{1}{2m}\left([P]^2 + m^2\omega_0^2[Q]^2\right)$$
$$= m\omega_0^2 \begin{bmatrix} q_{01}q_{10} & 0 & 0 & \cdots \\ 0 & q_{10}q_{01} + q_{12}q_{21} & 0 & \cdots \\ 0 & 0 & q_{21}q_{12} + q_{23}q_{32} & \cdots \\ \vdots & \vdots & \vdots & \ddots \end{bmatrix} \quad (2.521)$$

the momentum, square of coordinate, square of momentum and finally the total energy matrices are, respectively.

There is obvious now that the energy matrix has indeed the diagonal character, due to proper chose of the coordinate matrix where the absorption-emission cases were considered in the form known as the *selection rules* for transition probabilities between various vectorial sates, here for the harmonic oscillator systems; here it is the first (implicit) level of quantum theory in the Heisenberg matrix approach. Also note that the q's entering the total energy matrix are of general shape and the matrix remains the same at whatever time, i.e., expresses the stationary or eigen-values for energies of various $|n\rangle$ levels on the diagonal nn positions; this can be immediately seen once the substitution of above coordinate temporal solution $q_{nm}(t) = q_{nm}(0)\exp(i\omega_{nm}t)$ is employed to give $q_{nm}(t)q_{mn}(t) = q_{nm}(0)q_{mn}(0)$; thus the formalism is consistently with the major quantum percepts achieved so far.

However, the second (explicit) level of quantum theory in the Heisenberg matrix approach regards the evaluation of the diagonal components of the total energy matrix through employment of the commutation rule; in the matrix forms specialized to actual harmonic oscillator case it looks equivalently like:

$$[P][Q]-[Q][P]=-i\hbar[1]$$
$$\Leftrightarrow \frac{2m\omega_0}{i}\begin{bmatrix} q_{01}q_{10} & 0 & 0 & \cdots \\ 0 & q_{12}q_{21}-q_{10}q_{01} & 0 & \cdots \\ 0 & 0 & q_{23}q_{32}-q_{21}q_{12} & \cdots \\ \vdots & \vdots & \vdots & \ddots \end{bmatrix}$$
$$=\frac{\hbar}{i}\begin{bmatrix} 1 & 0 & 0 & \cdots \\ 0 & 1 & 0 & \cdots \\ 0 & 0 & 1 & \cdots \\ \vdots & \vdots & \vdots & \ddots \end{bmatrix} \quad (2.522)$$

The one-to-one equalization for all matrices' components leads with the coordinates' relationships

$$q_{01}q_{10}=\frac{\hbar}{2m\omega_0} \quad (2.523a)$$

$$q_{12}q_{21}-q_{10}q_{01}=\frac{\hbar}{2m\omega_0} \quad (2.523b)$$

$$q_{23}q_{32}-q_{21}q_{12}=\frac{\hbar}{2m\omega_0} \quad (2.523c)$$

...

which solved iteratively produces the general relationship:

$$q_{n,n+1}q_{n+1,n}=\frac{(n+1)\hbar}{2m\omega_0} \quad (2.524a)$$

having its counterparts (for $n \to n-1$)

$$q_{n-1,n} q_{n,n-1} = \frac{n\hbar}{2m\omega_0} \tag{2.524b}$$

These general solutions of the quantum commutation of coordinate-momentum matrices are now plugged into the correspondent *nn*-diagonal-$|n\rangle$ eigen-energy level to obtain:

$$\begin{aligned}[H_\omega]_{nn} &= E_n(\omega_0) = m\omega_0^2 \left(q_{n,n-1} q_{n-1,n} + q_{n,n+1} q_{n+1,n} \right) \\ &= \frac{\omega_0 \hbar}{2}(2n+1) = \omega_0 \hbar \left(n + \frac{1}{2} \right)_{n=0,1,2,\ldots}\end{aligned} \tag{2.525}$$

This is the final solution of the quantum eigen-energy problem of harmonic oscillator. There are some comments to be made about it; firstly, it is characterized by the so called zero-point energy $E_0 = \omega_0 \hbar / 2$ which is non-zero even on the state with zero quantum number; this has some additional "ontological" consequence regarding the creation and annihilation particles from the "quantum vacuum" – however this problem will be reload with the occasion of many-body quantum systems description and quantum information theory, in the forthcoming chapters. Additionally, the quantification of the harmonic oscillator levels include only the classical frequency (pulsation) and the Planck reduced constant, beside the quantification number; yet, the Heisenberg matrix formalism do not provide the eigen-functions as well, being from this point of view with less quantum information comparing the Schrödinger theory. Even more, beside the general difficult task to determine the appropriate coordinate matrix for a general quantum system (Hamiltonian) so that along the momentum one to produce diagonal Hamiltonian matrix (for direct identification of the eigen-energies as the diagonal elements of it) it may presents also the difficulty encountering when inverse of coordinate matrix are involved (as it is the central motion problems, and the problem of Hydrogen atom itself) through the special warns appearing when inverses of matrices are to be involved. For all these reasons, Heisenberg matrix quantum approach remains interesting only for systems formally equivalent with harmonic oscillatory ones, leaving space to Schrödinger theory as the general tool for treating a wide variety of systems based on special restrains imposed on the eigen-functions and energies, as will be in next sections exposed.

2.5 CONCLUSION

The main lessons to be kept for the further theoretical and applicative investigations of the quantum mechanical formalization that are approached in the present chapter pertain to the following:

- identifying the distribution nature of the wave function and, consequently the need for Green function as the quantum propagator/amplitude;
- employing the momentum and energy operators towards providing the Klein-Gordon and Schrödinger equations for bosons and fermions quantum evolution, respectively;
- understanding the quantum spin as the driving power of the time transformation in quantum equation such that to provide its homogeneous evolution with space coordinate transformation through such equation;
- writing the spin for bosons and fermions from Klein-Gordon and Schrödinger equations, respectively;
- dealing with quantum states, adjoint (Hermitic) operators and commutativity properties and operations;
- characterizing quantum systems by means of specific measurement uncertainty through the statistical concept of dispersion rewritten in quantum terms and average quantities (for coordinate and momentum operators);
- describing the quantum systems in relation with classical correspondences as certified by Ehrenfest theorem and equivalence between current and probability densities, so assuring the reliability of quantum description of nanoworld;
- learning the classical to quantum correspondence by employing the analytical theorems of classical mechanics as the Euler-Lagrange and Hamilton Jacobi equations towards the actual Schrodinger description of quantum particles by associated field;
- treating the quantum evolution by the concept of action and of its expansion, when generating the so called semi-classical (WKB:

Wentzel, Kramers and Brillouin) description of the quantum wave-packet and of the associated de Broglie wavelength;

- Solving conservation issue for density current probability as based on the internal symmetry of the quantum fields, thus opening the way for future breaking symmetry considerations in characterizing of chemical fields by quantum particles, bondons in special, see Volume III of this five volume work (Putz, 2016);
- formulating the formalized evolving quantum picture including the Heisenberg equation for operators through employing the Poisson parenthesis from classical mechanics;
- interpreting the quantum states by *bra-* and *-ket* Dirac vectors in a generalized Hilbert space of linear vectors, and of their algebraic relationships, among them and with the quantum operators;
- connecting the formalized quantum mechanics in Dirac notation with matrix representations of vectors (quantum states) and operators (quantum observables);
- developing the concept of quantum complete space of vectors and operators (*complete set of commutative operators*) with the illustration of the energy representation on the eigen-states/spectrum of a given evolving quantum system;
- finding applications of the matrix quantum mechanics by employing the matrix representations for coordinate and momentum, whose the preeminent (Heisenberg) illustration is exposed for the quantum oscillator with the allied (experimentally measured) eigen-values, yet without involving the non-substantial wave-function concept and realization.

KEYWORDS

- **Bra-ket formalism**
- **current density probability**
- **Ehrenfest theorem**
- **energy**
- **energy representation**
- **Euler-Lagrange formalism**

- **Hamilton formalism**
- **Hamilton-Jacobi formalism**
- **Hilbert space**
- **matrix quantum mechanics**
- **momentum**
- **operators**
- **quantum commutation**
- **Schrödinger equation**
- **spin**
- **WKB approximation**

REFERENCES

AUTHOR'S MAIN REFERENCES

Putz, M. V. (2016). *Quantum Nanochemistry. A Fully Integrated Approach: Vol. III. Quantum Molecules and Reactivity*. Apple Academic Press & CRC Press, Toronto-New Jersey, Canada-USA.

Putz, M. V., Lazea, M., Chiriac, A. (2010). *Introduction in Physical Chemistry. The Structure and Properties of Atoms and Molecules* (in Romanian), Mirton Publishing House, Timişoara.

FURTHER READINGS

Atkins, P. W. (1991). *Quanta: A Handbook of Concepts*, Oxford University Press, New York.

Barrett, J. (2001). *Structure and Bonding*, Royal Society of Chemistry, Cambridge.

Bohm, D. (1989). *Quantum Theory*, Dover, New York.

Burdett, J. K. (1996). *Chemical Bonding: A Dialogue*, Wiley, New York.

Christofferson, R. E. (1989). *Basic Principles and Techniques of Molecular Quantum Mechanics*, Springer, New York.

Clark, T., Koch, R. (1999). *The Chemist's Electronic Book of Orbitals*, Springer, Berlin.

Coulson, C. A. (1982). *The Shape and Structure of Molecules* (revised by, R. McWeeny), Oxford University press, New York.

Courant, D., Hilbert, D. (1953). *Methods of Mathematical Physics*, Vol 1, Interscience, New York.

Cox, P. A. (1996). *Introduction to Quantum Theory and Atomic Structure*, Oxford Chemistry Primers, Oxford University Press, New York.
Daudel, R., Leroy, G., Peeters, D., Sana, M. (1983). *Quantum Chemistry*, John Wiley & Sons, New York.
DeKock, R. L., Gray, H. B. (1989). *Chemical Structure and Bonding*, University Science Books, Mill Valley (CA).
Fischer, C. F. (1977). *The Hartree-Fock Method for Atoms*, Wiley, New York.
Grant, G. H., Richards, W. G. (1995). *Computational Chemistry*, Oxford Chemistry Primers, Oxford University Press, New York.
Hinchcliffe, A. (2000). *Modeling Molecular Structures*, Wiley, New York.
Jean, Y., Volatron, F., Burdett, J. K. (1993). *An Introduction to Molecular Orbitals*, Oxford University Press, New York.
Leach, A. R. (2000). *Molecular Modeling: Principles and Applications*, Longman, Harlow.
Levine, I. N. (2000). *Quantum Chemistry*, Prentice-Hall, Upper Saddle River (NJ).
Lowe, J. P. (1993). *Quantum Chemistry*, Academic Press, Boston.
McQuarrie, D. A. (1983). *Quantum Chemistry*, University Science Books, Mill Valley (CA).
Metiu, H. (2006), *Physical Chemistry-Quantum Mechanics*, Taylor & Francis, Boca Raton (FL).
Murrell, J. N., Kettle, S. F. A., Tedder, J. M. (1985). *The Chemical Bond*, Wiley, New York.
Pauling, L., Wilson, E. B. (1985). *Introduction to Quantum Mechanics with Applications to Chemistry*, Dover, New York.
Peebles, P. J. E. (1992). *Quantum Mechanics*, Princeton University Press, Princeton.
Pickett, W. E. (1989). *Pseudopotential Methods in Condensed Matter Applications*, North-Holland.
Schatz, G. C., Ratner, M. A. (1993). *Quantum Mechanics in Chemistry*, Ellis Horwood/Prentice Hall, Hemel Hempstead (UK).
Scott, P. R., Richards, W. G. (1994). *Energy Levels in Atoms and Molecules*, Oxford Chemistry Primers, Oxford University Press, New York.
Shriver, D. F., Atkins, P. W., Langford, C. H. (1994). *Inorganic Chemistry*, Oxford University Press, Oxford and, W. H. Freeman Co., New York.
Szabo, A., Ostlund, N. S. (1996). *Modern Quantum Chemistry: Introduction to Advanced Electronic Structure Theory*, Dover, New York.
Webster, B. (1990). *Chemical Bonding Theory*, Blackwell Scientific, Oxford.
Winter, M. J. (1993). *Chemical Bonding*, Oxford Chemistry Primers, Oxford University Press, New York.

CHAPTER 3

POSTULATES OF QUANTUM MECHANICS: BASIC APPLICATIONS

CONTENTS

ABSTRACT

The phenomenological quantum mechanics is practically "re-storied" in a more formalized way, under the so-called extended quantum postulates, while opening and unfolding more complex analysis of the Nature's phenomena, from nuclear, to atomic, to molecular, and to solid state and scattering observational quantum effects.

3.1 INTRODUCTION

Quantum theory developed into a universal theory of matter with special multidisciplinary treatment among the fundamental domain *Mathematical and Natural Sciences*, with objectives on fundamental research and application employing the inner of the matter structure in terms of fundamental particles (fermions and bosons), fields and forces, in isolate and interacting states. The spread of quantum theory influence is vast indeed and covers in a non-limitative way *the disciplines of physical-chemistry, chemical informatics, mathematical-chemistry, physical organic chemistry, nano-inorganic chemistry, biology-chemistry, biochemistry, bio-informatics, pharmaceutical chemistry, medical chemistry, ecotoxicology, geochemistry, QSAR,* etc., as many of them will be approached in the flowing of the present five-volume book. However, with the advent of nano-sciences, one can formulate a sort of general definition for the applied quantum phenomena as:

- the manifestation of *nano-structured* matter properties of natural systems in isolate state and in reciprocal and with environment interaction (including QS[A-activity/P-property/T-toxicity]R), explained and controlled by the *quantum postulates*, modeled with *physical-mathematical concepts* and applied by *computational or informational tools'* aid.

In this regard worth accounting for some of the current and probably future hot topics in frontier applied quantum theory, being in their own strategic studies able to revel new and fascinating manifested properties of mater from nano-to-macro scale, so offering new bridges from fundamental science to applied science (technology), namely:

- *The "bosonic" approach of the matter and the chemical bonding in special* (see the Volumes III and IV of the present five-fold book): as it is known that in this quantum state the matter can became condensed in limited spaces but with accumulating energy, on the nano-atomic level; happily, at the research group it has been developed the necessary know-howl for developing this research direction, by recent studies about the modeling of chemical bond as a quantum condensate (based on the bosonation electron model in chemical bond, with formation of quantum particle of the chemical bond – named *bondons*: on the theoretical level, this model was and is currently applied to extended nano-systems of graphene, silicene, germanene types, with description and prediction of the phase transfer for topological types defects (Stone-Wales); at the experimental level the research is seeking in observing the predicted phenomena by forming the equivalents bondons through unique experimental setup (now in project phase) with the aid of quantum optics and photonics integrated at the micro-electronic level, but with possible extensions at the micro-nano-bio-systems (MNBS), considered as being the key for future technologies (KET: Key Enabling Technologies) which include the quantum molecular based computer, molecular electronics (moletronics), etc., this way, considerably saving the non-renewable or hard-renewable natural resources (e.g., minerals, Cu, Al, Au, Ag, etc.) of Terra in general, but also of Romania in special, including the respective exploitation costs; this way, the afferent *research themes* are: (i) *Bondons' Theory as quantum particle of the molecular wave function*; (ii) *Bondonic characterization of the phase transitions in extended nano-systems (e.g., of graphenic and fullerenic types) with topological defects (e.g., by Stone-Wales rotations)*; (iii) *Spectral identification of corpuscular (bondonic) character of the chemical bond.*

- *Quantum modeling of chemical reactivity* (see later in Volume I/ Chapter 4 as well as in Volumes II and III of the present five-fold volume book): where the unified understanding, based on quantum-mechanics is targeted, eventually with the involvement of the bosonic-bondonic phenomenology, in explaining the reactivity mechanisms at the molecular level, respectively for atoms-in-molecules' aggregation and nano-composites; this way, the consecrate principle of electronegativity and chemical hardness, which stay at the base of explaining the chemical potentials' equalization of subsystems (e.g., the quantum atomic pool) in molecules (through, for instance, the frontiers delimited by the vanishing of the electronic density gradients in molecules, as happens in orbital hierarchy in atoms), but also respectively to the chemical reactions with the hard-and-soft-acids-and-bases paradigm, it is reformed and generalized from the combined perspective of electrophilicity (relating the activation energy) and of the chemical power (relating the maximum number of electrons interchanged in a chemical interaction, intra- or inter-molecular) concepts; this kind of studies create a physical-mathematical universal model in order to treat, e.g., the "chemical atom", i.e., the atom engaged in the chemical bond and prepared for reactivity – or further interaction; such approach permits the quantum control of a new projected molecules with specific properties of reactivity and specific response (on specific atoms or other molecules with certain "recognized as active/alert" molecular zones); this way, the "memory" effects are combined in this direction with modeling the quantum information, the quantum cryptography, with bohmian effects at large-distance interaction (about the electrons delocalization in polyenes and polymers, as example); this way it can be achieved a sort of "teleportation of the chemical information and of the chemical bond in general", phenomenon which is comprehensible from the perspective of up-named bondon-quantum particle of the chemical bond; the impact in economy and in quantum information transport, and so for the storage energy, in different nano-and mesoscopic processes is immediate, yet with applicability at systems which are still in study (fullerenes, endo-fullerenes, ionic liquids, composite systems of inorganic-organic type), etc.; in this context, the afferent research themes are: (i) *Electronegativity: the modern concept in Density Functional Theory; equalization and integration of atoms in molecules principles; topological coloring with electronegativity of extended nanosystems (polycyclic aromatic hydrocarbons-PAH, graphene, silicene etc.);* (ii) *Chemical hardness: companion of electronegativity; quantum observability problem; quantification of maximum hardness principle and in relation with the hard and soft acids and basis principles;* (iii) *Modeling and standardization of chemical reactions with min-max principles (and with the aromaticity ones) of electronegativity and chemical hardness;* (iv) *Unification of chemical reactivity principles: chemical action and chemical bond.*

- *Topological and algebraic description of the chemical–biological interaction and toxicity* (see the Volume V of the present fivefold book): corresponding with the "chemical life elixir dream" by designing of the new drugs with specific action, from active substances or pharmacophors and generic substances synthesized as a result of certain topological and computational predictions, this direction is developing models and algorithms for better understanding and controlling the mechanism of binding action of ligand (chemical substance, toxicant, respectively the "target" structure, meaning the structure which is chosen to be structurally optimized by the allosteric interaction mode) with receptor (of biological nature, organisms' sites, at the cellular level, which can be a biomolecule of enzyme type, a metabolic activator or acting as metabolic inhibitor); toxicity is this way characterized by the kind of bonding mechanism identified, direction in which, in the past few years, innovative algorithms for correlating of the ligand-receptor or substrate-enzyme interactions were successfully proposed, through reformulating the problem of quantity structure activity (biological) QSAR; by algebraic-orthogonal approach with Spectral-SAR variant, and ultimately with considering of semi-molecules, with simple conjugate bonds broken in such manner that can be able to form molecular chains with primary and/or secondary branches, more adapted to the one similar with "key-lock" bond mechanism in according with the Fisherian principle of the drug's action; this way, the essential step it was made in bringing from virtual a new molecule considered so far only with topo-computational value, being "decomposed" (by *SMILES –Simplified Molecular-Input Line-Entry System type*) on the level of "real" conceptual-interaction mechanism and bonding by lipo-cellular transduction under this fragmentary form; the unique but also new character of this approach opens promising premises for the future Quantum-SAR (Qu-SAR) and 3D-QSAR studies with high mechanistically (not necessary in the deterministic stricto-senso by merely as *chemical binding mechanisms*) prediction capacity, for a controlled design of the target molecule, with focused toxicological potential (e.g., low in toxicity value, for example, for alimentary additives or in cosmetic products, but with high toxic value in anti-HIV composition and for any other processes for cellular apoptosis in different degenerative diseases, as in arteriosclerosis, Alzheimer type, etc.), contributing to the so-called functional medicine by the proposed pharmacotoxicology and pharmacodynamics' conceptual-computational approach but also with synthesis perspectives of pharmaceutics' laboratory; for this direction, the afferent research themes list as: (i) *The Spectral-SAR method: algebraically approach of the statistic correlation structure-activity and structure-toxicity;* (ii) *Correspondent Qu-SAR Principles of the Organization for Co-operation and Economical Developing (OECD); modeling of ligand-receptor bond with Qu-SAR principles; applications to molecules of ecotoxicological interest (aliphatic structures, PAH, ionic liquids, etc.);* (iii) *SMILES modeling (Simplified Molecular-Input Line-Entry System) of the ligand-receptor bond; the virtual-real problem for the SMILES molecule in cellular transduction of toxicants;* (iv) *Topologic indices and molecular graphs: topological indices with quantum potential (Wiener type and equivalents); formulation of new indices and their correlation with the chemical reactivity and with bio-ecotoxicological activity.*

This way the quantum postulates stay at the foreground of all basic and advanced application, in physical sciences in general and in chemical bonding formation and evolution (transmission) in special in various forms of substance (atoms, molecules, solids, cells) and of its transformation. Accordingly, the in depth and illustrative presentation of quantum postulates (principles) is of first interest in creating a solid foundation for further developments either in fundamental or applicative natural sciences, as above described. They will be in the following exposed.

3.2 THE WAVE-FUNCTION CONTINUITY

Each particle, either in free movement or in bound state under a potential that do not depend explicitly on time may be associated with a wave-function ψ *which contains, in principle all, information of the system; moreover: it is a continuously function of coordinate (including for classical turning points)*

$$\lim_{x>a} \psi(x) = \lim_{x<a} \psi(x) \,, \forall a \in \Re \tag{3.1}$$

it is at least one fold derivable respecting coordinate(s) and with it as a continuous function of coordinate(s) as well:

$$\lim_{x>a} \partial_x \psi(x) = \lim_{x<a} \partial_x \psi(x), \forall a \in \Re \tag{3.2}$$

In following, the crucial quantum effect of tunneling may be explained only through employing the wave-function's shape and continuity constraints; then, an eminent application at nuclear level unfolds an important certification of the quantum mechanics' reliability.

3.2.1 QUANTUM TUNNELING AND THE GAMOW FACTOR

Having learned the basic types of wave-function either associated with a quantum particle evolving either within a classically allowed (E>V) or classically forbidden (E<V) potential regions, one may formally models in principle any kind of microscopic system, free or bounded, respectively.

However, to be more analytical, the quantum scenario of the Figure 3.1 is considered with a free particle in region I encountering the barrier potential with a finite height and length in region II, while since eventually passed through it (by the so-called tunneling effect) is regained on the region III again as free particle, yet with a decreased amplitude of the associate wave due to the dissipation effect in region II. Therefore, the wave functions on the regions may be written according with the semiclassical WKB result as:

$$\text{Region I: } \psi_I(x) = c_i \exp(ikx) + c_r \exp(-ikx) = \psi_i(x) + \psi_r(x) \tag{3.3}$$

$$\text{Region II: } \psi_{II}(x) = A\exp(Kx) + B\exp(-Kx) \tag{3.4}$$

$$\text{Region III: } \psi_{III}(x) = c_t \exp(ikx) = \psi_t(x) \tag{3.5}$$

with the notations:

$$k = \frac{p_{I,III}}{\hbar},\ p_{I,III} = \sqrt{2mE},\ K = \frac{p_{II}}{\hbar},\ p_{II} = \sqrt{2m(V_0 - E)} \tag{3.6}$$

Due to the fact that on region I we have both incident and reflected wave (as in classical picture), in region II the classically forbidden behavior may include stationary waves back and forth inside the barrier, wile the region III being without wave source at infinity hosts only the transmitted (advanced) wave coming from tunneling region II.

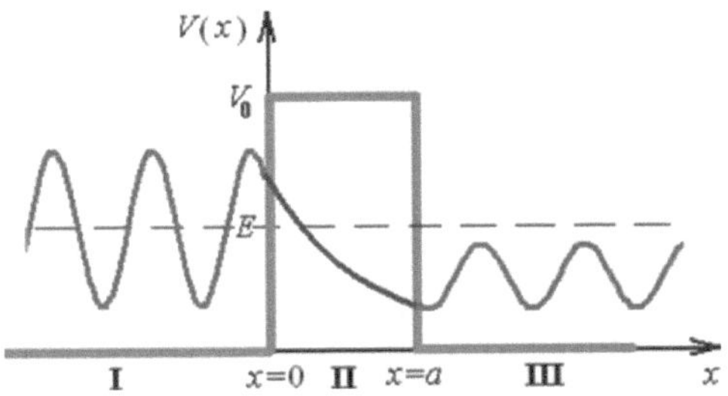

FIGURE 3.1 Typical regions with associated wave nature, as oscillating in regions I and III and as tunneling the potential barrier in II, for a quantum particle one-dimensional evolution produced at $-\infty$.

The wave-function coefficients may be determined from imposing the continuity conditions upon the genuine and first derivative functions on the regions' frontiers, equivalently with limiting or frontier conditions for solving the second order differential Schrödinger equation for each domain, respectively as

$$for\ \ x=0\ \begin{cases}\psi_I(x)\big|_{x=0}=\psi_{II}(x)\big|_{x=0}\\ \partial_x\psi_I(x)\big|_{x=0}=\partial_x\psi_{II}(x)\big|_{x=0}\end{cases} \tag{3.7}$$

$$for\ \ x=a\ \begin{cases}\psi_{II}(x)\big|_{x=a}=\psi_{III}(x)\big|_{x=a}\\ \partial_x\psi_{II}(x)\big|_{x=a}=\partial_x\psi_{III}(x)\big|_{x=a}\end{cases} \tag{3.8}$$

Explicitly we have:

$$for\ \ x=0\ \begin{cases}c_i+c_r=A+B\\ (c_i-c_r)i\dfrac{k}{K}=A-B\end{cases} \tag{3.9}$$

$$for\ \ x=a\ \begin{cases}A\exp(Ka)+B\exp(-Ka)=c_t\exp(ika)\\ A\exp(Ka)-B\exp(-Ka)=ic_t\dfrac{k}{K}\exp(ika)\end{cases} \tag{3.10}$$

Firstly, one may note that we have four equations for five coefficients to be found, thus we may at least deal with a parametric dependency in the general case. However, one may call the probability currents for the incident, reflected and transmitted waves (that can eventually be measured) through their standard definitions:

$$j_{i/r/t}=\frac{i\hbar}{2m}\left[\psi_{i/r/t}(x)\partial_x\psi^*_{i/r/t}(x)-\psi^*_{i/r/t}(x)\partial_x\psi_{i/r/t}(x)\right] \tag{3.11}$$

Therefore, we may measure the probability currents:

$$j_i=|c_i|^2\frac{1}{m},\quad j_r=|c_r|^2\frac{1}{m},\quad j_t=|c_t|^2\frac{1}{m} \tag{3.12}$$

fulfilling the conservation of particle (assuming no absorption process is happen within barrier)

$$j_i=j_r+j_t \tag{3.13}$$

rewritten as

$$1 = R + T \tag{3.14}$$

once the reflection and transmission factors are introduced:

$$R = \frac{j_r}{j_i} = \left|\frac{c_r}{c_i}\right|^2, \quad T = \frac{j_t}{j_i} = \left|\frac{c_t}{c_i}\right|^2 \tag{3.15}$$

From these, we are primarily interested in evaluating the transmission coefficient that defines the "power" or the quantum tunneling effect through the barrier. In this respect we are no longer bind to solve the continuity above equation for all involved coefficients but only for the ratio between that governing the transmitted to that carrying the incident waves. This may be indeed easily done through reconsidering the continuity equations with the new notation:

$$\alpha = \frac{k}{K} \tag{3.16}$$

while performing the sum end difference among equations of each set above to get two equivalent expressions for the coefficients A and B, respectively, as:

$$A = \frac{1}{2}\left[c_i(1 + i\alpha) + c_r(1 - i\alpha)\right] = \frac{1}{2}e^{-Ka}e^{ika}c_t(1 + i\alpha) \tag{3.17}$$

$$B = \frac{1}{2}\left[c_i(1 - i\alpha) + c_r(1 + i\alpha)\right] = \frac{1}{2}e^{Ka}e^{ika}c_t(1 - i\alpha) \tag{3.18}$$

This system contains only the incident, reflection and transmission coefficients; it may be further be simplified by multiplication of its first equation with $2(1 + i\alpha)/c_i$ and of the second with $-2(1 - i\alpha)/c_i$, following by summing up the results to get the transmission to incident coefficients successively as:

$$\frac{c_t}{c_i} = \frac{4i\alpha}{e^{ika}\left[e^{-Ka}(1 + i\alpha)^2 - e^{Ka}(1 - i\alpha)^2\right]}$$

$$= \frac{4i\alpha e^{-ika}}{(1 - \alpha^2)(e^{-Ka} - e^{Ka}) + 2i\alpha(e^{-Ka} + e^{Ka})}$$

$$= \frac{2i\alpha e^{-ika}}{(\alpha^2 - 1)\sinh(Ka) + 2i\alpha\cosh(Ka)}$$

$$= \frac{e^{-ika}}{\cosh(Ka) + \frac{i(1-\alpha^2)}{2\alpha}\sinh(Ka)} \tag{3.19}$$

where there were involved the trigonometric hyperbolic functions:

$$\sinh\alpha = \frac{e^{\alpha} - e^{-\alpha}}{2}; \quad \cosh\alpha = \frac{e^{\alpha} + e^{-\alpha}}{2} \tag{3.20}$$

Moreover, using their closure relationship

$$\cosh^2\alpha - \sinh^2\alpha = 1 \tag{3.21}$$

the transmission factor analytically unfolds like:

$$T = \left|\frac{C_t}{C_i}\right|^2 = \frac{1}{1 + \frac{(1+\alpha^2)^2}{4\alpha^2}\sinh^2(Ka)} \tag{3.22}$$

that in terms of energy and barrier potential cast as

$$T = \frac{1}{1 + \frac{\sinh^2\left(\frac{a}{\hbar}\sqrt{2m(V_0 - E)}\right)}{4\frac{E}{V_0}\left(1 - \frac{E}{V_0}\right)}} \tag{3.23}$$

One may immediately note that when the barrier is infinitely high we actually have no particle passing through the barrier:

$$\lim_{V_0\to\infty} T = \frac{1}{1+\infty} = 0 \tag{3.24}$$

However, at the classical turning point, when $E=V_0$, the transmission probability may be computed by using the sin-hyperbolic limit:

$$\lim_{x\to 0}\frac{\sinh x}{x} = \lim_{x\to 0}\frac{1}{x}\left(x + \frac{x^3}{3} + \frac{x^5}{5} + \ldots\right) = 1 \tag{3.25}$$

to yield:

$$\lim_{E \to V_0} T = \frac{1}{1 + \frac{m}{2\hbar^2} a^2 V_0^2} \tag{3.26}$$

meaning that it decreases as the volume of the barrier (aV_0) increases.

Finally, going to the general case of a variable barrier potential, the transmission factor, since carrying the probability meaning may be calculate for tunneling effects by its factorization to individual values corresponding with a discrimination of the in-out interval in rectangular consecutive barrier potentials, however with small (towards infinitesimally small) widths, see Figure 3.2. For instance, for the i-th such barrier of a tunneling phenomenon through a variable potential the transmission factor may be approximated by employing the formula

$$\sinh(K_i \Delta x_i) = \frac{1}{2}\left[\exp(K_i \Delta x_i) - \exp(-(K_i \Delta x_i))\right] \underset{\Delta x_i << 1}{\cong} \frac{\exp(K_i \Delta x_i)}{2} \tag{3.27}$$

to its working formulation:

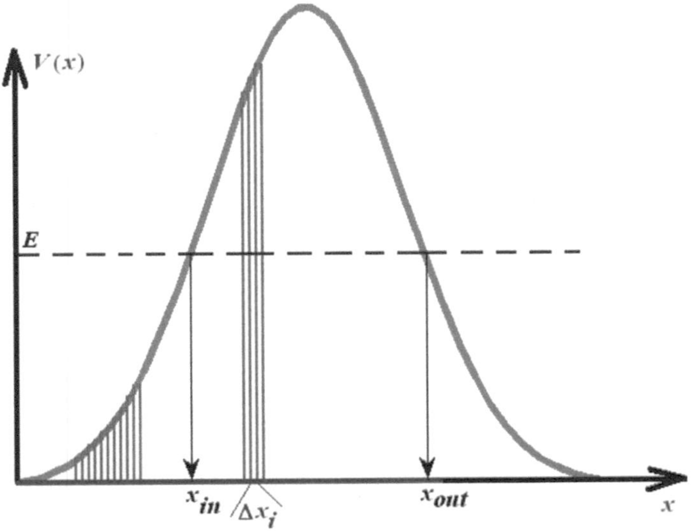

FIGURE 3.2 The factorization of a variable potential by rectangular barrier potentials of with finite widths.

$$T_i = \frac{1}{1+\dfrac{\exp\left(2\dfrac{\Delta x_i}{\hbar}\sqrt{2m(V_i - E)}\right)}{16\dfrac{E}{V_i}\left(1-\dfrac{E}{V_i}\right)}} \cong 16\frac{E}{V_i}\left(1-\frac{E}{V_i}\right)\exp\left(-2\frac{\Delta x_i}{\hbar}\sqrt{2m(V_i - E)}\right) \tag{3.28}$$

Now, since we have

$$16\frac{E}{V_i}\left(1-\frac{E}{V_i}\right) \in (0,4) \tag{3.29}$$

because the function $x(1-x)$ belongs to (0, 1/4) realm, a suitable probability significant transmission factor would be with the (Gamow) factor:

$$T_i \propto \exp\left(-2\frac{\Delta x_i}{\hbar}\sqrt{2m(V_i - E)}\right) \tag{3.30}$$

leading with the overall tunneling probability:

$$T = \prod_i T_i \tag{3.31}$$

or explicitly as

$$T \propto \exp\left(-\frac{2}{\hbar}\sum_i \Delta x_i \sqrt{2m(V_i - E)}\right) \tag{3.32}$$

or even more in the integral continuously framework:

$$T \propto \exp\left(-\frac{2}{\hbar}\int_{x_{in}}^{x_{out}} dx \sqrt{2m(V(x) - E)}\right) \tag{3.33}$$

where, the x_{in} and x_{out} are determined as the classical turning points where $E=V(x)$, see Figure 3.1.

Note that the present Gamow factor gives only the order of the transmission probability and is as good as the barrier is higher and shrink, i.e.,

$$p\Delta x = \sqrt{2m(V(x) - E)}\Delta x >> 1 \tag{3.34}$$

thus allowing for the (semi)classical observation of the effect. Consecrated applications of this formula were the study of the natural radioactivity (alpha particle emission by heavy nuclei) and the cold electronic emission (from metals under an intense electric field).

3.2.2 NUCLEAR SYSTEMS' LIFETIME BY ALFA DISINTEGRATION

Among the fundamental nuclear disintegration processes, the α particle (or ${}_{2}^{4}He$) emission by the heavy nuclei was discovered by K. Fajans, F. Soddy, and A. Russel in 1913 with the general scheme of reaction:

$${}_{Z}^{A}X \to {}_{Z-2}^{A-4}Y + {}_{2}^{4}He \tag{3.35}$$

with the famous Radium or Uranium transmutations:

$${}_{88}^{226}Ra \to {}_{86}^{222}Rn + {}_{2}^{4}He, \quad {}_{92}^{238}U_1 \to {}_{90}^{234}UX_1 + {}_{2}^{4}He \tag{3.36}$$

was brilliantly explained by the quantum tunneling effect, by G. Gamow (1928) and independently R. W. Gurney and E.U. Condon (1928), in a manner to be here revealed.

The physical phenomenon is illustrated in Figure 3.3, where the α particle with energy in orders of MeV once leaving the nucleus (A, Z) has still to pass through the Coulomb barrier

$$V(r) = \frac{2Z'e_0^2}{r}, \quad e_0^2 = \frac{e^2}{4\pi\varepsilon_0}, \quad Z' = Z - 2 \tag{3.37}$$

formed by its positive charge +2e and the remaining nucleus with the charge $+Z'e$ (with $Z'=Z-2$) as far as its energy is bellow this potential.

However, quantum tunneling is the main process as far as the alpha particle is localized between the nuclear radius' frontier R_n (as the first classical turning point):

$$R_n \cong r_0 A^{1/3}, \; r_0 = 1.2\times10^{-15}\,m = 1.2\,fermis(fm), \; 1\,fermi = 10^{-15}\,m \tag{3.38}$$

and the Coulomb escape classical turning point R_C:

$$E_\alpha = V(r)\big|_{r=R_c} \frac{2Z'e_0^2}{R_c} \Rightarrow R_c = \frac{2Z'e_0^2}{E_\alpha} \tag{3.39}$$

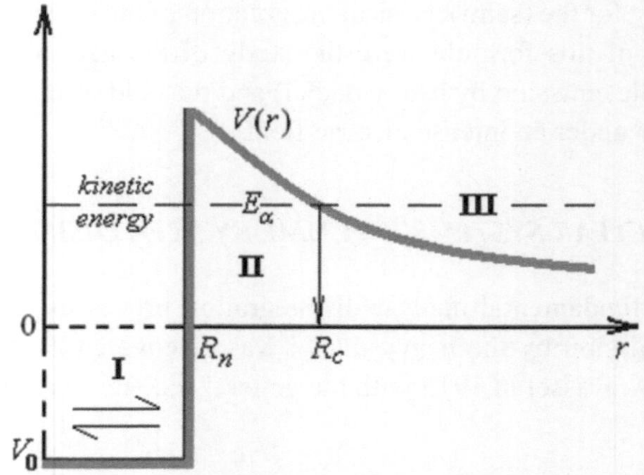

FIGURE 3.3 The phenomenology of α particle disintegration from a given heavy nucleus (region I, with very shrink radius R_n and very large potential depth V_0 compared with the rest of dimensions in the draw) through passing the barrier of region II towards its record (observation) in region III.

With these the tunneling factor and respectively the transition probability is given by

$$T_\alpha = \exp\left(-\frac{2}{\hbar}\int_{R_n}^{R_c} dr\sqrt{2m_\alpha\left(V(r)-E_\alpha\right)}\right) = \exp\left(-\frac{1}{\hbar}\sqrt{8m_\alpha E_\alpha}\int_{R_n}^{R_c} dr\sqrt{\frac{R_c}{r}-1}\right) \tag{3.40}$$

helping in estimating the disintegration life-time $\tau_{1/2}$ of the concerned nucleus by their reciprocal relationship:

$$\tau_{1/2} = \tau_0 \frac{1}{T_\alpha} \tag{3.41}$$

by means of the assumed uniform time of escaping of the alpha particle from the nuclear well, while understanding that the E energy is of dominant kinetic nature only outside of nucleus (and the Figure 3.3 should be considered symmetric in coordinate "r" so that fixing the $2R_0$ spanned distance of alpha particle before reaching the nucleus frontier):

$$\tau_0 = \frac{2R_n}{v_\alpha} = 2R_n\sqrt{\frac{m_\alpha}{2(E+V_0)}} \cong R_n\sqrt{\frac{2m_\alpha}{V_0}} \tag{3.42}$$

Therefore, the nuclear disintegration half-time (based on alpha particle tunneling) is written as:

$$\tau_{1/2} = r_0 A^{1/3} \sqrt{\frac{2m_\alpha}{V_0}} \frac{1}{T_\alpha} \tag{3.43}$$

with the Gamow tunneling factor for alpha particle to be in next determined from above integral. Basically, with the variable exchange

$$\frac{R_c}{r} = \frac{1}{\cos^2\theta}, \quad dr = -2R_c \sin\theta\cos\theta d\theta \tag{3.44}$$

$$\theta = \begin{cases} \theta_c = 0 & , r = R_c \\ \theta_n = \arccos\left(\sqrt{R_n / R_c}\right), & r = R_n \end{cases} \tag{3.45}$$

the involved integral successively becomes:

$$\begin{aligned} I_\alpha &= \int_{R_n}^{R_c} dr\sqrt{\frac{R_c}{r} - 1} = -2R_c \int_{\theta_n}^{0} d\theta \sin\theta\cos\theta\sqrt{\frac{1}{\cos^2\theta} - 1} = 2R_c \int_0^{\theta_n} \sin^2\theta\, d\theta \\ &= 2R_c\left(\frac{\theta_n}{2} - \frac{1}{2}\int_0^{\theta_n} \cos 2\theta\, d\theta\right) = R_c(\theta_n - \sin\theta_n \cos\theta_n) \\ &= R_c\left[\arccos\left(\sqrt{\frac{R_n}{R_c}}\right) - \sqrt{1 - \frac{R_n}{R_c}}\sqrt{\frac{R_n}{R_c}}\right] \end{aligned} \tag{3.46}$$

At this moment one should employ the fact that, usually, the escaping kinetic energy (at R_c) is lower than the corresponding potential (at R_n) as it may be immediately be illustrated for the paradigmatic Radium (Z=88, Z'=86, A=226, E_α=4.78 MeV) case

$$R_n({}^{226}_{88}Ra) = r_0(226)^{1/3} \cong 7.3\, fermis \tag{3.47}$$

$$R_c = 2Z'\left(\frac{e_0^2}{m_0c^2}\right)\left(\frac{m_0c^2}{E_\alpha}\right) = 172(2.8\, fm)\frac{0.5MeV}{4.78MeV} \cong 50\, fm \cong 7R_n \tag{3.48}$$

As such, there may be considered the first order series expansions in $R_n / R_c \cong 0$:

$$\arccos\left(\sqrt{\frac{R_n}{R_c}}\right) \cong \frac{\pi}{2} - \sqrt{\frac{R_n}{R_c}} \tag{3.49}$$

$$\sqrt{\frac{R_n}{R_c}}\sqrt{1-\frac{R_n}{R_c}} \cong \sqrt{\frac{R_n}{R_c}}\left(1-\frac{1}{2}\frac{R_n}{R_c}\right) \cong \sqrt{\frac{R_n}{R_c}} \tag{3.50}$$

the above integral approximates as:

$$I_\alpha \cong R_c\left[\frac{\pi}{2} - 2\sqrt{\frac{R_n}{R_c}}\right] \tag{3.51}$$

and the related Gamow α-tunneling factor:

$$T_\alpha = \frac{\exp\left(\frac{8}{\hbar}\sqrt{Z' e_0^2 m_\alpha r_0 A^{1/3}}\right)}{\exp\left(\frac{\pi Z' e_0^2}{\hbar}\sqrt{\frac{8m_\alpha}{E_\alpha}}\right)} \tag{3.52}$$

providing the half-life nuclear time as

$$\tau_{1/2} = r_0 A^{1/3}\sqrt{\frac{2m_\alpha}{V_0}}\exp\left(\frac{\pi Z' e_0^2}{\hbar}\sqrt{\frac{8m_\alpha}{E_\alpha}} - \frac{8}{\hbar}\sqrt{Z' e_0^2 m_\alpha r_0 A^{1/3}}\right) \tag{3.53}$$

Whishing to have a working formula, the last expression may be rewritten as:

$$\tau_{1/2}(s) = \frac{\frac{r_0}{c}\sqrt{2m_\alpha c^2(MeV)}\frac{A^{1/3}}{\sqrt{V_0(MeV)}}\exp\left(\pi\alpha\sqrt{\frac{8m_\alpha c^2}{(MeV)}}\frac{Z'}{\sqrt{E_\alpha(MeV)}}\right)}{\exp\left(\frac{8}{\hbar c}\sqrt{Z' e_0^2 r_0 A^{1/3} m_\alpha c^2(MeV)}\right)}$$

$$\cong \frac{0.33\cdot 10^{-23}\sqrt{2\cdot 3727.38}\frac{A^{1/3}}{\sqrt{V_0(MeV)}}\exp\left(\frac{7.297\cdot 10^{-3}}{\frac{\pi\sqrt{8\cdot 3727.38}}{\sqrt{E_\alpha(MeV)}}}Z'\right)}{\exp\left(\frac{8\cdot\sqrt{2.8\cdot 10^{-15}\cdot 0.5\cdot 10^{-15}\cdot 3727.38}}{1.0546\cdot 10^{-34}\cdot 3\cdot 10^{8}\cdot 6.24151\cdot 10^{12}}\sqrt{Z' A^{1/3}}\right)}$$

$$\cong \frac{28.4925\cdot 10^{-23}\frac{A^{1/3}}{\sqrt{V_0(MeV)}}\exp\left(\frac{3.95859(Z-2)}{\sqrt{E_\alpha(MeV)}}\right)}{\exp\left(2.92656\sqrt{(Z-2)A^{1/3}}\right)} \tag{3.54}$$

The present formula may be even more be specialized while choosing a sort of "universal" nuclear parameters; one may firstly consider the typical nuclear depth of barrier as $V_0 \approx 10^{10}\, MeV$. With this choice there is immediate to verify that the obtained formula works quite well for the extreme nuclear systems:

$$\tau_{1/2}\left({}^{226}_{88}Ra, E_\alpha = 4.78MeV\right) \cong 5.95\cdot 10^{12}\, s \approx 10^4\, years \tag{3.55}$$

$$\tau_{1/2}\left({}^{212}_{84}Po, E_\alpha = 8.78MeV\right) \cong 5.04\cdot 10^{-7}\, s \tag{3.56}$$

in reasonable agreement with the experiment. Even more, if one likes to obtain a single expression for all nuclear alpha disintegration, i.e., with the alpha particle energy as the single variable, the above employed depth of the nuclear well may be combined with the paradigmatic Radium characteristics (*Z*=88, *A*=226) to the working expression:

$$\ln \tau_{1/2}(s) \cong -126 + \frac{340.439}{\sqrt{E_\alpha(MeV)}} \tag{3.57}$$

whose representation against the various radioactive series respecting the alpha disintegrations shows quite a fine "fitting" agreement with the observed half-lifetime values, see Figure 3.4, thus making from the present approach a valuable one and confirming the exceptional reliability of the quantum mechanics, here through the continuity principle of underlining the tunneling effect.

However, despite the alpha tunneling was recognizing as a present effect even in every-day life, for example, in smoke detectors by the alpha emitter Americium-241 or as being related with the so-called "soft errors" in computer technology produced by the alpha emission by the radioactive elements contained in the packaging of semiconductor materials, the more general lesson in quantum formalism may be drawn from the continuity of the wave function and by its specialization through the tunneling effects: it is that the Gamow factor carriers the dominant effect in tunneling and that its exponent contains the main quantification information that can be detected since recalling by the semiclassical (WKB) wave-function formalism.

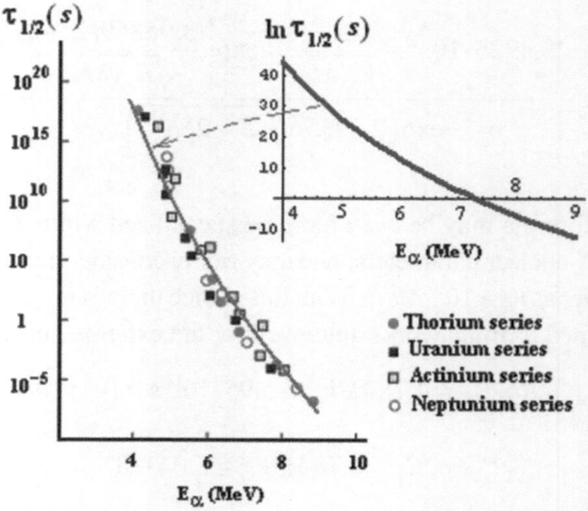

FIGURE 3.4 The "fitting" of the quantum tunneling effect based computed "universal" half-life times for nuclei emitting alpha particles with their observed values for representative radioactive series; adapted from (Rohlf, 1994; HyperPhysics, 2010).

Nevertheless, the special feature of semiclassical (Gamow or WKB) quantification as the link between the phenomenological Bohr quantization and the exact Schrödinger solution for given systems is to be in next sections explored.

3.3 EIGEN-ENERGIES AND EIGEN-FUNCTIONS

An observable system is represented by stationary wave-function $\psi(x)$ satisfying the analytical constraints:

- *it is a solution of the stationary Schrödinger equation:*

$$\widehat{H}\psi = E\psi \tag{3.58}$$

- *it is normalized:*

$$\int \psi^* \psi \, d\Gamma = 1 \tag{3.59}$$

- *it asymptotically vanishes when $x \to \infty$:*

$$\lim_{x\to\pm\infty} \psi(x) = \lim_{x\to\pm\infty} \psi^*(x) = 0 \tag{3.60}$$

This principle goes together with the fact that all functions that are solution to a given Schrödinger equation may be linearly combined (superimposed) in producing other proper (eigen) wave-functions of the system:

$$\exists \varphi_i : \widehat{H}\varphi_i = E\varphi_i \Rightarrow \exists \psi = \sum_i c_i \varphi_i, c_i \in \Im \mid \widehat{H}\psi = E\psi \tag{3.61}$$

Here follows illustration of this principle while solving three representative physical systems of matter: atomic hydrogen (and hydrogenic atoms), molecules in vibrational states and the solid state of free electrons (for clarifications on this linguistically paradox see Section 3.3.3).

3.3.1 ATOMIC HYDROGENIC STATES BY SOLVING LAGUERRE TYPE EQUATIONS

Aiming to provide the general eigen-values (energies) and eigen-functions (wave-functions) for the hydrogenic problem, one should consider that beside the presence of nucleus attractive potential, of a classical Coulombic nature, due to the spherical symmetry of the atoms the quantum centrifugal effects related has to be accounted as well. However, remembering the Heisenberg law rewritten in radial terms in Bohr atom $[rp = \hbar n,\ n = 1, 2, ...]$ one may use it to introduce the centrifugal quantum contribution through the kinetic energy in proper way, namely

$$\frac{p^2}{2m_0} = \frac{r^2 p^2}{2m_0 r^2} = \frac{(rp)^2}{2m_0 r^2} = \frac{\hbar^2 l(l+1)}{2m_0 r^2} = \frac{[\text{quantum kinetic moment}]^2}{2[\text{intertia moment}]} \tag{3.62}$$

where the kinetic moment was quantified not as direct generalization of $n^2 \to l^2$ but as $n^2 \to l(l+1)$ for the reasons that will be bellow revealed, together with the allowed values for the quantum index *l*, and justified from first principles in the Volume II of this five-volume set, dedicated to quantum atom detailed study. For the moment, one retain the effective potential for hydrogenic atoms with the general form

$$V_{eff}(r) = -\frac{Ze_0^2}{r} + \frac{\hbar^2 l(l+1)}{2m_0 r^2}, \quad e_0^2 = \frac{e^2}{4\pi\varepsilon_0} \tag{3.63}$$

In a way, one may interpret this potential as the sum of the classical Coulomb potential and the "orbital" (or quantum) kinetic moment energy or repulsive nature in the light of above considerations, see Figure 3.5. In fact, the kinetic energy of the total electronic motion in Hydrogen (and hydrogenic atoms) consists of two terms: one responsible for translation and other for spherical rotation (the so-called orbital motion).

Yet, Figure 3.5 displays a very interesting feature, namely there appears that the states with $l \neq 0$ are to some extent spatially mixed with those characterized by $l = 0$ generating that it can be called as "the spatially penetrating potentials" paradox; It will be solved in a way that will be revealed after the full analytical (radial) electronic motion in the central field will be solved out.

Therefore, the working Hydrogenic Schrödinger radial equation has the form

$$\left[-\frac{\hbar^2}{2m_0}\nabla^2 - \frac{Ze_0^2}{r} + \frac{\hbar^2 l(l+1)}{2m_0 r^2}\right]\psi_{nl}(r) = E_{nl}\psi_{nl}(r) \tag{3.64}$$

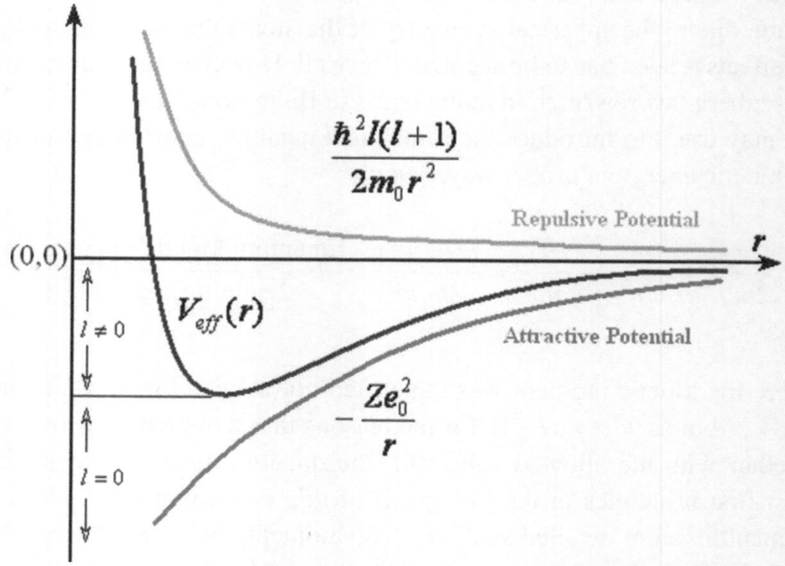

FIGURE 3.5 The existing potentials in hydrogenic atoms: attractive (Coulombic) and the repulsive (centrifugal) ones establishing the energetic regions governed by classical Coulombic ($l = 0$) or by quantum effective potential as the sum of both Coulombic and centrifugal (when $l \neq 0$).

while one still needs to express the Laplacian into spherical (radial restrained) form. This may be easily achieved once there are employed the definitions

$$\nabla^2 = \frac{\partial^2}{\partial x_1^2} + \frac{\partial^2}{\partial x_2^2} + \frac{\partial^2}{\partial x_3^2}, \quad r = \sqrt{x_1^2 + x_2^2 + x_3^2} \tag{3.65}$$

to systematically application of the chain-rule derivations:

$$\frac{\partial r}{\partial x_1} = \frac{x_1}{r} \tag{3.66a}$$

$$\frac{\partial f}{\partial x_1} = \frac{\partial f}{\partial r}\frac{\partial r}{\partial x_1} = \frac{x_1}{r}\frac{df}{dr} \tag{3.66b}$$

$$\frac{\partial^2 f}{\partial x_1^2} = \frac{\partial}{\partial x_1}\left(\frac{x_1}{r}\frac{df}{dr}\right) = \frac{1}{r}\frac{df}{dr} + \frac{x_1^2}{r}\frac{d}{dr}\left(\frac{1}{r}\frac{df}{dr}\right) \tag{3.66c}$$

from where by summing up in the 3D radial expression there is found that

$$\nabla^2 f = \frac{3}{r}\frac{df}{dr} + r\frac{d}{dr}\left(\frac{1}{r}\frac{df}{dr}\right) \tag{3.67a}$$

leading with the radial-spherical rewriting of the Laplacian operator:

$$\nabla_r^2 = \frac{2}{r}\frac{d}{dr} + \frac{d^2}{dr^2} = \frac{1}{r^2}\frac{d}{dr}\left(r^2\frac{d}{dr}\right) \tag{3.67b}$$

Now, under the appropriate substitution

$$\psi_{nl}(r) = \frac{U_{nl}(r)}{r} \tag{3.68}$$

the radial normalization condition takes its usual (wave-function square integrated) form,

$$1 = \int_0^\infty |\psi_{nl}(r)|^2 r^2 dr = \int_0^\infty |U_{nl}(r)|^2 dr \tag{3.69}$$

while the radial Schrödinger equation

$$\left[-\frac{\hbar^2}{2m_0}\left(\frac{2}{r}\frac{d}{dr}+\frac{d^2}{dr^2}\right)-\frac{Ze_0^2}{r}+\frac{\hbar^2 l(l+1)}{2m_0 r^2}\right]\psi_{nl}(r)=E_{nl}\psi_{nl}(r) \tag{3.70}$$

takes now the simpler form

$$\left[-\frac{\hbar^2}{2m_0}\frac{d^2}{dr^2}-\frac{Ze_0^2}{r}+\frac{\hbar^2 l(l+1)}{2m_0 r^2}\right]U_{nl}(r)=E_{nl}U_{nl}(r) \tag{3.71}$$

The last equation allows us in recognizing the asymptotic equation for sufficiently large radius distance ($r \to \infty$):

$$\frac{d^2}{dr^2}U_{nl}^{\infty}(r)=\left(\frac{\mathrm{A}}{2}\right)^2 U_{nl}^{\infty}(r) \tag{3.72}$$

with the bound state notation (discrete spectrum with negative eigen-energies)

$$\left(\frac{\mathrm{A}}{2}\right)^2=-\frac{2m_0 E_{nl}}{\hbar^2}=\frac{2m_0\left|E_{nl}\right|}{\hbar^2},\quad E_{nl}<0 \tag{3.73}$$

thus providing the general solution:

$$U_{nl}^{\infty}(r)=C_1 e^{-\mathrm{A}r/2}\underbrace{\left(+C_2 e^{\mathrm{A}r/2}\right)}_{\substack{\downarrow\\ 0\\ (by\ Postulate\ I)}} \tag{3.74}$$

Note that when the energy is positive the solution will be oscillating, as being associated with free particle, thus indicating the continuous part of the spectrum when the hydrogenic atom is ionized.

Yet, the asymptotic discussion allows for introducing the non-dimensional quantity:

$$x=\mathrm{A}r \tag{3.75}$$

and to perform the substitutions

$$r=\frac{x}{\mathrm{A}},\quad E_{nl}=-\frac{\mathrm{A}^2\hbar^2}{8m_0} \tag{3.76}$$

on to the original Schrödinger radial equation to rewrite

$$\left[-\frac{\hbar^2}{2m_0}\left(\frac{2}{x}\frac{d}{dx}+\frac{d^2}{dx^2}\right)-\frac{Ze_0^2}{x\text{A}}+\frac{\hbar^2 l(l+1)}{2m_0x^2}\right]\psi_{nl}(x)=-\frac{\hbar^2}{8m_0}\psi_{nl}(x) \quad (3.77)$$

or even more as:

$$x\frac{d^2\psi_{nl}}{dx^2}+2\frac{d\psi_{nl}}{dx}+\left[\frac{2m_0Ze_0^2}{\hbar^2\text{A}}-\frac{x}{4}-\frac{l(l+1)}{x}\right]\psi_{nl}=0 \quad (3.78)$$

This equation may be solved at once recognizing it is related with a form of the Laguerre's differential equation:

$$xy''+(1-x)y'+ny=0, \quad n\in\mathbf{N}^* \quad (3.79)$$

whose solution is the so-called *Laguerre polynomial* of n-degree:

$$y=L_n(x)=\sum_{\lambda=0}^{n}a_\lambda x^{\kappa+\lambda} \quad (3.80)$$

However, when substituted back in original Laguerre's equation one gets:

$$\sum_{\lambda=0}^{n}a_\lambda(\kappa+\lambda)^2x^{\kappa+\lambda-1}=\sum_{\lambda=0}^{n}a_\lambda(\kappa+\lambda-n)x^{\kappa+\lambda} \quad (3.81)$$

an expression that has to be fulfilled for every value of x, i.e., the coefficients of equally footing power to vanish; in these conditions, the lower power achieved in the left-hand-side for $\lambda=0$ should vanish as well, since it will have no correspondent in the right-hand-side part of equation, that is the so-called *indicial equation* is obtained

$$a_0\kappa^2=0\Rightarrow\kappa=0 \quad (3.82)$$

Moreover, the recurrence relation may be obtained setting the remaining above left hand side term (now starting to count from $\lambda=1$) with the substitution $\lambda\to\lambda+1$ in sum (and therefore subtracting one unit from the counter index) to obtain equality:

$$\sum_{\lambda=0}^{n}a_{\lambda+1}(\lambda+1)^2x^\lambda=\sum_{\lambda=0}^{n}a_\lambda(\lambda-n)x^\lambda \quad (3.83)$$

leaving with the relationship:

$$a_{\lambda+1} = \frac{\lambda - n}{(\lambda+1)^2} a_\lambda \tag{3.84}$$

and even more to the expanded (in n^{th} order) solution of the Laguerre equation:

$$y_n(x) = a_0 \begin{bmatrix} 1 - nx + \dfrac{n(n-1)}{(2!)^2} x^2 - ... + \\ (-1)^\lambda \dfrac{n(n-1)...(n-\lambda+1)}{(\lambda!)^2} x^\lambda + ... \dfrac{(-1)^n}{n!} x^n \end{bmatrix} \tag{3.85}$$

It eventually becomes the Laguerre's polynomial of degree n for appropriately choosing of $a_0 = 1$ for y_n, and $a_0 = n!$ for L_n such that to have the connection

$$\begin{aligned} L_n(x) &= n!\, y_n(x) \\ &= (-1)^n \left[x^n - \frac{n^2}{1!} x^{n-1} + \frac{n^2(n-1)^2}{2!} x^{n-2} - ... + (-1)^n n! \right] \end{aligned} \tag{3.86}$$

Now, if one like considering the differentiation of the original Laguerre's equation it is explicitly obtained

$$x\left(y^{(1)}\right)'' + (1+1-x)\left(y^{(1)}\right)' + (n-1)y^{(1)} = 0 \tag{3.87}$$

while after k-successive differentiations it yields

$$x\left(y^{(k)}\right)'' + (k+1-x)\left(y^{(k)}\right)' + (n-k)y^{(k)} = 0,\ k \geq 0 \tag{3.88}$$

thus having as the solution the *associate Laguerre polynomial* of degree n–k:

$$y^{(k)} = \frac{d^k}{dx^k} L_n \equiv L_n^k(x) \tag{3.89}$$

Even more, this last equation is equivalent with the following one

$$xY'' + 2Y + \left[n - \frac{k-1}{2} - \frac{x}{4} - \frac{k^2-1}{4x} \right] Y = 0 \tag{3.90}$$

under the transformation

$$Y = e^{-x/2} x^{(k-1)/2} L_n^k(x) \tag{3.91}$$

That consecrates the so-called *associated Laguerre function.*

There is now clear that our radial equation is of Laguerre form with the associate Laguerre function solution through the one-to-one correspondences, solved at once to:

$$l(l+1) = \frac{k^2 - 1}{4} \Rightarrow k = 2l + 1 \tag{3.92}$$

providing the number of possible angular degeneracies:

$$\frac{2m_0 Z e_0^2}{\hbar^2 \mathrm{A}} = n - \frac{k-1}{2} = n - l \tag{3.93}$$

It is this last relationship that through replacing A^2 it leads to the energy states of Hydrogenic atoms:

$$E_{n_*} = -\frac{1}{2} \underbrace{\frac{m_0 e_0^2}{\hbar^2}}_{1/a_0} \frac{Z^2 e_0^2}{n_*^2} = -\frac{1}{2} \frac{Z^2 e_0^2}{a_0 n_*^2}, \quad n_* = n - l \tag{3.94}$$

from where follows also the rule of allowed orbital quantum number values

$$l \in \{0, 1, \ldots, n-1\} \tag{3.95}$$

since the positive integer conditions of n and k are properly counted. Nevertheless, the result expresses the recovering of the Bohr type result for the eigen-energies, while the eigen-functions are to be written as:

$$\psi_{nl}(x) = C_{nl} e^{-x/2} x^l L_{n+l}^{2l+1}(x) \tag{3.96}$$

though the present prescriptions, by tacitly replacement of n notation with simple n for physical consistency, whereas the appeared constant to be determined by the radial normalization condition:

$$1=\int_0^\infty |\psi_{nl}(r)|^2 r^2 dr = \left(\frac{na_0}{2Z}\right)^3 \int_0^\infty |\psi_{nl}(x)|^2 x^2 dx$$
$$= \left(\frac{na_0}{2Z}\right)^3 C_{nl}^2 \int_0^\infty e^{-x} x^{2l} \left|L_{n+l}^{2l+1}(x)\right|^2 x^2 dx \tag{3.97}$$

when taking note by the equivalencies

$$r=\frac{x}{\mathrm{A}}=x\sqrt{\frac{\hbar^2}{8m_0|E_{nl}|}}=x\frac{n}{2Z}\underbrace{\frac{\hbar^2}{m_0 e_0^2}}_{\text{first Bohr radius}}=x\frac{n}{2Z}a_0 \Leftrightarrow x=\frac{2Z}{a_0 n}r \tag{3.98}$$

by employing the introduced notations and the already found results.

To this aim the integral representation of Laguerre polynomial is most useful; it is based on complex integral representation of the solution of the Laguerre original differential equation:

$$y_n(x)=\frac{1}{2\pi i}\oint \frac{z^{-n-1}}{1-z}\exp\left(\frac{-xz}{1-z}\right)dz \tag{3.99}$$

there is immediately to check that through considering its first and second differentiation with respect to x, respectively,

$$y_n(x)'=-\frac{1}{2\pi i}\oint \frac{z^{-n}}{(1-z)^2}\exp\left(\frac{-xz}{1-z}\right)dz \tag{3.100a}$$

$$y_n(x)''=\frac{1}{2\pi i}\oint \frac{z^{-n+1}}{(1-z)^3}\exp\left(\frac{-xz}{1-z}\right)dz \tag{3.100b}$$

they fulfill the Laguerre differential equation:

$$xy_n''+(1-x)y_n'+ny_n$$
$$=\frac{1}{2\pi i}\oint\left[\frac{xz^2}{(1-z)^2}-\frac{(1-x)z}{1-z}+n\right]\frac{z^{-n-1}}{1-z}\exp\left(\frac{-xz}{1-z}\right)dz$$
$$=-\frac{1}{2\pi i}\oint\frac{d}{dz}\left[\frac{z^{-n}}{1-z}\exp\left(\frac{-xz}{1-z}\right)\right]dz=0 \tag{3.101}$$

since the integrand function is single valued, i.e., taking the same value on initial and final point of the closed contour integral. With this one has automatically also the Laguerre polynomial complex integral representation:

$$L_n(x) = n!y_n(x) = \frac{n!}{2\pi i}\oint \frac{z^{-n-1}}{1-z}\exp\left(\frac{-xz}{1-z}\right)dz \quad (3.102)$$

Further on, employing the complex Laurent's theorem

$$f(z) = \sum_{\lambda} a_{\lambda}(z-z_0)^{\lambda} = \sum_{\lambda}\left[\frac{1}{2\pi i}\oint \frac{f(z)dz}{(z-z_0)^{\lambda+1}}\right](z-z_0)^{\lambda} \quad (3.103)$$

one may identically write that

$$\frac{1}{1-z}\exp\left(\frac{-xz}{1-z}\right) = \sum_{n=0}^{\infty} y_n z^n = \sum_{n=0}^{\infty}\frac{L_n(x)}{n!}z^n \quad (3.104)$$

This consequence of complex integral representation of Laguerre polynomial has another crucial implication when one deals with associated Laguerre function $L_n^k(x)$; it is obtained from k-differentiation (respecting x) of the Laguerre polynomial, having therefore the associate series resumming (not forget from previous discussion that $n \geq k$, otherwise the function $L_n^k(x)$ vanishes through k-derivations):

$$\frac{(-1)^k}{1-z}\left(\frac{z}{1-z}\right)^k \exp\left(\frac{-xz}{1-z}\right) = \sum_{n=k}^{\infty} y_n^{(k)} z^n = \sum_{n=k}^{\infty}\frac{L_n^k(x)}{n!}z^n \quad (3.105)$$

Such representation, the so-called *generating (Laguerre) function representation* since express one bi-variable function in a series of a function respecting one variable having the other as parameter, is most useful when integrals involving pairs of Laguerre function are to be evaluated. In general, integrals of the type

$$I_{nm} = \int_0^{\infty} e^{-x}x^{k-1}L_n^k(x)L_m^k(x)x^p dx \quad (3.106)$$

can be evaluated with the help of above associate Laguerre series resumming. In fact when two such series are firstly multiplied,

$$\sum_{n,m=k}^{\infty}\frac{z_1^n z_2^m}{n!m!}L_n^k(x)L_m^k(x)=\frac{(z_1z_2)^k}{(1-z_1)^{k+1}(1-z_2)^{k+1}}\exp\left(-\frac{xz_1}{1-z_1}-\frac{xz_2}{1-z_2}\right) \tag{3.107}$$

and then further convoluted with $e^{-x}x^{p+k-1}$ by integration with respecting x, while considering the exponents of the right hand side of last expression the decomposition

$$\frac{-xz}{1-z}=\frac{-x}{1-z}+x \tag{3.108}$$

and in the view of I_{nm} type integral, one gets the successive expressions:

$$\begin{aligned}\sum_{n,m=k}^{\infty}\frac{z_1^n z_2^m}{n!m!}I_{nm}&=\frac{(z_1z_2)^k}{\left[(1-z_1)(1-z_2)\right]^{k+1}}\int_0^\infty x^{k+p-1}\exp\left[-\left(\frac{1}{1-z_1}+\frac{1}{1-z_2}-1\right)x\right]\\&=\frac{(z_1z_2)^k}{\left[(1-z_1)(1-z_2)\right]^{k+1}}\underbrace{\int_0^\infty x^{k+p-1}\exp\left[-\frac{1-z_1z_2}{(1-z_1)(1-z_2)}x\right]}_{\substack{INTEGRAL\ of\ type\ \int_0^\infty x^\alpha e^{-\beta x}dx=\beta^{-\alpha-1}\alpha!\\(see\ Apendix\ )}}\\&=\frac{(z_1z_2)^k}{\left[(1-z_1)(1-z_2)\right]^{k+1}}\frac{(k+p-1)!}{(1-z_1z_2)^{k+p}}\left[(1-z_1)(1-z_2)\right]^{k+p}\\&=\frac{(z_1z_2)^k}{(1-z_1z_2)^{k+p}}\left[(1-z_1)(1-z_2)\right]^{p-1}(k+p-1)!\end{aligned} \tag{3.109}$$

Yet, the last expression may be even more transformed by using the inverse (or generalized) binomial theorem (see Appendix A.1.1)

$$\frac{1}{(1-z_1z_2)^{k+p}}=\sum_{\lambda=0}^{\infty}\binom{k+p+\lambda-1}{\lambda}(z_1z_2)^\lambda=\sum_{\lambda=0}^{\infty}\frac{(k+p+\lambda-1)!}{\lambda!(k+p-1)!}(z_1z_2)^\lambda \tag{3.110}$$

under the form:

$$\sum_{n,m=k}^{\infty}\frac{z_1^n z_2^m}{n!m!}I_{nm}=\left[(1-z_1)(1-z_2)\right]^{p-1}\sum_{\lambda=0}^{\infty}\frac{(k+p+\lambda-1)!}{\lambda!}(z_1z_2)^{k+\lambda} \quad (3.111)$$

In these conditions, for $n=m$ one has the identity

$$\sum_{n=k}^{\infty}(z_1z_2)^n I_{nm}=\sum_{\lambda=0}^{\infty}(n!)^2\frac{(k+p+\lambda-1)!}{\lambda!}\left[(1-z_1)(1-z_2)\right]^{p-1}(z_1z_2)^{k+\lambda} \quad (3.112)$$

from where follows that the typical integral I_{nn} is evaluated as:

$$I_{nn}=\int_0^{\infty}e^{-x}x^{k-1}\left|L_n^k(x)\right|^2 x^p dx=\sum_{\substack{CERTAIN\\ \lambda}}(n!)^2\left.\frac{(k+p+\lambda-1)!}{\lambda!}\right|_{\substack{for\\ given\\ p}} \quad (3.113)$$

such that

$$(z_1z_2)^n=\left[(1-z_1)(1-z_2)\right]^{p-1}(z_1z_2)^{k+\lambda} \quad (3.114)$$

as prescribed by the last series equality.

Going to exemplify the method we treat the first three cases:

- When $p=1$ we have:

$$\left[(1-z_1)(1-z_2)\right]_{p=1}^{p-1}(z_1z_2)^{k+\lambda}=(z_1z_2)^{k+\lambda}\underset{\lambda=n-k}{=}(z_1z_2)^n \quad (3.115)$$

and therefore the integration rule:

$$I_{1(nn)}=\int_0^{\infty}e^{-x}x^{k-1}\left|L_n^k(x)\right|^2 x^1 dx=(n!)^2\left.\frac{(k+p+\lambda-1)!}{\lambda!}\right|_{\substack{p=1\\ \lambda=n-k}}=\frac{(n!)^3}{(n-k)!} \quad (3.116)$$

that in terms of Hydrogenic principal and angular quantification numbers (n, l) it rewrites though applying the indicial substitutions:

$$\begin{cases} n \to n+l \\ k \to 2l+1 \end{cases} \tag{3.117}$$

leaving with the result:

$$I_1(n,l) = \int_0^\infty e^{-x} x^{2l} \left| L_{n+l}^{2l+1}(x) \right|^2 x^1 dx = \frac{[(n+l)!]^3}{(n-l-1)!} \tag{3.118}$$

- When $p = 2$ we have:

$$\left[(1-z_1)(1-z_2)\right]_{p=2}^{p-1} (z_1 z_2)^{k+\lambda} = (1 - z_1 - z_2 + z_1 z_2)(z_1 z_2)^{k+\lambda} \tag{3.119}$$

From which counts only the sum of those products that produce equal shooting in the powers of z_1 and z_2, namely

$$\underbrace{(z_1 z_2)^{k+\lambda}}_{\substack{for\ \lambda=n-k \\ \downarrow \\ (z_1 z_2)^n}} + \underbrace{(z_1 z_2)^{k+\lambda+1}}_{\substack{for\ \lambda=n-k-1 \\ \downarrow \\ (z_1 z_2)^n}} \tag{3.120}$$

from where the right values of λ are selected; therefore it leads with the integration rule:

$$I_{2(nn)} = \int_0^\infty e^{-x} x^{k-1} \left| L_n^k(x) \right|^2 x^2 dx = (n!)^2 \left[\left. \frac{(k+p+\lambda-1)!}{\lambda!} \right|_{\substack{p=2 \\ \lambda=n-k}} + \left. \frac{(k+p+\lambda-1)!}{\lambda!} \right|_{\substack{p=2 \\ \lambda=n-k-1}} \right]$$

$$= (n!)^2 \left[\frac{(n+1)!}{(n-k)!} + \frac{n!}{(n-k-1)!} \right] = \frac{(n!)^3}{(n-k)!}(2n-k+1) \tag{3.121}$$

which in the "language" of Hydrogenic integrations translates as:

$$I_2(n,l) = \int_0^{\infty} e^{-x} x^{2l} \left| L_{n+l}^{2l+1}(x) \right|^2 x^2 dx = 2n \frac{[(n+l)!]^3}{(n-l-1)!} \tag{3.122}$$

- When $p = 3$ we have the product:

$$\left[(1-z_1)(1-z_2)\right]_{p=3}^{p-1} (z_1 z_2)^{k+\lambda} = \left(1 - 2z_1 + z_1^2\right)\left(1 - 2z_2 + z_2^2\right)(z_1 z_2)^{k+\lambda} \tag{3.123}$$

that restricted to the significant terms, as above, it contains only the summation:

$$\underbrace{(z_1 z_2)^{k+\lambda}}_{\substack{for\ \lambda=n-k \\ \downarrow \\ (z_1 z_2)^n}} + 4\underbrace{(z_1 z_2)^{k+\lambda+1}}_{\substack{for\ \lambda=n-k-1 \\ \downarrow \\ (z_1 z_2)^n}} + \underbrace{(z_1 z_2)^{k+\lambda+2}}_{\substack{for\ \lambda=n-k-2 \\ \downarrow \\ (z_1 z_2)^n}} \tag{3.124}$$

producing the general integration rule

$$\begin{aligned}
I_{3(nn)} &= \int_0^{\infty} e^{-x} x^{k-1} \left| L_n^k(x) \right|^2 x^3 dx \\
&= (n!)^2 \left[\left. \frac{(k+p+\lambda-1)!}{\lambda!} \right|_{\substack{p=3 \\ \lambda=n-k}} + 4 \left. \frac{(k+p+\lambda-1)!}{\lambda!} \right|_{\substack{p=3 \\ \lambda=n-k-1}} + \left. \frac{(k+p+\lambda-1)!}{\lambda!} \right|_{\substack{p=3 \\ \lambda=n-k-2}} \right] \\
&= (n!)^2 \left[\frac{(n+2)!}{(n-k)!} + 4\frac{(n+1)!}{(n-k-1)!} + \frac{n!}{(n-k-2)!} \right] \\
&= \frac{(n!)^3}{(n-k)!} \left(6n^2 + 6n + k^2 - 3k - 6nk + 2 \right)
\end{aligned} \tag{3.125}$$

and with Hydrogenic specialization:

$$I_3(n,l) = \int_0^{\infty} e^{-x} x^{2l} \left| L_{n+l}^{2l+1}(x) \right|^2 x^3 dx = \frac{[(n+l)!]^3}{(n-l-1)!} \left[6n^2 - 2l(l+1) \right] \tag{3.126}$$

Now we are in position to calculate the normalization constant of the Hydrogenic wave-function, as well some of the radial mean values in direct manner. For instance, for the normalization constants we have:

$$1=\left(\frac{na_0}{2Z}\right)^3 C_{nl}^2\int_0^\infty e^{-x}x^{2l}\left|L_{n+l}^{2l+1}(x)\right|^2 x^2dx=\left(\frac{na_0}{2Z}\right)^3 C_{nl}^2 I_2(n,l)$$

$$=\left(\frac{na_0}{2Z}\right)^3 C_{nl}^2\, 2n\frac{\left[(n+l)!\right]^3}{(n-l-1)!} \tag{3.127}$$

that at once gives out:

$$C_{nl}=\left(\frac{2Z}{na_0}\right)^{3/2}\left\{\frac{(n-l-1)!}{2n\left[(n+l)!\right]^3}\right\}^{1/2} \tag{3.128}$$

and respectively, the Hydrogenic radial eigen-functions:

$$\psi_{nl}(r)=\left\{\frac{(n-l-1)!}{2n\left[(n+l)!\right]^3}\right\}^{1/2}\left(\frac{2Z}{na_0}\right)^{3/2}\left(\frac{2Z}{na_0}r\right)^{l} L_{n+l}^{2l+1}\left(\frac{2Z}{a_0 n}r\right)\exp\left(-\frac{Z}{n}\frac{r}{a_0}\right) \tag{3.129}$$

Yet, for individuating certain radial wave-functions for certain quantum numbers' combinations, one needs to better express the associate Laguerre functions. For that the next discussion regards the ways it may be unfolded. Firstly one may use the Laguerre generating function representation above to differentiate it n-times respecting z and then taking the $z\to 0$ limit that gives the Laguerre polynomial under the derivative form:

$$L_n(x)=e^x\lim_{z\to 0}\frac{\partial^n}{\partial z^n}\left[\frac{1}{1-z}\exp\left(\frac{-x}{1-z}\right)\right]=e^x\frac{d^n}{dx^n}\left(x^n e^{-x}\right) \tag{3.130}$$

as one may easily check out (eventually by induction). Then the Hydrogenic associate Laguerre functions are:

$$L_n^k(x)\Big|_{\substack{k=2l+1\\ n\to n+l}}=(-1)^k\frac{d^k}{dx^k}L_n(x)\Bigg|_{\substack{k=2l+1\\ n\to n+l}} \tag{3.131}$$

providing the working expression:

$$L_{n+l}^{2l+1}(x) = (-1)^{2l+1} \frac{d^{2l+1}}{dx^{2l+1}} \left[e^{x} \frac{d^{n+l}}{dx^{n+l}} \left(x^{n+l} e^{-x} \right) \right] \tag{3.132}$$

whose few results are summarized in the Table 3.1.

From Table 3.1 worth employing the ground state Hydrogenic wave-function, for $n = 1, l = 0$, that takes the simple analytical form:

$$\psi_{10}(r) = 2\left(\frac{Z}{a_0}\right)^{3/2} \exp\left(-Z\frac{r}{a_0}\right) \tag{3.133}$$

TABLE 3.1 Associate Laguerre Functions for the Main Hydrogenic Radial Wave-Functions

n	l	$L_{n+l}^{2l+1}(x)$
1	0	$L_1^1(x) = 1$
2	0	$L_2^1(x) = 2(2-x)$
2	1	$L_3^3(x) = 6$
3	0	$L_3^1(x) = 3\left(6-6x+x^2\right)$
3	1	$L_4^3(x) = 24(4-x)$
3	2	$L_5^5(x) = 120$
4	0	$L_4^1(x) = 4\left[24-(x-6)^2 x\right]$
4	1	$L_5^3(x) = 60\left[20+(x-10)x\right]$
4	2	$L_6^5(x) = 720(6-x)$
4	3	$L_7^7(x) = 5040$
5	0	$L_5^1(x) = 5\left\{120+x\left[-240+x(120+(x-20)x)\right]\right\}$
5	1	$L_6^3(x) = 120\left\{120-x\left[90+(x-18)x\right]\right\}$
5	2	$L_7^5(x) = 2520\left[42+(x-14)x\right]$
5	3	$L_8^7(x) = 40320(8-x)$
5	4	$L_9^9(x) = 362880$

In the same manner other wave-functions are determined, in a full normalized analytical mode.

Moreover, with the help of the Laguerre functions of Table 3.1, the full normalized expression of the associate radial wave functions may be analytically formulated; yet, the physical significance will be carried by the squared quantity $|r\psi_{nl}(r)|^2$, in radial normalization framework, as meaning the probability density the electronic existence will be encompassed by the concerned *nl* state (i.e., the so-called *orbital*). These quantities are represented in the Figures 3.6 and 3.7, giving the answer to the above formulated "spatially penetrating potentials" in the beginning of this section. The solution of this paradox relays in the actual observed spatially penetrating orbitals – a feature that really appears for any state with $n > 1$, i.e., displaying at least two degeneracies respecting the quantum (angular) number l. In other words, the mixture of potential states with $l = 0$ and $l \neq 0$ remarked at the beginning of the present analysis is here reflected in the penetrating features of the orbitals with non-zero angular number l respecting those characterized by $l = 0$.

This will have further consequence in adjusting of corresponding sub-orbital energies in various atomic structures, depending on the degree of these orbital penetrations, providing the successive energetic series of

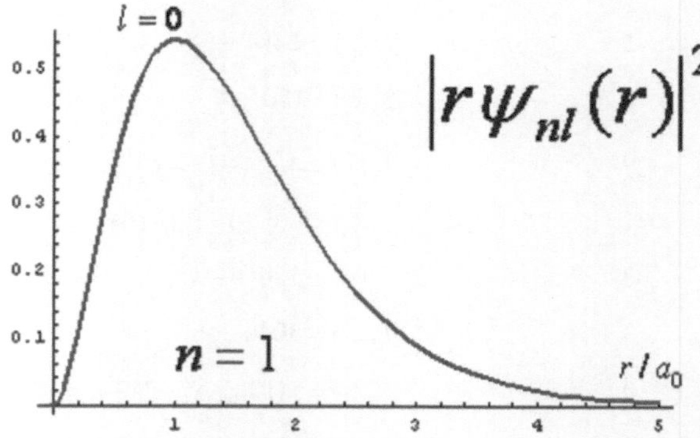

FIGURE 3.6 The paradigmatic representation of the electronic probability density of existence (wave-functions) for Hydrogenic atoms in the ground state, i.e., for the first level (or shell with n=1) with the respective sub-shell (with l=0), in accordance with the first quantum postulate of Section 3.2, i.e., displaying both continuity and vanishing behavior on the nucleus as well as to the asymptotic (long-range, at infinitum) existence distribution (naturally derived by the respective Laguerre polynomial of Table 3.1).

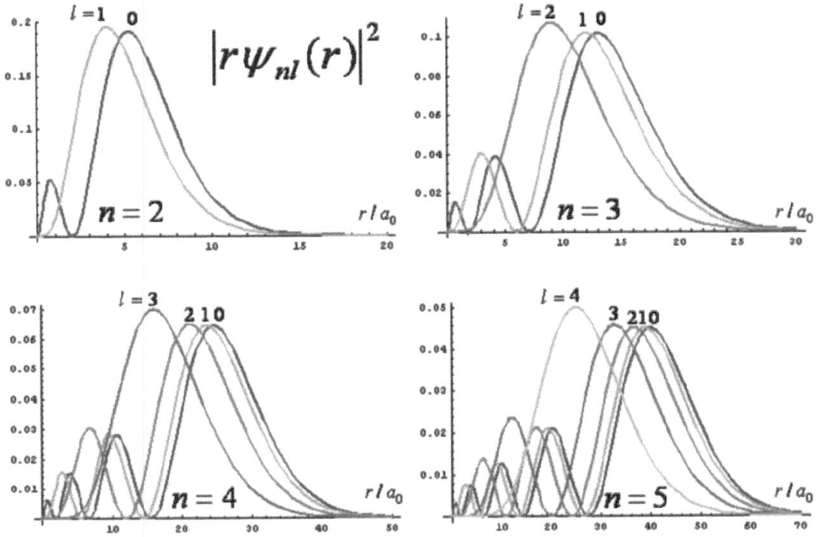

FIGURE 3.7 The representations of the electronic probability density of existence (wave-functions) for Hydrogenic atoms, for the superior (excited) levels (or shells, quantified by the number n) with the respective sub-shells (or orbitals, quantified by the mixed numbers nl), employing the derived radial wave functions in terms of respective Laguerre polynomials of Table 3.1.

shells n and sub-shells nl in the celebrated *atomic aufbau principle* (the atomic constitution). Moreover, one may still observe that the orbital with $l_{max} = n-1$ behaves as the Hydrogen first (ground state) orbital, supporting the idea of *chemical existence* in the *valence shell* of the atoms of the features similar to those specific for the Hydrogen atom, while the difference appears only by the shielding effect of the inner shells between the outermost level and the nucleus. These ideas will be reloaded in the next volume of this five-volume book, dedicated to quantum atom and periodicity.

Finally, one would like to use the developed rules of radial wave-function integration to compute other integrals of interests. For instance, one may wish to compute the quantum average (the observed) of the applied potential:

$$\langle V(r)\rangle_{\psi_{nl}} = \left\langle -\frac{Ze_0^2}{r}\right\rangle_{\psi_{nl}} = -Ze_0^2\left\langle r^{-1}\right\rangle_{\psi_{nl}} \tag{3.134}$$

therefore needing to compute the average

$$\left\langle r^{-1}\right\rangle_{\psi_{nl}} = \int_0^{\infty} |\psi_{nl}(r)|^2 r dr = \left(\frac{na_0}{2Z}\right)^2 \int_0^{\infty} |\psi_{nl}(x)|^2 x dx$$
$$= \left(\frac{na_0}{2Z}\right)^2 C_{nl}^2 I_1(n,l)$$
$$= \left(\frac{na_0}{2Z}\right)^2 \left(\frac{2Z}{na_0}\right)^3 \frac{(n-l-1)!}{2n[(n+l)!]^3} \frac{[(n+l)!]^3}{(n-l-1)!} = \frac{Z}{n^2 a_0} \quad (3.135)$$

that helps in concluding finding that:

$$\left\langle V(r)\right\rangle_{\psi_{nl}} = -\frac{Ze_0^2}{a_0 n^2} = 2E_n \quad (3.136)$$

in fully agreement with earlier introduced *virial theorem*.

Equally, one may be interested in evaluating the average of the distance of the Hydrogenic electron in a certain state, with direct specialization to the ground state as well. For that it is compulsory to evaluate the integral:

$$\left\langle r\right\rangle_{\psi_{nl}} = \int_0^{\infty} |\psi_{nl}(r)|^2 r^3 dr = \left(\frac{na_0}{2Z}\right)^4 \int_0^{\infty} |\psi_{nl}(x)|^2 x^3 dx = \left(\frac{na_0}{2Z}\right)^4 C_{nl}^2 I_3(n,l)$$
$$= \left(\frac{na_0}{2Z}\right)^4 \left(\frac{2Z}{na_0}\right)^3 \frac{(n-l-1)!}{2n[(n+l)!]^3} \frac{[(n+l)!]^3}{(n-l-1)!} \left[6n^2 - 2l(l+1)\right]$$
$$= \frac{a_0}{Z} \frac{3n^2 - l(l+1)}{2} \quad (3.137)$$

There is interesting to see that for the Hydrogen atom in its ground state, $Z = 1, n = 1, l = 0$, one has that the average of the distance of the electronic existence in this state looks like:

$$\left\langle r\right\rangle_{H,n=1,l=0} = \frac{3}{2} a_0 \quad (3.138)$$

thus being related with the first Bohr radius a_0 but not identical with it; from this point there is obvious that since the observed "orbit radius" (the average above) differs from the quantified "orbit radius" the notion of radius itself is less meaningful, or in other terms stands as a fuzzy (without precisely delimited frontier) quantity. The problem that arises from this is that the frontier of atoms thyself seems to not be properly quantified unless other quantities are introduced in relation with their valence or the outmost eigen-state. This will be achieved with the help of the most celebrated chemical notion of electronegativity that will be introduced latter in this volume, and applied on the atomic systems characterization in the Volume II of the present work (Putz, 2016).

Yet, we presented so fat the complete radial picture of eigen-values (quantified energies) and of the associated eigen-function (wave-functions) for the radial motion of electrons in central (spherical) potential, with results resembling the phenomenological ones derived by the Bohr theory, however opening the door to the electronic spatially characterization through the analytical systematically generation of their wave-functions.

3.3.2 MOLECULAR VIBRATIONAL STATES BY SOLVING HERMITE TYPE EQUATIONS

The vibrational states in molecules are modeled by electronic motion under the harmonic oscillator (1D) potential

$$V(x) = \frac{1}{2}k_{\omega}x^2 = \frac{1}{2}m\omega^2 x^2 \tag{3.139}$$

with the force constant

$$k_{\omega} = m\omega^2 \tag{3.140}$$

and the effective mass

$$m = \frac{m_1 m_2}{m_1 + m_2} \tag{3.141}$$

of a typical diatomic molecular oscillator, whose atomic masses m_1, m_2 are the expressed as atom gram quantities. For such physical-chemical system the associate Schrödinger equation looks like

$$\left(-\frac{\hbar^2}{2m_0}\frac{d^2}{dx^2}+\frac{1}{2}m\omega^2x^2\right)\psi_\omega(x)=E_\omega\psi_\omega(x) \tag{3.142}$$

that immediately rearranges as:

$$\frac{d^2}{dx^2}\psi_\omega(x)+\left(\alpha-\beta^2x^2\right)\psi_\omega(x)=0 \tag{3.143}$$

with

$$\alpha=\frac{2mE_\omega}{\hbar^2},\quad \beta=\frac{m\omega}{\hbar} \tag{3.144}$$

Further on, if one consider the substitution

$$x=\frac{\zeta}{\sqrt{\beta}} \tag{3.145}$$

the last equation becomes:

$$\frac{d^2\psi_\omega}{d\zeta^2}+\left[1-\zeta^2+\left(\frac{\alpha}{\beta}-1\right)\right]\psi_\omega=0 \tag{3.146}$$

that is recognized as the equation type for *Hermite's orthogonal functions*:

$$\psi''+\left(1-\zeta^2+2n\right)\psi=0,\quad n\in\mathbf{N} \tag{3.147}$$

Now, through the immediate correspondence of the two equations follows the identity:

$$\frac{\alpha}{\beta}-1=2n \tag{3.148}$$

which after the replacements of the shortcuts it leaves with the energy quantification

$$E_{\omega} = \hbar\omega\left(n + \frac{1}{2}\right), \quad n \in \mathbf{N} \tag{3.149}$$

for the harmonic oscillator.

However, for finding the associate eigen-functions, i.e., the wave-functions of the harmonic oscillator, one needs to solve the Hermite's orthogonal equation above; yet there is very interesting that it may be further simplified by the substitution(s)

$$\psi = v\exp\left(-\zeta^2 / 2\right) \tag{3.150a}$$

$$\psi'' = \left[\left(\zeta^2 - 1\right)v - 2\zeta v' + v''\right]\exp\left(-\zeta^2 / 2\right) \tag{3.150b}$$

to the simple *Hermite's differential equation*:

$$v'' - 2\zeta v' + 2nv = 0 \tag{3.151}$$

In the same way as proceeded with the Laguerre's polynomial, the Hermite equation may be solved by the specific complex integral representation:

$$v_n(\zeta) = \frac{1}{2\pi i}\oint z^{-n-1}\exp\left(\zeta^2 - (z-\zeta)^2\right)dz \tag{3.152}$$

that through its first and second derivatives with respecting ζ, respectively,

$$v_n(\zeta)' = \frac{1}{2\pi i}\oint 2z^{-n}\exp\left(\zeta^2 - (z-\zeta)^2\right)dz \tag{3.153a}$$

$$v_n(\zeta)'' = \frac{1}{2\pi i}\oint 4z^{-n+1}\exp\left(\zeta^2 - (z-\zeta)^2\right)dz \tag{3.153b}$$

shows that it fulfills the identity:

$$v'' - 2\zeta v' + 2nv = \frac{1}{2\pi i}\oint\left(4z^2 - 4\zeta z + 2n\right)z^{-n-1}\exp\left(\zeta^2 - (z-\zeta)^2\right)dz$$

$$= -\frac{2}{2\pi i}\oint \frac{d}{dz}\left[z^{-n}\exp\left(\zeta^2-(z-\zeta)^2\right)\right]dz = 0 \tag{3.154}$$

because of its integrant single value function shape that cancels it along the close contours, i.e., having the same values on the initial and final points.

Therefore, in the light of the Laurent theorem of residues there can be inferred that the complex integral solution of the Hermite equation may produce the *Hermite generating function* (with the same recipe as was previously done for generating Laguerre function):

$$\exp\left(\zeta^2-(z-\zeta)^2\right) = \sum_{n=0}^{\infty} v_n(\zeta)z^n = \sum_n \frac{H_n(\zeta)}{n!}z^n \tag{3.155}$$

where

$$H_n(\zeta) = n!v_n(\zeta) = \frac{n!}{2\pi i}\oint z^{-n-1}\exp\left(\zeta^2-(z-\zeta)^2\right)dz \tag{3.156}$$

stands as the complex representation for the Hermite polynomial. However, it may acquire also a workable expression by means of the n-th derivative of the above generating function expansion on both sides and then taking the $z \to 0$ limit; in these conditions one successively yields the Hermite polynomial or n-th degree:

$$H_n(\zeta) = \lim_{z\to 0} e^{\zeta^2}\frac{\partial^n}{\partial z^n}e^{-(z-\zeta)^2} = \lim_{z\to 0}(-1)^n e^{\zeta^2}\frac{\partial^n}{\partial \zeta^n}e^{-(z-\zeta)^2} = (-1)^n e^{\zeta^2}\frac{\partial^n}{\partial \zeta^n}e^{-\zeta^2} \tag{3.157}$$

whose the first ten realization are obtained systematically as:

$$H_0(\zeta) = 1 \tag{3.158a}$$

$$H_1(\zeta) = 2\zeta \tag{3.158b}$$

$$H_2(\zeta) = 4\zeta^2 - 2 \tag{3.158c}$$

$$H_3(\zeta) = 8\zeta^3 - 12\zeta \tag{3.158d}$$

$$H_4(\zeta)=16\zeta^4-48\zeta^2+12 \tag{3.158e}$$

$$H_5(\zeta)=32\zeta^5-160\zeta^3+120\zeta \tag{3.158f}$$

$$H_6(\zeta)=64\zeta^6-480\zeta^4+720\zeta^2-120 \tag{3.158g}$$

$$H_7(\zeta)=128\zeta^7-1344\zeta^5+3360\zeta^3-1680\zeta \tag{3.158h}$$

$$H_8(\zeta)=256\zeta^8-3584\zeta^6+13440\zeta^4-13440\zeta^2+1680 \tag{3.158i}$$

$$H_9(\zeta)=512\zeta^9-9216\zeta^7+48384\zeta^5-80640\zeta^3+30240\zeta \tag{3.158j}$$

$$H_{10}(\zeta)=1024\zeta^{10}-23040\zeta^8+161280\zeta^6-403200\zeta^4+30240\zeta^2-30240 \tag{3.158k}$$

Back to the quantum harmonic oscillator, the Hermite polynomials helps in writing its eigen-functions as:

$$\psi_n(\zeta)=C_n\frac{H_n(\zeta)}{n!}\exp\left(-\zeta^2/2\right) \tag{3.159}$$

with the amendment that the constants are still to be determined from the wave-function normalization condition in the *n*-th state:

$$1=\int_{-\infty}^{+\infty}\left|\psi_n(x)\right|^2 dx=\frac{C_n^2}{(n!)^2\sqrt{\beta}}\int_{-\infty}^{+\infty}\left|H_n(\zeta)\right|^2\exp\left(-\zeta^2\right)d\zeta \tag{3.160}$$

Thus, we have to evaluate the integrals of type:

$$J_0(n,m)=\int_{-\infty}^{+\infty}H_n(\zeta)H_m(\zeta)\exp\left(-\zeta^2\right)d\zeta \tag{3.161}$$

This can be done in the same way as proceed with the Laguerre's polynomials, i.e., by considering the product of two Hermite's generating functions:

$$\exp\left(\zeta^2-(z_1-\zeta)^2\right)\exp\left(\zeta^2-(z_2-\zeta)^2\right)=\left(\sum_n\frac{H_n(\zeta)}{n!}z_1^n\right)\left(\sum_m\frac{H_m(\zeta)}{m!}z_2^m\right) \tag{3.162}$$

followed by the multiplication of both sides with $\exp\left(-\zeta^2\right)$ and integrating to successively obtain:

$$\begin{aligned}
&\sum_{n,m}\left[\int_{-\infty}^{+\infty}H_n(\zeta)H_m(\zeta)\exp\left(-\zeta^2\right)d\zeta\right]\frac{z_1^n z_2^m}{n!m!}=\int_{-\infty}^{+\infty}\exp\begin{pmatrix}\zeta^2-(z_1-\zeta)^2\\-(z_2-\zeta)^2\end{pmatrix}d\zeta\\
&=\int_{-\infty}^{+\infty}\exp\left(-\zeta^2+2(z_1+z_2)\zeta-z_1^2-z_2^2\right)d\zeta\\
&=\exp\left[(z_1+z_2)^2-z_1^2-z_2^2\right]\underbrace{\int_{-\infty}^{+\infty}\exp\left[-(\zeta-(z_1+z_2))^2\right]d(\zeta-(z_1+z_2))}_{\text{of type }\int_{-\infty}^{+\infty}\exp\left(-ax^2\right)dx=\sqrt{\frac{\pi}{a}}}\\
&=\sqrt{\pi}\exp(2z_1z_2)\\
&=\sqrt{\pi}\sum_n\frac{(2z_1z_2)^n}{n!}\\
&=\sqrt{\pi}\sum_{n,m}\frac{2^n}{n!}z_1^n z_2^m\delta_{n,m}
\end{aligned} \tag{3.163}$$

from where the general rule of integration may be abstracted:

$$\int_{-\infty}^{+\infty}H_n(\zeta)H_m(\zeta)\exp\left(-\zeta^2\right)d\zeta=2^m m!\sqrt{\pi}\delta_{n,m} \tag{3.164}$$

with the actual specialization:

$$\int_{-\infty}^{+\infty}\left|H_n(\zeta)\right|^2\exp\left(-\zeta^2\right)d\zeta=2^n n!\sqrt{\pi} \tag{3.165}$$

This result allows the determination of the normalization constant above for the wave-function of the quantum harmonic oscillator,

$$C_n = \left(\frac{n!}{2^n} \sqrt{\frac{\beta}{\pi}} \right)^{1/2} \tag{3.166}$$

leading it with the complete form:

$$\psi_n(x) = \left(\frac{1}{2^n n!} \sqrt{\frac{m\omega}{\pi\hbar}} \right)^{1/2} H_n\left(x\sqrt{\frac{m\omega}{\hbar}} \right) \exp\left(-\frac{m\omega}{2\hbar} x^2 \right), n \in \mathbf{N} \tag{3.167}$$

Aiming to present an application of the given harmonic oscillator model for diatomic molecules, let's take the case of Iodine-Hydrogen system (HI) having the force constant

$$k_{HI} = 314.14 < N \cdot m^{-1} > \tag{3.168}$$

With the effective mass of the system computed through combining atomic gram masses with the Avogadro's number N_A:

$$m_{HI} = \frac{m_H m_I}{m_H + m_I} \frac{1}{N_A} \tag{3.169}$$

$$= \frac{(1.0079 \times 10^{-3} < kg \cdot mol^{-1} >)(126.905 \times 10^{-3} < kg \cdot mol^{-1} >)}{1.0079 \times 10^{-3} < kg \cdot mol^{-1} > + 126.905 \times 10^{-3} < kg \cdot mol^{-1} >}$$
$$\frac{1}{6.022045 \times 10^{23} mol^{-1}}$$
$$= 1.6605 \times 10^{-27} < kg > \tag{3.170}$$

the system's oscillating rate has the value:

$$\omega = \sqrt{\frac{k_{HI}}{m_{HI}}} = 43.4954 \cdot 10^{13} < s^{-1} > \tag{3.171}$$

it provides the eigen-energies (of vibrations on bonding)

$$E_{HI} = \left(n + \frac{1}{2} \right) (4.587 \cdot 10^{-20}) < J >, n \in \mathbf{N} \tag{3.172}$$

and the working eigen-functions' probability

$$|\psi_n(x)|^2 = \frac{1}{2^n n!}\sqrt{\frac{\beta}{\pi}}H_n^2\left(x\sqrt{\beta}\right)\exp\left(-\beta x^2\right),\ n \in \mathbf{N} \tag{3.173}$$

with the specialized parameter

$$\beta_{HI} = \frac{m_{HI}\omega_{HI}}{\hbar} = 6.84853 \cdot 10^{21} < m^{-2} > \tag{3.174}$$

In these conditions, noting that the equilibrium bonding distance for HI system is about:

$$R_{HI} \cong 1.89 \cdot 10^{-10} < m > \tag{3.175}$$

the graphical representations of the quantized probabilities for few states of this oscillating system are depicted in the Figure 3.8.

The analysis of the Figure 3.8 clearly illustrates that since in classical interpretation of the motion in the harmonic potential the system has its maximum probability to be found at the position x when its velocity is minimum, i.e., at the maximum distance (at the amplitude) allowed by the oscillation, while in quantum motion this is certainly not the case of the system in its vibrational ground state ($n = 1$) but only in the higher excited states (see the probability behavior for $n = 10$ and far above that) when the quantum probability of vibration becomes multiplied enough (by the quantum vibrational number) so that it shapes asymptotically to the classical potential of vibrational motion. Such behavior is nothing but the vibrational manifestation of the earlier discussed (Bohr) *correspondence principle* affirming that the *quantum motion approaches the classical one in the very high levels of quantification*.

Finally, seeing the vibrational ground state is so fragrant in contradiction with the classical oscillating behavior (i.e., having maximum of quantum probability where classically it is recorded zero probability) one may like to investigate the probability with which the vibrational system lying on its ground state

$$\psi_0(x) = \left(\frac{\beta}{\pi}\right)^{1/4}\exp\left(-\frac{\beta}{2}x^2\right) \tag{3.176}$$

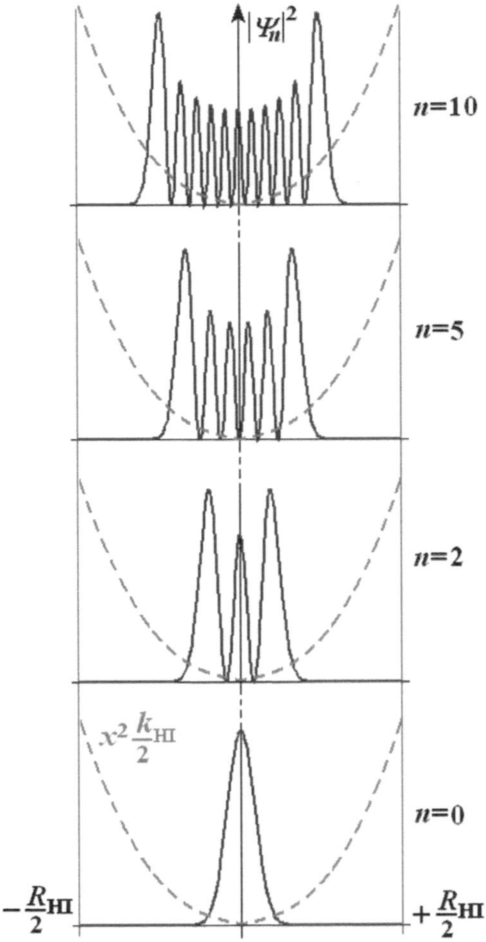

FIGURE 3.8 The quantum harmonic oscillator eigen-function probabilities' (density) representation (thick continuous curves) for ground state ($n = 0$), and few excited vibronic states ($n = 2$, 5, and 10) for the working case of HI molecule (respecting the coordinated centered on its mass center); the classical potential is as well illustrated (by the dashed curve in each instant) for facilitating the correspondence principle discussion.

is compressed within the classical domain; that is to impose the energetically condition:

$$\frac{1}{2}\hbar\omega = \frac{1}{2}m\omega^2 x_0^2 \tag{3.177}$$

from where the (semi)classical turning points are founded as:

$$x_0 = \pm\sqrt{\frac{\hbar}{m\omega}} = \pm\frac{1}{\sqrt{\beta}} \tag{3.178}$$

leaving with the probability calculation:

$$P_0(x_0) = \int_{-x_0}^{+x_0} \left|\psi_0(x)\right|^2 dx$$

$$= \left(\frac{\beta}{\pi}\right)^{1/2} \int_{-x_0}^{+x_0} \exp\left(-\beta x^2\right) dx = 2\left(\frac{\beta}{\pi}\right)^{1/2} \int_{0}^{|x_0|} \exp\left(-\beta x^2\right) dx$$

$$= 2\left(\frac{\beta}{\pi}\right)^{1/2} \int_{0}^{\left(1/\sqrt{\beta}\right)} \exp\left(-\beta x^2\right) dx \underset{z=x\sqrt{\beta}}{=} \frac{2}{\sqrt{\pi}} \int_{0}^{1} \exp\left(-z^2\right) dz \tag{3.179}$$

while recognizing the special error function result (see Appendix A.4):

$$P_0(x_0) = \mathrm{erf}(1) = 0.8427 \tag{3.180}$$

This result confirms the quantum behavior of the quantum motion that is not entirely encompassed by the (semi) classical turning points' domain, even its major part lays there, with "the rest" being dispersed by tunneling process (see the Postulate I discussion), being however a new manifestation of the indeterminacy (Heisenberg) relationship that impedes the position and momentum to be with the same precision simultaneously determined. Further aspects of the quantum vibrational motions are to be discussed in the next sections and chapter of this volume, whereas the application to various molecular systems will be systematically presented in the Volume 3 of this five-volume set.

3.3.3 SOLID STATE FREE ELECTRONIC STATES

Modeling of the solid state from an ordered chains and planes of atoms may be possible, in the first approximation, by considered each of the involved atoms as being represented by the *core* and *valence* states. As such the inner electrons, those close to the nucleus and those completing the atomic shells – form the core state or the *bulk solid state* ψ^{bulk}, while electrons on the incomplete atomic levels compose the *valence solid state* $\psi^{valence}$; however, being considered independent states, their scalar product hast to cancel over the entire solid state domain:

$$\left\langle \psi^{bulk} \middle| \psi^{valence} \right\rangle = 0 \tag{3.181}$$

This condition, considered in terms of associate potentials leaves with the idea that the associate potentials are *complementary*, i.e., when the bulk (electrostatic or Coulombic) potential is maximum the valence or orthogonalized potential reaches its minimum, and vice-versa. The result is the net uniform potential that models the so-called free electrons in solids, see Figure 3.9. Therefore, the free electronic model in solids has only formally the potential

$$V(x) = 0 \tag{3.182}$$

replaced in (1D) Schrödinger,

$$-\frac{\hbar^2}{2m_0}\frac{d^2}{dx^2}\psi_k(x) = E_k\psi_k(x) \tag{3.183}$$

while it is the result of atomic bulk-valence potential cancellation in rows and planes of a solid state of crystal type.

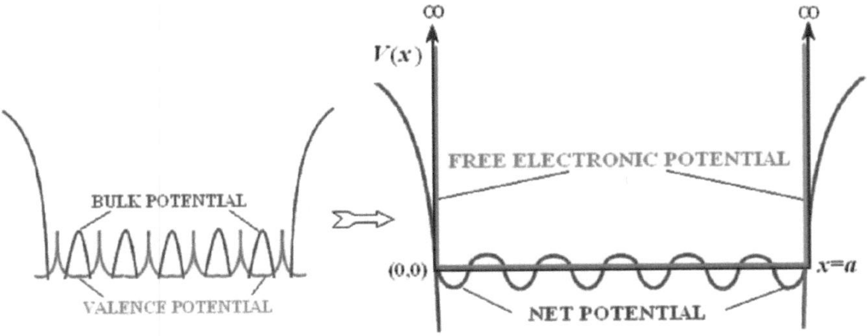

FIGURE 3.9 The resulting free electronic potential in solid state modeling (right) from superposition of the bulk (electrostatic or Coulombic) and the valence (orthogonal) potentials (in left) (Putz, 2006).

However, the actual solid state Schrödinger equation immediately rewrites as a simple differential equation with frontier conditions:

$$\begin{cases} \frac{d^2}{dx^2}\psi_k(x) + k\psi_k(x) = 0, \; k = \sqrt{\frac{2m_0 E_k}{\hbar^2}} \\ \psi_k(x)_{x=0} = \psi_k(x)_{x=a} = 0 \end{cases} \tag{3.184}$$

whose one general solution reads as:

$$\begin{aligned} \psi_k(x) &= A\exp(ikx) + B\exp(-ikx) \\ &= (A+B)\cos(kx) + i(A-B)\sin(kx) \\ &\equiv C_k^I \cos(kx) + C_k^{II} \sin(kx) \end{aligned} \tag{3.185}$$

When the first frontier constraint is employed one gets the value of the first constant above

$$0 = \psi_k(x)_{x=0} = C_k^I \tag{3.186}$$

that leaves the valid wave-function form to be:

$$\psi_k(x) = C_k^{II} \sin(kx) \tag{3.187}$$

which, under the remaining limiting constraint:

$$0 = \psi_k(x)_{x=a} = C_k^{II} \sin(ka) \tag{3.188}$$

provides the *k*-values

$$k = n_k \frac{\pi}{a} \tag{3.189}$$

and the implicit energy quantification

$$E_k = \frac{\hbar^2}{2m_0}k^2 = \frac{\pi^2\hbar^2}{2m_0 a^2}n_k^2 \tag{3.190a}$$

with the physical-mathematical acceptable quantum numbers:

$$n_k = 1, 2, 3, \ldots \tag{3.190b}$$

The fact that the state with $n = 0$ was excluded means, beyond it appears from the impossibility to have zero energy for a real system, the absence of the ground state in the free electronic solid state model. At this moment this results seems acceptable from the phenomenological argument that electrons having only kinetic energy their existence only on excited states appears plausible, although a more quantitative argument will be given through the variational quantum procedure applied on this model, as will be see in one of the next sections.

In next, the normalization of the k-wave-function adjusts the normalization constant as:

$$\begin{aligned} 1 &= \int_0^a |\psi_k(x)|^2 dx = \left(C_k^{II}\right)^2 \int_0^a \sin^2\left(\frac{n\pi x}{a}\right) dx \\ &= \left(C_k^{II}\right)^2 \int_0^a \frac{1}{2}\left[1 - \cos\left(2\frac{n\pi x}{a}\right)\right] dx \\ &= \left(C_k^{II}\right)^2 \left[\frac{x}{2} - \frac{\sin(2n\pi x/a)}{4(n\pi/a)}\right]_0^a = \left(C_k^{II}\right)^2 \frac{a}{2} \end{aligned} \tag{3.191}$$

resulting in

$$C_k^{II} = \sqrt{\frac{2}{a}} \tag{3.192}$$

and of the quantified stationary eigen-functions

$$\psi_k(x) = \sqrt{\frac{2}{a}} \sin\left(\frac{n\pi x}{a}\right) \tag{3.193}$$

or with its direct temporal (energy dependent) generalization:

$$\psi_k(x,t) = \sqrt{\frac{2}{a}} \sin\left(\frac{n\pi x}{a}\right) \exp\left(-i\frac{E_k t}{\hbar}\right) \tag{3.194}$$

The Figure 3.10 shows how the first few eigen-levels (energy and stationary wave-function) are displaced in the spectrum of free electronic movement within infinite wall cavity modeling the solid – crystal state. One may easily note the fragrant behavior according which the levels are more and more separated as increasing of the quantum number (or excited

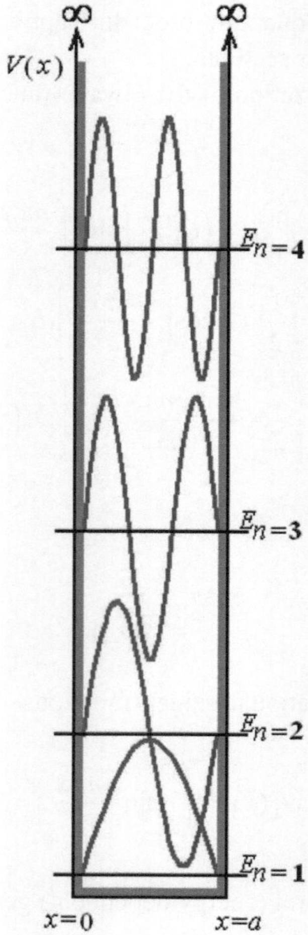

FIGURE 3.10 Illustration of the first four eigen-energy levels together with the associate stationary wave-function for the free electronic spectra in an infinite high well approximating the (1D) periodic solid state.

state), being this in a quite opposite way respecting the atomic spectra in which the levels were approaching more and more to the continuum level.

Such peculiar feature of the solid state gives it the special role it has in the quantum theory and physical-chemical applications. Yet, one may also with this simplified model checking out some interesting consequences.

An important one regards the application of the Heisenberg uncertainty principle:

$$\Delta\hat{x}\Delta\hat{p}_x \geq \frac{1}{2}\hbar \tag{3.195}$$

rewritten with the help of coordinate and momentum expectations:

$$\left[\left\langle \hat{x}^2\right\rangle - \left\langle \hat{x}\right\rangle^2\right]^{1/2}\left[\left\langle \hat{p}_x^2\right\rangle - \left\langle \hat{p}_x\right\rangle^2\right]^{1/2} \geq \frac{1}{2}\hbar \tag{3.196}$$

with the above general k-eigen functions each of the involved expected value becomes:

$$\begin{aligned}
\left\langle \hat{x}\right\rangle &= \int_0^a x\left|\psi_k(x)\right|^2 dx = \frac{2}{a}\int_0^a x\sin^2\left(\frac{n\pi x}{a}\right)dx \\
&= \frac{2}{a}\left[\frac{x^2}{4} - \frac{x\sin(2n\pi x/a)}{4(n\pi/a)} - \frac{\cos(2n\pi x/a)}{8(n\pi/a)^2}\right]_0^a \\
&= \frac{a}{2}
\end{aligned} \tag{3.197}$$

$$\begin{aligned}
\left\langle \hat{x}^2\right\rangle &= \int_0^a x^2\left|\psi_k(x)\right|^2 dx = \frac{2}{a}\int_0^a x^2\sin^2\left(\frac{n\pi x}{a}\right)dx \\
&= \frac{2}{a}\left[\frac{x^3}{6} - \left(\frac{x^2}{4(n\pi/a)} - \frac{1}{8(n\pi/a)^3}\right)\sin\left(\frac{n\pi x}{a}\right) - \frac{x\cos(2n\pi x/a)}{4(n\pi/a)^2}\right]_0^a \\
&= a^2\left(\frac{1}{3} - \frac{1}{2n^2\pi^2}\right)
\end{aligned} \tag{3.198}$$

$$\left\langle \hat{p}_x \right\rangle = \int_0^a \psi_k^*(x) \underbrace{\frac{\hbar}{i}\frac{\partial}{\partial x}}_{\hat{P}_x} \psi_k(x)dx = \frac{2}{a}\frac{\hbar}{i}\frac{n\pi}{a}\int_0^a \sin\left(\frac{n\pi x}{a}\right)\cos\left(\frac{n\pi x}{a}\right)dx$$

$$= -\frac{2}{a}\frac{\hbar}{i}\frac{n\pi}{a}\left[\frac{1}{2(n\pi x/a)}\cos^2\left(\frac{n\pi x}{a}\right)\right]_0^a = 0 \tag{3.199}$$

$$\left\langle \hat{p}_x^2 \right\rangle = -\hbar^2 \int_0^a \psi_k^*(x)\partial_x^2 \psi_k(x)dx = \frac{2}{a}\hbar^2\left(\frac{n\pi}{a}\right)^2 \underbrace{\int_0^a \sin^2\left(\frac{n\pi x}{a}\right)dx}_{a/2} = \hbar^2\left(\frac{n\pi}{a}\right)^2 \tag{3.200}$$

Now the uncertainty relation simplifies to:

$$n^2\pi^2 \geq 9 \tag{3.201}$$

that is valid for all positive integer quantum numbers, implicitly for all free electronic solid states. In other words, the uncertainty principle may be regarded as another reason the electrons in solid state are impeded to exist in fundamental (ground) state with $n = 0$.

From this conclusion one may inferred that since free electrons in solid may exist only on the exited stated they are also "free" to exist on any combination of them. Indeed, because the orthogonality condition among any two different eigen-functions, with $n_1 \neq n_2$,

$$\int_0^a \psi_k^*(x,n_1)\psi_k(x,n_2)dx = \frac{2}{a}\int_0^a \sin\left(\frac{n_1\pi x}{a}\right)\sin\left(\frac{n_2\pi x}{a}\right)dx$$

$$= \frac{1}{a}\left[\frac{\sin\left[(n_1 - n_2)\pi x/a\right]}{n_1 - n_2} - \frac{\sin\left[(n_1 + n_2)\pi x/a\right]}{n_1 + n_2}\right]_0^a = 0 \tag{3.202}$$

no matter if $n_2 = n_1 + 2s$ or $n_2 = n_1 + 2s + 1$ (with $s \in \mathrm{N}$), one may chose any linear combination, for instance one with the two-eigen-functions linear superposition,

$$\psi_{12}(x,t) = \frac{1}{\sqrt{2}}[\psi_1(x,t,n_1) + \psi_2(x,t,n_2)]$$
$$= \frac{1}{\sqrt{a}}\left[\sin\left(\frac{n_1\pi x}{a}\right)\exp\left(-i\frac{E_{n_1}t}{\hbar}\right) + \sin\left(\frac{n_2\pi x}{a}\right)\exp\left(-i\frac{E_{n_2}t}{\hbar}\right)\right] \quad (3.203)$$

stands as a valid solution of the Schrödinger equation for the system (since each individual – orthogonal eigen-wave fulfills it), respecting in addition the normalization condition as well:

$$\int_0^a \psi_{12}^*(x,t)\psi_{12}(x,t)dx = 1 \quad (3.204)$$

as there is immediately to check. Yet, this "collective" behavior of the free electrons in solids is at the foreground for quantum explanation of the superconducting and of other specific quantum properties, for example, the bondonic movement on graphenes, that will be unfold in more detail in Volume IV of this five-volume set.

3.4 SEMICLASSICAL QUANTIFICATION OF ENERGY

The eigen-energies of a system may be evaluated from the semiclassical action quantification, i.e., from equation

$$S(x_1, x_2, E) = \int_{x_1}^{x_2} \sqrt{2m(E - V(x))}dx = \left(n + \frac{1}{2}\right)\pi\hbar \quad (3.205)$$

with x_1, x_2 *being the classical-quantum turning points throughout fulfilling the condition:*

$$E = V(x) \quad (3.206)$$

This postulate tells in fact that the phase of the de Broglie packet is quantified in appropriate manner so that to provide the eigen-energies carried by it; even more, such quantification represent the natural axiological generalization of the historical Bohr and Bohr-Sommerfeld quantifications.

To derive this quantification condition one may recourse to the classical interference between a direct and the reflected wave on a certain point in space so that the stationary waves are produced (those that are appropriately to be quantified), see Figure 3.11.

There is evident from the representation of Figure 3.11 that the classical reflection is equivalent with a tunneling of the wall by the $\lambda / 2$ distance or inducing a delay of $-\pi$ in phase followed by turning the propagation direction. Therefore, the direct and reflected are respectively (in classically sense):

$$u_D = a\sin\left[\frac{2\pi}{\lambda}x - \omega t\right]; \; u_R = a\sin\left[\frac{2\pi}{\lambda}\left(2l - x + \frac{\lambda}{2}\right) - \omega t\right] \quad (3.207)$$

with their phase difference

$$\Delta\theta = \theta_R - \theta_D = \frac{2\pi}{\lambda}\left(2(l-x) + \frac{\lambda}{2}\right) \quad (3.208)$$

controlling the amplitude of the resulted interference wave

$$A = \sqrt{a^2 + a^2 + 2a^2\cos\Delta\theta} = \sqrt{2a^2\left[1 + \cos 2\frac{\pi}{\lambda}\left(2(l-x) + \frac{\lambda}{2}\right)\right]}$$

$$= \sqrt{4a^2\cos^2\left[\frac{\pi}{\lambda}\left(2(l-x) + \frac{\lambda}{2}\right)\right]} = 2a\cos\left[\frac{\pi}{\lambda}\left(2(l-x) + \frac{\lambda}{2}\right)\right] \quad (3.209)$$

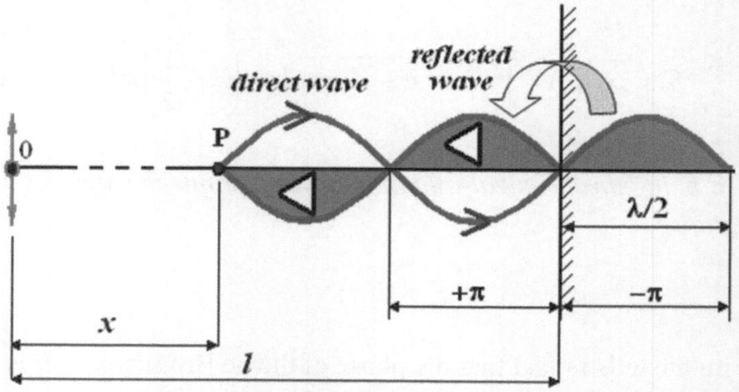

FIGURE 3.11 Schematic representation of the corpuscular-undulatory semi-classical quantification based on the interferential between the direct and reflected wave on an arbitrary point P inside of a cavity (potential) with a given width (see the text for details).

This result tells us that many things:

- the interference amplitude is stationary, i.e., do not depend on time
- the interference amplitude is modulated by the phase

$$\Theta = \pi\left(\frac{2(l-x)}{\lambda}+\frac{1}{2}\right) \tag{3.210}$$

that can be further re-arranged as

$$\Theta = \pi\left(n+\frac{1}{2}\right),\ n \in \mathbf{N} \vee \mathbf{N}^{*} \tag{3.211}$$

noting that when the $(l-x)$ distance is filled with stationary waves (with zero, one, two, etc.) nodes it has to fulfill the natural wave-length relationship, see Figure 3.12

$$l-x = n\frac{\lambda}{2} \tag{3.212}$$

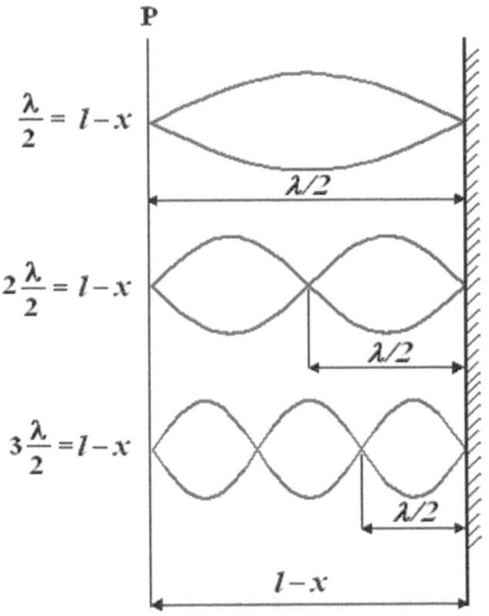

FIGURE 3.12 Illustration of the stationary waves' formation, in different modes, from the direct and reflected waves of Figure 3.11.

Finally, employing (postulating the equivalence of) the actual founded phase for *classical* stationary wave-condition with the *quantum* stationary action related phase of the de Broglie wave-packet

$$\frac{S(p(E))}{\hbar} = \Theta \tag{3.213}$$

in terms of the action functional

$$S(p(E)) = \int_{x_1}^{x_2} p dx = \int_{x_1}^{x_2} \sqrt{2m(E - V(x))} dx \tag{3.214}$$

the quantum-classical passage resulted as the present semi-classical quantification condition.

Application of the present semiclassical quantification method for recovering the consecrated energetic eigen-values for the simple systems of hydrogenic atoms, molecular vibrations and free electrons in solid state is to be in the next unfolded.

3.4.1 HYDROGENIC ATOMIC SYSTEMS

The fundamental problem of hydrogenic atom relies on considering the atomic effective (Coulombic + centrifugal contributions) potential

$$V_Z(r) = -\frac{Ze_0^2}{r} + \frac{c_l^2\hbar^2}{2m_0 r^2} \tag{3.215}$$

with orbital quantization compressed in orbital momentum constant c_l, onto the semiclassical action functional (integral)

$$\begin{aligned}\left(n + \frac{1}{2}\right)\pi\hbar &= \int_{r_{\min}}^{r_{\max}} \sqrt{2m_0\left(E_Z - V_Z(r)\right)} dr \\ &= \int_{r_{\min}}^{r_{\max}} \sqrt{2m_0 E_Z + \frac{2m_0 Ze_0^2}{r} - \frac{c_\theta^2\hbar^2}{r^2}} dr \\ &= \int_{r_{\min}}^{r_{\max}} \sqrt{-A + \frac{2B}{r} - \frac{C}{r^2}} dr \equiv \int_{r_{\min}}^{r_{\max}} f_Z(r) dr \end{aligned} \tag{3.216}$$

providing the following notation were introduced:

$$A = 2m_0|E_Z| = -2m_0 E_Z; \quad B = m_0 Z e_0^2; \quad C = c_l^2 \hbar^2 \tag{3.217}$$

Now, the main problem remains the evaluation of the formal action integral above; it is no trivial job, while an elegant and meaningful method appeals the complex integration technique through identifying the "poles" or "zeros" of the integrated function $f_Z(r)$ respecting the integrand r.

Actually, for the concerned function two poles are identified in the complex plane of Figure 3.13, producing the associated integrals to be solved according with the Cauchy residues theorem with application to the single or to multiple poles as well:

- one in the origin ($r \to 0$) as we have the integral:

$$J_I = \oint_I \underbrace{\frac{1}{r}\sqrt{-Ar^2 + 2Br - C}}_{f_I(r)} dr = +2\pi i \underset{r=0}{\mathbf{Rez}} f_I(r)$$
$$= 2\pi i \lim_{r\to 0}\left[r f_I(r)\right] = 2\pi i\sqrt{-C} = -2\pi\sqrt{C} \tag{3.218}$$

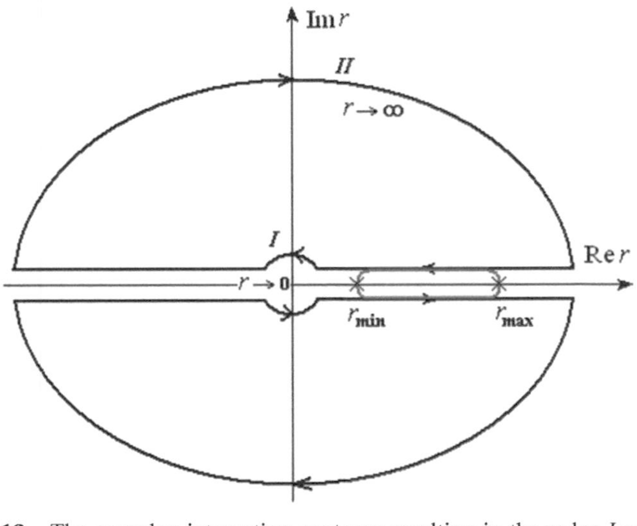

FIGURE 3.13 The complex integration contours resulting in the poles I and II and in the closed integral between r_{min} and r_{max} turning points of atomic hydrogenic action in the semiclassical quantification.

- one at infinity ($r \to \infty$) as we may transform the original integral through the variable change

$$r \to \frac{1}{r} \tag{3.219}$$

to the next one with multiple (double) pole at zero:

$$J_{II} = \oint_{II} \underbrace{\frac{(-1)}{r^2}\sqrt{-A+2Br-Cr^2}}_{f_{II}(r)}\, dr = -2\pi i \, \underset{r^2=0}{\mathbf{Rez}}\, f_{II}(r)$$
$$= -2\pi i \lim_{r\to 0}\left\{\frac{d}{dr}\left[r^2 f_{II}(r)\right]\right\} = -2\pi i\left(\frac{iB}{\sqrt{A}}\right) = 2\pi\frac{B}{\sqrt{A}} \tag{3.220}$$

With these, one can infer from the Figure 3.13 the following integral relationship holds (keeping in mind that the anti-clocking sense is the positive one, and the fact this was already counted in complex computation of integrals on the contours *I* and *II* before):

$$J_I + J_{II} + 2\int_{r_{\max}}^{r_{\min}} f_Z(r)dr = 0 \tag{3.221}$$

leaving with the general valuable result

$$\int_{r_{\min}}^{r_{\max}} \sqrt{-A+\frac{2B}{r}-\frac{C}{r^2}}\,dr = \pi\left(\frac{B}{\sqrt{A}}-\sqrt{C}\right) \tag{3.222}$$

Now we can immediately perform the semiclassical quantification for hydrogenic atoms through calling back the performed notations for *A*, *B* and *C*:

$$\left(n+\frac{1}{2}\right)\hbar = \frac{m_0 Z e_0^2}{\sqrt{2m_0\left|E_Z\right|}} - c_l\hbar \tag{3.223}$$

so that it can be rearranged in the usual Bohr-Schrödinger form:

$$\left|E_Z\right| = \frac{m_0^2 Z^2 e_0^4}{2\hbar^2 n_z^2} \tag{3.224}$$

by the formal quantum number re-labeling

$$n_Z = n + c_l + \frac{1}{2} \tag{3.225}$$

thus proofing the formal agreement of the semiclassical approach with the full quantum treatment. AT the same time it is here the opportunity to stress out that the phenomenological Bohr determination of atomic spectra appears as being really quantum and not semi-classical (as sometimes miss-believed) since providing the exact quantum number dependence; the "true" semiclassical undertake was here unveiled. Yet, the present semiclassical approach gives also the sing on the fact that the atomic quantum numbers of the ground state eigen-values cannot take the zero value in any circumstances, thus being of a certain value also in this respect.

3.4.2 *MOLECULAR VIBRATIONAL SYSTEMS*

In the vibrational molecular case the potential takes the usual 1D form (3.139), imposing the with the classical (to quantum) turning points

$$|x_\omega| = \sqrt{\frac{2E_\omega}{m\omega^2}} \tag{3.226}$$

while the eigen-energy enters the relationship

$$E_\omega = V_\omega(x_\omega) = \frac{1}{2}m\omega^2 x_\omega^2 \tag{3.227}$$

In these conditions, the semiclassical quantization postulate specializes successively as:

$$\begin{aligned}\left(n+\frac{1}{2}\right)\pi\hbar &= \int_{-x_\omega}^{+x_\omega}\sqrt{2m\left(E_\omega - V_\omega(x)\right)}dx \\ &= m\omega\int_{-x_\omega}^{+x_\omega}\sqrt{x_\omega^2 - x^2}\,dx = 2m\omega\int_{0}^{x_\omega}\sqrt{x_\omega^2 - x^2}\,dx\end{aligned} \tag{3.228}$$

that by means of the variable change

$$x = x_\omega \sin\vartheta,\ dx = x_\omega \cos\vartheta d\vartheta,\ x = x_\omega : \vartheta = \frac{\pi}{2} \tag{3.229}$$

it further becomes

$$\left(n+\frac{1}{2}\right)\pi\hbar = 2m\omega x_\omega^2 \int_0^{\pi/2} \cos^2\vartheta d\vartheta = m\omega x_\omega^2 \underbrace{\int_0^{\pi/2} (1+\cos 2\vartheta) d\vartheta}_{\pi/2} = \frac{\pi m\omega x_\omega^2}{2} \tag{3.230}$$

Finally, by considering the turning point – eigen-energy equation from above the semiclassical quantization rewrites as

$$\left(n+\frac{1}{2}\right)\pi\hbar = \frac{\pi E_\omega}{\omega} \tag{3.231}$$

yielding the well-known energy quantification of the vibrational (molecular) states in an exact manner, even through the application of the formal semiclassical quantification (3.149).

3.4.3 GENERALIZED $|x|^\alpha$ POTENTIAL SYSTEMS

The semiclassical quantification through the action integral (functional) gives the possibility of treating the free electrons in solid state in a more general context, i.e., through considering the 1D-potential

$$V_k(x) = k|x|^\alpha = \begin{cases} k & , \alpha \to 0 \ \ldots\ I: continuous\ \ spectra \\ k|x| & , \alpha = 1 \ \ldots\ II: linear\ \ potential \\ kx^2 & , \alpha = 2 \ \ldots\ III: harmonic\ \ potential \\ \infty & , \alpha \to \infty \ \ldots\ IV: solid\ \ well\ \ potential \end{cases} ,k \in \mathbf{R} \tag{3.232}$$

However, before specializing to all these cases the general approach will be firstly undertaken, noting that the classical-quantum turning points are obtained from the equality

$$E_k = V_k(x_k) \tag{3.233}$$

with the expressions:

$$|x_k| = \left(\frac{E_k}{k}\right)^{1/\alpha} \tag{3.234}$$

With this we have for quantization the chain equivalences:

$$\begin{aligned}\left(n+\frac{1}{2}\right)\pi\hbar &= \int_{-x_k}^{+x_k} \sqrt{2m(E_k - V_k(x))}dx \\ &= 2\int_0^{|x_k|} \sqrt{2m(E_k - V_k(x))}dx = 2\sqrt{2m}\int_0^{|x_k|} \sqrt{E_k\left(1-\frac{k|x|^\alpha}{E_k}\right)}dx\end{aligned} \tag{3.235}$$

suggesting therefore the variable substitution:

$$\frac{k|x|^\alpha}{E_k} = y = \begin{cases} 0 \ , \ x = 0 \\ 1 \ , \ x = |x_k| \end{cases} \tag{3.236}$$

$$|x| = k^{-1/\alpha} E_k^{1/\alpha} y^{1/\alpha}; \quad dx = \alpha^{-1} k^{-1/\alpha} E_k^{1/\alpha} y^{1/\alpha - 1} dy \tag{3.237}$$

with the help of which it transforms into

$$\left(n+\frac{1}{2}\right)\pi\hbar = 2\alpha^{-1} k^{-1/\alpha} E_k^{1/\alpha+1/2} \sqrt{2m_0} \int_0^1 y^{1/\alpha-1}(1-y)^{1/2} dy \tag{3.238}$$

involving an integral that is of Euler – β (or B) type (see Appendix A.2):

$$\beta(p,q) = \int_0^1 y^{p-1}(1-y)^{q-1} dy \tag{3.239}$$

for the individuation

$$p = \frac{1}{\alpha}, q = \frac{3}{2} \tag{3.240}$$

Such integrals may be analytically solved by making recourse to the Dirichlet relationship (see Appendix A.2):

$$\beta(p,q)=\frac{\Gamma(p)\Gamma(q)}{\Gamma(p+q)} \tag{3.241}$$

in terms of Gamma-Euler functions

$$\Gamma(p)=\int_0^\infty y^{p-1}e^{-y}dy \tag{3.242}$$

that nevertheless can be evaluated in the recursive manner by the rule:

$$\Gamma(p+1)=p\Gamma(p) \tag{3.243}$$

with

$$\Gamma\left(\frac{1}{2}\right)=\sqrt{\pi} \tag{3.244}$$

all in all there follows that the actual integral rearranges as

$$\int_0^1 y^{1/\alpha-1}(1-y)^{1/2}\,dy=\beta\left(\frac{1}{\alpha},\frac{3}{2}\right)=\frac{\sqrt{\pi}}{2}\frac{\Gamma(1/\alpha)}{\Gamma(1/\alpha+3/2)} \tag{3.245}$$

this way providing the above semiclassical quantification equation under the form:

$$\left(n+\frac{1}{2}\right)\sqrt{\pi}\hbar=\alpha^{-1}k^{-1/\alpha}E_k^{(2+\alpha)/2\alpha}\sqrt{2m}\frac{\Gamma(1/\alpha)}{\Gamma(1/\alpha+3/2)} \tag{3.246}$$

from where the eigen-energy is yielded:

$$E_{k,\alpha}=\left[\alpha\frac{\Gamma(1/\alpha+3/2)}{\Gamma(1/\alpha)}k^{1/\alpha}\sqrt{\frac{\pi}{2m_0}}\left(n+\frac{1}{2}\right)\hbar\right]^{\frac{2\alpha}{\alpha+2}} \tag{3.247}$$

We are going now to apply this result to various potential cases.

1. For the limit $\alpha\to 0$ the energy fills all the interval spectra above the potential:

$$E_{k,\alpha\to 0} \in [k, +\infty) \tag{3.248}$$

2. For the case $\alpha = 1$ the linear ("in V letter" potential) impose the corresponding eigen-energies of the form

$$E_{k,\alpha=1} = \left[\frac{3\pi}{4\sqrt{2m}}\left(n+\frac{1}{2}\right)k\hbar\right]^{\frac{2}{3}} \tag{3.249}$$

taking act of the Gamma-Euler particular values (see also the Appendix A.2):

$$\Gamma(1+3/2) = (3/2)\Gamma(3/2) = (3/2)(1/2)\Gamma(1/2) = \sqrt{\pi}/2 \tag{3.250a}$$

$$\Gamma(1) = 1 \tag{3.250b}$$

3. For the harmonic potential, with $\alpha = 2$, through replacing the involved new Gamma-Euler function

$$\Gamma(1/2+3/2) = \Gamma(2) = \Gamma(1+1) = \Gamma(1) = 1 \tag{3.251}$$

among the appropriate k-identification, namely

$$k = \frac{m\omega^2}{2} \tag{3.252}$$

the correct vibrational eigen-energy is getting out:

$$E_{k=m\omega^2/2,\alpha=2} = \left(n+\frac{1}{2}\right)\omega\hbar \tag{3.253}$$

4. For the free electrons in an infinite well model of solid state, i.e., for treating the infinite high potential from the limit $\alpha \to \infty$, one needs to note that since the recursive rule of the Gamma-Euler function written for inverse arguments there can be inferred that

$$\Gamma\left(\frac{1}{\alpha}\right)\xrightarrow{\alpha\to\infty}\alpha \tag{3.254}$$

from the equivalent limits:

$$\frac{1}{\lim\limits_{\alpha\to\infty}\Gamma(1/\alpha)}=\frac{\Gamma(1)}{\lim\limits_{\alpha\to\infty}\Gamma(1/\alpha)}=\lim_{\alpha\to\infty}\frac{\Gamma(1+1/\alpha)}{\Gamma(1/\alpha)}=\lim_{\alpha\to\infty}\frac{(1/\alpha)\Gamma(1/\alpha)}{\Gamma(1/\alpha)}=\frac{1}{\lim\limits_{\alpha\to\infty}\alpha} \tag{3.255}$$

This, along the other appearing limit:

$$\lim_{\alpha\to\infty}\Gamma\left(\frac{3}{2}+\frac{1}{\alpha}\right)=\Gamma\left(\frac{3}{2}\right)=\frac{\sqrt{\pi}}{2} \tag{3.256}$$

together the specific *k*-identification relating the width (say "*a*") of the well,

$$k=\left(\frac{1}{a}\right)^{\alpha} \tag{3.257}$$

there is immediate to obtain the formal quantified eigen-energies of free electrons in solids

$$E_{k=a^{-\alpha},\alpha\to\infty}=\frac{\pi^2\hbar^2}{2m_0a^2}n_*^2 \tag{3.258}$$

with the quantum numbers reshaped from the notation:

$$n_*^2=\left[\frac{1}{2}\left(n+\frac{1}{2}\right)\right]^2 \tag{3.259}$$

in order the traditional solid state formula to be formally regained.

However, worth remarking in the final that the semiclassical quantification, although always providing correct eigen-energies form, is only sometimes *exact*, as is the case of vibrational quantification, while requiring the quantum number re-notation for matching with the results given by the rigorous solution of the Schrödinger equation. Such limit is the reflection of the Bohr and Bohr-Sommerfeld semiclassical quantification treatment.

3.5 VARIATIONAL WAVE-FUNCTION AND ENERGIES

The stationary Schrödinger equation may be integrated to its eigen-values by means of the variational principle respecting the minimizing of the total (eigen) energy of the system:

$$\delta E = 0, E = \frac{\int \psi^* \widehat{H} \psi \, d\Gamma}{\int \psi^* \psi \, d\Gamma} \tag{3.260}$$

with ψ being a trial stationary wave-function suitable for the concerned system, in accordance with the previous enounced quantum postulates.

With this principle, actually all, natural systems may be quantum mechanically treated with the aid of trial-and-optimized wave-function to determine the correspondent eigen-energies. In next some of the most representative systems are to be accordingly treated from nuclear, to atomic, to molecular, and to solid-state level or matter's organization.

3.5.1 *HYDROGEN'S QUANTUM GROUND LEVEL*

Let be the stationary radial trial wave-function with two-parameters:

$$\psi_Z(C,\alpha,r) = C\exp(-\alpha r) \tag{3.261}$$

to be determined throughout imposing it the normalization and eigen-energy variational constraints.

Based on the general "Slater" type formula (see Appendix A.2)

$$I_k(m) = \int_0^\infty r^k e^{-mr} dr = \frac{k!}{m^{k+1}}, \quad \forall k \in \mathbf{N} \,\&\, m \in \mathbf{C}, \mathrm{Re}(m) > 0 \tag{3.262}$$

the wave-function radial normalization gives:

$$1 = \int_0^\infty \psi_Z^* \psi_Z r^2 dr = C^2 \int_0^\infty e^{-2\alpha r} r^2 dr = C^2 \frac{2!}{(2\alpha)^3} = \frac{C^2}{4\alpha^3} \tag{3.263}$$

from where the normalization constants is yield and the trial wave-function takes the intermediate form:

$$\psi_Z(\alpha,r) = 2\alpha^{3/2}\exp(-\alpha r) \tag{3.264}$$

to be further considered for employing variational principle on the eigen-energy:

$$E_Z(\alpha) = \int_0^\infty \psi_Z^*(\alpha,r)\widehat{H}\psi_Z(\alpha,r)r^2 dr \tag{3.265}$$

with the hydrogenic Hamiltonian

$$\widehat{H}_Z = -\frac{\hbar^2}{2m_0}\nabla^2 - \frac{Ze_0^2}{r},\ e_0^2 = \frac{e^2}{4\pi\varepsilon_0} \tag{3.266}$$

to be fully considered in radial-spherical coordinates; however, for the Laplacian term the involved integral can be easier evaluated through applying the Gauss surface-to-volume integral transformation (law) while counting the null contribution of the wave-function on the infinite expanded integrated surface; thus one can write:

$$0 = \int_{\Sigma\to\infty} \left(\psi^*\nabla\psi\right) d\Sigma_V = \int \nabla\left(\psi^*\nabla\psi\right) dV = \int \psi^*\nabla^2\psi dV + \int \nabla\psi^*\nabla\psi dV \tag{3. 267}$$

from where follows the relationship:

$$\int \psi^*\nabla^2\psi dV = -\int \left|\nabla\psi\right|^2 dV \tag{3.268}$$

that has the immediate correspondent in radial operators:

$$\int_0^\infty \psi^*\nabla_r^2\psi\, r^2 dr = -\int_0^\infty \left|\partial_r\psi\right|^2 r^2 dr \tag{3.269}$$

With this the above average energy integral becomes:

$$\begin{aligned}E_Z(\alpha) &= 2\frac{\alpha^5\hbar^2}{m_0}\int_0^\infty e^{-2\alpha r}r^2dr - 4\alpha^3 Ze_0^2\int_0^\infty e^{-2\alpha r}rdr \\ &= 2\frac{\alpha^5\hbar^2}{m_0}\frac{2!}{2^3\alpha^3} - 4\alpha^3 Ze_0^2\frac{1}{2^2\alpha^2} \\ &= \frac{\alpha^2\hbar^2}{2m_0} - \alpha Ze_0^2 \end{aligned} \tag{3.270}$$

while through the variational principle

$$0 = \partial_\alpha E_Z(\alpha) = \frac{\alpha\hbar^2}{m_0} - Ze_0^2 \tag{3.271}$$

one finally gets also the α-parameter with the form recuperating the inverse of first Bohr radius:

$$\alpha = \frac{Ze_0^2 m_0}{\hbar^2} = \frac{Z}{a_0} \tag{3.272}$$

with the help of which either the first radial wave-function expression

$$\psi_Z(r) = 2(Z/a_0)^{3/2}\exp(-Zr/a_0) \tag{3.273}$$

as well as the first Bohr-(eigen) energy (for the first quantum number in hydrogen atoms)

$$E_{Z,n=1} = -\frac{Z^2 e_0^4 m_0}{2\hbar^2} = -\frac{Z^2 e^4 m_0}{2(4\pi\varepsilon_0)^2\hbar^2} = -\frac{Z^2 e^4 m_0}{8\varepsilon_0^2 h^2} \tag{3.274}$$

are obtained, in fully agreement with previous phenomenological Bohr approach (1.82).

3.5.2 VIBRATIONAL GROUND LEVEL

The ω-vibrational state of a molecular 1D-system is described by the Hamiltonian:

$$\widehat{H}_{\omega} = -\frac{\hbar^2}{2m}\frac{\partial^2}{\partial x^2} + \frac{1}{2}m\omega^2 x^2 \tag{3.275}$$

while the stationary appropriate wave-function may be described by the two-parameters function:

$$\psi_{\omega}(c_1, c_2, x) = c_1 \exp\left(-c_2 x^2\right),\ c_1, c_2(>0) \in \mathbf{R} \tag{3.276}$$

so that carrying the geometry of the parabolic potential that determines it, with the two constants to be determined by the two quantum constrains of the direct normalization of the trial wave-function followed by application of the variational principle for the eigen-energy.

Starting with fulfilling the normalization condition for the trial wave-function, we have:

$$1 = \int_{-\infty}^{+\infty} \psi_{\omega}^{*}\psi_{\omega} dx = c_1^2 \int_{0}^{\infty} e^{-2c_2 x^2} dx = c_1^2 \sqrt{\frac{\pi}{2c_2}}$$

$$\Rightarrow c_1 = \left(\frac{2c_2}{\pi}\right)^{1/4} \tag{3.277}$$

where we have considered the 0th order Poisson integral (see Appendix A.2).

Going now to compute the trial vibrational energy, we calculate successively:

$$\widehat{H}_{\omega}\psi_{\omega} = -\frac{\hbar^2}{2m}\partial_x^2\left(c_1 e^{-c_2 x^2}\right) + \frac{c_1}{2}m\omega^2 x^2 e^{-c_2 x^2}$$

$$= c_1 c_2 \frac{\hbar^2}{m}\left(e^{-c_2 x^2} - 2c_2 x^2 e^{-c_2 x^2}\right) + \frac{c_1}{2}m\omega^2 x^2 e^{-c_2 x^2},$$

$$\psi_{\omega}^{*}\widehat{H}_{\omega}\psi_{\omega} = c_1^2 c_2 \frac{\hbar^2}{m}\left(e^{-2c_2 x^2} - 2c_2 x^2 e^{-2c_2 x^2}\right) + \frac{c_1^2}{2}m\omega^2 x^2 e^{-2c_2 x^2}$$

$$= c_1^2 c_2 \frac{\hbar^2}{m} e^{-2c_2 x^2} + \left(\frac{c_1^2}{2}m\omega^2 - 2c_1^2 c_2^2 \frac{\hbar^2}{m}\right) x^2 e^{-2c_2 x^2} \tag{3.278a}$$

$$E_\omega = \int_{-\infty}^{+\infty} \psi_\omega^* \widehat{H}_\omega \psi_\omega dx$$
$$= c_1^2 c_2 \frac{\hbar^2}{m}\sqrt{\frac{\pi}{2c_2}} + \left(\frac{c_1^2}{2} m\omega^2 - 2c_1^2 c_2^2 \frac{\hbar^2}{m}\right)\frac{1}{4c_2}\sqrt{\frac{\pi}{2c_2}}$$
$$= c_2 \frac{\hbar^2}{2m} + \frac{1}{c_2}\frac{m\omega^2}{8} \tag{3.278b}$$

where the 0th and 2nd order Poisson integrals were used (Appendix A.2) along the replacement of the above expression for c_1. Now, the variational principle on this energy leads with the c_2 result as well:

$$0 = \partial_{c_2} E_\omega(c_2) = \frac{\hbar^2}{2m} - \frac{1}{c_2^2}\frac{m\omega^2}{8},$$

$$\Rightarrow c_2 = \frac{m\omega}{2\hbar}, \quad \Rightarrow c_1 = \left(\frac{2c_2}{\pi}\right)^{1/4} = \left(\frac{m\omega}{\hbar\pi}\right)^{1/4} \tag{3.279}$$

so that the fundamental (ground state) energy

$$E_0(\omega) = \frac{1}{2}\hbar\omega \tag{3.280}$$

together with the associate eigen-function

$$\psi_0(\omega, x) = \left(\frac{m\omega}{\hbar\pi}\right)^{1/4} \exp\left(-\frac{m\omega}{2\hbar}x^2\right) \tag{3.281}$$

are furnished in fully agreement with the previous general quantum eigen-energies and -functions determinations for harmonic oscillator; thus proving also by this example the reliability of variational principle to provide both fundamental or ground state energy and its wave-function, in a consistent quantum mechanically manner.

3.5.3 GROUND STATE PARADOX OF FREE ELECTRONS IN SOLIDS

For the solid states the infinitely high potential barrier stands as a valid approximation for the electronic stationary behavior; however, this is equivalently to state that electrons are "free" outside the barrier, and evolving as stationary waves, with trial paradigmatic two-parameter trigonometric 1D form:

$$\psi_k(A,x) = A\sin(kx) \tag{3.282}$$

with the associate free electronic Hamiltonian:

$$\widehat{H}_k = -\frac{\hbar^2}{2m_0}\partial_x^2 \tag{3.283}$$

Note that if one would like to consider the Hamiltonian with some potential that mimics the infinite barrier in asymptotic limit, i.e., as is the case of $|x/a|^\alpha, \alpha \to \infty$, with a – the width of the free barrier, will soon conclude that the corresponding term in energy is not divergent in a very narrow domain, namely for $a \in [\lambda/2, \lambda/12]$ that is in obvious contradiction with forming of stationary waves in the well; therefore the only acceptable Hamiltonian in the case of infinite well is that restricted to the kinetic term only.

Unfolding the ordinary variational procedure, one starts with imposing the normalization condition on the trial wave-function:

$$\begin{aligned} 1 &= \int_0^a \psi_k^* \psi_k dx = A^2 \int_0^a \sin^2(kx) dx = \frac{A^2}{2}\int_0^a [1-\cos(2kx)] dx \\ &= \frac{A^2}{2}\left[a - \frac{1}{2k}\sin(2ka)\right] \end{aligned} \tag{3.284}$$

assisting the two constants' relationship in the form:

$$A = \sqrt{\frac{2}{a - \frac{1}{2k}\sin(2ka)}} = \sqrt{\frac{2}{a}}\sqrt{\frac{1}{1 - \frac{\sin(2ka)}{2ka}}} \tag{3.285}$$

On the other side, the energy computation with the A constant expression, i.e., in the normalization condition of the wave-function, yields the paradoxical result:

$$E_k = \int_0^a \psi_k^* \widehat{H}_k \psi_k dx$$
$$= A^2 k^2 \frac{\hbar^2}{2m_0} \int_0^a \sin^2(kx)dx = A^2 k^2 \frac{\hbar^2}{2m_0} \frac{1}{2}\left[a - \frac{1}{2k}\sin(2ka)\right] = \frac{\hbar^2 k^2}{2m_0} \quad (3.286)$$

since being in accordance with de Broglie quantization provides through variation

$$0 = \partial_k E_k = k\frac{\hbar^2}{m_0} \quad (3.287)$$

the solution

$$k = 0 \quad (3.288)$$

that produces the infinite amplitude in the above expression:

$$\lim_{k\to 0} A = \sqrt{\frac{2}{a}}\sqrt{\frac{1}{1 - \lim_{k\to 0}\frac{\sin(2ka)}{2ka}}} = \sqrt{\frac{2}{a}}\sqrt{\frac{1}{1-1}} = \infty \quad (3.289)$$

and consequently the "strange" couple of eigen-solutions:

$$\begin{cases} \psi_{k=0}(x) = \infty \cdot 0 = ? \\ E_{k=0} = 0 \end{cases} \quad (3.290)$$

leaving with the idea that electrons in the fundamental (ground) solid state are "hidden": they have no observable energy (or optimized –ground state energy) although they may have an un-determinate existence by means of the "associate" wave-function.

However, beside the fact that we made the first encounter with the "quantum hidden state" realization, the present paradox is solved by invoking other quantum postulate, namely that of wave-function continuity at the extremities of the infinite well:

$$\psi_k(x = a) = \psi_k(x > a) \quad (3.291)$$

that may be regarded also as a sort of wave-function variational principle

$$\delta\psi_k\big|_{x=a} = 0 \tag{3.292}$$

at the domain existence limit; explicitly looks as

$$A\sin(ka) = 0 \tag{3.293}$$

from where follows the entire spectra of k -quantification

$$k = \frac{\pi}{a}n,\ n = 1,2,... \tag{3.294}$$

with excluded zero (or ground state, or hidden state) solution above.

In these conditions the amplitude of the eigen-function will be proportional with the square root of the inverse of the well's width

$$A = \sqrt{\frac{2}{a}} \tag{3.295}$$

while the couple of finite and non-zero, eigen-solution of the electronic movement in the solid states (modeled as an infinite well) take the consecrated (already proved) forms:

$$\begin{cases} \psi_k(x) = \sqrt{\dfrac{2}{a}}\sin\left(n\pi\dfrac{x}{a}\right) \\ E_k = \dfrac{\hbar^2\pi^2}{2m_0a^2}n^2 \end{cases} \tag{3.296}$$

Yet, the case of solid states reveals the important idea that electronic behavior is at least forbidden in their truly ground state, or it happens in a hidden manner – this may be the quantum state that when approached to allow the super-conductivity phenomena; equivalently, one can say that since electrons in solid state are normally situated in "excited" states this may be the natural basic explanation for their propensity for conduction and metallic properties. All these ideas will be reloaded in a more depth and analytical regard in appropriate further sections and volume of this five-volume book.

Nevertheless, for the moment we remain with idea that, in a way or other, the variational principle (for energy or wave-function) are the necessary and sufficient requisites in order to solve the quantum eigen-problems for the ground or near ground sates.

3.6 CAUSAL QUANTUM EVOLUTION

Quantum states $|\ \rangle$ *may be transformed one into other by the causal effect of a quantum evolution unitary operator,*

$$\widehat{U}(\Delta t)\widehat{U}^{+}(\Delta t)=1 \tag{3.297}$$

with the property:

$$\widehat{U}(0)=\widehat{U}^{+}(0)=\hat{1} \tag{3.298}$$

employing the (unperturbed) evolution equation

$$|t+\Delta t\rangle=\widehat{U}(\Delta t)|t\rangle \tag{3.299}$$

This postulate allows the consistent formulation of the Schrödinger, Interaction, and the Heisenberg pictures, the treatment of time dependent perturbations, as well as the description of the quantum events by means of the so-called propagators (Green functions) linking them in a causal manner.

All these quantum aspects of evolution will be cast in the sequels.

3.6.1 SCHRLIDINGER'S PICTURE

We derive the Schrödinger equation in behalf of the evolution operator; for an infinitesimal small evolution time-interval ($\Delta t \to 0$) the evolution operator takes the form:

$$\widehat{U}(\Delta t)=\hat{1}-i\widehat{\omega}\Delta t \tag{3.300}$$

that is already recognized as being of a time-perturbation form respecting the unitary (self-symmetric or eigen-) state. Yet the perturbation term is driven by the $\widehat{\omega}$ operator that has to be hermitic

$$\hat{\omega}^{+} = \hat{\omega} \tag{3.301}$$

in order to assure the unitarity constraint of the evolution operator

$$\hat{1} = \hat{U}(\Delta t)\hat{U}^{+}(\Delta t) = \left(\hat{1} - i\hat{\omega}\Delta t\right)\left(\hat{1} + i\hat{\omega}^{+}\Delta t\right) \cong \hat{1} + \left(\hat{\omega}^{+} - \hat{\omega}\right)\Delta t \tag{3.302}$$

where the term of superior (second) order in time interval has been neglected.

The hermiticity of the $\hat{\omega}$ operator may relating it with the hermiticity of the Hamiltonian of the system by a conversion constant, here identified as the Planck constant,

$$\hat{H}_0 = \hbar\hat{\omega} \tag{3.303}$$

to yield for the evolution equation to

$$|t + \Delta t\rangle = \left(\hat{1} - \frac{i}{\hbar}\hat{H}_0\Delta t\right)|t\rangle = |t\rangle - \frac{i}{\hbar}\hat{H}_0\Delta t|t\rangle \tag{3.304}$$

or by a direct rearrangement to the form:

$$i\hbar\frac{|t + \Delta t\rangle - |t\rangle}{\Delta t} = \hat{H}_0|t\rangle \tag{3.305}$$

which within the limit $\Delta t \to 0$ regains the Schrödinger (temporal) equation

$$i\hbar\partial_t|t\rangle = \hat{H}_0|t\rangle \tag{3.306}$$

The consequences of this approach are multiple and fascinating:

- the (Bohr) stationary state:

$$\hat{H}_0|t_n\rangle = \hbar\hat{\omega}|t_n\rangle = \hbar\omega_n|t_n\rangle \equiv \varepsilon_n|t_n\rangle \tag{3.307}$$

carrying the structural information about their wave-corpuscular equivalence;

- the unperturbed state propagation as *harmonic plane waves*

$$\left|(t + \Delta t)_n\right\rangle = e^{-i\omega_n\Delta t}|t_n\rangle \tag{3.308}$$

for the general evolution operator is considered

$$\widehat{U}_0(\Delta t_n) = \exp\left(-i\widehat{\omega}_n \Delta t\right) = \exp\left(-\frac{i}{\hbar}\widehat{H}_0 \Delta t\right) \quad (3.309)$$

for whatever time interval of evolution, or with ending times explicitly as:

$$\widehat{U}_0(t, t_0) = e^{-\frac{i}{\hbar}\widehat{H}_0(t-t_0)} \quad (3.310)$$

- the evolution operator equation may be derived since replacing the specific state evolution expression

$$|t\rangle = \widehat{U}_0(t, t_0)|t_0\rangle \quad (3.311)$$

into the Schrödinger equation:

$$i\hbar\left[\partial_t \widehat{U}(t, t_0)\right]|t_0\rangle = \widehat{H}_0 \widehat{U}_0(t, t_0)|t_0\rangle \quad (3.312)$$

to get it out as:

$$i\hbar\partial_t \widehat{U}_0(t, t_0) = \widehat{H}_0 \widehat{U}_0(t, t_0) \quad (3.313)$$

This last equation opens the discussion about the evolution operator expression (or solution) if instead of a constant (stationary) Hamiltonian $\widehat{H}_0$ one has to treat a time-dependent Hamiltonian, $\widehat{H}(t)$. At this point two possibilities may be approached.

On is to consider the whole time-dependent Hamiltonian action through equation:

$$\partial_t \widehat{U}(t, t_0) = -\frac{i}{\hbar}\widehat{H}(t)\widehat{U}(t, t_0) \quad (3.314)$$

leading with the operatorial *formal* solution:

$$\widehat{U}(t, t_0) = \hat{1} - \frac{i}{\hbar}\int_{t_0}^{t} \widehat{H}(\tau)\widehat{U}(\tau, t_0)d\tau \quad (3.315)$$

in accord with the initial condition for the evolution operator condition:

$$\widehat{U}(t_0, t_0) = \hat{1} \quad (3.316)$$

Unfortunately this solution is not a solution, but another equation for evolution operator, this time of integral type. Yet, it allows the series expansion into the form:

$$\widehat{U}(t,t_0)=\hat{1}+\sum_{k=1}^{\infty}\widehat{U}^{(k)}(t,t_0) \tag{3.317}$$

with

$$\widehat{U}^{(k)}(t,t_0)=-\frac{i}{\hbar}\int_{t_0}^{t}\widehat{H}(\tau)\widehat{U}^{(k-1)}(\tau,t_0)d\tau \tag{3.318}$$

For exemplification, the two terms of the series looks like:

$$\widehat{U}^{(1)}(t,t_0)=-\frac{i}{\hbar}\int_{t_0}^{t}\widehat{H}(\tau)d\tau \tag{3.319a}$$

$$\widehat{U}^{(2)}(t,t_0)=-\frac{i}{\hbar}\int_{t_0}^{t}\widehat{H}(\tau)\widehat{U}^{(1)}(\tau,t_0)d\tau=\left(-\frac{i}{\hbar}\right)^2\int_{t_0}^{t}d\tau\widehat{H}(\tau)\int_{t_0}^{\tau}d\tau'\widehat{H}(\tau'),\quad \tau>\tau' \tag{3.319b}$$

…

from where appears the idea (the problem) of seeing the quantum evolution as a series of intermediate shorter evolutions, i.e., between a series of intermediate quantum events, that are linked in a given chronology (causality). Therefore, worth introducing the so-called *chronological product* among two time-dependent operators as:

$$\widehat{T}\left[\widehat{A}(t_1)\widehat{B}(t_2)\right]=\Theta(t_1-t_2)\widehat{A}(t_1)\widehat{B}(t_2)+\Theta(t_2-t_1)\widehat{B}(t_2)\widehat{A}(t_1) \tag{3.320}$$

in terms of the Heaviside-step function:

$$\Theta(\Delta t)=\begin{cases}1 & ,\ \Delta t>0\\ 0.5 & ,\ \Delta t=0\\ 0 & ,\ \Delta t<0\end{cases} \tag{3.321}$$

With the help of chronological product also the problem of non-commutativity of observables (Hamiltonians) at different times is covered. Returning to the evolution operator of the second order, the double integral has to be counted out in the following manner: since the integration is made upon two directions the result has to be divided by (2!) in order to not over-count the integral contributions; thus the second order contribution result looks like:

$$\hat{U}^{(2)}(t,t_0)=\frac{1}{2!}\left(-\frac{i}{\hbar}\right)^2\int_{t_0}^{t}\int_{t_0}^{\tau}\hat{T}\left[\hat{H}(\tau)\hat{H}(\tau')\right]d\tau d\tau' \quad (3.322)$$

while the general series unfolds as:

$$\hat{U}(t,t_0)=\hat{1}+\sum_{k=1}^{\infty}\frac{1}{k!}\left(-\frac{i}{\hbar}\right)^k\int_{t_0}^{t}...\int_{t_0}^{\tau_k}\hat{T}\left[\hat{H}(\tau_1)...\hat{H}(\tau_k)\right]d\tau_1...d\tau_k \quad (3.323)$$

or under symbolic resumation:

$$\hat{U}(t,t_0)=\hat{T}\exp\left(-\frac{i}{\hbar}\int_{t_0}^{t}\hat{H}(\tau)d\tau\right) \quad (3.324)$$

The problem with this way of treating the evolution operator solution for the time-dependent Hamiltonian, although elegant, is little practicable since the theory work only if the series is entirely considered; if only few terms are considered then, practically an infinity of terms from the global Hamiltonian are omitted and the description blows up!

The second way of treating the temporal Hamiltonians is doing perturbations over the un-perturbed (stationary) Hamiltonian, producing a quantum evolution as depicted in Figure 3.14.

Actually, the problem is to describe the perturbed wave-function $|\psi(t)\rangle$ from the information contained within the unperturbed state $|t\rangle$, having both time-evolution. To begin, the evolution of the stationary states (under the Hamiltonian $\hat{H}_0$) is considered as represented on the ortho-normalized

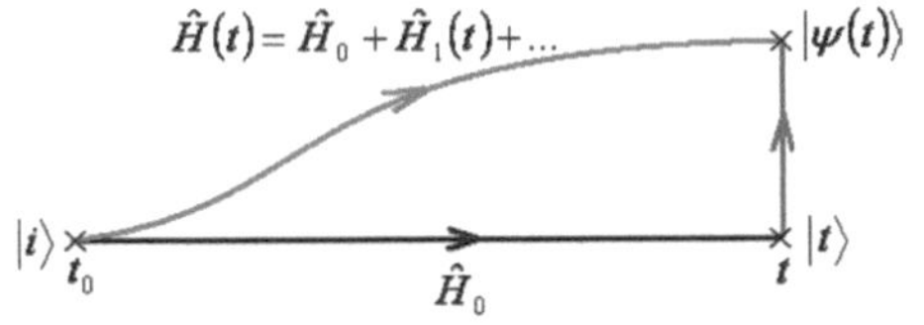

FIGURE 3.14 Illustration of schematic relationships between states and Hamiltonians in unperturbed and perturbed quantum evolutions.

set of (continuous) states, say $\left\{|a\rangle, \widehat{H}_0|a\rangle = E(a)|a\rangle\right\}$, so that we may write for an arbitrary time that its state is expanded as:

$$
\begin{aligned}
|t\rangle &= \int_{\{|a\rangle\}} |a\rangle\langle a|t\rangle d\mu(a) \\
&= \int_{\{|a\rangle\}} |a\rangle\langle a|\widehat{U}_0(t,t_0)|t_0\rangle d\mu(a) \\
&= \int_{\{|a\rangle\}} |a\rangle\langle a|e^{-\frac{i}{\hbar}\widehat{H}_0(t-t_0)}|t_0\rangle d\mu(a) \\
&= \int_{\{|a\rangle\}} |a\rangle e^{-\frac{i}{\hbar}E(a)(t-t_0)}\langle a|t_0\rangle d\mu(a)
\end{aligned}
\tag{3.325}
$$

from where there follows at once (comparing the start and end of the equality chain) the Fourier coefficients are

$$\langle a|t\rangle = e^{-\frac{i}{\hbar}E(a)(t-t_0)}\langle a|t_0\rangle \tag{3.326}$$

In the same manner for the perturbed wave-function evolution the correspondent representation will be:

$$|\psi(t)\rangle = \int_{\{|a\rangle\}} |a\rangle\langle a|\psi(t)\rangle d\mu(a) \tag{3.327}$$

with the formal Fourier coefficients:

$$\langle a|\psi(t)\rangle = e^{-\frac{i}{\hbar}E(a)(t-t_0)}c(a,t) \tag{3.328}$$

while searching for the coefficients $c(a,t)$. For that, let's employ the Schrödinger equation

$$i\hbar\partial_t|\psi(t)\rangle = \left[\widehat{H}_0 + \widehat{H}_1(t)\right]|\psi(t)\rangle \tag{3.329}$$

to the equivalent forms:

$$i\hbar\partial_t\langle a|\psi(t)\rangle = \langle a|\left[\widehat{H}_0 + \widehat{H}_1(t)\right]|\psi(t)\rangle$$

$$\Leftrightarrow\left[i\hbar\left(-\frac{i}{\hbar}E(a)\right)c(a,t)+i\hbar\partial_t c(a,t)\right]e^{-\frac{i}{\hbar}E(a)(t-t_0)}$$
$$=\underbrace{\langle a|\widehat{H}_0}_{\langle a|E(a)}|\psi(t)\rangle+\langle a|\widehat{H}_1(t)|\psi(t)\rangle$$

$$\Leftrightarrow\left[E(a)c(a,t)+i\hbar\partial_t c(a,t)\right]e^{-\frac{i}{\hbar}E(a)(t-t_0)}$$
$$=E(a)\underbrace{\langle a|\psi(t)\rangle}_{e^{-\frac{i}{\hbar}E(a)(t-t_0)}c(a,t)}+\langle a|\widehat{H}_1(t)|\psi(t)\rangle \tag{3.330a}$$

leaving with:

$$i\hbar\partial_t c(a,t)=e^{-\frac{i}{\hbar}E(a)(t-t_0)}\langle a|\widehat{H}_1(t)\hat{1}_{\{|a'\rangle\}}|\psi(t)\rangle$$
$$=e^{-\frac{i}{\hbar}E(a)(t-t_0)}\int_{\{|a'\rangle\}}\langle a|\widehat{H}_1(t)|a'\rangle\underbrace{\langle a'|\psi(t)\rangle}_{e^{-\frac{i}{\hbar}E(a')(t-t_0)}c(a',t)}d\mu(a') \tag{3.330b}$$

that means the integral-differential equation:

$$i\hbar\partial_t c(a,t)=\int_{\{|a'\rangle\}}e^{-\frac{i}{\hbar}[E(a')-E(a)](t-t_0)}\langle a|\widehat{H}_1(t)|a'\rangle c(a',t)d\mu(a') \tag{3.331}$$

with the initial condition

$$c(a,t_0)=\langle a|t_0\rangle \tag{3.332}$$

obvious from above construction and Figure 3.14.

As before was the previous case for the evolution operator equation, the solution writes as:

$$c(a,t)=\langle a|t_0\rangle-\frac{i}{\hbar}\int_{t_0}^{t}\left\{\int_{\{|a'\rangle\}}e^{-\frac{i}{\hbar}[E(a')-E(a)](\tau-t_0)}\langle a|\widehat{H}_1(\tau)|a'\rangle c(a',\tau)d\mu(a')\right\}d\tau \tag{3.333}$$

With this solution, one may now express the Fourier coefficients of the perturbed state:

$$\langle a|\psi(t)\rangle = e^{-\frac{i}{\hbar}E(a)(t-t_0)}c(a,t)$$

$$= \underbrace{e^{-\frac{i}{\hbar}E(a)(t-t_0)}\langle a|t_0\rangle}_{\langle a|t\rangle}$$

$$-\frac{i}{\hbar}\int_{t_0}^{t}\int_{\{|a'\rangle\}} e^{-\frac{i}{\hbar}[E(a')-E(a)](\tau-t_0)} e^{-\frac{i}{\hbar}E(a)(t-t_0)}\langle a|\widehat{H}_1(\tau)|a'\rangle \overbrace{e^{\frac{i}{\hbar}E(a')(\tau-t_0)}\langle a'|\psi(\tau)\rangle}^{c(a',\tau)} d\mu(a')d\tau$$

(3.334)

that gives

$$\langle a|\psi(t)\rangle = \langle a|t\rangle - \frac{i}{\hbar}\int_{t_0}^{t}\int_{\{|a'\rangle\}} e^{\frac{i}{\hbar}E(a)(\tau-t)}\langle a|\widehat{H}_1(\tau)|a'\rangle\langle a'|\psi(\tau)\rangle d\mu(a')d\tau$$

(3.335)

or, through restricting the unitary operator $\hat{1}_{\{|a'\rangle\}}$ and taking out the "bra" state $\langle a|$ one remains with the searched expression for the perturbed wave-function:

$$|\psi(t)\rangle = |t\rangle - \frac{i}{\hbar}\int_{t_0}^{t} e^{\frac{i}{\hbar}\widehat{H}_0(\tau-t_0)}\widehat{H}_1(\tau)|\psi(\tau)\rangle d\tau \tag{3.336}$$

where the time interval was "adjusted" from $(\tau - t) \to (\tau - t_0)$ both for being consistent with the integration limits (the causality or chronology assumed) and for paralleling the $c(a,t)$ above solution.

Last case here is to consider also a stationary perturbation, i.e.,

$$\widehat{H}_1 \neq f(t) \tag{3.337}$$

In these conditions, one return to the $c(a,t)$ problem to particularize its solutions recursively; yet one restricts itself to the first order and made the calculations with the hope of "guessing" some global recipe:

$$c^{(0)}(a,t) = \langle a|t_0\rangle;$$

$$c^{(1)}(a,t) = \langle a|t_0\rangle - \frac{i}{\hbar}\int_{\{|a'\rangle\}}\langle a|\widehat{H}_1|a'\rangle \underbrace{c(a',\tau)}_{\downarrow\; c^{(0)}(a',t)=\langle a'|t_0\rangle} d\mu(a')\int_{t_0}^{t} e^{-\frac{i}{\hbar}[E(a')-E(a)](\tau-t_0)}d\tau$$

$$=\langle a|t_0\rangle-\frac{i}{\hbar}\left(\frac{\hbar}{-i}\right)\int_{\{|a'\rangle\}}\langle a|\widehat{H}_1|a'\rangle\langle a'|t_0\rangle d\mu(a')\frac{e^{-\frac{i}{\hbar}[E(a')-E(a)](t-t_0)}-1}{E(a')-E(a)} \tag{3.338}$$

with the help of which one gets immediately, as before:

$$\begin{aligned}\langle a|\psi(t)\rangle^{(1)} &= e^{-\frac{i}{\hbar}E(a)(t-t_0)}c^{(1)}(a,t)\\ &= \underbrace{e^{-\frac{i}{\hbar}E(a)(t-t_0)}\langle a|t_0\rangle}_{\langle a|t\rangle}+\int_{\{|a'\rangle\}}\langle a|\widehat{H}_1|a'\rangle\underbrace{e^{-\frac{i}{\hbar}E(a')(t-t_0)}\langle a'|t_0\rangle}_{\langle a'|t\rangle}\\ &\quad \frac{1-e^{\frac{i}{\hbar}[E(a')-E(a)](t-t_0)}}{E(a')-E(a)}d\mu(a')\\ &= \langle a|t\rangle+\int_{\{|a'\rangle\}}\langle a|\widehat{H}_1|a'\rangle\langle a'|t\rangle\frac{1-e^{\frac{i}{\hbar}[E(a')-E(a)\pm i\eta](t-t_0)}}{E(a')-E(a)\pm i\eta}d\mu(a')\end{aligned} \tag{3.339}$$

where in the last expression the so-called "Feynman integral prescription" was applied for rising or decreasing the energy spectrum with an infinitesimally small imaginary quantity $\pm i\eta$ so that the divergences due to poles $E(a)=E(a')$ be avoided. From here on the convention will be that the "plus" sign (+) be attributed to the so-called *retarded* solution ($t_0\to-\infty$), while that with "minus" sign (−) is to be the *advanced* evolution ($t_0\to+\infty$); for both work-frames we have:

$$\underbrace{e^{\frac{i}{\hbar}[E(a')-E(a)](t-t_0)}}_{\substack{\downarrow\\ 0\\ \textit{for INFINITE arguments}}}\times\underbrace{e^{\frac{i}{\hbar}[\pm i\eta](t-t_0)}}_{\substack{\exp\left(-\eta\frac{t-t_0}{\hbar}\right)\ \to\ 0\ \textit{for}\ t-t_0\to+\infty\\ \exp\left(+\eta\frac{t-t_0}{\hbar}\right)\ \to\ 0\ \textit{for}\ t-t_0\to-\infty}}=0\times0=0 \tag{3.340}$$

so we get the first order result for Fourier coefficients

$$\langle a|\psi^{(\pm)}(t)\rangle^{(1)}=\langle a|t\rangle+\int_{\{|a'\rangle\}}\langle a|\frac{\widehat{H}_1}{E(a')-E(a)\pm i\eta}|a'\rangle\langle a'|t\rangle d\mu(a') \tag{3.341}$$

from where the respective first order perturbed state' wave function reads:

$$\left|\psi^{(\pm)}(t)\right\rangle^{(1)} = \hat{1}_{\{|a'\rangle\}}|t\rangle + \int_{\{|a'\rangle\}} \frac{1}{E(a') - \widehat{H}_0 \pm i\eta} \widehat{H}_1 |a'\rangle\langle a'|t\rangle d\mu(a') \tag{3.342}$$

with the immediate step in its generalization in terms of the so-called *Born series*:

$$\left|\psi^{(\pm)}(t)\right\rangle = \int_{\{|a'\rangle\}} \underbrace{\left[|a'\rangle + \frac{1}{E(a') - \widehat{H}_0 \pm i\eta} \widehat{H}_1 |a'\rangle + \ldots \right]}_{BORN\ SERIES} \langle a'|t\rangle d\mu(a') \tag{3.343}$$

This important result may be rewritten in two important equivalent forms; one is through the notation:

$$\left|\psi^{(\pm)}(t)\right\rangle = \widehat{\Omega}^{(\pm)} |t\rangle \tag{3.344}$$

with the *Born operator*

$$\begin{aligned} \widehat{\Omega}^{(\pm)} &= \int_{\{|a'\rangle\}} \left[\hat{1} + \frac{1}{E(a') - \widehat{H}_0 \pm i\eta} \widehat{H}_1 + \ldots \right] |a'\rangle\langle a'| d\mu(a') \\ &= \hat{1} + \frac{1}{E(a') - \widehat{H}_0 \pm i\eta} \widehat{H}_1 + \ldots \end{aligned} \tag{3.345}$$

whose schematic representation is given in the Figure 3.15; while the other regards the resumation under the self-consistent equation celebrated as the *Lippmann-Schwinger equation*:

$$\left|\psi^{(\pm)}(t)\right\rangle = \int_{\{|a'\rangle\}} \left|\psi^{(\pm)}(a')\right\rangle \langle a'|t\rangle d\mu(a') \tag{3.346}$$

with

$$\begin{aligned} \left|\psi^{(\pm)}(a')\right\rangle &= |a'\rangle + \frac{1}{E(a') - \widehat{H}_0 \pm i\eta} \widehat{H}_1 |a'\rangle + \ldots \\ &= \left[\hat{1} + \frac{1}{E(a') - \widehat{H}_0 \pm i\eta} \widehat{H}_1 + \ldots \right] |a'\rangle \\ &= \widehat{\Omega}^{(\pm)} |a'\rangle \end{aligned} \tag{3.347}$$

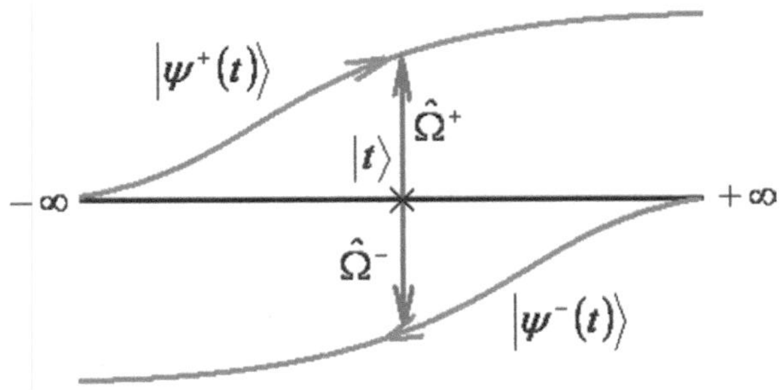

FIGURE 3.15 Representation of the retarded (+) and advanced (–) states obtained from a non-perturbed state by a stationary series of perturbations.

In either case the limiting condition looks like:

$$\lim_{t\to\mp\infty}\left|\psi^{(\pm)}(t)\right\rangle = |t\rangle \tag{3.348}$$

in accordance with the introduced retarded-advanced conventions and with the Figure 3.15.

3.6.2 UNITARY ("U") PICTURE. HEISENBERG PICTURE

A transformed quantum picture is said that one obtained by means of a unitary time-dependent operator, i.e., with ordinary property

$$\widehat{U}(t)\widehat{U}^{+}(t)=1$$

$$\underset{\partial_t}{\Leftrightarrow}\left[\partial_t\widehat{U}(t)\right]\widehat{U}^{+}(t)=-\widehat{U}(t)\left[\partial_t\widehat{U}^{+}(t)\right] \tag{3.349}$$

along the Hamiltonian related derivative properties

$$\partial_t\widehat{U}(t)=-\frac{i}{\hbar}\widehat{H}\widehat{U}(t),\quad \partial_t\widehat{U}^{+}(t)=\frac{i}{\hbar}\widehat{H}\widehat{U}^{+}(t) \tag{3.350}$$

performed upon the Schrödinger picture, producing for operators and states the new objects:

$$\widehat{A}_U(t)=\widehat{U}(t)\widehat{A}\widehat{U}^{+}(t) \tag{3.351}$$

$$|\phi_U(t)\rangle=\widehat{U}(t)|\phi\rangle \tag{3.352}$$

Note that in Schrödinger picture operators are not time-dependent (if not otherwise specified, as was the earlier case with Hamiltonian interaction and perturbation discussion), while in U-picture they necessarily become so. Actually the usefulness of the "U" transformation is exactly that to obtaining the time-dependent operators from stationary ones, along the time-dependent quantum states; this may be translated in the ancient Greek Parmenides' "*pantha rei*" philosophy – "all is flowing" (that is true in the "U picture" where both operators and states are time-dependent). Nevertheless, here we will explore a general and then a specialized transformation of the Schrödinger picture.

Say we are referring to the state $|t\rangle$ fulfilling the direct and conjugate Schrödinger equations:

$$i\hbar\partial_t|t\rangle = \widehat{H}|t\rangle \tag{3.353a}$$

$$-i\hbar\partial_t\langle t| = \langle t|\widehat{H} \tag{3.353b}$$

Then, its projector (called also as the *statistical operator*)

$$\Delta_t = |t\rangle\langle t| \equiv \widehat{W}(t) \tag{3.354}$$

obeys the equivalences

$$\begin{aligned} i\hbar\partial_t\widehat{W}(t) &= i\hbar\left[\left(\partial_t|t\rangle\right)\langle t| + |t\rangle\left(\partial_t\langle t|\right)\right] \\ &= \widehat{H}|t\rangle\langle t| - |t\rangle\langle t|\widehat{H} = \widehat{H}\widehat{W}(t) - \widehat{W}(t)\widehat{H} \end{aligned} \tag{3.355}$$

to the operatorial equation:

$$i\hbar\partial_t\widehat{W}(t) = \left[\widehat{H}, \widehat{W}(t)\right] \tag{3.356}$$

from where the stationary states are recovered through the commutation condition

$$\left[\widehat{H}, \widehat{W}(t)\right] = 0 \Rightarrow \partial_t\widehat{W}(t) = 0 \tag{3.357}$$

as there are the states at thermodynamic equilibrium, for example, those written as

$$\widehat{W} = Z^{-1}\exp\left(-\beta\widehat{H}\right) \cong Z^{-1}\left[1-\left(\beta\widehat{H}\right)+\frac{1}{2!}\left(\beta\widehat{H}\right)^2+...\right] \quad (3.358)$$

this way justifying the statistical operator name, where Z stands for partition function and $\beta = 1/k_B T$ the inverse of thermal energy, in the base of operatorial expansion and of self-commutativity of Hamiltonian at whatever power

$$\left[\widehat{H},\widehat{H}^n\right] = \widehat{H}^{n+1} - \widehat{H}^{n+1} = 0 \ \forall n \in \mathbf{N} \quad (3.359)$$

Now, within U-picture, the (time dependent) statistical operator, the (time-dependent) Hamiltonian and a general state become:

$$\widehat{W}_U(t) = \widehat{U}(t)\widehat{W}\widehat{U}^+(t) \quad (3.360)$$

$$\widehat{H}_U(t) = \widehat{U}(t)\widehat{H}\widehat{U}^+(t) \quad (3.361)$$

$$|t_U\rangle = \widehat{U}(t)|t\rangle \quad (3.362)$$

Through employing the derivative properties of the *unitary operator in U-picture* and the Schrödinger picture of the statistical operator equation, it rewrites successively in the "U"-picture:

$$i\hbar\partial_t\widehat{W}_U(t) = i\hbar\partial_t\left[\widehat{U}(t)\widehat{W}\widehat{U}^+(t)\right]$$

$$= i\hbar\left[\partial_t\widehat{U}(t)\right]\underbrace{\hat{1}_U}_{\widehat{U}^+\widehat{U}}\widehat{W}\widehat{U}^+(t) + i\hbar\widehat{U}(t)\widehat{W}\underbrace{\hat{1}_U}_{\widehat{U}^+\widehat{U}}\left[\partial_t\widehat{U}^+(t)\right] + \widehat{U}(t)\underbrace{\left[i\hbar\partial_t\widehat{W}\right]}_{\left[\widehat{H},\widehat{W}\right]}\widehat{U}^+(t)$$

$$= i\hbar\left[\partial_t\widehat{U}(t)\right]\widehat{U}^+(t)\underbrace{\widehat{U}(t)\widehat{W}\widehat{U}^+(t)}_{\widehat{W}_U} + i\hbar\underbrace{\widehat{U}(t)\widehat{W}\widehat{U}^+(t)}_{\widehat{W}_U}\underbrace{\widehat{U}(t)\left[\partial_t\widehat{U}^+(t)\right]}_{-\left[\partial_t\widehat{U}(t)\right]\widehat{U}^+(t)}$$

$$+\left[\widehat{U}(t)\widehat{H}\underbrace{\hat{1}_U}_{\widehat{U}^+\widehat{U}}\widehat{W}\widehat{U}^+(t)-\widehat{U}(t)\widehat{W}\underbrace{\hat{1}_U}_{\widehat{U}^+\widehat{U}}\widehat{H}\widehat{U}^+(t)\right]$$

$$=i\hbar\left[\left(\partial_t\widehat{U}(t)\right)\widehat{U}^+(t),\widehat{W}_U\right]+\left[\begin{array}{c}\underbrace{\widehat{U}(t)\widehat{H}\widehat{U}^+(t)}_{\widehat{H}_U}\underbrace{\widehat{U}(t)\widehat{W}\widehat{U}^+(t)}_{\widehat{W}_U}\\ -\underbrace{\widehat{U}(t)\widehat{W}\widehat{U}^+(t)}_{\widehat{W}_U}\underbrace{\widehat{U}(t)\widehat{H}\widehat{U}^+(t)}_{\widehat{H}_U}\end{array}\right]$$

$$=\left[i\hbar\left(\partial_t\widehat{U}(t)\right)\widehat{U}^+(t),\widehat{W}_U\right]+\left[\widehat{H}_U,\widehat{W}_U\right];$$

$$i\hbar\partial_t\widehat{W}_U(t)=\left[\widehat{H}_U+i\hbar\left(\partial_t\widehat{U}(t)\right)\widehat{U}^+(t),\widehat{W}_U\right] \tag{3.363}$$

When comparing this U-equation for statistical operator with the corresponding one within Schrödinger picture, one can immediately identify that the "rôle" of the Schrödinger Hamiltonian is played formally by:

$$\widehat{H}=\widehat{H}_U+i\hbar\left(\partial_t\widehat{U}(t)\right)\widehat{U}^+(t) \tag{3.364}$$

from where there is immediately get also the interesting relationship:

$$i\hbar\left(\partial_t\widehat{U}(t)\right)\widehat{U}^+(t)=\widehat{H}-\widehat{H}_U \tag{3.365}$$

that, through right multiplication with the unitary operator, and using its basic definition, it provides a sort of Schrödinger equation for the unitary operator:

$$i\hbar\left(\partial_t\widehat{U}(t)\right)=\left(\widehat{H}-\widehat{H}_U\right)\widehat{U}(t); \tag{3.366}$$

the last equation has the philosophical meaning that the unitary operator itself "travels" upon a Schrödinger type equation to transform the Schrödinger picture into other one, showing this way with necessity more fluidic (causal) features; moreover the "passage" between the two pictures is assured by their Hamiltonian difference.

Let's investigate the U-transformation for a general Schrödinger operator that is not Hamiltonian, say $\widehat{A}$; within Schrödinger picture it fulfills the conservation equation:

$$i\hbar\left(\partial_t \widehat{A}\right) = 0 \tag{3.367}$$

since in this picture operators are generally considered as non-time-dependent. Yet, with the U-picture it becomes, repeating in an analogue manner as before for the statistical operator:

$$i\hbar\partial_t \widehat{A}_U(t) = i\hbar\partial_t\left[\widehat{U}(t)\widehat{A}\widehat{U}^+(t)\right]$$
$$= i\hbar\left[\partial_t \widehat{U}(t)\right]\widehat{A}\widehat{U}^+(t) + i\hbar\widehat{U}(t)\widehat{A}\left[\partial_t \widehat{U}^+(t)\right] + \widehat{U}(t)\underbrace{\left[i\hbar\partial_t \widehat{A}\right]}_{0}\widehat{U}^+(t);$$
$$i\hbar\partial_t \widehat{A}_U(t) = \left[\left(\partial_t \widehat{U}(t)\right)\widehat{U}^+(t), \widehat{A}_U\right] \tag{3.368}$$

Putting side-by-side the two U-transformations, for statistical operator and for the general operator looks synthetically like:

$$i\hbar\partial_t \widehat{W}_U(t) = \left[\widehat{H}, \widehat{W}_U\right] \tag{3.369}$$

$$i\hbar\partial_t \widehat{A}_U(t) = \left[\widehat{H} - \widehat{H}_U, \widehat{A}_U\right] \tag{3.370}$$

from where for the specialization

$$\widehat{H} = \widehat{H}_U \tag{3.371}$$

they behave as

$$i\hbar\partial_t \widehat{W}_U(t) = \left[\widehat{H}_U, \widehat{W}_U\right] \tag{3.372}$$

$$i\hbar\partial_t \widehat{A}_U(t) = 0 \tag{3.373}$$

that recovers exactly the Schrödinger picture since the unitary operator is now reduced to the constant:

$$i\hbar\left(\partial_t\widehat{U}(t)\right)=0\Rightarrow\widehat{U}(t)=ct=\hat{1} \tag{3.374}$$

being this the only solution fulfilling all of the unitary operator constraints (norm, initial conditions, etc.).

With this appears the question of the U-transformation of the *unitary operator in Schrödinger picture* (3.310); for doing this one has to employ the evolution equations in Schrödinger- and U-pictures, respectively as:

$$|t\rangle=\widehat{U}(t,t_0)|t_0\rangle \tag{3.375a}$$

$$|t_U\rangle=\widehat{U}(t)|t\rangle \tag{3.375b}$$

to combine them in the transformation:

$$\widehat{U}_U(t,t_0)|t_0\rangle=|t_U\rangle=\widehat{U}(t)|t\rangle=\widehat{U}(t)\widehat{U}(t,t_0)|t_0\rangle \tag{3.376}$$

or in the even more general one

$$\widehat{U}_U(t,t_0)\left|(t_0)_U\right\rangle=|t_U\rangle=\widehat{U}(t)\widehat{U}(t,t_0)|t_0\rangle=\widehat{U}(t)\widehat{U}(t,t_0)\widehat{U}^+(t_0)\left|(t_0)_U\right\rangle \tag{3.377}$$

Releasing with the (expected) U-transformation for Schrödinger unitary operator:

$$\widehat{U}_U(t,t_0)=\widehat{U}(t)\widehat{U}(t,t_0)\widehat{U}^+(t_0) \tag{3.378}$$

Next, let's see its equation of motion throughout performing the appropriate derivative following the above exposed line of derivations:

$$\begin{aligned}i\hbar\partial_t\widehat{U}_U(t,t_0)&=i\hbar\partial_t\left[\widehat{U}(t)\widehat{U}(t,t_0)\widehat{U}^+(t_0)\right]\\&=i\hbar\left[\partial_t\widehat{U}(t)\right]\underbrace{\hat{1}_U}_{\widehat{U}^+\widehat{U}}\widehat{U}(t,t_0)\widehat{U}^+(t_0)+i\hbar\widehat{U}(t)\underbrace{\left[\partial_t\widehat{U}(t,t_0)\right]}_{-\frac{i}{\hbar}\widehat{H}\widehat{U}(t,t_0)}\widehat{U}^+(t_0)\end{aligned}$$

$$= i\hbar\left[\partial_t \widehat{U}(t)\right]\widehat{U}^+(t)\underbrace{\widehat{U}(t)\widehat{U}(t,t_0)\widehat{U}^+(t_0)}_{\widehat{U}_U(t,t_0)} + \widehat{U}(t)\widehat{H}\underbrace{\hat{1}_U}_{\widehat{U}^+\widehat{U}}\widehat{U}(t,t_0)\widehat{U}^+(t_0)$$

$$= i\hbar\left[\partial_t \widehat{U}(t)\right]\widehat{U}^+(t)\widehat{U}_U(t,t_0) + \underbrace{\widehat{U}(t)\widehat{H}\widehat{U}^+(t)}_{\widehat{H}_U}\underbrace{\widehat{U}(t)\widehat{U}(t,t_0)\widehat{U}^+(t_0)}_{\widehat{U}_U(t,t_0)}$$

$$= \underbrace{\left\{i\hbar\left[\partial_t \widehat{U}(t)\right]\widehat{U}^+(t) + \widehat{H}_U\right\}}_{\widehat{\mathcal{H}}}\widehat{U}_U(t,t_0)$$

(3.379)

Therefore, we form the U-transformed equation for the Schrödinger unitary operator:

$$\begin{cases} i\hbar\partial_t \widehat{U}_U(t,t_0) = \widehat{H}(t)\widehat{U}_U(t,t_0) \\ \widehat{U}_U(t_0,t_0) = \hat{1} \end{cases} \quad (3.380)$$

from where the whole previous section problem the time-dependent Hamiltonian arises; it may be eventually resumed with the general solution:

$$\widehat{U}_U(t,t_0) = \widehat{T}\exp\left(-\frac{i}{\hbar}\int_{t_0}^{t}\widehat{H}(\tau)d\tau\right) \quad (3.381)$$

or through Born series when expanded $\widehat{H}$ in time-perturbation terms.

Finally, the particular *Heisenberg picture* may be gained fro the U-picture for the special choice of the unitary operator:

$$\widehat{U}_{Hei}(t) = \widehat{U}(t_0,t) = e^{\frac{i}{\hbar}\widehat{H}(t-t_0)} = \widehat{U}^+(t,t_0) \quad (3.382)$$

that is expected to produce the same effect on Schrödinger picture as the classical inverse (vectorial) composed (relative) motion, i.e., through passing from the inertial (Laboratory) to the non-inertial (Mass-Center) system. The first effect of such transformation regards the state vectors that now behave like:

$$|t_I\rangle = \widehat{U}_{Hei}(t)|t\rangle = \widehat{U}_{Hei}(t)\widehat{U}(t,t_0)|t_0\rangle = \underbrace{\widehat{U}^+(t,t_0)\widehat{U}(t,t_0)}_{\hat{1}}|t_0\rangle = |t_0\rangle \quad (3.383)$$

so that producing no movement or evolution of them; in Parmenides' Greek philosophical paradigm "the river was stopped, being transformed into a lake". This may be further checked through computing the actual Hamiltonian:

$$\begin{aligned}
\widehat{H}_{Hei} &= \widehat{H}_{Hei} + i\hbar(\partial_t \widehat{U}_{Hei}(t))\widehat{U}^{+}_{Hei}(t) \\
&= \widehat{H}_{Hei} - i\hbar \widehat{U}_{Hei}(t)\partial_t \widehat{U}^{+}_{Hei}(t) \\
&= \widehat{H}_{Hei} - i\hbar \widehat{U}(t_0,t)\partial_t \widehat{U}^{+}(t_0,t) \\
&= \widehat{H}_{Hei} - i\hbar \widehat{U}^{+}(t,t_0)\underbrace{\partial_t \widehat{U}(t,t_0)}_{-\frac{i}{\hbar}\widehat{H}\widehat{U}(t,t_0)} \\
&= \widehat{H}_{Hei} - \widehat{U}^{+}(t,t_0)\widehat{H}\widehat{U}(t,t_0) \\
&= \widehat{H}_{Hei} - \widehat{U}(t_0,t)\widehat{H}\widehat{U}^{+}(t_0,t) \\
&= \widehat{H}_{Hei} - \underbrace{\widehat{U}_{Hei}(t)\widehat{H}\widehat{U}^{+}_{Hei}(t)}_{\widehat{H}_{Hei}};
\end{aligned}$$

$$\widehat{H}_{Hei} = \hat{0} \tag{3.384}$$

Replacing this result in above U-equations for statistical operator and the working operator, we find out that within Heisenberg picture we have stationary projectors or statistical operators:

$$i\hbar\partial_t \widehat{W}_{Hei}(t) = \left[\hat{0}, \widehat{W}_U\right] = \hat{0} \Rightarrow \widehat{W}_{Hei}(t) = |t_0\rangle\langle t_0| = ct. \tag{3.385}$$

while the general operators are the only objects in the Hilbert space that are still evolving, however upon the resulting equation:

$$i\hbar\partial_t \widehat{A}_{Hei}(t) = \left[\hat{0} - \widehat{H}_{Hei}, \widehat{A}_{Hei}\right] = -\left[\widehat{H}_{Hei}, \widehat{A}_{Hei}\right] \tag{3.386}$$

In ontological terms, within Heisenberg picture the observables (that corresponds with the quantum operators) are those that are in moving, or in Parmenides' picture: "we are measuring or observing (or fishing) the quantum phenomena through moving on the lake of stationary quantum states (eigen-states)".

3.6.3 QUANTUM TRANSITIONS: INTERACTION PICTURE

Having discussed the quantum evolution in terms of unitary and statistical operators we can make the further step towards describing quantum transitions. The starting point stays, as already custom with, on the unitary evolution action on given initial state:

$$|t\rangle = \widehat{U}(t,t_0)|t_0\rangle \tag{3.387}$$

that support the writing of the statistical operator at some evolution time and state:

$$\widehat{W}(t) = |t\rangle\langle t| = \widehat{U}(t,t_0)\underbrace{|t_0\rangle\langle t_0|}_{\widehat{W}(t_0)}\widehat{U}^+(t,t_0) = \widehat{U}(t,t_0)\widehat{W}(t_0)\widehat{U}^+(t,t_0) \tag{3.388}$$

With the help of this temporal statistic operator one may write the *average of an observable* $\hat{A}$, in a given representation $\hat{1}_{\{|n\rangle\}} = \sum_n |n\rangle\langle n|$, on selected state $|t\rangle$ of a measurement as:

$$\begin{aligned}
\left\langle \hat{A} \right\rangle_t &= \langle t|\hat{A}|t\rangle = \langle t|\hat{1}_{\{|n\rangle\}}\hat{A}|t\rangle \\
&= \sum_n \langle t|n\rangle\langle n|\hat{A}|t\rangle = \sum_n \langle n|\hat{A}\underbrace{|t\rangle\langle t|}_{\widehat{W}(t)}|n\rangle \\
&= \sum_n \langle n|\hat{A}\widehat{W}(t)|n\rangle = \mathrm{Tr}\left(\hat{A}\widehat{W}(t)\right);
\end{aligned}$$

$$\left\langle \hat{A} \right\rangle_t = \mathrm{Tr}\left(\hat{A}\widehat{W}(t)\right) \tag{3.389}$$

introducing the so-called *trace functional* with the definition:

$$\mathrm{Tr}\left(\hat{\bullet}\right)_{\{|n\rangle\}} = \sum_n \langle n|\hat{\bullet}|n\rangle \tag{3.390}$$

with the following elementary properties:

- $\mathrm{Tr}\left(\hat{\bullet}\right)$ is invariant respecting the spectral basis of representation:

$$\begin{aligned}
\mathrm{Tr}\left(\hat{A}\right)_{\{|a\rangle\}} &= \sum_a \langle a|\hat{A}|a\rangle = \sum_a \langle a|\hat{A}\hat{1}_{\{|b\rangle\}}|a\rangle \\
&= \sum_a\sum_b \langle a|\hat{A}|b\rangle\langle b|a\rangle = \sum_b\sum_a \langle b|a\rangle\langle a|\hat{A}|b\rangle
\end{aligned}$$

$$= \sum_b \langle b|\hat{1}_{\{|a\rangle\}} \hat{A}|b\rangle = \sum_b \langle b|\hat{A}|b\rangle;$$

$$\mathrm{Tr}\left(\hat{A}\right)_{\{|a\rangle\}} = \mathrm{Tr}\left(\hat{A}\right)_{\{|b\rangle\}} \tag{3.391}$$

and the proof can be extended also to the continuum spectrum (with associate closure relation), as well between any representations.

- $\mathrm{Tr}\left(\hat{\bullet}\right)$ absorbs the multiplication with a constant:

$$\alpha \mathrm{Tr}\left(\hat{A}\right) = \mathrm{Tr}\left(\alpha \hat{A}\right), \alpha \in \mathrm{C} \tag{3.392}$$

- $\mathrm{Tr}\left(\hat{\bullet}\right)$ is distributive along the operatorial summation:

$$\mathrm{Tr}\left(\hat{A}+\hat{B}\right) = \mathrm{Tr}\left(\hat{A}\right) + \mathrm{Tr}\left(\hat{B}\right) \tag{3.393}$$

- $\mathrm{Tr}\left(\hat{\bullet}\right)$ is invariant under commutativity of operators:

$$\mathrm{Tr}\left(\hat{A}\hat{B}\right) = \mathrm{Tr}\left(\hat{B}\hat{A}\right) \Leftrightarrow \mathrm{Tr}\left(\left[\hat{A},\hat{B}\right]\right) = 0 \tag{3.394a}$$

as can be immediately shown:

$$\begin{aligned}
\mathrm{Tr}\left(\hat{A}\hat{B}\right) &= \sum_n \langle n|\hat{A}\hat{B}|n\rangle = \sum_n \langle n|\hat{A}\hat{1}_{\{|k\rangle\}} \hat{B}|n\rangle = \sum_n \sum_k \langle n|\hat{A}|k\rangle\langle k|\hat{B}|n\rangle \\
&= \sum_k \sum_n \langle k|\hat{B}|n\rangle\langle n|\hat{A}|k\rangle = \sum_k \langle k|\hat{B}\hat{1}_{\{|n\rangle\}} \hat{A}|k\rangle \\
&= \sum_k \langle k|\hat{B}\hat{A}|k\rangle = \mathrm{Tr}\left(\hat{B}\hat{A}\right)
\end{aligned} \tag{3.394a}$$

- $\mathrm{Tr}\left(\hat{\bullet}\right)$ is normalized for the statistical operator at any time, due to the Parceval closure relation

$$\mathrm{Tr}\left(\widehat{W}(t)\right) = \sum_t \langle t|\widehat{W}(t)|t\rangle = \sum_t \langle t|t\rangle\langle t|t\rangle = \sum_t \left|\langle t|t\rangle\right|^2 = 1 \tag{3.395}$$

Other properties will be unfolded later with the occasion of density matrix quantum formalism. For the moment we use this operatorial property to make the remark that while fixing the *initial* and *final* quantum states through the statistical operator and projector, respectively as:

$$\widehat{W}_i = |t_0\rangle\langle t_0| \; \dots \; evolution \; \dots \; \widehat{\Lambda}_f = |f\rangle\langle f| \tag{3.396}$$

we have that

$$\mathrm{Tr}\left(\widehat{W}_i \widehat{\Lambda}_f\right) = 0 \tag{3.397}$$

since the initial and final states are considered orthogonal (because are independent), $\langle t_0 | f\rangle = 0$, while when considered the time-dependent statistical operator $\widehat{W}(t)$ the so-called quantum *transition probability* is defined as:

$$\begin{aligned}\wp_{t_0 \to t} &\equiv \wp(t,t_0) = \mathrm{Tr}\left(\widehat{W}(t)\widehat{\Lambda}_f\right)\\ &= \mathrm{Tr}\left(\widehat{U}(t,t_0)\widehat{W}_i\widehat{U}^+(t,t_0)\widehat{\Lambda}_f\right) \underset{t=t_0}{\to} 0\end{aligned} \tag{3.398}$$

with the help of which the *rate of transition* may be introduced as well by

$$\begin{aligned}\mathbf{R}(t,t_0) &= \partial_t \wp(t,t_0)\\ &= \mathrm{Tr}(\underbrace{\left[\partial_t \widehat{U}(t,t_0)\right]}_{-\frac{i}{\hbar}\widehat{H}\widehat{U}}\widehat{W}_i\widehat{U}^+(t,t_0)\widehat{\Lambda}_f) + \mathrm{Tr}(\widehat{U}(t,t_0)\widehat{W}_i\underbrace{\left[\partial_t \widehat{U}^+(t,t_0)\right]}_{\frac{i}{\hbar}\widehat{U}^+\widehat{H}}\widehat{\Lambda}_f)\\ &= -\frac{i}{\hbar}\left[\mathrm{Tr}\left(\widehat{H}\widehat{U}(t,t_0)\widehat{W}_i\widehat{U}^+(t,t_0)\widehat{\Lambda}_f\right) - \mathrm{Tr}\left(\widehat{U}(t,t_0)\widehat{W}_i\widehat{U}^+(t,t_0)\widehat{H}\widehat{\Lambda}_f\right)\right]\\ &= -\frac{i}{\hbar}\left[\mathrm{Tr}\left(\widehat{U}(t,t_0)\widehat{W}_i\widehat{U}^+(t,t_0)\widehat{\Lambda}_f\widehat{H}\right) - \mathrm{Tr}\left(\widehat{U}(t,t_0)\widehat{W}_i\widehat{U}^+(t,t_0)\widehat{H}\widehat{\Lambda}_f\right)\right]\\ &= -\frac{i}{\hbar}\mathrm{Tr}\left(\widehat{U}(t,t_0)\widehat{W}_i\widehat{U}^+(t,t_0)\left[\widehat{\Lambda}_f,\widehat{H}\right]\right);\\ \mathbf{R}(t,t_0) &= \frac{i}{\hbar}\mathrm{Tr}\left(\widehat{U}(t,t_0)\widehat{W}_i\widehat{U}^+(t,t_0)\left[\widehat{H},\widehat{\Lambda}_f\right]\right)\end{aligned} \tag{3.399}$$

Worth remarking that, the statistical operator takes the crucial role in carrying the entirely quantum evolution, from initial state to the end. In this respect, there seems it may hidden an even more crucial information that links the stationary (equilibrium) states with those being engaged in (temporal) evolution.

Therefore, the statistical operator may be firstly written in a way characterizing the stationary (equilibrium playing the role of some initial state), as earlier introduced

$$\widehat{W}(\beta) = Z_{QS}^{-1} \exp\left(-\beta \widehat{H}\right) \tag{3.400}$$

from where, in order the unitary (normalization) condition for its trace to be respected,

$$1 = \mathrm{Tr}\left(\widehat{W}(\beta)\right) \tag{3.401}$$

one gets the operational definition for the *quantum statistical partition function*:

$$Z_{QS} = \mathrm{Tr}\left(e^{-\beta \widehat{H}}\right) = \sum_n \langle n | e^{-\beta \widehat{H}} | n \rangle \tag{3.402}$$

in terms of the trace function, computed on the spectrum of the Hamiltonian involved.

Secondly, for time-dependent statistical operator, a similar relation may be heuristically write down as being in relation with the unitary operator

$$\widehat{W}(t) = Z_{QM}^{-1} \widehat{U}(t,t_0) = Z_{QM}^{-1} \exp\left(-\frac{i}{\hbar} \widehat{H}(t - t_0)\right) \tag{3.403}$$

that leaves, in the virtue of the same arguments as above, with the definition of the actual *quantum mechanically partition function*:

$$Z_{QM} = \mathrm{Tr}\left(e^{-\frac{i}{\hbar}\widehat{H}(t-t_0)}\right) = \sum_n \langle n | e^{-\frac{i}{\hbar}\widehat{H}(t-t_0)} | n \rangle \tag{3.404}$$

Now, since the two statistical operators represent the same reality, since (in principle) any instantaneous state may be prepared as to become the initial state for further evolution (in this so-called *adiabatic quantum evolution*) their partition function should be equalized; that produces the so-called *Wick rotation*:

$$t - t_0 = -i\hbar\beta = -\frac{i\hbar}{k_B T} \tag{3.405}$$

that beyond of its mathematical (real-to-imaginary) continuation feature expresses the natural intrinsic reciprocal relationship between the time and temperature in quantum evolution. Yet this can be put in phenomenological relation with the evolution of the Universe as a whole, see Figure 3.16.

For instance if one imagine the birth of the Universe from the time approaching the zero moment, but from it's *a priori* side $t \le t_0 \to 0$ one gets that such moment correspond with an entirely excited picture of the Universe, with an infinite temperature, from where the natural (spontaneous and stimulated) emissions justify the earlier light sea coming from the "birth" of Universe. Instead, for very large times $t_0 < t \to \infty$ Universe cools down to its 0K ground state, from where, no emission is possible nor any absorption (if it is an isolated entity, as "Universe"), in accordance with the entropic principle as well. In between these extremes, both the quantum evolutions as well as the emission-absorption processes - the "Life" at various non-extreme temperatures takes place. There is therefore remarkably how the quantum postulates imply consistent ideas about the ontology of the Universe itself, being thus more than a set of (mathematical) rules but carrying also a lot of physics inside!

Returning to the transition probability it may be further evaluated as:

$$\wp_{i\to f}(t,t_0) = \mathrm{Tr}\left(\hat{\Lambda}_f \hat{U}(t,t_0)\hat{W}_i \hat{U}^+(t,t_0)\right)$$
$$= \sum_a \langle a| \underbrace{\hat{\Lambda}_f}_{|f\rangle\langle f|} \hat{U}(t,t_0) \underbrace{\hat{W}_i}_{|t_0\rangle\langle t_0|} \hat{U}^+(t,t_0)|a\rangle$$

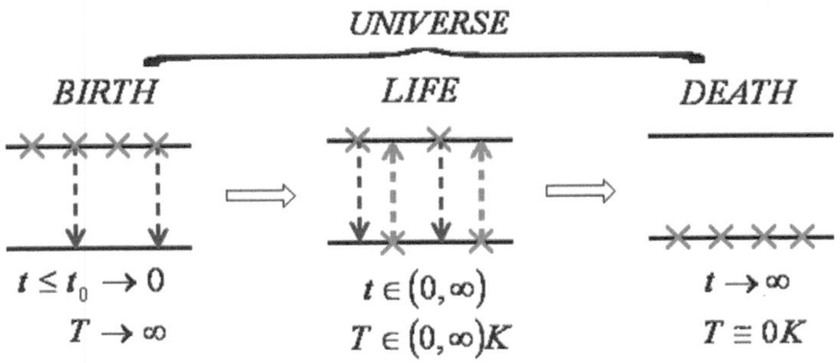

FIGURE 3.16 The quantum evolution of the Universe from the quantum transition perspective, from birth to death, being characterized either by time and temperature changes based on the Wick rotation rule.

$$
\begin{aligned}
&= \sum_a \langle a | \underbrace{\widehat{\Lambda}_f}_{|f\rangle\langle f|} \widehat{U}(t,t_0) \underbrace{\widehat{W}_i}_{|t_0\rangle\langle t_0|} \widehat{U}^+(t,t_0) | a \rangle \\
&= \sum_a \langle a | f \rangle \langle f | \widehat{U}(t,t_0) | t_0 \rangle \langle t_0 | \widehat{U}^+(t,t_0) | a \rangle \\
&= \sum_a \langle f | \widehat{U}(t,t_0) | t_0 \rangle \langle t_0 | \widehat{U}^+(t,t_0) | a \rangle \langle a | f \rangle \\
&= \langle f | \widehat{U}(t,t_0) | t_0 \rangle \langle t_0 | \widehat{U}^+(t,t_0) \hat{1}_{\{|a\rangle\}} | f \rangle;
\end{aligned}
$$

$$
\wp_{i\to f}(t,t_0) = \left| \langle f | \widehat{U}(t,t_0) | t_0 \rangle \right|^2 \tag{3.406}
$$

representing the transition probability from the initial $|t_0\rangle$ to the final state $|f\rangle$ as the square of the amplitude $\langle f|\widehat{U}(t,t_0)|t_0\rangle$ of this transition.

The practical implementation of this formula is made through the so-called *interaction picture – constructed as the Heisenberg picture of the non-perturbed problem*, i.e., having for the U-transformation the Heisenberg particularization:

$$
\widehat{U}(t) \to \widehat{U}_{Hei}(t) = \widehat{U}_0(t_0,t) = e^{\frac{i}{\hbar}\widehat{H}_0(t-t_0)} \tag{3.407}
$$

With this, the evolution operator in the interaction picture is specialized from its general formulation

$$
\widehat{U}_{Int}(t,t_0) = \widehat{T}\exp\left(-\frac{i}{\hbar}\int_{t_0}^{t} \widehat{H}_{Int}(\tau)d\tau \right) \tag{3.408}
$$

to its first order cut as:

$$
\widehat{U}^{(1)}_{Int}(t,t_0) = \hat{1} - \frac{i}{\hbar}\int_{t_0}^{t} \widehat{H}_{Int}(\tau)d\tau \tag{3.409}
$$

with the interaction Hamiltonian expressed as well in the first order of the time-dependent perturbation:

$$
\begin{aligned}
\widehat{H}_{Int}(\tau) \to \widehat{H}^{(1)}_{Int} &= \widehat{H}^{(1)}_{Hei} + i\hbar\left(\partial_t \widehat{U}_{Hei}(t)\right)\widehat{U}^+_{Hei}(t) \\
&= \widehat{U}_{Hei}(t)\widehat{H}^{(1)}\widehat{U}^+_{Hei}(t) + i\hbar\left(\partial_t \widehat{U}_{Hei}(t)\right)\widehat{U}^+_{Hei}(t)
\end{aligned}
$$

$$
\begin{aligned}
&= \widehat{U}_{Hei}(t)\left(\widehat{H}_0 + \widehat{H}_1(t)\right)\widehat{U}^+_{Hei}(t) + i\hbar\underbrace{\left(\partial_t \widehat{U}_{Hei}(t)\right)}_{\frac{i}{\hbar}\widehat{H}_0\widehat{U}_{Hei}(t)}\widehat{U}^+_{Hei}(t)\\
&= \underbrace{\widehat{U}_{Hei}(t)\widehat{H}_0}_{\widehat{H}_0\widehat{U}_{Hei}-\left[\widehat{H}_0,\widehat{U}_{Hei}\right]} \widehat{U}^+_{Hei}(t) + \widehat{U}_{Hei}(t)\widehat{H}_1(t)\widehat{U}^+_{Hei}(t) - \widehat{H}_0\underbrace{\widehat{U}_{Hei}(t)\widehat{U}^+_{Hei}(t)}_{\hat{1}}\\
&= \widehat{H}_0\underbrace{\widehat{U}_{Hei}(t)\widehat{U}^+_{Hei}(t)}_{\hat{1}} - \underbrace{\left[\widehat{H}_0,\widehat{U}_{Hei}\right]}_{\hat{0}}\widehat{U}^+_{Hei}(t) + \widehat{U}_{Hei}(t)\widehat{H}_1(t)\widehat{U}^+_{Hei}(t) - \widehat{H}_0;
\end{aligned}
$$

$$
\begin{aligned}
\widehat{H}^{(1)}_{Int} &= \widehat{U}_{Hei}(t)\widehat{H}_1(t)\widehat{U}^+_{Hei}(t)\\
&= e^{\frac{i}{\hbar}\widehat{H}_0(t-t_0)}\widehat{H}_1(t)e^{-\frac{i}{\hbar}\widehat{H}_0(t-t_0)}
\end{aligned}
\tag{3.410}
$$

In these conditions the transition amplitude becomes in interaction picture:

$$
\begin{aligned}
\wp^{(1)}_{Int}(t,t_0) &= \left|\left\langle f\left|\widehat{U}^{(1)}_{Int}(t,t_0)\right|t_0\right\rangle\right|^2\\
&= \left|\left\langle f\right|\left[\hat{1} - \frac{i}{\hbar}\int_{t_0}^{t}\widehat{H}^{(1)}_{Int}(\tau)d\tau\right]\left|t_0\right\rangle\right|^2\\
&= \left|\underbrace{\langle f|t_0\rangle}_{0} - \frac{i}{\hbar}\left\langle f\right|\left[\int_{t_0}^{t}\widehat{H}^{(1)}_{Int}(\tau)d\tau\right]\left|t_0\right\rangle\right|^2\\
&= \frac{1}{\hbar^2}\left|\int_{t_0}^{t}\left\langle f\left|\widehat{H}^{(1)}_{Int}(\tau)\right|t_0\right\rangle d\tau\right|^2\\
&= \frac{1}{\hbar^2}\left|\int_{t_0}^{t}\left\langle f\left|e^{\frac{i}{\hbar}\widehat{H}_0(\tau-t_0)}\widehat{H}_1(\tau)e^{-\frac{i}{\hbar}\widehat{H}_0(\tau-t_0)}\right|t_0\right\rangle d\tau\right|^2\\
&= \frac{1}{\hbar^2}\left|\int_{t_0}^{t}e^{\frac{i}{\hbar}E_f(\tau-t_0)}\left\langle f\left|\widehat{H}_1(\tau)\right|t_0\right\rangle e^{-\frac{i}{\hbar}E_0(\tau-t_0)}d\tau\right|^2\\
&= \frac{1}{\hbar^2}\left|\int_{t_0}^{t}e^{\frac{i}{\hbar}(E_f-E_0)\tau}\left\langle f\left|\widehat{H}_1(\tau)\right|t_0\right\rangle d\tau\right|^2;
\end{aligned}
$$

$$\wp_{Int}^{(1)}(t,t_0)=\frac{1}{\hbar^2}\left|\int_{t_0}^{t} e^{i\omega_{f0}\tau}\left\langle f\middle|\widehat{H}_1(\tau)\middle|t_0\right\rangle d\tau\right|^2, \quad \omega_{f0}=\frac{E_f-E_0}{\hbar} \tag{3.411}$$

The last step regards the effective computation of this expression for a periodic perturbation

$$\widehat{H}_1(\tau)=\underbrace{\widehat{F}\exp(i\omega\tau)}_{EMISSION}+\underbrace{\widehat{F}^{+}\exp(-i\omega\tau)}_{ABSORPTION} \tag{3.412}$$

chosen so that to fulfill also the hermiticity condition for a quantum Hamiltonian, while accounting either for emission and absorption processes in a quantum transition; this stands also for an indirect justification for that the two processes are intimately related. Yet, we can consider for convenience just one part of the above Hamiltonian to study emission and absorption separately.

As such, let's consider the emission process only; thus we successively get for transition probability:

$$\begin{aligned}
\wp_{Int/Emis}^{(1)}(t,t_0=0)&=\frac{\left|\left\langle f\middle|\widehat{F}\middle|t_0\right\rangle\right|^2}{\hbar^2}\left|\int_{t_0=0}^{t}\exp\left[i\left(\omega_{f0}+\omega\right)\tau\right]d\tau\right|^2\\
&=\frac{1}{\hbar^2}\frac{\left|\left\langle f\middle|\widehat{F}\middle|t_0\right\rangle\right|^2}{\left(\omega_{f0}+\omega\right)^2}\left|\exp\left[i\left(\omega_{f0}+\omega\right)t\right]-1\right|^2\\
&=\frac{1}{\hbar^2}\frac{\left|\left\langle f\middle|\widehat{F}\middle|t_0\right\rangle\right|^2}{\left(\omega_{f0}+\omega\right)^2}\left\{2-\exp\left[i\left(\omega_{f0}+\omega\right)t\right]-\exp\left[-i\left(\omega_{f0}+\omega\right)t\right]\right\}\\
&=\frac{2}{\hbar^2}\frac{\left|\left\langle f\middle|\widehat{F}\middle|t_0\right\rangle\right|^2}{\left(\omega_{f0}+\omega\right)^2}\left\{1-\cos\left[\left(\omega_{f0}+\omega\right)t\right]\right\}
\end{aligned} \tag{3.413}$$

while for the rate of transition we have:

$$\lim_{t\to\infty}\mathbf{R}_{Int/Emis}^{(1)}(t,t_0=0)=\lim_{t\to\infty}\left[\partial_t\wp_{Int/Emis}^{(1)}(t,t_0=0)\right]$$

$$=\frac{2\pi}{\hbar^2}\left|\left\langle f\left|\widehat{F}\right|t_0\right\rangle\right|^2\lim_{t\to\infty}\frac{\sin\left[\left(\omega_{f0}+\omega\right)t\right]}{\pi\left(\omega_{f0}+\omega\right)}$$

$$=\frac{2\pi}{\hbar^2}\left|\left\langle f\left|\widehat{F}\right|t_0\right\rangle\right|^2\delta\left(\omega_{f0}+\omega\right);$$

$$\lim_{t\to\infty}\mathbf{R}^{(1)}_{Int/Emis}(t,t_0=0)=\frac{2\pi}{\hbar}\left|\left\langle f\left|\widehat{F}\right|t_0\right\rangle\right|^2\delta\left(E_f-E_0+\hbar\omega\right) \tag{3.414}$$

where the delta-Dirac sinus-representation,

$$\delta\left(\omega\right)=\frac{1}{2\pi}\int_{-\infty}^{+\infty}e^{i\omega t}dt=\lim_{T\to\infty}\frac{1}{2\pi}\int_{-T}^{+T}e^{i\omega t}dt=\lim_{T\to\infty}\frac{e^{i\omega T}-e^{-i\omega T}}{2\pi i\omega}=\lim_{T\to\infty}\frac{1}{\pi}\frac{\sin\left(\omega T\right)}{\omega} \tag{3.415}$$

as well as the normalization-multiplication rule:

$$\delta\left(ax\right)=\frac{1}{|a|}\delta\left(x\right) \tag{3.416}$$

were considered.

In the similar manner, for absorption process one yield:

$$\lim_{t\to\infty}\mathbf{R}^{(1)}_{Int/Abs}(t,t_0=0)=\frac{2\pi}{\hbar}\left|\left\langle f\left|\widehat{F}^{+}\right|t_0\right\rangle\right|^2\delta\left(E_f-E_0-\hbar\omega\right) \tag{3.417}$$

From the condition both rate expression be non-zero there follows the Bohr transition rules (his second postulate); therefore they resulted as belonging to the present quantum evolution – interaction picture in the fist order of the periodic perturbation. Such output widely justifies why in practice the first order perturbation in most cases enough for realistically treating the quantum evolution; note that higher orders of perturbation will have higher orders of the Planck constant at denominator with the consequence of drastically diminishing the nominal result – corresponding with a significant veiling of the observed/measured phenomenon.

3.6.4 GREEN FUNCTION AND ITS CAUSAL PROPERTIES

Back to the very quantum mechanically problem: the wave-function; how it is intimately related with the quantum evolution between two sets of events, prepared or observed as "initial-final", characterized by the coordinates (x_1, t_1) and (x_2, t_2)? The answer came from the generalization of the *Huygens' optical principle*, according which the wave-front's amplitude on the event point (x_2, t_2) is determined (thus causally) by the sum of all amplitudes generated by the wave oscillations on the earlier event (x_1, t_1). In terms of the wave-function this is transposed as follows (written in 1D for the shake of simplicity):

$$\psi(x_2,t_2) = i\int G(x_2,t_2;x_1,t_1)\psi(x_1,t_1)dx_1 \ , \ t_2 > t_1 \tag{3.418}$$

Worth commenting upon this important relationship: firstly it introduces the so-called *Green function* with the role of bridging quantum events at the level of wave-functions; it was also earlier introduced with the same occasion of introducing wave-function formalism, however featuring here in a more formal and general fashion; it is as well related with the causality of events so that the above relation may be even more formally rewritten as:

$$\Theta(t_2 - t_1)\psi(x_2,t_2) = i\int G^+(x_2,t_2;x_1,t_1)\psi(x_1,t_1)dx_1 \tag{3.419}$$

thus explicitly specifying the causal ordering by the Heaviside-step function, while in this case the *retarded Green function* $G^+(x_2,t_2;x_1,t_1)$ was considered; In the case the *advanced Green function* $G^-(x_2,t_2;x_1,t_1)$ is concerned the causal equation becomes:

$$\Theta(t_1 - t_2)\psi(x_2,t_2) = -i\int G^-(x_2,t_2;x_1,t_1)\psi(x_1,t_1)dx_1 \tag{3.420}$$

Both Green functions have the property of factorization or composing across the time slicing, for example, for the time ordering $t_2 > t_1 > t_0$:

$$\begin{aligned}
\psi(x_2,t_2) &= i\int G^+(x_2,t_2;x_1,t_1)\left[\psi(x_1,t_1)\right]dx_1 \\
&= i\int G^+(x_2,t_2;x_1,t_1)\left[i\int G^+(x_1,t_1;x_0,t_0)\psi(x_0,t_0)dx_0\right]dx_1 \\
&= i\int\left[i\int G^+(x_2,t_2;x_1,t_1)G^+(x_1,t_1;x_0,t_0)dx_1\right]\psi(x_0,t_0)dx_0 \\
&\equiv i\int G^+(x_2,t_2;x_0,t_0)\left[\psi(x_0,t_0)\right]dx_0 \\
&\Rightarrow G^+(x_2,t_2;x_0,t_0) = i\int G^+(x_2,t_2;x_1,t_1)G^+(x_1,t_1;x_0,t_0)dx_1
\end{aligned} \tag{3.421}$$

and similarly for the time ordering $t_2 < t_1 < t_0$

$$G^-(x_2,t_2;x_0,t_0) = -i\int G^-(x_2,t_2;x_1,t_1)G^-(x_1,t_1;x_0,t_0)dx_1 \tag{3.422}$$

from where one may inferred also the formal identity (their reciprocal conjugated conversion)

$$G^+(x_b,t_b;x_a,t_a) = \left[G^-(x_a,t_a;x_b,t_b)\right]^* \tag{3.423}$$

while their mixed product generates the delta-Dirac representation, e.g., for time ordering $t_2 > t_1 < t_0$:

$$\begin{aligned}
\psi(x_2,t_2) &= i\int G^+(x_2,t_2;x_1,t_1)\left[\psi(x_1,t_1)\right]dx_1 \\
&\underset{t_1<t_0}{=} i\int G^+(x_2,t_2;x_1,t_1)\left[-i\int G^-(x_1,t_1;x_0,t_0)\psi(x_0,t_0)dx_0\right]dx_1 \\
&= \int\left[\int G^+(x_2,t_2;x_1,t_1)G^-(x_1,t_1;x_0,t_0)dx_1\right]\psi(x_0,t_0)dx_0 \\
&\equiv \int\delta(x_0-x_2)\left[\psi(x_0,t_0)\right]dx_0 \\
&\Rightarrow \delta(x_0-x_2) = \int G^+(x_2,t_2;x_1,t_1)G^-(x_1,t_1;x_0,t_0)dx_1
\end{aligned} \tag{3.424}$$

and similarly for the time ordering $t_2 < t_1 > t_0$

$$\delta(x_0-x_2) = \int G^-(x_2,t_2;x_1,t_1)G^+(x_1,t_1;x_0,t_0)dx_1 \tag{3.425}$$

However, from now on we will consider only retarded Green function if not otherwise indicated; yet, when do so, we practically consider it equivalent with the so-called propagator of quantum effects:

$$G^+(x_2,t_2;x_1,t_1) \equiv (x_2,t_2|x_1,t_1) \tag{3.426}$$

Another meaningful observation is that the Green function convolution (integration) with (source) wave-function is done upon the coordinate only in the basic definition, while the entirely time evolution is contained within the Green function. In relation with this aspect one may establish also the Green function connection with the time evolution operator, through remembering that

$$|t_2\rangle = \widehat{U}(t_2,t_1)|t_1\rangle \tag{3.427}$$

which may equivalently rewritten as in coordinate representation to recover the wave-function relationship:

$$\begin{aligned}\psi(x_2,t_2) &= \langle x_2|t_2\rangle \\ &= \langle x_2|\widehat{U}(t_2,t_1)\hat{1}_{\{|x_1\rangle\}}|t_1\rangle \\ &= \int \langle x_2|\widehat{U}(t_2,t_1)|x_1\rangle \underbrace{\langle x_1|t_1\rangle}_{\psi(x_1,t_1)} dx_1 \\ &= \int \langle x_2|\widehat{U}(t_2,t_1)|x_1\rangle \psi(x_1,t_1)dx_1\end{aligned} \tag{3.428}$$

from where, by analogy with the above Green function definition there follows the connection:

$$iG(x_2,t_2;x_1,t_1) = \langle x_2|\widehat{U}(t_2,t_1)|x_1\rangle \tag{3.429}$$

Therefore one may interpret the appearance of the Green function as related with the matrix elements of the evolution operator in the coordinate representation in linking wave-function causality. Moreover, the above relation with evolution operator allows the formulation of the spectral representation for the Green function

$$\begin{aligned}G(x_2,t_2;x_1,t_1) &= -i\langle x_2|\hat{1}_{\{|\varphi_E\rangle\}}\widehat{U}(t_2,t_1)\hat{1}_{\{|\varphi_E\rangle\}}|x_1\rangle \\ &= -i\sum_E \langle x_2|\varphi_E\rangle \underbrace{\langle \varphi_E|e^{-\frac{i}{\hbar}\widehat{H}t_2}}_{\exp\left(-\frac{i}{\hbar}Et_2\right)\langle\varphi_E|} \underbrace{e^{\frac{i}{\hbar}\widehat{H}t_1}|\varphi_E\rangle}_{|\varphi_E\rangle\exp\left(-\frac{i}{\hbar}Et_1\right)} \langle \varphi_E|x_1\rangle \\ &= -i\sum_E \varphi_E^*(x_2)\varphi_E(x_1)\exp\left[-\frac{i}{\hbar}E(t_2-t_1)\right]\end{aligned} \tag{3.430}$$

when the closure relation is inserted in the evolution operator matrix elements in terms of energy eigen-function $\varphi_E(x)$ from the spectrum of the systems Hamiltonian.

In next, giving its intimacy with wave-function one may whish to establishing the Green function equation, i.e., the analogues of that specific to Schrödinger wave-function equation in coordinate representation:

$$i\hbar\partial_t\psi(x,t) = \widehat{H}(t)\psi(x,t) \tag{3.431}$$

this mat be achieved through applying the operatorial action $[i\hbar\partial_t - \widehat{H}(t)]$ over the, let's say, the retarded Green function to obtain successively:

$$\begin{aligned}
& i\int\left[i\hbar\partial_{t_2} - \widehat{H}(t_2)\right]G^{+}(x_2,t_2;x_1,t_1)\psi(x_1,t_1)dx_1 \\
&= \left[i\hbar\partial_{t_2} - \widehat{H}(t_2)\right]\Theta(t_2-t_1)\psi(x_2,t_2) \\
&= i\hbar\partial_t\left[\Theta(t_2-t_1)\psi(x_2,t_2)\right] - \Theta(t_2-t_1)\widehat{H}(t)\psi(x_2,t_2) \\
&= i\hbar\psi(x_2,t_2)\underbrace{\partial_{t_2}\Theta(t_2-t_1)}_{\delta(t_2-t_1)} + \Theta(t_2-t_1)\underbrace{i\hbar\partial_{t_2}\psi(x_2,t_2)}_{\widehat{H}(t_2)\psi(x_2,t_2)} - \Theta(t_2-t_1)\widehat{H}(t_2)\psi(x_2,t_2) \\
&= i\hbar\psi(x_2,t_2)\delta(t_2-t_1)
\end{aligned} \tag{3.432}$$

so that in order the end sides of this equality chain to hold for any wave-function arguments there results that:

$$\left[i\hbar\partial_{t_2} - \widehat{H}(t_2)\right]G^{+}(x_2,t_2;x_1,t_1) = \hbar\delta(x_2-x_1)\delta(t_2-t_1) \tag{3.433}$$

which obviously fulfills the above expressing and thus constitutes as the Green function equation, corresponding to the wave-function Schrödinger equation; note that in this derivation the delta-Dirac function representation as the derivative of the Heaviside step function was employed base don the complex integral representation of the last:

$$\begin{aligned}
\frac{d}{dt}\Theta(t) &= -\frac{1}{2\pi i}\lim_{\varepsilon\to 0}\int_{-\infty}^{+\infty}\frac{d}{dt}\frac{e^{-i\omega t}}{\omega+i\varepsilon}d\omega = \frac{1}{2\pi}\int_{-\infty}^{+\infty}\lim_{\varepsilon\to 0}\left[\frac{\omega e^{-i\omega t}}{\omega+i\varepsilon}\right]d\omega \\
&= \frac{1}{2\pi}\int_{-\infty}^{+\infty}e^{-i\omega t}d\omega = \delta(t)
\end{aligned} \tag{3.434}$$

There is clear now that the Green function formalism do not automatically solve the initial Schrödinger problem, but replaces it with a more general one when also the causality of events counts. The passage from the retarded to advanced Green function equation can be easily made though the previously stipulated recipe.

Another important consequence of the Green function equation resides in the fact it introduces both the time- and coordinate- integration through the associate closure relation:

$$1=\int_{-\infty}^{+\infty}\int_{-\infty}^{+\infty}\delta(x_2-x_1)\delta(t_2-t_1)dx_2dt_2$$
$$=\frac{1}{\hbar}\int_{-\infty}^{+\infty}\int_{-\infty}^{+\infty}dx_2dt_2\left[i\hbar\partial_{t_2}-\widehat{H}(t_2)\right]G^{+}(x_2,t_2;x_1,t_1) \quad (3.435)$$

the actual discussion opens the possibility of unifying the time and coordinate integration measure into a single space-time one:

$$ds=dxdt,\qquad s=(x,t) \quad (3.436a)$$

$$\delta(s_2-s_1)=\delta(x_2-x_1)\delta(t_2-t_1) \quad (3.436b)$$

so that the Green function normalization relation simplifies as

$$1=\int_{-\infty}^{+\infty}\delta(s_2-s_1)ds_2=\frac{1}{\hbar}\int_{-\infty}^{+\infty}ds_2\left[i\hbar\partial_{t_2}-\widehat{H}(t_2)\right]G^{+}(s_2;s_1) \quad (3.437)$$

associate with the equation

$$\left[i\hbar\partial_{t_2}-\widehat{H}(t_2)\right]G^{+}(s_2;s_1)=\hbar\delta(s_2-s_1) \quad (3.438)$$

These forms may be further employed for the case the Hamiltonian of the system is considered as being composed from the free- and (time dependent) perturbed part:

$$\widehat{H}(s)=\widehat{H}_0+\widehat{V}(s) \quad (3.439)$$

in which case they are individuated for the *free Green function* $G_0^+(s_2;s_1)$:

$$1_{G_0^+} = \int_{-\infty}^{+\infty} \delta(s_2 - s_1) ds_2 = \frac{1}{\hbar}\int_{-\infty}^{+\infty} ds_2 \left[i\hbar\partial_{t_2} - \widehat{H}_0 \right] G_0^+(s_2;s_1) \quad (3.440a)$$

$$\hbar\delta(s_2 - s_1) = \left[i\hbar\partial_{t_2} - \widehat{H}_0 \right] G_0^+(s_2;s_1) \quad (3.440b)$$

while for the perturbed Hamiltonian we use them formally to consecutively get:

$$\begin{aligned}
&\left[i\hbar\partial_{t_2} - \widehat{H}_0 + \widehat{V}(s_2) \right] G^+(s_2;s_1) = \hbar\delta(s_2 - s_1) \\
&\Leftrightarrow \left[i\hbar\partial_{t_2} - \widehat{H}_0 \right] G^+(s_2;s_1) = \hbar\delta(s_2 - s_1) + \widehat{V}(s_2)G^+(s_2;s_1) \\
&= \left[i\hbar\partial_{t_2} - \widehat{H}_0 \right] G_0^+(s_2;s_1) + \left[1_{G_0^+} \right] \widehat{V}(s_2)G^+(s_2;s_1) \\
&= \left[i\hbar\partial_{t_2} - \widehat{H}_0 \right] G_0^+(s_2;s_1) + \int \delta(s_2 - s)\widehat{V}(s)G^+(s;s_1) ds \\
&= \left[i\hbar\partial_{t_2} - \widehat{H}_0 \right] G_0^+(s_2;s_1) + \frac{1}{\hbar}\int \left[i\hbar\partial_{t_2} - \widehat{H}_0 \right] G_0^+(s_2;s)\widehat{V}(s)G^+(s;s_1) ds
\end{aligned} \quad (3.441)$$

leaving with the identity among the free and full Green functions:

$$G^-(s_2;s_1) = G_0^+(s_2;s_1) + \int G_0^+(s_2;s)\widehat{\widehat{V}}(s)G^+(s;s_1) ds \quad (3.442)$$

with the notation:

$$\widehat{\widehat{V}}(s) = \frac{1}{\hbar}\widehat{V}(s) \quad (3.443)$$

as corresponding to the *Lippmann-Schwinger equation for the Green functions or propagators*. It emphasizes on the role the free Green function plays in determining the Green function of the perturbed systems. Moreover, seeing the potential appearance as a scattering process that affects (perturb) the free motion, the Lippmann-Schwinger equation may be unfolded through its iterative application generating the so-called *Dyson series* quoting the single, double, and multiple scatterings:

$$G^+(s_2;s_1) = G_0^+(s_2;s_1)$$

$$+\int G_0^+(s_2;s)\hat{\hat{V}}(s)\left[G_0^+(s;s_1)+\int G_0^+(s;s')\hat{\hat{V}}(s')G^+(s';s_1)ds'\right]ds$$
$$=G_0^+(s_2;s_1)+\int G_0^+(s_2;s)\hat{\hat{V}}(s)G_0^+(s;s_1)ds$$
$$+\int G_0^+(s_2;s)\hat{\hat{V}}(s)\left[\int G_0^+(s;s')\hat{\hat{V}}(s')G^+(s';s_1)ds'\right]ds$$
$$=G_0^+(s_2;s_1)+\int G_0^+(s_2;s)\hat{\hat{V}}(s)G_0^+(s;s_1)ds$$
$$+\int G_0^+(s_2;s)\hat{\hat{V}}(s)\left[\int G_0^+(s;s')\hat{\hat{V}}(s')\begin{bmatrix}G_0^+(s';s_1)+\\ \int G_0^+(s';s'')\hat{\hat{V}}(s'')G^+(s'';s_1)ds''\end{bmatrix}ds'\right]ds$$
$$=G_0^+(s_2;s_1)+\int G_0^+(s_2;s)\hat{\hat{V}}(s)G_0^+(s;s_1)ds$$
$$+\int\int G_0^+(s_2;s)\hat{\hat{V}}(s)G_0^+(s;s')\hat{\hat{V}}(s')G_0^+(s';s_1)ds'ds+\ldots$$

(3.444)

There is immediate that having this Dyson series of Green function the wave function solution of the perturbed systems may be written according with the Huygens' quantum principle as:

$$\psi^{(+)}(s_2)=i\int G^+(s_2;s_1)\psi(s_1)ds_1$$
$$=i\int\left[G_0^+(s_2;s_1)+\int G_0^+(s_2;s)\hat{\hat{V}}(s)G^+(s;s_1)ds\right]\psi(s_1)ds_1$$
$$=i\int G_0^+(s_2;s_1)\psi(s_1)ds_1+\int G_0^+(s_2;s)\hat{\hat{V}}(s)\left[i\int G^+(s;s_1)\psi(s_1)ds_1\right]ds;$$
$$\psi^{(+)}(s_2)=\underbrace{\phi(s_2)}_{\substack{free\ or\ plane\\ wave}}+\underbrace{\int G_0^+(s_2;s)\hat{\hat{V}}(s)\psi^{(+)}(s)ds}_{\substack{scattered\\ wave}}$$

(3.445)

telling that it may be regarded as the superposition of two waves: one unperturbed, free or plane wave $\varphi(s_2)$, and one scattered, however written in terms of the same unknown general wave-function. It represents the actual form (with Green function) for the earlier perturbed quantum state expansion by the Born series, from where follows the identifications:

$$\hat{V}\equiv\hat{H}_1 \quad (3.446)$$

$$G_0^+ \equiv \frac{\hbar}{E - E_0 + i\eta} \tag{3.447}$$

while the last will be proved bellow within a direct analytical context. Nevertheless, the analogy allows resuming the perturbed wave-function *in terms of plane waves*:

$$\psi^{(+)}(s_2) = i\int G^+(s_2;s_1)\phi(s_1)ds_1 \tag{3.448}$$

with the help of which one may construct also the so-called *scattering matrix* (*S-matrix*) through representing it on some final (free or plane) state $|\phi_f(x_2,t_2)\rangle$ by means of bra-ket convolution:

$$\begin{aligned}
S_{fi} &= \langle\phi_f(s_2)|\psi_i^{(+)}(s_2)\rangle \\
&= \int\phi_f^*(s_2)\psi_i^{(+)}(s_2)ds_2 \\
&= -i\int\int\phi_f^*(s_2)G^+(s_2;s_1)\phi_i(s_1)ds_2ds_1 \\
&= -i\int\int\phi_f^*(s_2)G_0^+(s_2;s_1)\phi_i(s_1)ds_2ds_1 \\
&\quad -i\int\int\phi_f^*(s_2)\left[\int G_0^+(s_2;s)\hat{\hat{V}}(s)G_0^+(s;s_1)ds\right]\phi_i(s_1)ds_2ds_1 \\
&\quad -i\int\int\phi_f^*(s_2)\left[\int\int \begin{matrix} G_0^+(s_2;s)\hat{\hat{V}}(s)G_0^+(s;s') \\ \hat{\hat{V}}(s')G_0^+(s';s_1)ds'ds \end{matrix}\right]\phi_i(s_1)ds_2ds_1 - \ldots
\end{aligned} \tag{3.449}$$

that may be simplified recognizing that some of involved integrals may be rearranged so that to systematically apply the Huygens rule, for instance:

$$\int G_0^+(s_2;s_1)\phi_i(s_1)ds_1 \underset{t_2>t_1}{=} -i\phi_i(s_2) \tag{3.450a}$$

$$\int\phi_f^*(s_2)G_0^+(s_2;s)ds_2 \underset{t_2>t}{=} +i\phi_f^*(s),\ \int G_0^+(s;s_1)\phi_i(s_1)ds_1 \underset{t>t_1}{=} -i\phi_i(s) \tag{3.450b}$$

this way producing the result:

$$\begin{aligned}
S_{fi} &= \delta(k_f - k_i)\delta(\omega_f - \omega_i) \\
&\quad -i\int\phi_f^*(s)\hat{\hat{V}}(s)\phi_i(s)ds \\
&\quad +i\int\int\phi_f^*(s)\hat{\hat{V}}(s)G_0^+(s;s')\hat{\hat{V}}(s')\phi_i(s')dsds' - \ldots
\end{aligned} \tag{3.451}$$

in terms of perturbation potentials and free Green functions at various scattering events, in various scattering modes, i.e., no scattering-the first term, single scattering – the second term, double scattering –the third terms, etc., averaged over the initial and final (ad infinitum, in past and future times) plane waves:

$$\phi(s)_{s=(x,t)} = \frac{1}{\sqrt{2\pi\hbar}}\exp\left[i(kx-\omega t)\right] = \frac{1}{\sqrt{2\pi\hbar}}\exp\left[\frac{i}{\hbar}(px-Et)\right] \tag{3.452}$$

However for that the present theory be complete a special discussion on the free Green function and of its analytical forms is compulsory; it will be undertaken in what follows.

3.6.5 FREE PARTICLE'S PROPAGATOR

There are many ways for evaluation and equivalent forms of free Green function. Here, we will start with evaluating its Fourier transformation

$$G_0^+(x_2,t_2;x_1,t_1) = \iint \frac{dpdE}{(2\pi\hbar)^2} G_0^+(p,E)\exp\left[\frac{i}{\hbar}p(x_2-x_1)\right]\exp\left[-\frac{i}{\hbar}E(t_2-t_1)\right] \tag{3.453}$$

through the successive identities, starting with the associate Schrödinger like equation:

$$\begin{aligned}
\hbar\delta(x_2-x_1)\delta(t_2-t_1) &= \left(i\hbar\partial_{t_2} - \widehat{H}_0\right)G_0^+(x_2,t_2;x_1,t_1) \\
&= \left(i\hbar\partial_{t_2} + \frac{\hbar^2}{2m}\partial_{x_2}^2\right)G_0^+(x_2,t_2;x_1,t_1) \\
&= \iint \frac{dpdE}{(2\pi\hbar)^2}\left(E - \frac{p^2}{2m}\right)G_0^+(p,E)\exp\left[\frac{i}{\hbar}p(x_2-x_1)\right] \\
&\quad \times \exp\left[-\frac{i}{\hbar}E(t_2-t_1)\right]
\end{aligned} \tag{3.454}$$

while from the delta Dirac definition we have

$$\hbar\delta(x_2-x_1)\delta(t_2-t_1)=\hbar\iint\frac{dpdE}{(2\pi\hbar)^2}\exp\left[\frac{i}{\hbar}p(x_2-x_1)\right]\exp\left[-\frac{i}{\hbar}E(t_2-t_1)\right]$$

(3.455)

From the need the two expressions be equivalent we find the momentum-energy retarded free Green function as:

$$G_0^+(p,E)=\frac{\hbar}{E-\frac{p^2}{2m}+i\eta} \tag{3.456}$$

where the analytic continuation follows the Feynman prescription in order to avoid energy singularity through the integration. This expression is in accordance with that one found from previous U-picture intervening in the Born series for treating the temporal perturbation in Lippmann-Schwinger equation.

With this the space-time free Green function becomes:

$$G_0^+(x_2,t_2;x_1,t_1)=\hbar\int_{-\infty}^{+\infty}\frac{dp}{2\pi\hbar}\exp\left[\frac{i}{\hbar}p(x_2-x_1)\right]\int_{-\infty}^{+\infty}\frac{1}{2\pi\hbar}\frac{\exp\left[-\frac{i}{\hbar}E(t_2-t_1)\right]}{E-\frac{p^2}{2m}+i\eta}dE$$

$$=\hbar\int_{-\infty}^{+\infty}\frac{dp}{2\pi\hbar}\exp\left[\frac{i}{\hbar}p(x_2-x_1)\right]\int_{-\infty}^{+\infty}\frac{1}{2\pi\hbar}\frac{\exp\left[-\frac{i}{\hbar}\left(E'+\frac{p^2}{2m}\right)(t_2-t_1)\right]}{E'+i\eta}dE'$$

$$=i\int_{-\infty}^{+\infty}\frac{dp}{2\pi\hbar}\exp\left[\frac{i}{\hbar}p(x_2-x_1)-\frac{i}{\hbar}\frac{p^2}{2m}(t_2-t_1)\right]\underbrace{\int_{-\infty}^{+\infty}\frac{1}{2\pi i}\frac{\exp\left[-\frac{i}{\hbar}E'(t_2-t_1)\right]}{E'+i\eta}dE'}_{-\Theta\left(\frac{t_2-t_1}{\hbar}\right)}$$

$$=-i\int_{-\infty}^{+\infty}\frac{dp}{2\pi\hbar}\exp\left\{\frac{i}{\hbar}\left[p(x_2-x_1)-\frac{p^2}{2m}(t_2-t_1)\right]\right\}\Theta(t_2-t_1);$$

$$G_0^+(x_2,t_2;x_1,t_1)=-i\Theta(t_2-t_1)\int\phi_p(x_2,t_2)\phi_p^*(x_1,t_1)dp$$

(3.457)

in terms of the (free) plane wave $\phi_p(x,t)$, and where the integral representation of the step function was employed along its multiplication property

$$\Theta(\alpha \bullet) = \Theta(\bullet),\ \alpha > 0 \tag{3.458}$$

Remarkably, this expression is in accordance with the spectral expansion of the general Green function, however here adapted to the free motion case. This is another confirmation for that the momentum-energy free Green function takes indeed the above $G_0^+(p,E)$ expression. However, going further with the free Green function evaluation in terms of plane wave, one yields the following transformations:

$$G_0^+(x_2,t_2;x_1,t_1) = -\frac{i}{2\pi\hbar}\int_{-\infty}^{+\infty} dp \exp\left\{-\frac{i}{\hbar}\left[\frac{p^2}{2m}(t_2-t_1) - p(x_2-x_1)\right]\right\}\Theta(t_2-t_1)$$

$$= -\frac{i}{2\pi\hbar}\int_{-\infty}^{+\infty} dp \exp\left\{-\frac{i}{\hbar}\left[\frac{p\sqrt{t_2-t_1}}{\sqrt{2m}} - \frac{\sqrt{2m}(x_2-x_1)}{2\sqrt{t_2-t_1}}\right]^2 + \frac{i}{\hbar}\frac{m(x_2-x_1)^2}{2(t_2-t_1)}\right\}\Theta(t_2-t_1)$$

$$= -\frac{i}{2\pi\hbar}\int_{-\infty}^{+\infty} dp \exp\left\{-\frac{i}{\hbar}\frac{t_2-t_1}{2m}\left[p - \frac{m(x_2-x_1)}{t_2-t_1}\right]^2 + \frac{i}{\hbar}\frac{m(x_2-x_1)^2}{2(t_2-t_1)}\right\}\Theta(t_2-t_1)$$

$$= -\frac{i}{2\pi\hbar}\exp\left[\frac{i}{\hbar}\frac{m(x_2-x_1)^2}{2(t_2-t_1)}\right]\int_{-\infty}^{+\infty} dp \exp\left\{-i\frac{t_2-t_1}{2m\hbar}\underbrace{\left[p - \frac{m(x_2-x_1)}{t_2-t_1}\right]}_{\xi}^2\right\}\Theta(t_2-t_1)$$

$$= -\frac{i}{2\pi\hbar}\exp\left[\frac{i}{\hbar}\frac{m(x_2-x_1)^2}{2(t_2-t_1)}\right]\underbrace{\int_{-\infty}^{+\infty} d\xi \exp\left\{-i\underbrace{\frac{t_2-t_1}{2m\hbar}}_{a}\xi^2\right\}}_{\int_{-\infty}^{+\infty} d\xi \exp(-ia\xi^2) = \sqrt{\frac{\pi}{ia}}}\Theta(t_2-t_1)$$

$$= -\frac{i}{2\pi\hbar}\sqrt{\frac{2\pi m\hbar}{i(t_2-t_1)}}\exp\left[\frac{i}{\hbar}\frac{m(x_2-x_1)^2}{2(t_2-t_1)}\right]\Theta(t_2-t_1);$$

$$G_0^+(x_2,t_2;x_1,t_1) = -i\sqrt{\frac{m}{2\pi i\hbar(t_2-t_1)}}\exp\left[i\frac{m(x_2-x_1)^2}{2\hbar(t_2-t_1)}\right]\Theta(t_2-t_1) \tag{3.459}$$

while the advanced free Green function will read

$$G_0^-(x_2,t_2;x_1,t_1)=i\sqrt{\frac{m}{2\pi i\hbar(t_2-t_1)}}\exp\left[i\frac{m(x_2-x_1)^2}{2\hbar(t_2-t_1)}\right]\Theta(t_1-t_2) \quad (3.460)$$

Finally, worth noting that the 3D generalization is straightforward and the result may be at once written down from the 1D analysis by the propagator:

$$G_0^+(\mathbf{r}_2,t_2;\mathbf{r}_1,t_1)=-i\left[\frac{m}{2\pi i\hbar(t_2-t_1)}\right]^{3/2}\exp\left[i\frac{m|\mathbf{r}_2-\mathbf{r}_1|^2}{2\hbar(t_2-t_1)}\right]\Theta(t_2-t_1) \quad (3.461)$$

The free propagator plays an important role in path integral formulation of quantum mechanics and will be reloaded with that occasion soon below. Yet, it remains the problem of assigning the free Green function the stationary expression, i.e., when is not viewed as a propagator, a matter that will be as well unfolded later when describing the scattering process as a measurement tool for quantum phenomena, see the last postulate of quantum mechanics in this chapter.

3.7 STATIONARY PERTURBATIONS

The physical reality shows that systems, even at microscopic level, are not isolated, but subjects of various perturbations. While temporal perturbations were treated within causal motion by the preceding postulate, the stationary ones are the subject of the present discourse:

The stationary perturbations are those that do not affect the system structure but only its eigen-spectrum (eigen-energies and eigen-functions).

Analytically, this postulate may be formulated through the coupling Hamiltonian:

$$\widehat{H}(\lambda)=\widehat{H}_0+\lambda\widehat{H}_1,\quad \lambda\in[0,1] \quad (3.462)$$

whose associate eigen-problem

$$\widehat{H}(\lambda)|\mathrm{E}(\lambda)\rangle=\mathrm{E}(\lambda)|\mathrm{E}(\lambda)\rangle \quad (3.463)$$

has the eigen-solutions

$$|\mathrm{E}(\lambda)\rangle = \underbrace{|0\rangle}_{\substack{ISOLATED\\STATE}} + \lambda \underbrace{|1\rangle}_{\substack{FIRST\\CORRECTED\\STATE}} + \lambda^2 \underbrace{|2\rangle}_{\substack{SECOND\\CORRECTED\\STATE}} + ... \tag{3.464}$$

$$\mathrm{E}(\lambda) = \underbrace{\mathrm{E}^{(0)}}_{\substack{UNPERTURBED\\ENERGY}} + \lambda \underbrace{\mathrm{E}^{(1)}}_{\substack{FIRST\\ORDER\\ENERGY}} + \lambda^2 \underbrace{\mathrm{E}^{(2)}}_{\substack{SECOND\\ORDER\\ENERGY}} + ... \tag{3.465}$$

as eigen-state and eigen-energy λ-(coupling)expansions.

However, the problem is to express the corrected terms of perturbation is terms of eigen-energies and wave functions of the non-perturbed (isolated) system, assumed with the complete determined solution. Yet, note that the perturbed eigen-function are not per se normalized, the normalization procedure being reload for each order the preservative approximation is considered.

The general algorithm of finding eigen-energies and states as well as some basic atomic, molecular and free solid states applications follows.

3.7.1 GENERAL PERTURBATION ALGORITHM

Let's consider the non-degenerate non-perturbed discrete (stationary) solved problem

$$\widehat{H}_0|\varepsilon_k\rangle = \varepsilon_k|\varepsilon_k\rangle \tag{3.466}$$

whose eigen-states made the ortho-normalized basis:

$$\langle\varepsilon_j|\varepsilon_k\rangle = \delta_{jk} \tag{3.467}$$

$$\hat{1}_{\{|\varepsilon_k\rangle\}} = \sum_{\substack{k=1\\k\in\mathrm{N}}}^{\infty}|\varepsilon_k\rangle\langle\varepsilon_k| \tag{3.468}$$

In these conditions, the perturbed eigen-states are generically written as a superposition of all non-perturbed eigen-energies:

$$|\mathrm{E}(\lambda)\rangle = \sum_k c_k(\lambda)|\varepsilon_k\rangle \tag{3.469}$$

while the perturbation itself is comprised in the coefficients' expansion upon the above recipe:

$$c_k(\lambda) = c_k^{(0)} + \lambda c_k^{(1)} + \lambda^2 c_k^{(2)} + \ldots \tag{3.470}$$

With this specialization, the perturbed eigen-problem equivalently becomes:

$$\begin{aligned}
&\widehat{H}(\lambda)\left|\mathrm{E}(\lambda)\right\rangle = \mathrm{E}(\lambda)\left|\mathrm{E}(\lambda)\right\rangle \\
&\Leftrightarrow \left(\widehat{H}_0 + \lambda \widehat{H}_1\right)\sum_k c_k(\lambda)\left|\varepsilon_k\right\rangle = \mathrm{E}(\lambda)\sum_k c_k(\lambda)\left|\varepsilon_k\right\rangle \\
&\Leftrightarrow \sum_k c_k(\lambda)\underbrace{\widehat{H}_0\left|\varepsilon_k\right\rangle}_{\varepsilon_k\left|\varepsilon_k\right\rangle} + \lambda\sum_k c_k \widehat{H}_1\left|\varepsilon_k\right\rangle = \mathrm{E}(\lambda)\sum_k c_k(\lambda)\left|\varepsilon_k\right\rangle \\
&\overset{\left\langle\varepsilon_j\right|}{\Leftrightarrow} \sum_k c_k(\lambda)\varepsilon_k \underbrace{\left\langle\varepsilon_j\middle|\varepsilon_k\right\rangle}_{\delta_{ik}} + \lambda\sum_k c_k\left\langle\varepsilon_j\middle|\widehat{H}_1\middle|\varepsilon_k\right\rangle = \mathrm{E}(\lambda)\sum_k c_k(\lambda)\underbrace{\left\langle\varepsilon_j\middle|\varepsilon_k\right\rangle}_{\delta_{jk}} \\
&\Leftrightarrow c_j(\lambda)\left[\mathrm{E}(\lambda) - \varepsilon_j\right] = \lambda\sum_k\left\langle\varepsilon_j\middle|\widehat{H}_1\middle|\varepsilon_k\right\rangle c_k \\
&\Leftrightarrow \left[c_j^{(0)} + \lambda c_j^{(1)} + \lambda^2 c_j^{(2)} + \ldots\right]\left[\mathrm{E}^{(0)} - \varepsilon_j + \lambda\mathrm{E}^{(1)} + \lambda^2\mathrm{E}^{(2)} + \ldots\right] \\
&= \lambda\sum_k\left\langle\varepsilon_j\middle|\widehat{H}_1\middle|\varepsilon_k\right\rangle\left[c_k^{(0)} + \lambda c_k^{(1)} + \lambda^2 c_k^{(2)} + \ldots\right]
\end{aligned} \tag{3.471}$$

from where, by equal power of coefficients one successively gets the cut-offs:

Order (0):

$$c_j^{(0)}\left[\mathrm{E}^{(0)} - \varepsilon_j\right] = 0 \tag{3.472}$$

Order (1):

$$\left[\mathrm{E}^{(0)} - \varepsilon_j\right]c_j^{(1)} + \mathrm{E}^{(1)}c_j^{(0)} = \sum_k\left\langle\varepsilon_j\middle|\widehat{H}_1\middle|\varepsilon_k\right\rangle c_k^{(0)} \tag{3.473}$$

Order (2):

$$\left[\mathrm{E}^{(0)} - \varepsilon_j\right]c_j^{(2)} + \mathrm{E}^{(1)}c_j^{(1)} + \mathrm{E}^{(2)}c_j^{(0)} = \sum_k\left\langle\varepsilon_j\middle|\widehat{H}_1\middle|\varepsilon_k\right\rangle c_k^{(1)} \tag{3.474}$$

…

Order (*p*):

$$\left[\mathrm{E}^{(0)}-\varepsilon_j\right]c_j^{(p)}+\mathrm{E}^{(1)}c_j^{(p-1)}+\ldots+\mathrm{E}^{(p)}c_j^{(0)}=\sum_k\left\langle\varepsilon_j\middle|\widehat{H}_1\middle|\varepsilon_k\right\rangle c_k^{(p)} \tag{3.475}$$

Let's now analyze each order in perturbation, based on the above separate, however somehow iterative, equations.

Order (0): The solution of this (unperturbed) problem is immediate:

$$\mathrm{E}^{(0)}=\varepsilon_n \tag{3.476}$$

recovering the whole isolated energy spectrum, while for the wave-function reads as:

$$\left|\mathrm{E}^{(0)}\right\rangle=\sum_k c_k^{(0)}\left|\varepsilon_k\right\rangle\overset{!}{=}\left|\varepsilon_n\right\rangle \tag{3.477}$$

from where there follows the necessary identity:

$$c_k^{(0)}=\delta_{kn} \tag{3.478}$$

so that the 0th order equation is verified as:

$$c_j^{(0)}\left[\mathrm{E}^{(0)}-\varepsilon_j\right]=0\Leftrightarrow\delta_{jn}\left[\varepsilon_n-\varepsilon_j\right]=0 \tag{3.479}$$

Order (1): Here, apart of employing the results of the order (0) perturbation analysis, two cases are distinguished, namely one in which the associate equation is specialized for some $j=n$ in the non-perturbed discrete spectrum that gives:

$$\underbrace{\left[\varepsilon_n-\varepsilon_n\right]}_{0}c_n^{(1)}+\mathrm{E}^{(1)}\underbrace{\delta_{nn}}_{1}=\sum_k\left\langle\varepsilon_n\middle|\widehat{H}_1\middle|\varepsilon_k\right\rangle\delta_{kn} \tag{3.480}$$

releasing the first order energy perturbation

$$\mathrm{E}^{(1)}=\left\langle\varepsilon_n\middle|\widehat{H}_1\middle|\varepsilon_n\right\rangle \tag{3.481}$$

as the average of the perturbation Hamiltonian over the non-perturbed eigen-states, while emphasizing on impossibility of $c_n^{(1)}$ evaluation since canceling its energy multiplication, but assumed with indeterminate expression:

$$c_n^{(1)} = \delta_{jn} \mathrm{Z}^{(1)} \tag{3.482}$$

Instead, for the case in which $j \neq n$ the corrected energy vanishes while allowing the determination of the first order perturbation coefficient:

$$\left[\varepsilon_n - \varepsilon_j\right] c_{j\neq n}^{(1)} + \mathrm{E}^{(1)} \underbrace{\delta_{j\neq n}}_{0} = \sum_k \left\langle \varepsilon_j \middle| \widehat{H}_1 \middle| \varepsilon_k \right\rangle \delta_{kn} \tag{3.483}$$

$$c_{j\neq n}^{(1)} = \frac{\left\langle \varepsilon_j \middle| \widehat{H}_1 \middle| \varepsilon_n \right\rangle}{\varepsilon_n - \varepsilon_j} \tag{3.484}$$

Combining both cases the first order perturbation coefficient of the perturbed wave-function looks like:

$$c_j^{(1)} = \delta_{jn} \mathrm{Z}^{(1)} + \left(1 - \delta_{jn}\right) \frac{\left\langle \varepsilon_j \middle| \widehat{H}_1 \middle| \varepsilon_n \right\rangle}{\varepsilon_n - \varepsilon_j} \tag{3.485}$$

Order (2): The same procedure as for the previous order applies, however with a supplemented degree of complication since considering the results and cases raised from lower orders. As such, for the $j = n$ case the original equation of second order perturbation unfolds as:

$$\begin{aligned} &\underbrace{\left[\varepsilon_n - \varepsilon_n\right]}_{0} c_j^{(2)} + \mathrm{E}^{(1)} \underbrace{c_{j=n}^{(1)}}_{\mathrm{Z}^{(1)}} + \mathrm{E}^{(2)} \underbrace{\delta_{nn}}_{1} = \sum_k \left\langle \varepsilon_n \middle| \widehat{H}_1 \middle| \varepsilon_k \right\rangle \overbrace{c_k^{(1)}}^{k=n \quad k\neq n} \\ &\Leftrightarrow \mathrm{E}^{(2)} = \underbrace{-\mathrm{E}^{(1)} \mathrm{Z}^{(1)} + \underbrace{\left\langle \varepsilon_n \middle| \widehat{H}_1 \middle| \varepsilon_n \right\rangle}_{\mathrm{E}^{(1)}} \underbrace{c_{k=n}^{(1)}}_{\mathrm{Z}^{(1)}}}_{0} + \sum_k \left\langle \varepsilon_n \middle| \widehat{H}_1 \middle| \varepsilon_k \right\rangle \underbrace{c_{k\neq n}^{(1)}}_{\frac{\left\langle \varepsilon_k \middle| \widehat{H}_1 \middle| \varepsilon_n \right\rangle}{\varepsilon_n - \varepsilon_k}} \end{aligned} \tag{3.486}$$

until the expression of the second order energy perturbation:

$$\mathrm{E}^{(2)} = \sum_{k\neq n} \frac{\left| \left\langle \varepsilon_k \middle| \widehat{H}_1 \middle| \varepsilon_n \right\rangle \right|^2}{\varepsilon_n - \varepsilon_k} \tag{3.487}$$

while leaving, as before, the $j = n$ second order coefficient of wave-function expansion as undetermined:

$$c_n^{(2)} = \delta_{jn} \mathrm{Z}^{(2)} \tag{3.488}$$

Analogously, the $j \neq n$ case leaves with the second order coefficient determination while canceling the associate energy:

$$\left[\varepsilon_n - \varepsilon_j\right] c_{j\neq n}^{(2)} + \underbrace{\mathrm{E}^{(1)}}_{\left\langle \varepsilon_n \middle| \widehat{H}_1 \middle| \varepsilon_n \right\rangle} \underbrace{c_{j\neq n}^{(1)}}_{\frac{\left\langle \varepsilon_j \middle| \widehat{H}_1 \middle| \varepsilon_n \right\rangle}{\varepsilon_n - \varepsilon_j}} + \mathrm{E}^{(2)} \underbrace{\delta_{n\neq j}}_{0} = \sum_k \left\langle \varepsilon_j \middle| \widehat{H}_1 \middle| \varepsilon_k \right\rangle \underbrace{c_k^{(1)}}_{k=n \quad k\neq n}$$

$$\Rightarrow c_{j\neq n}^{(2)} = \frac{1}{\varepsilon_n - \varepsilon_j} \left\{ \begin{array}{l} -\left\langle \varepsilon_n \middle| \widehat{H}_1 \middle| \varepsilon_n \right\rangle \dfrac{\left\langle \varepsilon_j \middle| \widehat{H}_1 \middle| \varepsilon_n \right\rangle}{\varepsilon_n - \varepsilon_j} + \left\langle \varepsilon_j \middle| \widehat{H}_1 \middle| \varepsilon_n \right\rangle \underbrace{c_{k=n}^{(1)}}_{\mathrm{Z}^{(1)}} \\ + \sum\limits_{k\neq n} \left\langle \varepsilon_j \middle| \widehat{H}_1 \middle| \varepsilon_k \right\rangle \underbrace{c_{k\neq n}^{(1)}}_{\frac{\left\langle \varepsilon_k \middle| \widehat{H}_1 \middle| \varepsilon_n \right\rangle}{\varepsilon_n - \varepsilon_k}} \end{array} \right\}$$

$$= \mathrm{Z}^{(1)} \frac{\left\langle \varepsilon_j \middle| \widehat{H}_1 \middle| \varepsilon_n \right\rangle}{\varepsilon_n - \varepsilon_j} - \frac{\left\langle \varepsilon_j \middle| \widehat{H}_1 \middle| \varepsilon_n \right\rangle \left\langle \varepsilon_n \middle| \widehat{H}_1 \middle| \varepsilon_n \right\rangle}{\left(\varepsilon_n - \varepsilon_j\right)^2} + \sum_{k\neq n} \frac{\left\langle \varepsilon_j \middle| \widehat{H}_1 \middle| \varepsilon_k \right\rangle \left\langle \varepsilon_k \middle| \widehat{H}_1 \middle| \varepsilon_n \right\rangle}{\left(\varepsilon_n - \varepsilon_j\right)\left(\varepsilon_n - \varepsilon_k\right)} \quad (3.489)$$

Combining both cases we can write for the second order coefficient of perturbed wave-function the general expression:

$$c_j^{(2)} = \delta_{jn} \mathrm{Z}^{(2)} + \left(1 - \delta_{jn}\right) \mathrm{Z}^{(1)} \frac{\left\langle \varepsilon_j \middle| \widehat{H}_1 \middle| \varepsilon_n \right\rangle}{\varepsilon_n - \varepsilon_j}$$
$$+ \left(1 - \delta_{jn}\right) \left[\sum_{k\neq n} \frac{\left\langle \varepsilon_j \middle| \widehat{H}_1 \middle| \varepsilon_k \right\rangle \left\langle \varepsilon_k \middle| \widehat{H}_1 \middle| \varepsilon_n \right\rangle}{\left(\varepsilon_n - \varepsilon_j\right)\left(\varepsilon_n - \varepsilon_k\right)} - \frac{\left\langle \varepsilon_j \middle| \widehat{H}_1 \middle| \varepsilon_n \right\rangle \left\langle \varepsilon_n \middle| \widehat{H}_1 \middle| \varepsilon_n \right\rangle}{\left(\varepsilon_n - \varepsilon_j\right)^2} \right] \quad (3.490)$$

Now, worth making the observation according which the corrections $\mathrm{Z}^{(1)}$ and $\mathrm{Z}^{(2)}$ are not entering the perturbed energies corrections, thus may be principally set as being equal with zero (0) since they do not affect the perturbed spectra. Moreover, there can be easily proved that such choice is equivalent with condition that perturbed states are orthogonal on the non-perturbed eigen-states: if one defined the "p" order states as:

$$\left| p \right\rangle = \sum_k c_k^{(p)} \left| \varepsilon_k \right\rangle \quad (3.491)$$

and the "p" undetermined correction coefficient as:

$$Z^{(p)} = c_{k=n}^{(p)} = \sum_k c_k^{(p)} \delta_{nk} = \sum_k c_k^{(p)} \langle \varepsilon_n | \varepsilon_k \rangle = \langle \varepsilon_n | \underbrace{\sum_k c_k^{(p)} | \varepsilon_k \rangle}_{|p\rangle} = \langle \varepsilon_n | p \rangle \quad (3.492)$$

there is immediate that the condition:

$$0 = Z^{(p)} = \langle \varepsilon_n | p \rangle, \ \forall |\varepsilon_n\rangle \in \left\{ \widehat{H}_0 \right\}_{spectra} \ \& |p\rangle \in \left\{ \widehat{H}_1 \right\}_{spectra} \quad (3.493)$$

leaves with the physical condition that the Hilbert (sub)spaces of the isolated and perturbation Hamiltonians are orthogonal, $\mathcal{H}_{\{\widehat{H}_0\}} \perp \mathcal{H}_{\{\widehat{H}_1\}}$, thus allowing their direct product to be reproduce the whole-problem spectra (levels and states) of the perturbed system:

$$\mathcal{H}_{\{\widehat{H}_0 + \lambda \widehat{H}_1\}} = \mathcal{H}_{\{\widehat{H}_0\}} \otimes \mathcal{H}_{\{\widehat{H}_1\}} \quad (3.494)$$

With this remarkable result, the full perturbed ($\lambda = 1$) energy and wave function may be written as the series:

$$E_n(\lambda = 1) = \varepsilon_n + \langle \varepsilon_n | \widehat{H}_1 | \varepsilon_n \rangle + \sum_{k \neq n} \frac{\left| \langle \varepsilon_k | \widehat{H}_1 | \varepsilon_n \rangle \right|^2}{\varepsilon_n - \varepsilon_k} + \ldots \quad (3.495)$$

$$|E_n(\lambda = 1)\rangle = |\varepsilon_n\rangle + \sum_{k \neq n} |\varepsilon_k\rangle \frac{\langle \varepsilon_k | \widehat{H}_1 | \varepsilon_n \rangle}{\varepsilon_n - \varepsilon_k}$$
$$+ \sum_{k \neq n} |\varepsilon_k\rangle \left[\sum_{j \neq n} \frac{\langle \varepsilon_k | \widehat{H}_1 | \varepsilon_j \rangle \langle \varepsilon_j | \widehat{H}_1 | \varepsilon_n \rangle}{(\varepsilon_n - \varepsilon_k)(\varepsilon_n - \varepsilon_j)} - \frac{\langle \varepsilon_k | \widehat{H}_1 | \varepsilon_n \rangle \langle \varepsilon_n | \widehat{H}_1 | \varepsilon_n \rangle}{(\varepsilon_n - \varepsilon_k)^2} \right] + \ldots \quad (3.496)$$

while, usually, in practice, there are retained only the expansion until the second order in energy and the first order in wave-function, respectively. Even so, the calculations imply the evaluation of all matrix elements $\langle \varepsilon_k | \widehat{H}_1 | \varepsilon_n \rangle$, being a non-trivial job unless some of them are identically null (as is the case of harmonic oscillator, see the Section 2.4.6).

Other special appearances of the stationary perturbations are to be exposed in what following for some paradigmatic physical situations.

3.7.2 THE ZEEMAN EFFECT

The problem of degenerate levels requires the reformulation of the perturbation upon the isolated systems by considering them characterized by the vector states $\left|E^{(0)},a\right\rangle$, in terms of the multiplication factor a equaling the degree of degeneration $g(E^{(0)})$ of a given level from the non-perturbed spectrum:

$$\hat{H}_0\left|E^{(0)},a\right\rangle = E^{(0)}\left|E^{(0)},a\right\rangle \tag{3.497}$$

Moreover, since this degeneration it can be considered itself as constituting a base defining a sub-Hilbert space the entire non-perturbed eigenstate my be unfolded on it as

$$|0\rangle = \hat{1}_{\{|E^{(0)},a\rangle\}}|0\rangle = \sum_{a}^{g(E^{(0)})}\left|E^{(0)},a\right\rangle\left\langle E^{(0)},a\middle|0\right\rangle \tag{3.498}$$

of which Fourier coefficients $\left\langle E^{(0)},a\middle|0\right\rangle$ are to be determined out through the perturbative treatment.

As such, one has to reconsider the perturbation of the isolated eigenproblem, according with the general perturbation recipe, to be equivalently cast as:

$$\begin{aligned}
&\left(\hat{H}_0 + \lambda\hat{H}_1\right)|E(\lambda)\rangle = E(\lambda)|E(\lambda)\rangle \\
&\Leftrightarrow \left(\hat{H}_0 + \lambda\hat{H}_1\right)\left(|0\rangle + \lambda|1\rangle + \lambda^2|2\rangle + \ldots\lambda^p|p\rangle + \ldots\right) \\
&= \left(E^{(0)} + \lambda E^{(1)} + \lambda^2 E^{(2)} + \ldots\lambda^p E^{(p)} + \ldots\right)\left(|0\rangle + \lambda|1\rangle + \lambda^2|2\rangle + \ldots\lambda^p|p\rangle + \ldots\right) \\
&\Leftrightarrow \hat{H}_0|0\rangle + \lambda\left[\hat{H}_0|1\rangle + \hat{H}_1|0\rangle\right] + \lambda^2\left[\hat{H}_0|2\rangle + \hat{H}_1|1\rangle\right] \\
&+\ldots\lambda^p\left[\hat{H}_0|p\rangle + \hat{H}_1|p-1\rangle\right] + \ldots \\
&= E^{(0)}|0\rangle + \lambda\left[E^{(0)}|1\rangle + E^{(1)}|0\rangle\right] + \lambda^2\left[E^{(0)}|2\rangle + E^{(1)}|1\rangle + E^{(2)}|0\rangle\right] \\
&+\ldots\lambda^p\left[E^{(0)}|p\rangle + E^{(1)}|p-1\rangle + \ldots + E^{(p)}|0\rangle\right] + \ldots
\end{aligned} \tag{3.499}$$

in terms of the perturbed states:

$$\left|p\right\rangle = \sum_k c_k^{(p)} \left|\varepsilon_n\right\rangle \tag{3.500}$$

written from the un-perturbed eigen-states, see the previous section.

The identification of equal orders of perturbation coupling releases with the respective eigen-problems:
Order 0:

$$\widehat{H}_0\left|0\right\rangle = \mathrm{E}^{(0)}\left|0\right\rangle \tag{3.501}$$

is automatically satisfied by the unperturbed system (and spectra).
Order 1:

$$\left(\widehat{H}_0 - \mathrm{E}^{(0)}\right)\left|1\right\rangle + \left(\widehat{H}_1 - \mathrm{E}^{(1)}\right)\left|0\right\rangle = 0 \tag{3.502}$$

Order 2:

$$\left(\widehat{H}_0 - \mathrm{E}^{(0)}\right)\left|2\right\rangle + \left(\widehat{H}_1 - \mathrm{E}^{(1)}\right)\left|1\right\rangle - \mathrm{E}^{(2)}\left|0\right\rangle = 0 \tag{3.503}$$

...

Order p:

$$\left(\widehat{H}_0 - \mathrm{E}^{(0)}\right)\left|p\right\rangle + \left(\widehat{H}_1 - \mathrm{E}^{(1)}\right)\left|p-1\right\rangle - \mathrm{E}^{(2)}\left|p-2\right\rangle - ... - \mathrm{E}^{(p)}\left|0\right\rangle = 0 \tag{3.504}$$

While the restriction to the first order will define the so-called the first Zeeman approximation, for all perturbed states may be imposed the condition that their eigen-vectors be orthogonal on those (and on any) of the unperturbed system:

$$\left\langle p \middle| \mathrm{E}^{(0)}, a\right\rangle = 0 \tag{3.505}$$

as prescribed by the general perturbation algorithm (see the previous section). Yet, even without this condition the Zeeman approximation leads though left composition with degenerate states $\left\langle \mathrm{E}^{(0)}, a\right|$ with the equivalent forms:

$$\left\langle \mathrm{E}^{(0)}, a\right|\left(\widehat{H}_0 - \mathrm{E}^{(0)}\right)\left|1\right\rangle + \left\langle \mathrm{E}^{(0)}, a\right|\left(\widehat{H}_1 - \mathrm{E}^{(1)}\right)\left|0\right\rangle = 0$$

$$\Leftrightarrow \underbrace{\left(\mathrm{E}^{(0)}-\mathrm{E}^{(0)}\right)\left\langle \mathrm{E}^{(0)},a\middle|1\right\rangle}_{0}+\left\langle \mathrm{E}^{(0)},a\middle|\widehat{H}_1\middle|0\right\rangle-\mathrm{E}^{(1)}\left\langle \mathrm{E}^{(0)},a\middle|0\right\rangle=0$$

$$\Leftrightarrow \left\langle \mathrm{E}^{(0)},a\middle|\widehat{H}_1\hat{1}_{\{|\mathrm{E}^{(0)},a'\rangle\}}\middle|0\right\rangle-\mathrm{E}^{(1)}\left\langle \mathrm{E}^{(0)},a\middle|\hat{1}_{\{|\mathrm{E}^{(0)},a'\rangle\}}\middle|0\right\rangle=0$$

$$\Leftrightarrow \sum_{a'}^{g\left(\mathrm{E}^{(0)}\right)}\left\langle \mathrm{E}^{(0)},a\middle|\widehat{H}_1\middle|\mathrm{E}^{(0)},a'\right\rangle\left\langle \mathrm{E}^{(0)},a'\middle|0\right\rangle$$

$$-\mathrm{E}^{(1)}\sum_{a'}^{g\left(\mathrm{E}^{(0)}\right)}\underbrace{\left\langle \mathrm{E}^{(0)},a\middle|\mathrm{E}^{(0)},a'\right\rangle}_{\delta_{aa'}}\left\langle \mathrm{E}^{(0)},a'\middle|0\right\rangle=0$$

$$\Rightarrow \sum_{a'}^{g\left(\mathrm{E}^{(0)}\right)}\left[\left\langle \mathrm{E}^{(0)},a\middle|\widehat{H}_1\middle|\mathrm{E}^{(0)},a'\right\rangle-\delta_{aa'}\mathrm{E}^{(1)}\right]\left\langle \mathrm{E}^{(0)},a'\middle|0\right\rangle=0 \tag{3.506}$$

that is nothing else than a $g(\mathrm{E}^{(0)})$ dimensional system for the searched Fourier coefficients in the base of the unperturbed degenerate vectors $\left|\mathrm{E}^{(0)},a\right\rangle$. The system admits non-zero solution only if exists an unperturbed vector, which for certain value of $\mathrm{E}^{(1)}$ the determinant of the homogeneous system vanishes:

$$\det\left[\left\langle \mathrm{E}^{(0)},a\middle|\widehat{H}_1\middle|\mathrm{E}^{(0)},a'\right\rangle-\delta_{aa'}\mathrm{E}^{(1)}\right]=0 \tag{3.507}$$

This is an algebraic equation of order $g(\mathrm{E}^{(0)})$ that may have as many distinct solutions or with some of them still equal (still degenerate), say:

$$\underset{\substack{\downarrow\\ g_1}}{\mathrm{E}_1^{(1)}},\underset{\substack{\downarrow\\ g_2}}{\mathrm{E}_2^{(1)}},\ldots,\underset{\substack{\downarrow\\ g_s}}{\mathrm{E}_s^{(1)}} \tag{3.508}$$

while still preserving the summation condition according which their partial degeneracies has to sum up into the total degeneration number:

$$g_1+g_2\ldots+g_s=g(\mathrm{E}^{(0)}) \tag{3.509}$$

There will be said that the perturbation takes off the degeneration of the level $\mathrm{E}^{(0)}$ if $s>1$, that is the unperturbed level is split out into its multiplicity or on some part of it. This phenomenon is called the *Zeeman effect* and, most remarkably, appears even in the first order of perturbation, as above described. If we have $s = g(\mathrm{E}^{(0)})$ there it is said that the degeneration is completely removed, through splitting the unperturbed level in $g(\mathrm{E}^{(0)})$ levels with one multiplicity. Of course, when $1 < s < g(\mathrm{E}^{(0)})$ the degeneration is only partially removed (or solved) by applied perturbation.

However, once the system is solved and the perturbed energies determined in terms of the matrix elements $\langle \mathrm{E}^{(0)}, a | \hat{H}_1 | \mathrm{E}^{(0)}, a' \rangle$ the recorded energy of a given level is then given by shifting the unperturbed spectrum with the computed correction:

$$\mathrm{E}_i = \mathrm{E}^{(0)} + \mathrm{E}_i^{(1)} \tag{3.510}$$

Note that the one-to-one correspondence between the first order correction energies and the matrix elements of the perturbation Hamiltonian in the basis of the degenerate unperturbed system,

$$\mathrm{E}_a^{(1)} \leftrightarrow \langle \mathrm{E}^{(0)}, a | \hat{H}_1 | \mathrm{E}^{(0)}, a' = a \rangle \tag{3.511}$$

may take place only if the perturbation Hamiltonian commutes with the set of operators whose eigen-values generates the degeneration of $\mathrm{E}^{(0)}$, say

$$\left[\hat{H}_1, \hat{A}_i \right] = 0, \ \forall i \tag{3.512}$$

meaning that they share the same eigen-states basis for the eigen-problem

$$\hat{H}_1 | \mathrm{E}^{(0)}, a \rangle = \mathrm{E}_a^{(1)} | \mathrm{E}^{(0)}, a \rangle \tag{3.513}$$

Indeed in such situation one has:

$$\langle \mathrm{E}^{(0)}, a | \hat{H}_1 | \mathrm{E}^{(0)}, a' \rangle = \mathrm{E}_{a'}^{(1)} \underbrace{\langle \mathrm{E}^{(0)}, a | \mathrm{E}^{(0)}, a' \rangle}_{\delta_{aa'}} = \mathrm{E}_a^{(1)} \tag{3.514}$$

and the above determinant equation rearranges in the diagonal manner as

$$\det\left[\left(\mathrm{E}_{a'}^{(1)} - \mathrm{E}^{(1)} \right) \delta_{aa'} \right] = 0 \tag{3.515}$$

leaving with the simple polynomial:

$$\prod_{i=1}^{g\left(\mathrm{E}^{(0)}\right)} \left(\mathrm{E}_{a_i}^{(1)} - \mathrm{E}^{(1)} \right) = 0 \tag{3.516}$$

and therefore with the complete removal of the degeneracy through the $g(\mathrm{E}^{(0)})$ different solutions for the first order energy correction:

$$\mathrm{E}_1^{(1)} = \mathrm{E}_{a_1}^{(1)}, \ldots, \mathrm{E}_i^{(1)} = \mathrm{E}_{a_i}^{(1)}, \ldots, \mathrm{E}_{g\left(\mathrm{E}^{(0)}\right)}^{(1)} = \mathrm{E}_{a_{g\left(\mathrm{E}^{(0)}\right)}}^{(1)} \tag{3.517}$$

This procedure can be repeated when the second order of energy correction is considered, however with the caution that the degeneracy of the first order corrections may appear as well, while in most cases the first order algorithm is enough for modeling the Zeeman's effect.

3.7.3 NUCLEAR ISOTROPIC CORRECTIONS ON HYDROGENIC ATOMS

Another interesting effect is that produced on the ground state (in the first instance) energy of the hydrogenic atoms by the fact the nucleus is no more considered as a point-like, but having its charge distributed either (say uniformly) on its volume or on its surface. In what follows we will explore for both the uniformly volume and surface nuclear distributed charge how much such physical model perturbations will affect the hydrogenic ground state energy

$$\varepsilon_0^Z = -\frac{Ze_0^2}{2a_0}, \; e_0^2 = \frac{e^2}{4\pi\varepsilon_0} \tag{3.518}$$

within the radial normalized ground state wave function

$$\psi_{10}(r) = \langle r|\psi_{10}\rangle = 2\left(\frac{Z}{a_0}\right)^{3/2} \exp\left(-Z\frac{r}{a_0}\right) \tag{3.519}$$

Note that considering the involved parameters as belonging to the realm:

- Nuclear radius $R \cong 10^{-15}[m]$;
- First Bohr radius $a_0 \cong 0.5 \cdot 10^{-10}[m] = 5 \cdot 10^{-11}[m]$
- Hydrogenic average atomic number $Z \cong 10$

we have in practice that for values around the nuclear frontier we may consider that

$$\frac{Zr}{a_0} \to \frac{ZR}{a_0} \cong \frac{10 \cdot 10^{-15}}{5 \cdot 10^{-11}} = 2 \cdot 10^{-4} \Rightarrow \exp\left(-\frac{ZR}{a_0}\right) \cong e^0 = 1 \tag{3.520}$$

thus limiting the above wave-function to the working one

$$\tilde{\psi}_{10}(r) = 2\left(\frac{Z}{a_0}\right)^{3/2} \tag{3.521}$$

Above this the main problem is to establish the form of the (radial) perturbation Hamiltonian $\widehat{H}_1(r)$ with the help of which the first order correction energy writes generally as:

$$\begin{aligned}
\mathrm{E}_1^{(1)} &= \langle \psi_{10} | \widehat{H}_1(r) | \psi_{10} \rangle \\
&= \langle \psi_{10} | \underbrace{\hat{1}_{\{|r\rangle\}}}_{\int_0^\infty r^2 |r\rangle\langle r| dr} \widehat{H}_1(r) | \psi_{10} \rangle = \int_0^\infty r^2 \langle \psi_{10} | r \rangle \underbrace{\langle r | \widehat{H}_1(r) | \psi_{10} \rangle}_{H_1(r)\langle r|} dr \\
&= \int_0^\infty r^2 H_1(r) \underbrace{\langle \psi_{10} | r \rangle}_{\psi_{10}^*(r)} \underbrace{\langle r | \psi_{10} \rangle}_{\psi_{10}(r)} dr = \int_0^\infty r^2 H_1(r) |\psi_{10}(r)|^2 \, dr
\end{aligned} \tag{3.522}$$

with the actual working specialization:

$$\tilde{\mathrm{E}}_1^{(1)} = \int_0^\infty r^2 H_1(r) |\tilde{\psi}_{10}(r)|^2 \, dr = 4\left(\frac{Z}{a_0}\right)^3 \int_0^\infty r^2 H_1(r) dr \tag{3.523}$$

Now, since the perturbation Hamiltonian stands in fact as an interaction potential (energy) $H_1(r)$ created by a charge field it has to be firstly related with the *potential function* $\varphi(r)$ by the definition relations

$$\begin{aligned}
H(r) &= (-e)\varphi(r) \\
&= H_0(r) + H_1(r) \\
&= -\frac{Ze_0^2}{r} + H_1(r)
\end{aligned} \tag{3.524}$$

Next, the potential function is at its turn to be determined though the electric field – gradient of potential relationship (derived from the force-potential one):

$$\vec{E} = -grad\varphi \tag{3.525}$$

which in radial coordinates rewrites:

$$E_r = -\frac{d\varphi}{dr} \Leftrightarrow d\varphi = -E_r dr \tag{3.526}$$

from where it unfolds as

$$\varphi(r) = -\int E_r dr + ct \tag{3.527}$$

The remaining radial electric field is to be derived in terms of charge source and the geometrical conditions through the *Gauss fundamental law* of electrostatics in integral version as well:

$$\oint \vec{E} d\vec{S} = \frac{Q}{\varepsilon_0} \tag{3.528}$$

that in spherical symmetry (as assumed fro nucleus either in volume or surface charge distribution), i.e., $\left|\vec{E}\right| = E_r = ct$, simply becomes

$$E_r \underbrace{\oint \vec{n}_r d\vec{S}}_{4\pi r^2} = \frac{Q}{\varepsilon_0} \tag{3.529}$$

recovering the classical Coulomb field law:

$$E_r = \frac{1}{4\pi\varepsilon_0}\frac{Q}{r^2} \tag{3.530}$$

Now, depending on the type of charge distribution one has to follow the reverse step until recognizing the perturbation Hamiltonian and to use it in calculating the isotropic nuclear charge correction(s) on the (ground states) unperturbed energies of hydrogenic systems.

1. Let's start with the case of uniformly (isotropic) nuclear charge distribution in volume. Yet, we have to consider the situations $r < R$ and $r \geq R$. The last situation is immediately recognized from the Coulomb law on the basis that all charge sources "behind" the action horizon may be treated as "point like sources":

$$E^V_{r\geq R} = \frac{1}{4\pi\varepsilon_0}\frac{Q}{r^2} \tag{3.531}$$

The electric field inside the nuclear volume has to take into account that while delimiting an inner volume with radius $r < R$ it contains at its turn the charge proportional with the volume encompassed, whose proportionality constant is represented by the uniformly volume distributed charge density:

$$\rho_Q = \frac{3}{4\pi}\frac{Q}{R^3} \tag{3.532}$$

with this we have the effective inner charge

$$Q_{r<R} = \rho_Q \frac{4}{3}\pi r^3 = Q\frac{r^3}{R^3} \tag{3.533}$$

that provides the inner electric field as well:

$$E_{r<R}^V = \frac{1}{4\pi\varepsilon_0}\frac{Q_{r<R}}{r^2} = \frac{1}{4\pi\varepsilon_0}\frac{Q}{R^3}r \tag{3.534}$$

Once having the electric field the stage of potential function may be undertaken. Firstly, for the outside nuclear filed we obtain:

$$\varphi_{r\geq R}^V(r) = -\int E_{r\geq R}^V dr + ct_{r\geq R}$$

$$= -\frac{Q}{4\pi\varepsilon_0}\int\frac{1}{r^2}dr + ct_{r\geq R} = \frac{Q}{4\pi\varepsilon_0 r} + ct_{r\geq R} \tag{3.535}$$

with the associate constant found through the asymptotic (natural) condition

$$\varphi_{r\geq R}(r \to \infty) = 0 \Leftrightarrow ct_{r\geq R} = 0 \tag{3.536}$$

thus leaving with the working potential function:

$$\varphi_{r\geq R}^V(r) = \frac{Q}{4\pi\varepsilon_0 r} \tag{3.537}$$

Analogously, for the inner potential function we firstly get:

$$\varphi_{r<R}^V(r) = -\int E_{r<R}^V dr + ct_{r<R}$$

$$= -\frac{1}{4\pi\varepsilon_0}\frac{Q}{R^3}\int r dr + ct_{r<R} = -\frac{1}{4\pi\varepsilon_0}\frac{Q}{R^3}\frac{r^2}{2} + ct_{r<R} \tag{3.538}$$

while the actual constant of integration is searched through the boundary equivalent potential condition on the nuclear surface:

$$\varphi_{r<R}^V(r = R) = \varphi_{r\geq R}^V(r = R) \Rightarrow ct_{r<R} = \frac{3}{2}\frac{1}{4\pi\varepsilon_0}\frac{Q}{R} \tag{3.539}$$

the resulting inner potential function looks therefore like:

$$\varphi^{V}_{r<R}(r)=\frac{Q}{4\pi\varepsilon_0}\frac{1}{R}\left(\frac{3}{2}-\frac{r^2}{2R^2}\right) \tag{3.540}$$

Both inner and outside potential function solutions may be reunited with the help of Heaviside step function

$$\varphi^{V}(r)=\frac{Q}{4\pi\varepsilon_0 r}\Theta(r-R)+\frac{Q}{4\pi\varepsilon_0}\frac{1}{R}\left(\frac{3}{2}-\frac{r^2}{2R^2}\right)\Theta(R-r) \tag{3.541}$$

that may be further rearranged employing the step function property:

$$\Theta(\bullet)+\Theta(-\bullet)=1 \tag{3.542}$$

to be cast as

$$\varphi^{V}(r)=\frac{Q}{4\pi\varepsilon_0 r}+\frac{Q}{4\pi\varepsilon_0}\left(\frac{3}{2R}-\frac{r^2}{2R^3}-\frac{1}{r}\right)\Theta(R-r) \tag{3.543}$$

Now, identifying the nuclear charge

$$Q=Ze \tag{3.544}$$

and writing the total Hamiltonian as above defined in terms of potential function,

$$\begin{aligned}H^{V}(r)&=(-e)\varphi^{V}(r)\\&=\underbrace{-\frac{Ze^2}{4\pi\varepsilon_0 r}}_{H_0(r)}+\left[\frac{Ze^2}{4\pi\varepsilon_0 r}-\frac{Ze^2}{4\pi\varepsilon_0 R}\left(\frac{3}{2}-\frac{r^2}{2R^2}\right)\right]\Theta(R-r)\end{aligned} \tag{3.545}$$

from where identifying the perturbation contribution

$$H_1^{V}(r)=\left[\frac{Ze_0^2}{r}-\frac{Ze_0^2}{R}\left(\frac{3}{2}-\frac{r^2}{2R^2}\right)\right]\Theta(R-r) \tag{3.546}$$

In these conditions there is now immediate to evaluate the hydrogenic ground state energy correction due to nuclear spatially extent interaction:

$$\tilde{E}_1^{(1)V}=4\left(\frac{Z}{a_0}\right)^3\int_0^{\infty}r^2H_1^{V}(r)dr$$

$$=4\left(\frac{Z}{a_0}\right)^3 \underbrace{\int_0^\infty dr\Theta(R-r)r^2}_{\substack{\downarrow \\ \int_0^R dr...}}\left[\frac{Ze_0^2}{r}-\frac{Ze_0^2}{R}\left(\frac{3}{2}-\frac{r^2}{2R^2}\right)\right]$$

$$=4\left(\frac{Z}{a_0}\right)^3 Ze_0^2\int_0^R\left(r-\frac{3r^2}{2R}+\frac{r^4}{2R^2}\right)dr;$$

$$\tilde{\mathrm{E}}_1^{(1)V}=\frac{2}{5}\frac{Z^4e_0^2}{a_0^3}R^2=-\frac{4}{5}\left(\frac{R}{a_0}\right)^2 Z^2\varepsilon_0^Z \tag{3.547}$$

a result showing that the correction is indeed no significant since

$$\frac{4}{5}\left(\frac{R}{a_0}\right)^2 Z^2\cong\frac{4}{5}\frac{10^{-30}}{25\cdot 10^{-22}}10^2\cong 3\cdot 10^{-8}\cong 0 \tag{3.548}$$

as the actual atomic-nucleus environment implies.

2. Let's see whether the situation changes in the situation the nuclear charge is uniformly localized on its (assumed spherical) surface only. In this situation the potential function may be directly abstracted from the previous calculations to have the branches:

$$\varphi^S(r)=\begin{cases}\dfrac{Q}{4\pi\varepsilon_0}\dfrac{1}{R}, & r\leq R\\[2ex] \dfrac{Q}{4\pi\varepsilon_0}\dfrac{1}{r}, & r>R\end{cases} \tag{3.549}$$

that can be reunited in a single expression due to the step function:

$$\varphi^S(r)=\frac{Q}{4\pi\varepsilon_0 r}\underbrace{\Theta(r-R)}_{1-\Theta(R-r)}+\frac{Q}{4\pi\varepsilon_0}\frac{1}{R}\Theta(R-r)$$

$$=\frac{Q}{4\pi\varepsilon_0 r}+\frac{Q}{4\pi\varepsilon_0}\left(\frac{1}{R}-\frac{1}{r}\right)\Theta(R-r) \tag{3.550}$$

providing as above the associated nuclear surface containing Hamiltonian (with $Q=Ze$):

$$H^S(r)=(-e)\varphi^S(r)$$

$$= \underbrace{-\frac{Ze_0^2}{r}}_{H_0(r)} - Ze_0^2\left(\frac{1}{R} - \frac{1}{r}\right)\Theta(R-r) \tag{3.551}$$

Once the surface perturbation Hamiltonian being identified as:

$$H_1^S(r) = Ze_0^2\left(\frac{1}{r} - \frac{1}{R}\right)\Theta(R-r) \tag{3.552}$$

the first order correction energy to the hydrogenic ground state yields, like before, through shrinking the infinite interval of radial integration to that recommended by the Heaviside step function $\Theta(R-r)$, i.e., $0 \le r \le R$, to the equivalent expressions:

$$\begin{aligned}\tilde{E}_1^{(1)S} &= 4\left(\frac{Z}{a_0}\right)^3 \int_0^\infty r^2 H_1^S(r)dr \\ &= 4\left(\frac{Z}{a_0}\right)^3 Ze_0^2 \int_0^R \left(r - \frac{r^2}{R}\right)dr;\end{aligned}$$

$$\tilde{E}_1^{(1)S} = \frac{2}{3}\frac{Z^4 e_0^2}{a_0^3}R^2 = -\frac{4}{3}\left(\frac{R}{a_0}\right)^2 Z^2 \varepsilon_0^Z \tag{3.553}$$

a result very similar with that obtained in the volume charge distribution approach, with the same insignificant effective contribution to the hydrogenic ground state energy, since caring the same 10^{-8} order as before although with a smooth higher multiplication factor:

$$\frac{\tilde{E}_1^{(1)S}}{\tilde{E}_1^{(1)V}} = \frac{4}{3}\frac{5}{4} = \frac{5}{3} \cong 1.67 > 1 \tag{3.554}$$

Similar correction cases, either in higher orders or on excited hydrogenic states, may be considered following the same line of analysis however with the same principally conclusion that they do not in fact contribute to the real shift (or perturbation) of the hydrogenic atoms within the point like nucleus framework; this leads with the idea that indeed, for atomic and supra-atomic systems the point like hypothesis of the nucleus finely works and will be in next assumed as such (for instance when treating the chemical bonding by means of the Dirac theory, see next chapters).

3.7.4 HARMONIC OSCILLATOR PERTURBATIONS

The perturbation algorithm for energy corrections may be applied for any completely solved isolated system, of which, apart of hydrogenic ones, the harmonic oscillator represents an important application for modeling molecular open states. There will be considered two cases of perturbations: one having the symmetric coordinate perturbation, while the second is linear in coordinate with a constant that may represent some external electric intensity.

1. Starting with the harmonic oscillator with the unperturbed Hamiltonian

$$\widehat{H}_0 = \frac{\hat{p}^2}{2m} + \frac{k_{\omega_0}}{2}\hat{x}^2, \quad k_{\omega_0} = m\omega_0^2 \tag{3.555}$$

with the associate spectrum

$$\varepsilon_n = \hbar\omega_0\left(n + \frac{1}{2}\right), n \in \mathrm{N} \tag{3.556}$$

one searches for its correction produced by the existence of the perturbation Hamiltonian:

$$\widehat{H}_1 = \frac{b}{2}\hat{x}^2, \ b \in \mathrm{R} \tag{3.557}$$

The direct way of determining the spectrum corrections relays on composing the total Hamiltonian as

$$\widehat{H} = \widehat{H}_0 + \widehat{H}_1 = \frac{\hat{p}^2}{2m} + \frac{k_{\omega_0} + b}{2}\hat{x}^2 \tag{3.558}$$

and to identify from it the new oscillating frequency (due to inclusion of perturbation) throughout

$$k_{\omega_0} + b = m\omega^2$$

$$\Rightarrow \omega = \sqrt{\frac{k_{\omega_0} + b}{m}} = \underbrace{\sqrt{\frac{k_{\omega_0}}{m}}}_{\omega_0}\sqrt{1 + \frac{b}{k_{\omega_0}}} = \omega_0\sqrt{1 + \frac{b}{k_{\omega_0}}} \tag{3.559}$$

which can be expanded in series up to the desired order, say two, in terms of "*b*" perturbation, using the series expansion according with the McLaurin formula

$$\sqrt{1+\bullet} \cong 1+\frac{\bullet}{2}-\frac{\bullet^2}{8} \tag{3.560}$$

to obtain the perturbed frequency:

$$\omega \cong \omega_0\left(1+\frac{b}{2k_{\omega_0}}-\frac{b^2}{8k_{\omega_0}^2}\right) \tag{3.561}$$

and then by means of direct substitution back on the original spectra the perturbed energy is provided

$$\mathrm{E}_n \cong \varepsilon_n+\frac{b}{2k_{\omega_0}}\varepsilon_n-\frac{b^2}{8k_{\omega_0}^2}\varepsilon_n \tag{3.562}$$

from where the first and second energy corrections are individually identified as:

$$\mathrm{E}_n^{(1)}=\frac{b}{2k_{\omega_0}}\varepsilon_n \tag{3.563}$$

$$\mathrm{E}_n^{(2)}=-\frac{b^2}{8k_{\omega_0}^2}\varepsilon_n \tag{3.564}$$

Yet, there is noted that in fact no use of the perturbation algorithm was made in deriving these corrections; therefore some cross-check is require. This can be done either using the properties of Hermite polynomials in employing the wave-functions of harmonic oscillator, or, more elegantly, through further employing the quantum information contained into the Heisenberg matrix approach. For the future purposes the second way is here unfolded. We have to actually compute the matrices elements $\langle n|\widehat{H}_1|n\rangle$ and $\langle n'|\widehat{H}_1|n\rangle$, $n,n'\in \mathrm{N}$ for computing the first and second order energy corrections, respectively. That means, we have in fact to compute the quantities $\langle n|\hat{x}^2|n\rangle$ and $\langle n'|\hat{x}^2|n\rangle$, leading with idea that the coordinate matrix $[x]$ has to be somehow further exploited from the Heisenberg matrix theory of harmonic oscillator. In this respect one may observe two important things with the coordinate $[Q]$ matrix derived in Section 2.4.6:

- The matrix is symmetrical due to the symmetry of the harmonic potential, thus allows identification

$$q_{n-1,n} = q_{n,n-1} \tag{3.565}$$

with the help of which they can be computed from their combined relationship as:

$$q_{n-1,n} = q_{n,n-1} = \frac{\alpha}{\sqrt{2}}\sqrt{n}, \quad \alpha = \sqrt{\frac{\hbar}{m\omega_0}} \tag{3.566}$$

to provide the rewritten of the harmonic oscillator coordinate matrix as:

$$[x] = \frac{\alpha}{\sqrt{2}}\begin{bmatrix} 0 & 1 & 0 & \cdots \\ 1 & 0 & \sqrt{2} & \cdots \\ 0 & \sqrt{2} & 0 & \cdots \\ \vdots & \vdots & \vdots & \ddots \end{bmatrix} \tag{3.567}$$

- This matrix may be seen as being composed by two different matrices,

$$[x] = \frac{\alpha}{\sqrt{2}}\left([a^+] + [a]\right) \tag{3.568}$$

one for the upper diagonal and other for the down diagonal rows of $\sqrt{n}$ quantum numbers; accordingly they may be eventually called annihilation $[a]$ and creation $[a^+]$ matrices (with associate operators, of course) for the reason bellow revealed, however displaying like:

$$[a] = \begin{bmatrix} 0 & 1 & 0 & \cdots \\ 0 & 0 & \sqrt{2} & \cdots \\ 0 & 0 & 0 & \cdots \\ \vdots & \vdots & \vdots & \ddots \end{bmatrix}, \quad [a^+] = \begin{bmatrix} 0 & 0 & 0 & \cdots \\ 1 & 0 & 0 & \cdots \\ 0 & \sqrt{2} & 0 & \cdots \\ \vdots & \vdots & \vdots & \ddots \end{bmatrix} \tag{3.569}$$

In terms of operators, the coordinate operator relationship with annihilation and creation operators reads

$$\hat{x} = \frac{\alpha}{\sqrt{2}}\left(\hat{a}^{+} + \hat{a}\right) \tag{3.570}$$

while for the matrix elements there is found through inspection of above matrices the operatorial rules:

$$\langle n'|\hat{a}|n\rangle = \sqrt{n}\delta_{n',n-1} = \sqrt{n}\langle n'|n-1\rangle \tag{3.571}$$

$$\langle n'|\hat{a}^{+}|n\rangle = \sqrt{n+1}\delta_{n',n+1} = \sqrt{n+1}\langle n'|n+1\rangle \tag{3.572}$$

assuming the quantum numbers' combinations $(n'=0, n=1)$ and $(n'=1, n=0)$ so that at extreme the annihilation operator to act over the state with $n=1$ (i.e., to have something the "annihilate") while the creation operator applies on the state with $n=0$ (i.e., to have to create something from " nothing"). Moreover, these combinations generates the first matrix element (equal to "1") above and bellow the diagonal in matrices of annihilation and creation operators (when the matrices structures are defined as beginning with 0th line and 0th column, respectively); as such, these matrices are said to be written in the particle's "number" representation, since rooting on the quantum numbers. More details and extensions of these ideas are to be presented with occasion of many-body quantum systems discussion, latter on. For the moment one retains the annihilation and creation operatorial actions:

$$\hat{a}|n\rangle = \sqrt{n}|n-1\rangle \tag{3.573}$$

$$\hat{a}^{+}|n\rangle = \sqrt{n+1}|n+1\rangle \tag{3.574}$$

abstracted from above matrices' elements.

For the shake of completeness, note that in the same manner the analysis of the Heisenberg momentum matrix provides the particle's representation as:

$$[p] = \frac{i\beta}{\sqrt{2}}\begin{bmatrix} 0 & -1 & 0 & \cdots \\ 1 & 0 & -\sqrt{2} & \cdots \\ 0 & \sqrt{2} & 0 & \cdots \\ \vdots & \vdots & \vdots & \ddots \end{bmatrix} = \frac{i\beta}{\sqrt{2}}\left([a^{+}]-[a]\right),\ \beta = \sqrt{\hbar m\omega_0} \tag{3.575}$$

with the correspondent operatorial connection with creation and annihilation operators:

$$\hat{p} = \frac{\alpha}{\sqrt{2}}\left(\hat{a}^{+} - \hat{a}\right) \tag{3.576}$$

Now, having rewritten the coordinate matrix elements with the help of creation and annihilation matrix elements as well will highly help our perturbation calculation since preserving the absolute generality (i.e., not restraining the checking only to the ground state for instance) and involving the so-called "shifted diagonalization" or delta-Kronecker rules that elegantly select the mixed contributing states to the energy.

Let's proceed with the first order correction to successively get:

$$\begin{aligned}
\mathrm{E}_n^{(1)} &= \langle n|\hat{H}_1|n\rangle = \frac{b}{2}\langle n|\hat{x}^2|n\rangle = \frac{b}{2}\langle n|\left[\frac{\alpha}{\sqrt{2}}\left(\hat{a}^{+} + \hat{a}\right)\right]^2|n\rangle \\
&= \frac{b\alpha^2}{4}\langle n|\left(\hat{a}^{+}\hat{a}^{+} + \hat{a}^{+}\hat{a} + \hat{a}\hat{a}^{+} + \hat{a}\hat{a}\right)|n\rangle \\
&= \frac{b\alpha^2}{4}[\langle n|\hat{a}^{+}\underbrace{\hat{a}^{+}|n\rangle}_{\sqrt{n+1}|n+1\rangle} + \langle n|\hat{a}^{+}\underbrace{\hat{a}|n\rangle}_{\sqrt{n}|n-1\rangle} + \langle n|\hat{a}\underbrace{\hat{a}^{+}|n\rangle}_{\sqrt{n+1}|n+1\rangle} + \langle n|\hat{a}\underbrace{\hat{a}|n\rangle}_{\sqrt{n}|n-1\rangle}] \\
&= \frac{b\alpha^2}{4}[\sqrt{n+1}\langle n|\underbrace{\hat{a}^{+}|n+1\rangle}_{\sqrt{n+2}|n+2\rangle} + \sqrt{n}\langle n|\underbrace{\hat{a}^{+}|n-1\rangle}_{\sqrt{n}|n\rangle} \\
&\quad + \sqrt{n+1}\langle n|\underbrace{\hat{a}|n+1\rangle}_{\sqrt{n+1}|n\rangle} + \sqrt{n}\langle n|\underbrace{\hat{a}|n-1\rangle}_{\sqrt{n-1}|n-2\rangle}] \\
&= \frac{b\alpha^2}{4}[\sqrt{(n+1)(n+2)}\underbrace{\langle n|n+2\rangle}_{0} + n\underbrace{\langle n|n\rangle}_{1} \\
&\quad + (n+1)\underbrace{\langle n|n\rangle}_{1} + \sqrt{n(n-1)}\underbrace{\langle n|n-2\rangle}_{0}] \\
&= \frac{b\alpha^2}{2}\left(n + \frac{1}{2}\right) \\
&= \frac{b\alpha^2}{2\hbar\omega_0}\underbrace{\hbar\omega_0\left(n + \frac{1}{2}\right)}_{\varepsilon_n} = \frac{b}{2\hbar\omega_0}\frac{\hbar}{m\omega_0}\varepsilon_n = \frac{b}{2k_{\omega_0}}\varepsilon_n
\end{aligned} \tag{3.577}$$

thus regaining exactly the same expression as from direct series expansion, nevertheless proofing both the perturbation as well as annihilation-creation operatorial formalisms.

Going to checkout the second order correction we have to evaluate:

$$\mathrm{E}_n^{(2)} = \sum_{n' \neq n} \frac{\left|\langle n'|\widehat{H}_1|n\rangle\right|^2}{\varepsilon_n - \varepsilon_{n'}} \tag{3.578}$$

that it reduces to the matrix element $\langle n'|\widehat{H}_1|n\rangle$ calculation, which develops as above to yield:

$$\begin{aligned}\langle n'|\widehat{H}_1|n\rangle &= \frac{b}{2}\langle n'|\hat{x}^2|n\rangle \\ &= \frac{b\alpha^2}{4}[\sqrt{(n+1)(n+2)}\underbrace{\langle n'|n+2\rangle}_{\delta_{n',n+2}} + n\underbrace{\langle n'|n\rangle}_{0} \\ &+(n+1)\underbrace{\langle n'|n\rangle}_{0} + \sqrt{n(n-1)}\underbrace{\langle n'|n-2\rangle}_{\delta_{n',n-2}}]\end{aligned} \tag{3.579}$$

thus selecting only the non-vanishing states $n' = n+2$ and $n' = n-2$ for which we separately have:

$$\left|\langle n+2|\widehat{H}_1|n\rangle\right|^2 = \frac{b^2\alpha^4}{16}(n+1)(n+2) \tag{3.580}$$

$$\left|\langle n-2|\widehat{H}_1|n\rangle\right|^2 = \frac{b^2\alpha^4}{16}n(n-1) \tag{3.581}$$

Therefore we still need to evaluate the differences:

$$\varepsilon_n - \varepsilon_{n+w} = \hbar\omega_0\left(n+\frac{1}{2}\right) - \hbar\omega_0\left(n+w+\frac{1}{2}\right) = -\hbar\omega_0 w = \begin{cases} 2\hbar\omega_0 \ , w=-2 \\ -2\hbar\omega_0 \ , w=2 \end{cases} \tag{3.582}$$

so that the second order energy correction becomes:

$$\mathrm{E}_n^{(2)} = \frac{\left|\langle n-2|\widehat{H}_1|n\rangle\right|^2}{\varepsilon_n - \varepsilon_{n-2}} + \frac{\left|\langle n+2|\widehat{H}_1|n\rangle\right|^2}{\varepsilon_n - \varepsilon_{n+2}}$$

$$=\frac{b^2\alpha^4}{16}\left[\frac{n(n-2)}{2\hbar\omega_0}+\frac{(n+1)(n+2)}{-2\hbar\omega_0}\right]$$

$$=-\frac{b^2\alpha^4}{8\hbar\omega_0}\left(n+\frac{1}{2}\right)$$

$$=-\frac{b^2\alpha^4}{8\hbar^2\omega_0^2}\underbrace{\hbar\omega_0\left(n+\frac{1}{2}\right)}_{\varepsilon_n}=-\frac{b^2}{8\hbar^2\omega_0^2}\frac{\hbar^2}{m^2\omega_0^2}\varepsilon_n=-\frac{b^2}{8k_{\omega_0}^2}\varepsilon_n \tag{3.583}$$

with an identical overlap with the result of the series expansion, thus confirming once more the reliability of the perturbation and annihilation-creation algorithms.

2. When the perturbation over the harmonic motion is considered linear in coordinate, i.e., when

$$\widehat{H}_1=b\hat{x},\ b\in\mathrm{R} \tag{3.584}$$

the problem seems to be even more simpler under the total Hamiltonian influence

$$\widehat{H}=\widehat{H}_0+\widehat{H}_1=\frac{\hat{p}^2}{2m}+\frac{k_{\omega_0}}{2}\hat{x}^2+b\hat{x}$$

$$=\frac{\hat{p}^2}{2m}+\frac{k_{\omega_0}}{2}\left(\hat{x}^2+\frac{2b}{k_{\omega_0}}\right)$$

$$=\underbrace{\frac{\hat{p}^2}{2m}+\frac{k_{\omega_0}}{2}\underbrace{\left(\hat{x}+\frac{b}{k_{\omega_0}}\right)}_{\hat{x}^{\#}}{}^2}_{\widehat{H}_0^{\#}}-\frac{b^2}{2k_{\omega_0}}=\widehat{H}_0^{\#}-\frac{b^2}{2k_{\omega_0}} \tag{3.585}$$

however appearing the problem whether the new coordinate operator $\hat{x}^{\#}$ maintains the same commutation relationship with momentum, a crucial matter in preserving the unperturbed harmonic spectrum. One has:

$$\left[\hat{p},\hat{x}^{\#}\right]=\left[\hat{p},\hat{x}+\frac{b}{k_{\omega_0}}\right]=\underbrace{\left[\hat{p},\hat{x}\right]}_{-i\hbar}+\underbrace{\left[\hat{p},\frac{b}{k_{\omega_0}}\right]}_{0}=-i\hbar \tag{3.586}$$

thus assuring the maintenance of the harmonic spectrum even with the shifter coordinate unperturbed Hamiltonian,

$$\widehat{H}_0^{\#}|n\rangle=\varepsilon_n|n\rangle \tag{3.587}$$

while the perturbed spectra is obtained employing the energy eigen-value problem for the whole Hamiltonian:

$$\widehat{H}|n\rangle=\left(\widehat{H}_0^{\#}-\frac{b^2}{2k_{\omega_0}}\right)|n\rangle=\left(\varepsilon_n-\frac{b^2}{2k_{\omega_0}}\right)|n\rangle\overset{!}{=}\mathrm{E}_n|n\rangle \tag{3.588}$$

from where immediately follows:

$$\mathrm{E}_n=\varepsilon_n-\frac{b^2}{2k_{\omega_0}} \tag{3.589}$$

as the corrected energy of the harmonic oscillator with linear coordinate perturbation.

The only remaining point is the assessment of the order in which this correction appears, and this will be done through searching this result with the help of annihilation-creation approach. As such, for the first order correction we obtain:

$$\begin{aligned}
\mathrm{E}_n^{(1)}&=\langle n|\widehat{H}_1|n\rangle=b\langle n|\hat{x}|n\rangle=\frac{b\alpha}{\sqrt{2}}\langle n|\left(\hat{a}^{+}+\hat{a}\right)|n\rangle\\
&=\frac{b\alpha}{\sqrt{2}}[\langle n|\underbrace{\hat{a}^{+}|n\rangle}_{\sqrt{n+1}|n+1\rangle}+\langle n|\underbrace{\hat{a}|n\rangle}_{\sqrt{n}|n-1\rangle}]\\
&=\frac{b\alpha}{\sqrt{2}}[\sqrt{n+1}\underbrace{\langle n|n+1\rangle}_{0}+\sqrt{n}\underbrace{\langle n|n-1\rangle}_{0}];\\
\mathrm{E}_n^{(1)}&=0
\end{aligned} \tag{3.590}$$

For the second order we need the non-diagonal matrix elements of the above terms

$$\langle n'|\hat{x}|n\rangle = \frac{b\alpha}{\sqrt{2}}[\sqrt{n+1}\underbrace{\langle n'|n+1\rangle}_{\delta_{n',n+1}} + \sqrt{n}\underbrace{\langle n'|n-1\rangle}_{\delta_{n',n-1}}] \tag{3.591}$$

that select the proper states with $n'=n+1$ and $n'=n-1$, and, as in the previous analysis we found out:

$$\left|\langle n+1|\widehat{H}_1|n\rangle\right|^2 = \frac{b^2\alpha^2}{2}(n+1) \tag{3.592}$$

$$\left|\langle n-1|\widehat{H}_1|n\rangle\right|^2 = \frac{b^2\alpha^2}{2}n \tag{3.593}$$

$$\varepsilon_n - \varepsilon_{n+w} = -\hbar\omega_0 w = \begin{cases} \hbar\omega_0 \ , \ w=-1 \\ -\hbar\omega_0 \ , \ w=1 \end{cases} \tag{3.594}$$

so that we write

$$\begin{aligned} E_n^{(2)} &= \frac{\left|\langle n-1|\widehat{H}_1|n\rangle\right|^2}{\varepsilon_n - \varepsilon_{n-1}} + \frac{\left|\langle n+1|\widehat{H}_1|n\rangle\right|^2}{\varepsilon_n - \varepsilon_{n+1}} \\ &= \frac{b^2\alpha^2}{2}\left[\frac{n}{\hbar\omega_0} - \frac{n+1}{\hbar\omega_0}\right] = -\frac{b^2\alpha^2}{2}\frac{1}{\hbar\omega_0} = -\frac{b^2}{2\hbar\omega_0}\frac{\hbar}{m\omega_0} = -\frac{b^2}{2k_{\omega_0}} \end{aligned} \tag{3.595}$$

thus identifying the present correction as being that of the second order.

We let for the reader to practice the similar perturbation problems in momentum, with the perturbation Hamiltonian taking the forms $\widehat{H}_1 = \left\{b\hat{p}^2, b\hat{p}^4, b\hat{p}\hat{x}^2\hat{p}\right\}$, $b \in \mathrm{R}$, chosen so that to be hermitic, by using the annihilation-creation representation for the momentum operator and the associate rules.

3.7.5 QUASI-FREE ELECTRONIC MODEL OF SOLIDS

As we saw previously, see Sections 3.3.3, 3.4.3, and 3.5.3, in the solid state environment there seems to exists two equivalent types of wave-function solutions, due to the large extension of the medium that allows the back and forth propagation of the electrons in a given eigen-state.

In free electrons approximation we actually consider the effective cancellation of the bulk and valence potentials for whose their states assumed as independent, i.e., orthogonal. Let's reformulate here the free solution so that to allow more realistic potential of the system. Firstly, let's separate the forth (+) and back (–) propagations by the normalized wave functions:

$$\Psi_k^{(\pm)}(x) = \frac{1}{\sqrt{L}} \exp(\pm ikx) \tag{3.596}$$

thus fulfilling the crystal domain (L) integration constraint:

$$\int_0^L \left[\Psi_k^{(\pm)}(x)\right]^* \Psi_k^{(\pm)}(x)dx = 1 \tag{3.597}$$

Yet, the boundary conditions, at the extremes of the crystal domain provide the geometric quantification of the wave-vector, as:

$$\Psi_k^{(\pm)}(x=0) = \Psi_k^{(\pm)}(x=L) \Leftrightarrow 1 = \exp(ikL)$$
$$\Rightarrow k = n\frac{2\pi}{L}, n \in \mathrm{N}^* \tag{3.598}$$

However, due to the unit cell periodicity assumed (since observed) for a solid crystal, the whole domain (in 1D direction here) may be considered portioned to become:

$$L = a\mathrm{N},\ \mathrm{N} \in \mathrm{N}^* \tag{3.599}$$

with a & N being the unit cell (1D) length and their number along the crystal, respectively. With this assumption the whole analysis may be restrained on the unit-cell level, so that the quantified wave vector reads:

$$k_n = n\frac{2\pi}{a}, n \in \mathrm{N}^* \tag{3.600}$$

Nevertheless, this restriction has a fundamental consequence in establishing the so-called first *Brillouin* zone as the first (the so-called "ground") quantified interval for the wave vector (of wave function or packet), in a symmetric way so that comprising back and forth propagations:

$$k_{n=1}^{Brillouin} = \left[-\frac{\pi}{a}, +\frac{\pi}{a} \right] \tag{3.601}$$

Finally, at the edges of this interval the above wave-functions are superimposed either as sum and differences to provide the normalized free electronic solution of motion in the unit cell of a periodic solid crystal:

$$\Psi_{e}^{(0)}(x) = \frac{1}{\sqrt{2}} \left[\Psi_{k=\pi/a}^{(+)}(x) + \Psi_{k=\pi/a}^{(-)}(x) \right] = \sqrt{\frac{2}{L}} \cos\left(\frac{\pi x}{a} \right) \tag{3.602}$$

$$\Psi_{0}^{(0)}(x) = \frac{1}{i\sqrt{2}} \left[\Psi_{k=\pi/a}^{(+)}(x) - \Psi_{k=\pi/a}^{(-)}(x) \right] = \sqrt{\frac{2}{L}} \sin\left(\frac{\pi x}{a} \right) \tag{3.603}$$

One may immediately check that both these solutions satisfy the normalization constraint:

$$\int_0^L \left[\Psi_{0}^{(0)}(x) \right]^2 dx = \int_0^L \left[\Psi_{e}^{(0)}(x) \right]^2 dx = 1 \tag{3.604}$$

since employing the natural number nature of the quantity L/a, see above. Moreover, these expressions represent a generalization of the previously considered single wave-function corresponding to the first eigen-value (see Section 3.3.3)

$$\varepsilon_{n=1} = \frac{\hbar^2}{2m_0} k_{n=1}^2 \tag{3.605}$$

in at least two ways:

- Firstly, the previously "sin" solution now is normalized to the whole crystal domain although with the behavior reduced to the single unit-cell; the consistency is checked out through normalization condition along the entire crystal;
- Secondly, the previously "sin" solution is now accompanied by the "cosine" solution, equally valid, and corresponding to the same energy quantization; thus, the two eigen-functions represent a *degenerate case of free electronic motion in solid*!

Such behavior of the free wave-function of electrons in solid greatly helps in establishing the form of the crystal bulk potential as well, considering it not entirely cancelled by that specific to the valence state. In other words, while representing the 1D atomic row in a crystal as in Figure 3.17, there is evident that the electronic crystal (perturbation) potential may be modeled in the same way as done for the wave-function, i.e., through considering it as a superposition (however here only as a sum) of the forth and back (Fourier) contributions:

$$V(x) = \sum_n \left[V_n \exp(i2\pi nx/a) + V_{-n}\exp(-i2\pi nx/a)\right] = 2\sum_n V_n \cos(2\pi nx/a) \tag{3.606}$$

Note that this is a potential than acts in principle over entire spectra of electrons in crystal; thus, the perturbation of the "degenerate ground" level of the crystal looks like:

$$\widehat{H}_1 = 2V_1 \cos(2\pi x/a) \tag{3.607}$$

This perturbation acts at once on both degenerate wave-functions of the ground level in a unit cell, producing the correction energies:

$$\mathrm{E}_{1[0]}^{(1)} = \lim_{L\to a}\int_0^L \Psi_0^{(0)}(x)\widehat{H}_1\Psi_0^{(0)}(x)dx$$

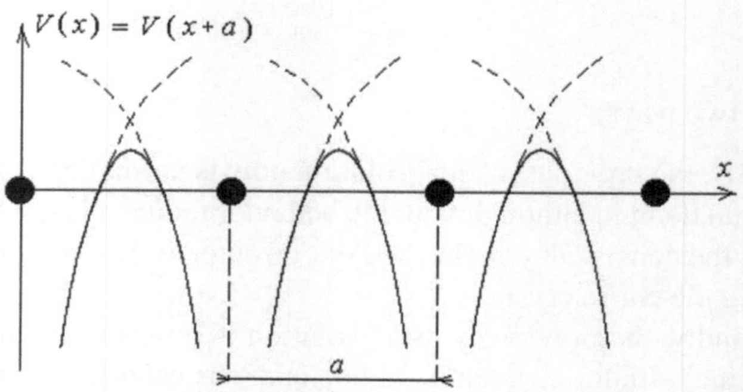

FIGURE 3.17 The resulted 1D- crystal potential from combined atomic bulk Coulombic contributions (Putz, 2006).

$$
\begin{aligned}
&= \frac{4V_1}{a}\lim_{L\to a}\int_0^L \cos\left(\frac{2\pi x}{a}\right)\sin^2\left(\frac{\pi x}{a}\right)dx \\
&= \frac{4V_1}{a}\lim_{L\to a}\left[-\frac{L}{4}+\frac{a}{4\pi}\sin\left(\frac{2\pi L}{a}\right)-\frac{a}{16\pi}\sin\left(\frac{4\pi L}{a}\right)\right]; \\
\mathrm{E}_{1[0]}^{(1)} &= -V_1; \qquad (3.608)
\end{aligned}
$$

$$
\begin{aligned}
\mathrm{E}_{1[e]}^{(1)} &= \lim_{L\to a}\int_0^L \Psi_e^{(0)}(x)\widehat{H}_1\Psi_e^{(0)}(x)dx \\
&= \frac{4V_1}{a}\lim_{L\to a}\int_0^L \cos\left(\frac{2\pi x}{a}\right)\cos^2\left(\frac{\pi x}{a}\right)dx \\
&= \frac{4V_1}{a}\lim_{L\to a}\left[\frac{L}{4}+\frac{a}{4\pi}\sin\left(\frac{2\pi L}{a}\right)+\frac{a}{16\pi}\sin\left(\frac{4\pi L}{a}\right)\right]; \\
\mathrm{E}_{1[e]}^{(1)} &= V_1 \qquad (3.609)
\end{aligned}
$$

Eventually, their difference:

$$
\mathrm{E}_{1[e]}^{(1)} - \mathrm{E}_{1[0]}^{(1)} = 2V_1 \qquad (3.610)
$$

corresponds to the gap produced by the perturbation potential of the bulk electrons in crystal, see Figure 3.18.

This results says that from above two free wave functions that once considered previously as "sin" plays the role of the lower "ground" state, while that with "cos" is the "excited" one in the crystal configuration; moreover, under the potential perturbation they separate in what is usually known in solid state theory as being the "valence" and "conduction" bands. We success therefore to construct this more realistic picture of the solid state crystals with the useful tool as stationary perturbation' algorithm is.

3.8 QUANTUM MEASUREMENTS

What means "to measure"? One possible answer is to create a deliberate interaction between the measured object and the measurement apparatus.

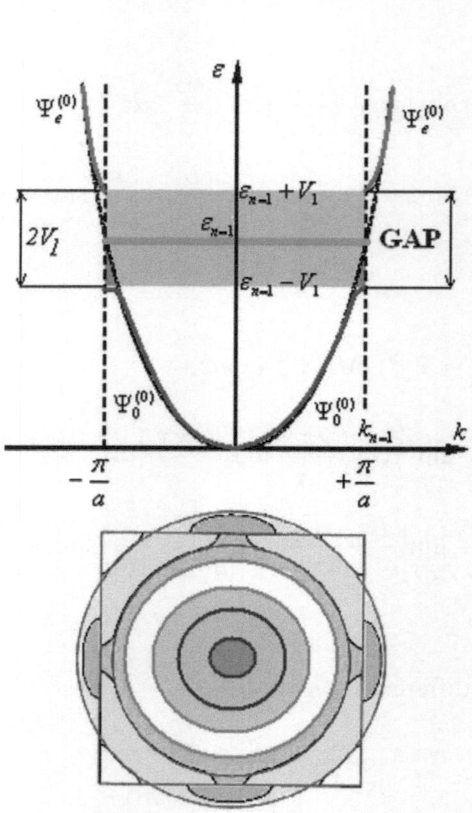

FIGURE 3.18 Illustration of the removed degeneracy of the free electronic wave-functions in the first Brillouin zone of a solid crystal by the perturbation potential generating the potential gap separating the valence and the (excited) conducting states, along the projection on the wave-vector plane (lower draw). See the text for details (Putz, 2006).

Then appears another question: "what is a measurement apparatus"? The answer is: that one that ideally obeys the physical laws. Yet, the question is "what is measurement process" – to which one can specify that the interaction has to be enough significant so that a changing in measured state to be recorded; from this follows that the measured states are dynamical states. In conclusion: "what and how much can be measured"?

For better answering let's analyze the experiment in which the resistance of an electric circuit is measured through measuring the current intensity I and the nominal tension U. Classically, one is interested in intrinsic value of the resistance and not the state of the resistor; in this case the experiment may be repeated at whatever moment of time by switching off the current, while on reloading the new experiment appears as being independent respecting the previous one; moreover, at each new experiment different quantities may be evaluated, while their set may be attributed to the same system even obtained at different moments of time.

Instead, in quantum mechanics the state is important in measurement, while all is fluctuating due to the undulatory nature of the states' components (electrons, atoms); as a consequence the influence of the measuring apparatus becomes partially out of control; as such the quantum mechanics may predict the maximal precision of an ideal experiment (beyond of any subjective or accidental errors). Thus, quantum measurements do not measure one system but two systems (object and the apparatus) in interaction, i.e., merely measures the interaction itself or the dynamical state of the concerned system to be measured.

Finally, what can we measure in quantum mechanics? The states themselves looses their causal evolution when interacting with an apparatus while achieving a sort of deterministic evolution; therefore, the measurement is even not on the states but on operators that have an intrinsic role on the system structure; from this point we recover the phenomenological idea that those operators that commutes with Hamiltonian of the system are observables and can be averaged on certain states to be measured (observed). Yet, other quantities may be measured on perturbed states of the systems under investigation.

These brings us to postulate that *quantum measuring is realized through stationary observables averaged on eigen-states or through recording perturbations of free or isolated causal (dynamic) states for a system.*

This postulate is in what follows unfolded, exemplified and applied on fundamental quantum mechanically (previously exposed) concepts.

3.8.1 CLASSICAL VS. QUANTUM ELECTRIC RESISTANCE

As stated in introduction, measuring the electric resistance is different within classical or quantum framework; here we illustrate this statement by treating both cases for this experimental issue.

1. The classical framework assumes electrons as particles forming the current density $\vec{j}$ under the external applied electric field $\vec{E}$ between

points "1" and "2" delimiting the frontier of the measured resistor for its resistance R, see Figure 3.19.

Their reciprocal relationship may be found out through the following:

$$\vec{j} = \frac{i_{12}}{S}\vec{n}_j = \frac{1}{S}\frac{\Delta q_{12}}{\Delta t_{12}}\vec{n}_j = \frac{(-e)n_e LS}{S\Delta t_{12}}\vec{n}_j = -en_e \underbrace{\frac{L}{\Delta t_{12}}}_{\langle v \rangle}\vec{n}_j$$

$$= -en_e \langle v \rangle \vec{n}_j = -en_e \frac{0 + a_e\tau}{2}\vec{n}_j$$

$$= -en_e\left(-\frac{e\vec{E}}{2m_0}\right)\tau = \frac{e^2 n_e \tau}{2m_0}\vec{E} \tag{3.611}$$

thus leaving with the equivalent expressions:

$$\vec{j} = \sigma_R \vec{E} \tag{3.612}$$

$$\vec{E} = \rho_R \vec{j} \tag{3.613}$$

involving the conductivity σ_R and resistivity ρ_R, constants respectively:

$$\sigma_R = \frac{e^2 n_e \tau}{2m_0} \tag{3.614}$$

$$\rho_R = \frac{1}{\sigma_R} \tag{3.615}$$

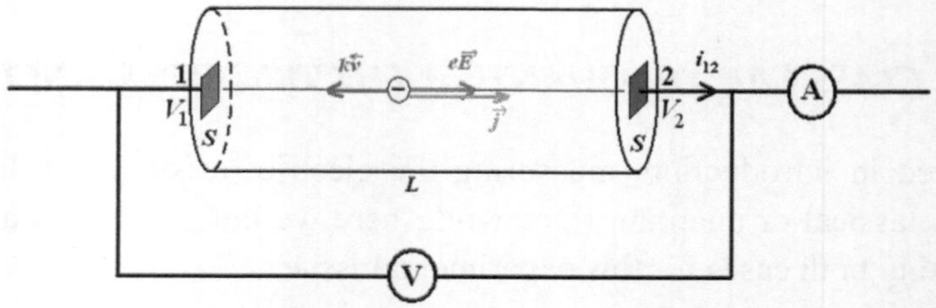

FIGURE 3.19 Sketch of the apparatus measuring the classical electrical resistance, where the electron is viewed as a moving particle through resistor.

in terms of electronic density n_e, the average time τ between two consecutive scatterings, beside fundamental constants as electronic charge and mass, e, m_0.

Yet, the resistance is to be determined by employing one of these electric field-current density equations to be multiplied by the elemental translation in resistor, dl,

$$\vec{E}d\vec{l} = \rho_R \vec{j}d\vec{l} = \rho_R \frac{i}{S}\vec{n}_j d\vec{l} \tag{3.616}$$

and then to integrate for the constant current this expression between the two ends,

$$\underbrace{\underbrace{\int_1^2 \vec{E}d\vec{l}}_{V_1-V_2}}_{u} = i_{12} \underbrace{\int_1^2 \frac{\rho_R}{S}\vec{n}_j d\vec{l}}_{R} \tag{3.617}$$

to get the local Ohm law

$$u_{12} = i_{12}R \tag{3.618}$$

linking the registered electric current and the measured tension at the end of the resistor by means of the resistance:

$$R = \int_1^2 \frac{\rho_R}{S}\vec{n}_j d\vec{l} \tag{3.619}$$

Note that for constant section and homogeneous matrix of resistor its resistance simply becomes:

$$R = \rho_R \frac{L}{S} \tag{3.620}$$

Worth noting that this classical description (sometimes known as Drude's theory) predicts that for ballistic trajectories of electrons in resistor, i.e., for movement without scattering with the resistor's matrix, the average time between two successive scatterings becomes infinity, thus predicting infinity conductivity, zero resistivity, and thus no resistance recorded:

$$\tau \to \infty \Rightarrow \sigma_R \to \infty \Rightarrow \rho_R \to 0 \Rightarrow R \to 0 \tag{3.621}$$

In next will be proved that this description is inconsistent with the quantum nature of the electron, and that a resistance "quanta" is to be observed or measure in any conditions, even when no scattering processes are involved.

2. When considering undulatory nature of the electron, the working definition of its current has no surface involvement, while its density within the resistor writes as:

$$n_e^q = \frac{2\pi}{L} f(E) \tag{3.622}$$

in terms of the its (first) quantized wave-vector, assumed for the resistor's crystal as, see Section 2.4.6.5,

$$k = \frac{2\pi}{L} \tag{3.623}$$

distributed on energy states according with the function $f(E)$. Moreover, the electronic velocity is as well written within undulatory framework as, see Section 1.3.2,

$$v^q = \frac{1}{\hbar}\frac{\Delta E}{\Delta k} \tag{3.624}$$

in terms of energy consumed ΔE, in accordance with the temporal Heisenberg principle,

$$\Delta E \Delta t \cong \hbar \tag{3.625}$$

However, being about the waves, one has to account for various modes of vibrations so that the summation over all possible wave-vectors has to be solved out, and this is done through performing the integral transformation:

$$\sum_{k=0}^{\infty} \bullet \rightarrow 2(spin)\frac{L}{2\pi}\int_0^{\infty} \bullet dk = \frac{L}{\pi}\int_0^{\infty} \bullet dk \tag{3.626}$$

with these the quantum electronic current intensity successively reads:

$$i^q = -e n_e^q v^q$$

$$= -e\frac{2\pi}{L} f(E) \frac{1}{\hbar} \frac{\Delta E}{\Delta k}$$

$$= -e\frac{2\pi}{L\hbar} \sum_{k=0}^{\infty} f(E) \frac{\partial E}{\partial k}$$

$$= -e\frac{2\pi}{L\hbar} \frac{L}{\pi} \int_{0}^{\infty} f(E) \frac{\partial E}{\partial k} dk;$$

$$i^q(\varepsilon) = -\frac{2e}{\hbar} \int_{\varepsilon}^{\infty} f(E) dE \tag{3.627}$$

Now, considering the distribution function approximated by the Heaviside step function,

$$f(E) \cong \Theta(\varepsilon - E) \tag{3.628}$$

and considering the two eigen-energies from where (ε_1) and to where (ε_2) the electron is accommodate with the measurement apparatus, see Figure 3.20, one has for the resulted recorded quantum current:

$$i_{12}^q = i^q(\varepsilon_1) - i^q(\varepsilon_2) = -\frac{2e}{\hbar} \int_{\varepsilon_1}^{\varepsilon_2} f(E) dE$$

$$\cong -\frac{2e}{\hbar} \int_{\varepsilon_1}^{\varepsilon_2} \Theta(\varepsilon - E) dE = -\frac{2e}{\hbar} (\varepsilon_2 - \varepsilon_1) = \frac{2e^2}{\hbar} \underbrace{\frac{\varepsilon_1 - \varepsilon_2}{e}}_{u_{12}^q};$$

$$i_{12}^q = \frac{2e^2}{\hbar} u_{12}^q \tag{3.629}$$

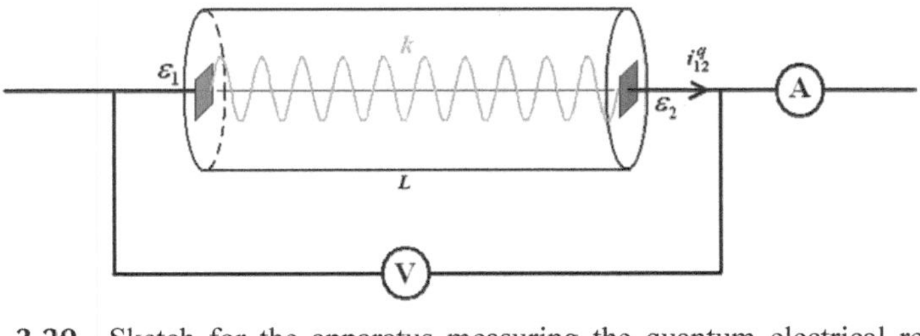

FIGURE 3.20 Sketch for the apparatus measuring the quantum electrical resistance, where the electron is viewed as a traveling wave through resistor.

The simple comparison of this expression with the local Ohm law, one gets the quantum resistance with the form:

$$R^q = \frac{\hbar}{2e^2} \cong 12.9[k\Omega] \tag{3.630}$$

known as the *contact resistance*, while playing the role of a quanta for electrical resistance. Overall, there is clear that while no scattering process was involved or considered, the measured resistance is non-zero, in contrast with earlier classical prediction.

There is therefore this proof of how much the quantum nature may influence our ordinary perception or classical measurements, in a way that may affect either our fundamental knowledge as well as practical (nano) applications. Further systematic of what can we observed out from the quantum world into average measurements is in next exposed.

3.8.2 QUANTUM CONSERVATION LAWS

Being given a stationary operator $\widehat{O}$ it fulfills the Heisenberg equation (see Sections 2.3.5 and 2.4.6):

$$\frac{d\widehat{O}}{dt} = \frac{1}{i\hbar}\left[\widehat{O},\widehat{H}\right] \tag{3.631}$$

while if it further commutes with the Hamiltonian as well,

$$\left[\widehat{O},\widehat{H}\right] = 0 \tag{3.632}$$

there follows that its averaged quantity on a given (or prepared through an apparatus) state is conserved:

$$\frac{d}{dt}\left\langle\widehat{O}\right\rangle_\psi = \frac{d}{dt}\langle\psi|\widehat{O}|\psi\rangle = \langle\psi|\underbrace{\frac{d}{dt}\widehat{O}}_{0}|\psi\rangle = 0;$$

$$\Rightarrow \left\langle\widehat{O}\right\rangle_\psi = const. \tag{3.633}$$

The first eminent example of this result is the Hamiltonian itself for a system: as far it does not depend explicitly on time, i.e., through the potential energy, it fulfills the conditions:

$$\frac{\partial \widehat{H}}{\partial t} = 0, \ \left[\widehat{H}, \widehat{H}\right] = 0 \tag{3.634}$$

immediately results that for any eigen state the eigen-energy is a constant (observable) quantity:

$$\left\langle \widehat{H} \right\rangle_{\psi} = \langle \psi | \underbrace{\widehat{H} | \psi \rangle}_{E|\psi\rangle} = E \underbrace{\langle \psi | \psi \rangle}_{1} = E \overset{!}{=} const. \tag{3.635}$$

consecrating the *energy conservation law* in the quantum mechanical frame.

Going to analyze the momentum operatorial behavior, it influences the translation of systems through the unitary translational operator, in the same way as introduced the time evolution operator:

$$\widehat{D}(\Delta x) = \exp\left[-\frac{i}{\hbar}\Delta x \widehat{p}\right] \tag{3.636}$$

in terms of the coordinate shift and the conjugate momentum on the movement direction, with the seminal property on states, see the Section 2.4.4:

$$\widehat{D}(\Delta x)|x\rangle = |x + \Delta x\rangle \tag{3.637}$$

Since the translational operator depends on momentum as well, it satisfies the specific functional commutation with coordinate (see Section 2.4.4), namely

$$\left[\widehat{x}, \widehat{D}\right] = i\hbar \partial_{\widehat{p}} \widehat{D} \tag{3.638}$$

there results the equivalent expressions

$$\Leftrightarrow \widehat{x}\widehat{D} - \widehat{D}\widehat{x} = i\hbar\left(-\frac{i}{\hbar}\Delta x\right)\widehat{D}$$

$$\Leftrightarrow \widehat{x}\widehat{D} = \widehat{D}\left(\widehat{x} + \Delta x\right)$$

$$\Leftrightarrow \hat{D}^{+}\hat{x}\hat{D} = \underbrace{\hat{D}^{+}\hat{D}}_{\hat{1}}\left(\hat{x} + \Delta x\right)$$

$$\Leftrightarrow \hat{x} + \Delta x = \hat{D}^{+}\hat{x}\hat{D} \tag{3.639}$$

until the unitary transformation of the coordinate into the translated one is reached.

Now being this unitary transformation established as that governing the translations, if one considers infinitesimal movement Δx of the measurement apparatus on x direction, the working operator looks like:

$$\hat{D}(\Delta x) \cong 1 - \frac{i}{\hbar}\Delta x\hat{p} \tag{3.640}$$

in the first order of the coordinate displacements approximation, while in the same framework, the condition that this movement leaves the Hamiltonian of the system invariant produces the chain relations

$$\begin{aligned}
\hat{H}(0) &= \hat{H}(\Delta x) \\
&= \hat{D}^{+}\hat{H}(0)\hat{D} \\
&\cong \left(1 + \frac{i}{\hbar}\Delta x\hat{p}\right)\hat{H}(0)\left(1 - \frac{i}{\hbar}\Delta x\hat{p}\right) \\
&= \hat{H}(0) - \frac{i}{\hbar}\Delta x\left[\hat{H}(0), \hat{p}\right] + \underbrace{(\Delta x)^2}_{\cong 0}\frac{\hat{p}\hat{H}(0)\hat{p}}{\hbar^2} \\
&= \hat{H}(0) - \frac{i}{\hbar}\Delta x\left[\hat{H}(0), \hat{p}\right]
\end{aligned} \tag{3.641}$$

from where the condition

$$\left[\hat{p}, \hat{H}\right] = 0 \tag{3.642}$$

that, along the fact that momentum do not explicitly depends of time, certifies, in fact, the *momentum conservation law* along the translation (quantum processes):

$$\left\langle \hat{p} \right\rangle_{\psi} = const. \tag{3.643}$$

Let's now consider that the apparatus performs a rotation of some angle (say$\Delta\alpha$) around the $0z$ axis in the (x, y) plane so that a given state suffers the rotation operator action according with the (already custom) rule:

$$\hat{R}(\Delta\alpha)|\psi(0)\rangle = |\psi(\Delta\alpha)\rangle \tag{3.644}$$

as was previously case of the translation operator. Yet one has to more analytically specify the form of rotation operator in order to search for the observable conserved upon rotation of measurement apparatus.

In this regard, basing on the Figure 3.21 one has for an axial rotation the coordinate transformations

$$x_2 = x_1 \cos\theta + y_1 \sin\theta \tag{3.645a}$$

$$y_2 = -x_1 \sin\theta + y_1 \cos\theta \tag{3.645b}$$

from where through employing the infinitesimal rotation angle,

$$\theta \cong \Delta\alpha \to 0 \tag{3.646}$$

they rewrite as

$$x_2 = x_1 + y_1\Delta\alpha \tag{3.647a}$$

$$y_2 = y_1 - x_1\Delta\alpha \tag{3.647b}$$

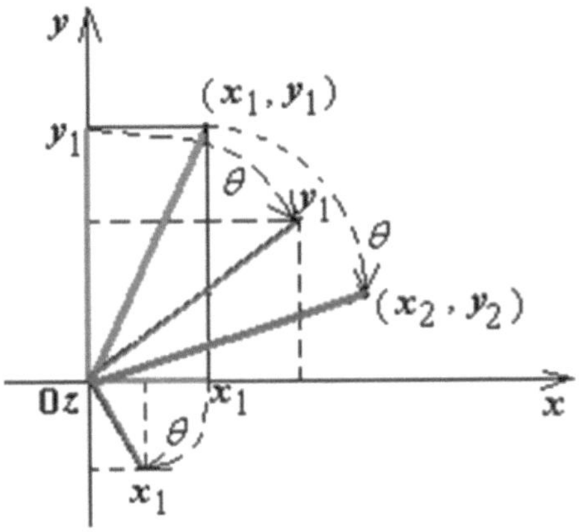

FIGURE 3.21 Basic illustration for the planar coordinates change upon perpendicular axis rotation.

Since the limiting trigonometric expressions:

$$\sin\Delta\alpha \cong \Delta\alpha ; \cos\Delta\alpha \cong 1 \tag{3.648}$$

These results, tells us that the effect of the $0z$ axial rotation is equivalent with two combined linear translations, one with $y\Delta\alpha$ and the other with $-x\Delta\alpha$ on $0x$ and $0y$ directions, respectively:

$$|\psi(\Delta\alpha_{0z})\rangle = |\psi(x + y\Delta\alpha_{0z}, y - x\Delta\alpha_{0z})\rangle = \widehat{R}(\Delta\alpha_{0Z})|\psi(x,y)\rangle \tag{3.649}$$

The passage for determining the form of associate rotation operator is made by *recognizing* the respective translation operators, in the first order approximation for the rotation angle $\Delta\alpha$, while specifying the conjugates of the momenta's with the directions suffering the movement

$$\widehat{D}_x \cong 1 - \frac{i}{\hbar}\widehat{y}\Delta\alpha_{0z}\widehat{p}_x \tag{3.650a}$$

$$\widehat{D}_y^+ \cong 1 + \frac{i}{\hbar}\widehat{x}\Delta\alpha_{0z}\widehat{p}_y \tag{3.650b}$$

With their help the rotation operator may be now expressed as:

$$\widehat{R}(\Delta\alpha_{0z}) = \widehat{D}_x\widehat{D}_y^+ \cong \left(1 - \frac{i}{\hbar}\widehat{y}\Delta\alpha_{0z}\widehat{p}_x\right)\left(1 + \frac{i}{\hbar}\widehat{x}\Delta\alpha_{0z}\widehat{p}_y\right)$$

$$= 1 - \frac{i}{\hbar}\left(\widehat{y}\widehat{p}_x - \widehat{x}\widehat{p}_y\right)\Delta\alpha_{0z} + \frac{\widehat{y}\widehat{p}_x\widehat{x}\widehat{p}_y}{\hbar^2}\underbrace{(\Delta\alpha_{0z})^2}_{\cong 0};$$

$$\widehat{R}(\Delta\alpha_{0z}) = 1 - \frac{i}{\hbar}\left(\widehat{y}\widehat{p}_x - \widehat{x}\widehat{p}_y\right)\Delta\alpha_{0z} \tag{3.651}$$

However, a conveyable expression can be obtained if we consider the concerned rotation as "active", i.e., moving the (quantum) vectorial basis of the system instead of considering the movement of the apparatus as was until now assumed; still, the two movements are equivalent until a minus sign to the rotation angle,

$$\Delta\alpha_{0z}^{Active/System} = -\Delta\alpha_{0z}^{Pasive/Apparatus} \tag{3.652}$$

since they are in opposite directions (as they are inverse composed, in the same manner the inertial and non-inertial systems are considered). In these conditions the rotation operator appears as written in terms of the kinetic momentum on $0z$ direction

$$\widehat{R}\left(\Delta\alpha_{0z}^{Active}\right)=1-\frac{i}{\hbar}\underbrace{\left(\widehat{x}\widehat{p}_y-\widehat{y}\widehat{p}_x\right)}_{\widehat{L}_z}\Delta\alpha_{0z}^{Active}=1-\frac{i}{\hbar}\widehat{L}_z\Delta\alpha_{0z}^{Active} \quad (3.653)$$

while recognizing it from its the classical (vectorial product) definition:

$$\mathbf{L}=\mathbf{x}\times\mathbf{p}=\begin{vmatrix}\mathbf{i} & \mathbf{j} & \mathbf{k}\\ x & y & z\\ p_x & p_y & p_z\end{vmatrix}=\mathrm{i}\underbrace{\left(yp_z-zp_y\right)}_{L_x}+\mathrm{j}\underbrace{\left(zp_x-xp_z\right)}_{L_y}+\mathrm{k}\underbrace{\left(xp_y-yp_x\right)}_{L_z} \quad (3.654)$$

From now on, one can easily repeat the procedure unfolded for translation operator, to successively have for preserving the Hamiltonian of the system upon infinitesimal (active) rotation

$$\begin{aligned}\widehat{H}(0)&=\widehat{H}(\Delta\alpha)=\widehat{R}^{+}\widehat{H}(0)\widehat{R}\\&\cong\left(1-\frac{i}{\hbar}\widehat{L}_z\Delta\alpha_{0z}^{Active}\right)\widehat{H}(0)\left(1+\frac{i}{\hbar}\widehat{L}_z\Delta\alpha_{0z}^{Active}\right)\\&=\widehat{H}(0)-\frac{i}{\hbar}\Delta\alpha_{0z}^{Active}\left[\widehat{H}(0),\widehat{L}_z\right]+\underbrace{\left(\Delta\alpha_{0z}^{Active}\right)^2}_{\cong 0}\frac{\widehat{L}_z\widehat{H}(0)\widehat{L}_z}{\hbar^2}\\&=\widehat{H}(0)-\frac{i}{\hbar}\Delta\alpha_{0z}^{Active}\left[\widehat{H}(0),\widehat{L}_z\right]\end{aligned} \quad (3.655)$$

leaving with commutator condition:

$$\left[\widehat{L}_z,\widehat{H}(0)\right]=0 \quad (3.656)$$

In similar manner there be found that the rotations along the other directions are quantified by associate quantum kinetic moments that commute with Hamiltonian of the system:

$$\left[\widehat{L}_x,\widehat{H}(0)\right]=0,\quad \left[\widehat{L}_y,\widehat{H}(0)\right]=0 \quad (3.657)$$

Moreover, having these relationships, there can be immediately shown that they hold also for the squares of kinetic moments, on individual directions

$$\left[\hat{L}_x^2,\hat{H}(0)\right]=\left[\hat{L}_x\hat{L}_x,\hat{H}(0)\right]=\hat{L}_x\underbrace{\left[\hat{L}_x,\hat{H}(0)\right]}_{0}+\underbrace{\left[\hat{L}_x,\hat{H}(0)\right]}_{0}\hat{L}_x=0 \quad (3.658a)$$

$$\left[\hat{L}_y^2,\hat{H}(0)\right]=0;\quad \left[\hat{L}_z^2,\hat{H}(0)\right]=0 \quad (3.658b)$$

as well on their summed total kinetic momentum:

$$\left[\hat{L}^2,\hat{H}(0)\right]=\left[\hat{L}_x^2+\hat{L}_y^2+\hat{L}_z^2,\hat{H}(0)\right]=0 \quad (3.659)$$

Now, since all these moments are not explicitly time-dependent there follows that both the total kinetic moment as well as its projections are constants of the quantum movements and observable in average over the eigen-states:

$$\left\langle \hat{L}^2\right\rangle_\psi = const.;\quad \left\langle \hat{L}_{x,y,z}\right\rangle_\psi = const. \quad (3.660)$$

However, worth finally draw attention on a very interesting property of the rotational operator, while generalizing it as:

$$\hat{R}(\theta)=\exp\left[-\frac{i}{\hbar}\hat{L}_z\theta\right] \quad (3.661)$$

for the active rotation along the $0z$ axis, for instance. The operator $\hat{L}_z$ is called the generator of the *Lie group of rotations* in this case, since the rotation operator fulfills the basic group properties. Beside, it enters as the coupling constant in the equation:

$$\frac{d\hat{R}(\theta)}{d\theta}=-\frac{i}{\hbar}\hat{L}_z\hat{R}(\theta) \quad (3.662)$$

with the boundary condition $\hat{R}(0)=\hat{1}$. This equation has the important consequence that allows the finding of the $0z$ kinetic momentum eigen-equation, since observing that the effect of the rotation operator on an arbitrary eigen-state $|\Upsilon\rangle$ expanded on the basis of the eigen-states $|m\rangle$ of the operator $\hat{L}_z$ we have:

$$\hat{R}(\theta)|\Upsilon\rangle = \sum_m |m\rangle e^{-\frac{i}{\hbar}\theta m} \tag{3.663}$$

with m the eigen-values of the Hermitic operator $\hat{L}_z$. Thus, through combining the last two identities we get two equivalent expressions for the same operatorial action, namely:

$$\hat{L}_z\hat{R}(\theta)|\Upsilon\rangle = \sum_m \hat{L}_z|m\rangle e^{-\frac{i}{\hbar}\theta m} \tag{3.664}$$

$$\hat{L}_z\hat{R}(\theta)|\Upsilon\rangle = \frac{\hbar}{-i}\frac{d}{d\theta}\left[\hat{R}(\theta)|\Upsilon\rangle\right] = \frac{\hbar}{-i}\sum_m |m\rangle\frac{d}{d\theta}e^{-\frac{i}{\hbar}\theta m} = \sum_m m|m\rangle e^{-\frac{i}{\hbar}\theta m} \tag{3.665}$$

from where we immediately identify the $\hat{L}_z$ eigen equation:

$$\hat{L}_z|m\rangle = m|m\rangle \tag{3.666}$$

This equation is most valuable when solving the angular motion of electron in atomic systems, driven by the quantified realm of the eigen-values m, as expected.

3.8.3 QUANTUM ELASTIC SCATTERING ON FIXED TARGET

One of the common quantum experiments consists in investigating the states of a system through elastic scattering on it by an income wave, say with momentum **p**, while observing the divergent scattered wave from the target $|0\rangle$ on a certain direction, say fixed by the versor $\mathbf{n} \in \Delta\Omega$, see Figure 3.22.

Additionally, there is assumed that during the scattering process neither modification of the inner structure of the system investigated nor other binding phenomena are produced. The measured quantity is the so-called *effective crossing section* $\Delta\sigma$, usually defined by the number of particle

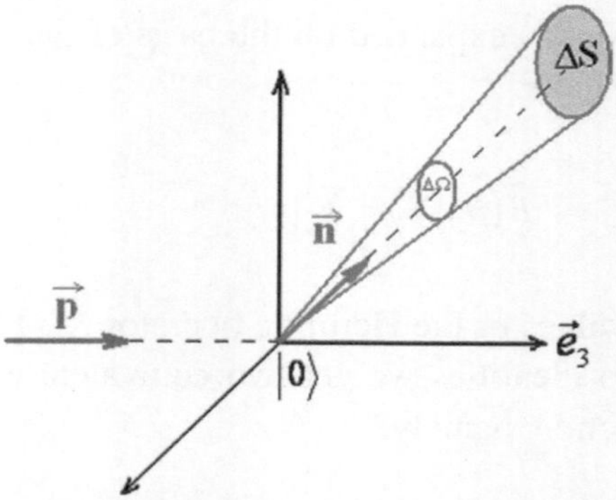

FIGURE 3.22 Sketch of the elastic scattering and observation of the emergent wave on certain direction in space.

detected in the angular angle $\Delta\Omega$ in a given time Δt from the number of the particles that crosses the unit aria placed in the target zone perpendicular on the incident fascicle direction (being independent on the density of incoming particles, which can be projected even one after another towards target):

$$\Delta\sigma = \frac{\Delta N}{\frac{\delta N}{\delta S}} \tag{3.667}$$

Alternatively, it may be expressed with the help of quantum probability of detection the emergent particles coming on detector through scattering from the incident ones on target, $\wp_{i\to f}$, respecting the incident flux of particle on target Φ_{in}:

$$\Delta\sigma = \frac{\wp_{i\to f}}{\Phi_{in}} \tag{3.668}$$

that can be further viewed as the integral of the effective differential cross section $d\sigma / d\Omega$:

$$\Delta\sigma = \int_{\Delta\Omega} \left(\frac{d\sigma}{d\Omega}\right)_{(\mathbf{p},\mathbf{n})} d\Omega \tag{3.669}$$

Worth noting that this picture is valid as far the time of interaction is significantly less than the time of preparation (of incident flux) or of the time of measurement (of emergent flux). However, as the strategy of work we are searching for the *asymptotic form* of the scattered wave, Ψ_{em}, since it is observed at large distances from the target; from it the density of current probability will be computed, $\mathbf{j}_{em}$, and then integrated to yield the transition probability:

$$\wp_{i\to f} = \int_{\Delta S}\int_{-\infty}^{+\infty} \mathbf{j}_{em} dt d\mathbf{S} \tag{3.670}$$

Going to unfold the above "measurement program" the first stage is in characterizing the free particle traveling toward its target. It is supposed to have an average (known but not as 100% determined) momentum p and characterized by a wave function in momentum representation in the very moment of scattering:

$$\langle \mathbf{P}|0\rangle = \varphi(\mathbf{P}-\mathbf{p}) \tag{3.671}$$

with the properties: (i) it becomes Dirac delta function when the momentum is precisely determined; (ii) has in any case a clear sharp maximum; (iii) it is an even function $|\varphi(-\mathbf{P})| = |\varphi(\mathbf{P})|$ so that the momentum to be well defined through the average on the unperturbed state:

$$\begin{aligned}
\langle 0|\hat{\mathbf{p}}|0\rangle &= \langle 0|\hat{1}_{\{|\mathbf{P}\rangle\}}\hat{\mathbf{p}}|0\rangle = \int_{-\infty}^{+\infty} \langle 0|\mathbf{P}\rangle \underbrace{\langle \mathbf{P}|\hat{\mathbf{p}}|0\rangle}_{\mathbf{P}\langle\mathbf{P}|} d\mathbf{P} \\
&= \int_{-\infty}^{+\infty} \mathbf{P}|\langle \mathbf{P}|0\rangle|^2 d\mathbf{P} = \int_{-\infty}^{+\infty} \mathbf{P}|\varphi(\mathbf{P}-\mathbf{p})|^2 d\mathbf{P} \\
&\overset{\mathbf{P}-\mathbf{p}=\mathbf{P}'}{=} \int_{-\infty}^{+\infty} (\mathbf{P}'+\mathbf{p})|\varphi(\mathbf{P}')|^2 d\mathbf{P}' = \mathbf{p}\underbrace{\int_{-\infty}^{+\infty} |\varphi(\mathbf{P}')|^2 d\mathbf{P}'}_{\langle 0|0\rangle=1} + \underbrace{\int_{-\infty}^{+\infty} \mathbf{P}'|\varphi(\mathbf{P}')|^2 d\mathbf{P}'}_{\substack{0 \\ by\ (iii)\ above}} = \mathbf{p}
\end{aligned} \tag{3.672}$$

With this we can turn to the perturbation process that is described for momentum states of the Lippmann-Schwinger expansion, see Sections 3.6.3:

$$|\Psi_{\mathbf{P}}^{(+)}\rangle = |\mathbf{P}\rangle + \frac{1}{E(\mathbf{P}) - \widehat{H}_0 + i\eta} V(\mathbf{x})|\Psi_{\mathbf{P}}^{(+)}\rangle \tag{3.673}$$

which in coordinate representation becomes:

$$\begin{aligned}\left\langle \mathbf{x}\middle|\Psi_{\mathbf{P}}^{(+)}\right\rangle &= \langle \mathbf{x}|\mathbf{P}\rangle + \langle \mathbf{x}|\frac{1}{E(\mathbf{P})-\widehat{H}_0+i\eta}\hat{1}_{\{|\mathbf{x}\rangle\}}V(\mathbf{x})\left|\Psi_{\mathbf{P}}^{(+)}\right\rangle \\ &= \langle \mathbf{x}|\mathbf{P}\rangle + \int d\mathbf{y}\underbrace{\langle \mathbf{x}|\underbrace{\frac{1}{E(\mathbf{P})-\widehat{H}_0+i\eta}}_{G^{(+)}}|\mathbf{y}\rangle}_{g^{(+)}(\mathbf{x}-\mathbf{y})}\underbrace{\langle \mathbf{y}|V(\mathbf{x})}_{V(\mathbf{y})\langle\mathbf{y}|}\left|\Psi_{\mathbf{P}}^{(+)}\right\rangle \\ &= \langle \mathbf{x}|\mathbf{P}\rangle + \int d\mathbf{y}g^{(+)}(\mathbf{x}-\mathbf{y})V(\mathbf{y})\left\langle \mathbf{y}\middle|\Psi_{\mathbf{P}}^{(+)}\right\rangle \end{aligned} \tag{3.674}$$

Before going further we have to evaluate the space averaged Green function $g^{(+)}(\mathbf{x}-\mathbf{y})$ involved; for that we firstly rewrite it as:

$$\begin{aligned} g^{(+)}(\mathbf{x}-\mathbf{y}) &= \langle \mathbf{x}|G^{(+)}\hat{1}_{\{|\mathbf{P}'\rangle\}}|\mathbf{y}\rangle = \int \langle \mathbf{x}|G^{(+)}|\mathbf{P}'\rangle\langle \mathbf{P}'|\mathbf{y}\rangle d\mathbf{P}' \\ &= \int \langle \mathbf{x}|\frac{1}{E(\mathbf{P})-\widehat{H}_0+i\eta}|\mathbf{P}'\rangle\langle \mathbf{P}'|\mathbf{y}\rangle d\mathbf{P}' \\ &\overset{\widehat{H}_0|\mathbf{P}'\rangle=E(\mathbf{P}')|\mathbf{P}'\rangle}{=} \int \frac{1}{E(\mathbf{P})-E(\mathbf{P}')+i\eta}\langle \mathbf{x}|\mathbf{P}'\rangle\langle \mathbf{P}'|\mathbf{y}\rangle d\mathbf{P}' \end{aligned} \tag{3.675}$$

By recognizing the generating distribution functions for the (de Broglie) free wave-packets, see Section 2.2.2,

$$\langle \mathbf{x}|\mathbf{P}'\rangle = u_{\mathbf{P}'}(\mathbf{x}) = \frac{1}{(2\pi\hbar)^{3/2}}e^{\frac{i}{\hbar}\mathbf{P}'\cdot\mathbf{x}} \tag{3.676a}$$

$$\langle \mathbf{P}'|\mathbf{y}\rangle = \bar{u}_{\mathbf{P}'}(\mathbf{y}) = \frac{1}{(2\pi\hbar)^{3/2}}e^{-\frac{i}{\hbar}\mathbf{P}'\cdot\mathbf{y}} \tag{3.676b}$$

together with (de Broglie) energy-momentum relationships:

$$E(\mathbf{P}) = \frac{P^2}{2m} = \frac{\hbar^2}{2m}k^2 \tag{3.677a}$$

$$E(\mathbf{P}') = \frac{P'^2}{2m} = \frac{\hbar^2}{2m}k'^2 \tag{3.677b}$$

the last expression of the space averaged Green function successively expresses as (note that in 3D we also have $\mathbf{P'} = \hbar\mathbf{k'}$ while for integrand we put $d\mathbf{P'} = \hbar^3 d\mathbf{k'}$):

$$
\begin{aligned}
g^{(+)}(\mathbf{x}-\mathbf{y}) &= \hbar^3 \int \frac{1}{\frac{\hbar^2}{2m}\left(k^2 - k'^2\right) + i \underbrace{\eta}_{\eta'\hbar^2/2m}} \frac{1}{(2\pi\hbar)^3} e^{i\mathbf{k'}\cdot(\mathbf{x}-\mathbf{y})} d\mathbf{k'} \\
&= -\frac{1}{(2\pi)^3}\frac{2m}{\hbar^2}\int \frac{1}{k'^2 - k^2 - i\eta'} e^{i\mathbf{k'}\cdot(\mathbf{x}-\mathbf{y})} d\mathbf{k'} \\
&= -\frac{1}{(2\pi)^3}\frac{2m}{\hbar^2}\underbrace{\int_0^{2\pi} d\varphi}_{2\pi}\int_0^{\pi} d\theta \int_0^{\infty} dk' k'^2 \sin\theta \frac{1}{k'^2 - k^2 - i\eta'} e^{ik'|\mathbf{x}\text{-}\mathbf{y}|\cos\theta} \\
&= -\frac{1}{(2\pi)^2}\frac{2m}{\hbar^2}\int_0^{\infty} dk' \frac{k'^2}{k'^2 - k^2 - i\eta'} \underbrace{\int_0^{\pi} d\theta \sin\theta \exp\left(ik'|\mathbf{x}\text{-}\mathbf{y}|\underbrace{\cos\theta}_{\xi}\right)}_{\int_{-1}^{+1} d\xi \exp\left(ik'\cdot|\mathbf{x}\text{-}\mathbf{y}|\xi\right)} \\
&= -\frac{1}{(2\pi)^2}\frac{2m}{\hbar^2}\int_0^{\infty} dk' \frac{k'^2}{k'^2 - k^2 - i\eta'} \underbrace{\int_{-1}^{+1} d\xi \exp\left(ik'|\mathbf{x}\text{-}\mathbf{y}|\xi\right)}_{\frac{\exp\left(ik'|\mathbf{x}\text{-}\mathbf{y}|\right)-\exp\left(-ik'|\mathbf{x}\text{-}\mathbf{y}|\right)}{ik'|\mathbf{x}\text{-}\mathbf{y}|}} \\
&= -\frac{1}{(2\pi)^2}\frac{2m}{\hbar^2}\frac{1}{i|\mathbf{x}\text{-}\mathbf{y}|}\int_0^{\infty} dk' \frac{k'}{k'^2 - k^2 - i\eta'}\left[\exp\left(ik'|\mathbf{x}\text{-}\mathbf{y}|\right) - \exp\left(-ik'|\mathbf{x}\text{-}\mathbf{y}|\right)\right] \\
&= -\frac{1}{(2\pi)^2}\frac{2m}{\hbar^2}\frac{1}{i|\mathbf{x}\text{-}\mathbf{y}|}\left\{
\begin{array}{l}
\displaystyle\int_0^{\infty} dk' \frac{k'\exp\left(ik'|\mathbf{x}\text{-}\mathbf{y}|\right)}{k'^2 - k^2 - i\eta'} \\
\displaystyle - \int_0^{(-)\infty} d(-)k' \frac{(-)k'\exp\left(-i(-)k'|\mathbf{x}\text{-}\mathbf{y}|\right)}{k'^2 - k^2 - i\eta'}
\end{array}\right\} \\
&= -\frac{1}{(2\pi)^2}\frac{2m}{\hbar^2}\frac{1}{i|\mathbf{x}\text{-}\mathbf{y}|}\left\{\int_0^{\infty} dk' \frac{k'\exp\left(ik'|\mathbf{x}\text{-}\mathbf{y}|\right)}{k'^2 - k^2 - i\eta'} + \int_{-\infty}^{0} dk' \frac{k'\exp\left(ik'|\mathbf{x}\text{-}\mathbf{y}|\right)}{k'^2 - k^2 - i\eta'}\right\} \\
&= -\frac{1}{(2\pi)^2}\frac{2m}{\hbar^2}\frac{1}{i|\mathbf{x}\text{-}\mathbf{y}|}\int_{-\infty}^{\infty} dk' \frac{k'\exp\left(ik'|\mathbf{x}\text{-}\mathbf{y}|\right)}{k'^2 - k^2 - i\eta'}
\end{aligned}
\tag{3.678}
$$

This result requires only the one step of complex integration, while the denominator Feynman recipe tells us that this integration has to be done around the pole $k'_1 = +k$, since the equation $k'^2 = k^2 + i\eta'$ suggests a slightly movement with $+i\eta'$, $\eta' \cong 0^+$, toward higher half complex plane, while avoiding the pole through rotation in the local trigonometric direction:

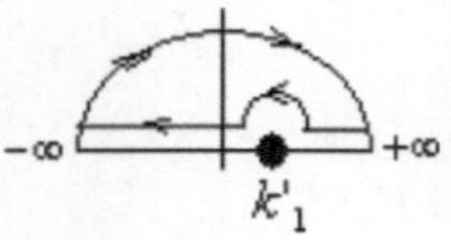

The application of the residue (Cauchy) theorem leaves with:

$$g^{(+)}(\mathbf{x}-\mathbf{y}) = -\frac{1}{(2\pi)^2}\frac{2m}{\hbar^2}\frac{1}{i|\mathbf{x}\text{-}\mathbf{y}|}(+2\pi i)\left[(k'-k)\frac{k'\exp(ik'|\mathbf{x}\text{-}\mathbf{y}|)}{(k'-k)(k'+k)}\right]_{k'=k}$$
$$= -\frac{1}{4\pi}\frac{2m}{\hbar^2}\frac{\exp(ik|\mathbf{x}\text{-}\mathbf{y}|)}{|\mathbf{x}\text{-}\mathbf{y}|} \tag{3.679}$$

$$g^{(+)}(\mathbf{x}-\mathbf{y}) = -\frac{1}{4\pi}\frac{2m}{\hbar^2}\frac{\exp\left(\frac{i}{\hbar}P|\mathbf{x}\text{-}\mathbf{y}|\right)}{|\mathbf{x}\text{-}\mathbf{y}|} \tag{3.680}$$

This expression may be still transformed for acquiring the asymptotic information, through modeling the distance $|\mathbf{x}\text{-}\mathbf{y}|$ following the prescription: $\mathbf{x} = r\mathbf{n}$, $|\mathbf{x}| = r \to \infty$, $\forall\mathbf{y}$; accordingly one gets:

$$|\mathbf{x}\text{-}\mathbf{y}| = \sqrt{(\mathbf{x}\text{-}\mathbf{y})^2} = \sqrt{\mathbf{x}^2 + \mathbf{y}^2 - 2\mathbf{x}\cdot\mathbf{y}}$$
$$= |\mathbf{x}|\sqrt{1 + \frac{\mathbf{y}^2}{|\mathbf{x}|^2} - \frac{2\mathbf{x}\cdot\mathbf{y}}{|\mathbf{x}|^2}} = r\sqrt{1 + \frac{\mathbf{y}^2}{r^2} - \frac{2\mathbf{n}\cdot\mathbf{y}}{r}}$$
$$\cong r\left[1 + \frac{1}{2}\left(\frac{\mathbf{y}^2}{r^2} - \frac{2\mathbf{n}\cdot\mathbf{y}}{r}\right)\right] = r - \mathbf{n}\cdot\mathbf{y} + \underbrace{\frac{\mathbf{y}^2}{r}}_{\cong 0};$$
$$|\mathbf{x}\text{-}\mathbf{y}| \cong r - \mathbf{n}\cdot\mathbf{y} \tag{3.681}$$

While noting this distance appears at the denominator of the space averaged (retarded) Green function, we can employ once more the asymptotic contribution:

$$\frac{1}{|\mathbf{x}-\mathbf{y}|} \cong \frac{1}{r-\mathbf{n}\cdot\mathbf{y}} = \frac{1}{r\left(1-\frac{\mathbf{n}\cdot\mathbf{y}}{r}\right)} = \frac{1}{r}\left(1-\frac{\mathbf{n}\cdot\mathbf{y}}{r}\right)^{-1}$$

$$\cong \frac{1}{r}\left(1+\frac{\mathbf{n}\cdot\mathbf{y}}{r}\right) = \frac{1}{r} + \underbrace{\frac{\mathbf{n}\cdot\mathbf{y}}{r^2}}_{\cong 0};$$

$$\frac{1}{|\mathbf{x}-\mathbf{y}|} \cong \frac{1}{r} \tag{3.682}$$

With this the space averaged retarded Green function takes the simplified (further workable) form:

$$g^{(+)}(\mathbf{x}-\mathbf{y}) = -\frac{1}{4\pi}\frac{2m}{\hbar^2}\frac{\exp\left[\frac{i}{\hbar}P(r-\mathbf{n}\cdot\mathbf{y})\right]}{r} \tag{3.683}$$

Worth noting that this represent in fact the asymptotic solution of the free motion (Schrödinger type) equation:

$$\left[\frac{\hbar^2}{2m}\left(\partial_{\mathbf{x}}^2 + k^2\right)\right] g^{(+)}(\mathbf{x}-\mathbf{y}) = \delta^3(\mathbf{x}-\mathbf{y}) \tag{3.684}$$

representing the stationary version of the more general (time dependent ones) presented in Section 3.6.4. However, its validity can be proofed quite easily by employing the Schrödinger equation

$$\left[-\frac{\hbar^2}{2m}\partial_{\mathbf{x}}^2 + V(\mathbf{x})\right]\left|\Psi_{\mathbf{P}}^{(+)}\right\rangle = E(\mathbf{P})\left|\Psi_{\mathbf{P}}^{(+)}\right\rangle \tag{3.685}$$

consecutively rearranged as:

$$f^{(+)}(\mathbf{n},\mathbf{P}) = -\sqrt{\frac{2\pi}{\hbar}} m \int \underbrace{e^{-\frac{i}{\hbar}P(\mathbf{n}\cdot\mathbf{y})}}_{(2\pi\hbar)^{3/2}\langle P_{\mathbf{n}}|\mathbf{y}\rangle} \underbrace{V(\mathbf{y})\langle\mathbf{y}|}_{\langle\mathbf{y}|V(\mathbf{x})}\left|\Psi_{\mathbf{P}}^{(+)}\right\rangle d\mathbf{y}$$

$$= -(2\pi)^2\hbar m \int \langle P_{\mathbf{n}}|\mathbf{y}\rangle\langle\mathbf{y}|V(\mathbf{x})\left|\Psi_{\mathbf{P}}^{(+)}\right\rangle d\mathbf{y}$$

$$\Leftrightarrow \frac{\hbar^2}{2m}\left(\partial_{\mathbf{x}}^2 + k^2\right)\left\langle \mathbf{x} \middle| \Psi_{\mathbf{P}}^{(+)} \right\rangle = \int \delta^3(\mathbf{x}-\mathbf{y}) V(\mathbf{y}) \left\langle \mathbf{y} \middle| \Psi_{\mathbf{P}}^{(+)} \right\rangle d\mathbf{y} \tag{3.686}$$

until the formal "extraction" of the retarded wave-function:

$$\left\langle \mathbf{x} \middle| \Psi_{\mathbf{P}}^{(+)} \right\rangle = \int \underbrace{\left[\frac{\hbar^2}{2m}\left(\partial_{\mathbf{x}}^2 + k^2\right)\right]^{-1} \delta^3(\mathbf{x}-\mathbf{y})}_{g^{(+)}(\mathbf{x}-\mathbf{y})} V(\mathbf{y}) \left\langle \mathbf{y} \middle| \Psi_{\mathbf{P}}^{(+)} \right\rangle d\mathbf{y} \tag{3.687}$$

from where the above Green function equation of the time-independent free particle. Nevertheless, this last wave-function represents only the scattered resulting wave-function when the free Green function is convoluted with the potential interaction in the Lippmann-Schwinger perturbation way; it has to be further supplemented with the free (incoming) traveling wave-packet in order to realistically model the scattering process, as previously exposed. With these, the asymptotic scattered wave-function looks like:

$$\begin{aligned}
\left\langle \mathbf{x} \middle| \Psi_{\mathbf{P}}^{(+)} \right\rangle &= \langle \mathbf{x} | \mathbf{P} \rangle + \int g^{(+)}(\mathbf{x}-\mathbf{y}) V(\mathbf{y}) \left\langle \mathbf{y} \middle| \Psi_{\mathbf{P}}^{(+)} \right\rangle d\mathbf{y} \\
&= \frac{1}{(2\pi\hbar)^{3/2}} e^{\frac{i}{\hbar}\mathbf{P}\cdot\mathbf{x}} - \frac{1}{4\pi}\frac{2m}{\hbar^2}\frac{e^{\frac{i}{\hbar}Pr}}{r} \int e^{-\frac{i}{\hbar}P(\mathbf{n}\cdot\mathbf{y})} V(\mathbf{y}) \left\langle \mathbf{y} \middle| \Psi_{\mathbf{P}}^{(+)} \right\rangle d\mathbf{y} \\
&= \frac{1}{(2\pi\hbar)^{3/2}} \left[e^{\frac{i}{\hbar}\mathbf{P}\cdot\mathbf{x}} - \frac{(2\pi\hbar)^{3/2}}{4\pi}\frac{2m}{\hbar^2}\frac{e^{\frac{i}{\hbar}Pr}}{r} \int e^{-\frac{i}{\hbar}P(\mathbf{n}\cdot\mathbf{y})} V(\mathbf{y}) \left\langle \mathbf{y} \middle| \Psi_{\mathbf{P}}^{(+)} \right\rangle d\mathbf{y} \right] \\
&\equiv \frac{1}{(2\pi\hbar)^{3/2}} \left[e^{\frac{i}{\hbar}\mathbf{P}\cdot\mathbf{x}} + f^{(+)}(\mathbf{n},\mathbf{P}) \frac{e^{\frac{i}{\hbar}Pr}}{r} \right]
\end{aligned} \tag{3.688}$$

where we have introduced the *scattering amplitude* with the equivalent forms

$$\begin{aligned}
f^{(+)}(\mathbf{n},\mathbf{P}) &= -\sqrt{\frac{2\pi}{\hbar}} m \int \underbrace{e^{-\frac{i}{\hbar}P(\mathbf{n}\cdot\mathbf{y})}}_{(2\pi\hbar)^{3/2}\langle P_{\mathbf{n}} | \mathbf{y} \rangle} \underbrace{V(\mathbf{y})\langle \mathbf{y} |}_{\langle \mathbf{y} | V(\mathbf{x})} \left| \Psi_{\mathbf{P}}^{(+)} \right\rangle d\mathbf{y} \\
&= -(2\pi)^2 \hbar m \int \langle P_{\mathbf{n}} | \mathbf{y} \rangle \langle \mathbf{y} | V(\mathbf{x}) \left| \Psi_{\mathbf{P}}^{(+)} \right\rangle d\mathbf{y}
\end{aligned}$$

$$= -(2\pi)^2 \hbar m \langle P_{\mathbf{n}} | V(\mathbf{x}) | \Psi_{\mathbf{P}}^{(+)} \rangle \tag{3.689}$$

until the last one that does not depend by a certain representation, being thus an invariant of the scattering.

The scattering process is not entirely described until the full temporal perturbation is considered, see Section 3.6.1:

$$\begin{aligned}
|\Psi^{(+)}(t)\rangle &= \widehat{\Omega}^{(+)} |t\rangle \\
&= \widehat{\Omega}^{(+)} e^{-\frac{i}{\hbar}\widehat{H}_0 t} \widehat{1}_{\{|\mathbf{P}\rangle\}} |0\rangle \\
&= \int \widehat{\Omega}^{(+)} e^{-\frac{i}{\hbar}\widehat{H}_0 t} |\mathbf{P}\rangle \underbrace{\langle \mathbf{P}|0\rangle}_{\varphi(\mathbf{P}-\mathbf{p})} d\mathbf{P} \\
&= \int \underbrace{\widehat{\Omega}^{(+)} |\mathbf{P}\rangle}_{|\Psi_{\mathbf{P}}^{(+)}\rangle} e^{-\frac{i}{\hbar}E(\mathbf{P})t} \varphi(\mathbf{P}-\mathbf{p}) d\mathbf{P};
\end{aligned}$$

$$|\Psi^{(+)}(t)\rangle = \int |\Psi_{\mathbf{P}}^{(+)}\rangle e^{-\frac{i}{\hbar}E(\mathbf{P})t} \varphi(\mathbf{P}-\mathbf{p}) d\mathbf{P} \tag{3.690}$$

The perturbed evolution state can now be spatially represented as:

$$\begin{aligned}
\langle \mathbf{x} | \Psi^{(+)}(t)\rangle &= \int \langle \mathbf{x} | \Psi_{\mathbf{P}}^{(+)}\rangle e^{-\frac{i}{\hbar}E(\mathbf{P})t} \varphi(\mathbf{P}-\mathbf{p}) d\mathbf{P} \\
&= \frac{1}{(2\pi\hbar)^{3/2}} \int \left[e^{\frac{i}{\hbar}\mathbf{P}\cdot\mathbf{x}} + f^{(+)}(\mathbf{n},\mathbf{P}) \frac{e^{\frac{i}{\hbar}Pr}}{r} \right] e^{-\frac{i}{\hbar}E(\mathbf{P})t} \varphi(\mathbf{P}-\mathbf{p}) d\mathbf{P} \\
&= \underbrace{\frac{1}{(2\pi\hbar)^{3/2}} \int \varphi(\mathbf{P}-\mathbf{p}) e^{-\frac{i}{\hbar}\left[E(\mathbf{P})t - \mathbf{P}\cdot\mathbf{x}\right]} d\mathbf{P}}_{I_1} \\
&\quad + \underbrace{\frac{1}{(2\pi\hbar)^{3/2}} \frac{1}{r} \int f^{(+)}(\mathbf{n},\mathbf{P}) \varphi(\mathbf{P}-\mathbf{p}) e^{-\frac{i}{\hbar}\left[E(\mathbf{P})t - Pr\right]} d\mathbf{P}}_{I_2}
\end{aligned} \tag{3.691}$$

In next we will evaluate these integrals separately. For the first integral we will consider the variable change $\mathbf{P'} = \mathbf{P} - \mathbf{p}$ to get for the exponent:

$$\begin{aligned}
E(\mathbf{P})t - \mathbf{P}\cdot\mathbf{x} &= E(\mathbf{P'}+\mathbf{p})t - (\mathbf{P'}+\mathbf{p})\cdot\mathbf{x} \\
&= \frac{(\mathbf{P'}+\mathbf{p})^2}{2m}t - (\mathbf{P'}+\mathbf{p})\cdot\mathbf{x} = \frac{\mathbf{P'}^2}{2m}t + \underbrace{\frac{\mathbf{p}^2}{2m}}_{E(\mathbf{p})}t + \mathbf{P'}\underbrace{\frac{\mathbf{p}}{m}}_{\mathbf{v}}t - (\mathbf{P'}+\mathbf{p})\cdot\mathbf{x} \\
&= \left[E(\mathbf{p})t - \mathbf{p}\cdot\mathbf{x}\right] + \frac{\mathbf{P'}^2}{2m}t - \mathbf{P'}\cdot(\mathbf{x}-\mathbf{v}t)
\end{aligned} \tag{3.692}$$

while employing the fact of knowing the incident momentum, i.e., $\mathbf{P'}^2 = (\mathbf{P}-\mathbf{p})^2 \to 0$, we have:

$$\begin{aligned}
I_1 &= \frac{1}{(2\pi\hbar)^{3/2}} e^{-\frac{i}{\hbar}[E(\mathbf{P})t-\mathbf{p}\cdot\mathbf{x}]} \int \varphi(\mathbf{P'}) \underbrace{e^{-\frac{i}{\hbar}\frac{\mathbf{P'}^2}{2m}t}}_{\cong 1} e^{\frac{i}{\hbar}\mathbf{P'}\cdot(\mathbf{x}-\mathbf{v}t)} d\mathbf{P'} \\
&= e^{-\frac{i}{\hbar}[E(\mathbf{P})t-\mathbf{p}\cdot\mathbf{x}]} \underbrace{\frac{1}{(2\pi\hbar)^{3/2}} \int \varphi(\mathbf{P'}) e^{\frac{i}{\hbar}\mathbf{P'}\cdot(\mathbf{x}-\mathbf{v}t)} d\mathbf{P'}}_{\substack{\text{Fourier transformation} \\ \chi(\mathbf{x}-\mathbf{v}t)}};
\end{aligned}$$

$$I_1 = \chi(\mathbf{x}-\mathbf{v}t) e^{-\frac{i}{\hbar}[E(\mathbf{P})t-\mathbf{p}\cdot\mathbf{x}]} \tag{3.693}$$

For the second integral above, within the same variable changing and approximation as done for the first integral, we can write in the first instance that:

$$\begin{aligned}
P &= \sqrt{\mathbf{P}^2} = \sqrt{(\mathbf{P'}+\mathbf{p})^2} = \sqrt{\mathbf{P'}^2 + \mathbf{p}^2 + 2\mathbf{P'p}} = p\sqrt{1 + \frac{\mathbf{P'}^2}{p^2} + \frac{2\mathbf{P'p}}{p^2}} \\
&\cong p\left[1 + \frac{1}{2}\left(\frac{\mathbf{P'}^2}{p^2} + \frac{2\mathbf{P'p}}{p^2}\right)\right] = p + \underbrace{\frac{\mathbf{P'}^2}{2p}}_{\cong 0} + \frac{\mathbf{P'}}{p}\underbrace{\mathbf{p}}_{p\mathbf{e}_3} \cong p + \mathbf{P'}\cdot\mathbf{e}_3
\end{aligned} \tag{3.694}$$

while the exponent becomes:

$$\begin{aligned}E(\mathbf{P})t - Pr &= \frac{\mathbf{P'}^2}{2m}t + \frac{\mathbf{p}^2}{2m}t + \mathbf{P'}\cdot\mathbf{v}t - pr - \mathbf{P'}\cdot\mathbf{e}_3 r \\ &= \frac{\mathbf{P'}^2}{2m}t + \left(\frac{\mathbf{p}^2}{2m}t - pr\right) - \mathbf{P'}\cdot\left(\mathbf{e}_3 r - \mathbf{v}t\right)\end{aligned} \tag{3.695}$$

so that the whole integral takes the form:

$$\begin{aligned}I_2 &= \frac{1}{(2\pi\hbar)^{3/2}}\frac{1}{r}e^{-\frac{i}{\hbar}\left(\frac{\mathbf{p}^2}{2m}t-pr\right)}\int f^{(+)}(\mathbf{n},\underbrace{\mathbf{P'}}_{\cong 0}+\mathbf{p})\varphi(\mathbf{P'})\underbrace{e^{-\frac{i}{\hbar}\frac{\mathbf{P'}^2}{2m}t}}_{\cong 1}e^{\frac{i}{\hbar}\mathbf{P'}\cdot(\mathbf{e}_3 r-\mathbf{v}t)}d\mathbf{P'} \\ &= \frac{1}{r}e^{-\frac{i}{\hbar}\left(\frac{\mathbf{p}^2}{2m}t-pr\right)}f^{(+)}(\mathbf{n},\mathbf{p})\underbrace{\frac{1}{(2\pi\hbar)^{3/2}}\int\varphi(\mathbf{P'})e^{\frac{i}{\hbar}\mathbf{P'}\cdot(\mathbf{e}_3 r-\mathbf{v}t)}d\mathbf{P'}}_{\substack{\text{Fourier transformation} \\ \chi(\mathbf{e}_3 r-\mathbf{v}t)}} \\ &= \frac{1}{r}e^{-\frac{i}{\hbar}\left(\frac{\mathbf{p}^2}{2m}t-pr\right)}f^{(+)}(\mathbf{n},\mathbf{p})\underbrace{\chi(\underbrace{\mathbf{e}_3 r-\underbrace{\mathbf{v}}_{v\mathbf{e}_3}t}_{\mathbf{e}_3(r-vt)})}_{\chi(0,0,r-vt)};\end{aligned}$$

$$I_2 = \frac{1}{r}e^{-\frac{i}{\hbar}\left(\frac{\mathbf{p}^2}{2m}t-pr\right)}f^{(+)}(\mathbf{n},\mathbf{p})\chi(0,0,r-vt) \tag{3.696}$$

All in all, recalling of the relationship $\mathbf{x} = r\mathbf{n}$ we have the complete scattered wave-function (in coordinate representation) unfolded as:

$$\begin{aligned}\left\langle\mathbf{x}\middle|\Psi^{(+)}(t)\right\rangle &\cong \underbrace{\chi(\mathbf{n}r-\mathbf{v}t)e^{-\frac{i}{\hbar}[E(\mathbf{P})t-(\mathbf{p}\cdot\mathbf{n})r]}}_{\text{plane wave}} \\ &\quad \underbrace{+f^{(+)}(\mathbf{n},\mathbf{p})\chi(0,0,r-vt)\frac{1}{r}e^{-\frac{i}{\hbar}\left(\frac{\mathbf{p}^2}{2m}t-pr\right)}}_{\text{spherical wave}} \\ &= \langle\mathbf{x}|t\rangle + \Psi_{em}(\mathbf{x},t)\end{aligned} \tag{3.697}$$

leaving with the idea that the scattering process has as the effect both the propagation of the incident unperturbed wave-function on which there is superimposed the scattered wave-function with a spherical nature, see Figure 3.23.

We can further proceed with computation of the probability of the current density carried by the emergent wave-function Ψ_{em}, see Section 2.2.9,

$$\mathbf{j}_{em}(\mathbf{x},t)=\frac{i\hbar}{2m}\left(\Psi_{em}\vec{\nabla}\Psi^{*}_{em}-\Psi^{*}_{em}\vec{\nabla}\Psi_{em}\right) \tag{3.698}$$

as being that registered by the measurement apparatus. For evaluating this expression let's rewrite the emergent wave-function as:

$$\Psi_{em}(\mathbf{x},t)=f^{(+)}(\mathbf{n},\mathbf{p})e^{-\frac{i}{\hbar}E(\mathbf{p})t}h(r) \tag{3.699}$$

$$h(r)=\chi(0,0,r-vt)\frac{1}{r}e^{\frac{i}{\hbar}pr} \tag{3.700}$$

from where there is clear that performing the space derivatives of emergent function is transferred to radial derivative of the function $h(r)$:

$$\vec{\nabla}=\mathbf{e}_1\frac{\partial}{\partial x_1}+\mathbf{e}_2\frac{\partial}{\partial x_2}+\mathbf{e}_3\frac{\partial}{\partial x_3}\equiv\mathbf{e}_i\partial_i,\ i=\overline{1,3} \tag{3.701a}$$

$$\vec{\nabla}h(r)=\mathbf{e}_i\partial_i h(r)=\mathbf{e}_i\left[\partial_r h(r)\right]\partial_i r \tag{3.701b}$$

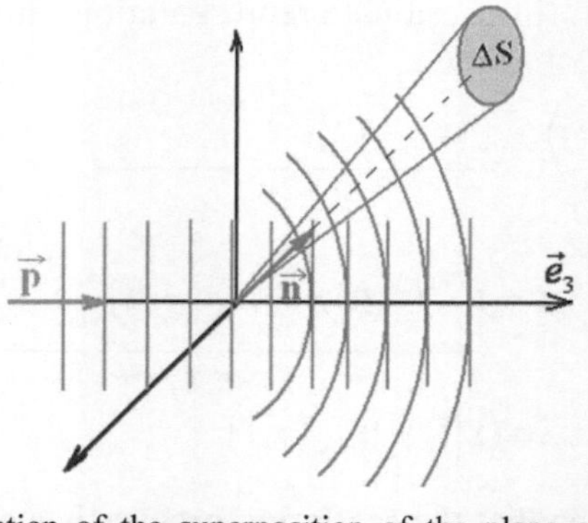

FIGURE 3.23 Illustration of the superposition of the plane and spherical waves in observation of the quantum elastic scattering.

Thus, we have, respectively:

$$\partial_i r = \partial_i \sqrt{x_1^2 + x_2^2 + x_3^2} = \frac{x_i}{r} \tag{3.702a}$$

$$\Rightarrow \mathbf{e}_i \partial_i r = \frac{\mathbf{e}_i x_i}{r} = \frac{\mathbf{x}}{r} = \mathbf{n} \quad \Rightarrow \vec{\nabla} h(r) = [\partial_r h(r)]\mathbf{n} \tag{3.702b}$$

$$\partial_r h(r) = \left([\partial_r \chi]\frac{1}{r} - \frac{\chi}{r^2} + i\frac{p\chi}{r\hbar} \right) e^{\frac{i}{\hbar}pr} \tag{3.702c}$$

$$\partial_r h^*(r) = \left([\partial_r \chi]\frac{1}{r} - \frac{\chi}{r^2} - i\frac{p\chi}{r\hbar} \right) e^{-\frac{i}{\hbar}pr} \tag{3.702d}$$

While for the probability of current density components we have:

$$\begin{aligned}
\Psi_{em}\vec{\nabla}\Psi^*_{em} &= f^{(+)}(\mathbf{n},\mathbf{p}) e^{-\frac{i}{\hbar}E(\mathbf{p})t} h(r) \left[f^{(+)}(\mathbf{n},\mathbf{p}) \right]^* e^{\frac{i}{\hbar}E(\mathbf{p})t} \vec{\nabla} h^*(r) \\
&= \left| f^{(+)}(\mathbf{n},\mathbf{p}) \right|^2 h(r) \left[\partial_r h^*(r) \right] \mathbf{n} \\
&= \left| f^{(+)}(\mathbf{n},\mathbf{p}) \right|^2 \frac{\chi}{r} e^{\frac{i}{\hbar}pr} \left([\partial_r \chi]\frac{1}{r} - \frac{\chi}{r^2} - i\frac{p\chi}{r\hbar} \right) e^{-\frac{i}{\hbar}pr} \mathbf{n};
\end{aligned}$$

$$\Psi_{em}\vec{\nabla}\Psi^*_{em} = \left| f^{(+)}(\mathbf{n},\mathbf{p}) \right|^2 \frac{\chi}{r} \left([\partial_r \chi]\frac{1}{r} - \frac{\chi}{r^2} - i\frac{p\chi}{r\hbar} \right) \mathbf{n} \tag{3.703}$$

and analogous for the other component:

$$\Psi^*_{em}\vec{\nabla}\Psi_{em} = \left| f^{(+)}(\mathbf{n},\mathbf{p}) \right|^2 \frac{\chi}{r} \left([\partial_r \chi]\frac{1}{r} - \frac{\chi}{r^2} + i\frac{p\chi}{r\hbar} \right) \mathbf{n} \tag{3.704}$$

so that for the registered current in the asymptotic zone of the measurement apparatus we get:

$$\begin{aligned}
\mathbf{j}_{em}(\mathbf{x},t) &= -\frac{i\hbar}{2m} \left| f^{(+)}(\mathbf{n},\mathbf{p}) \right|^2 \frac{\chi}{r} 2i \frac{p\chi}{r\hbar} \mathbf{n} \\
&= \left| f^{(+)}(\mathbf{n},\mathbf{p}) \right|^2 \frac{\chi^2}{r^2} \underbrace{\frac{p}{m}}_{v} \mathbf{n};
\end{aligned}$$

$$\mathbf{j}_{em}\left(\mathbf{x},t\right)=v\mathbf{n}\left|f^{(+)}(\mathbf{n},\mathbf{p})\right|^2\frac{\chi^2(0,0,r-vt)}{r^2} \tag{3.705}$$

Note the already interesting aspect of this result that do not explicitly depend on Planck constant, being therefore of classical (direct observable) nature. Now we are approaching the information recorded by the detector apparatus. The probability that one particle to be registered ever (as times goes from $-\infty$ to $+\infty$) is therefore computed as follows:

$$\begin{aligned}
\vec{\Phi}_{em} &= \int_{-\infty}^{+\infty}\mathbf{j}_{em}\left(\mathbf{x},t\right)dt \\
&= \int_{-\infty}^{+\infty} v\mathbf{n}\left|f^{(+)}(\mathbf{n},\mathbf{p})\right|^2\frac{1}{r^2}\chi^2(0,0,\underbrace{r-vt}_{z})dt \\
&= \int_{+\infty}^{-\infty} v\mathbf{n}\left|f^{(+)}(\mathbf{n},\mathbf{p})\right|^2\frac{1}{r^2}\chi^2(0,0,z)\frac{(-)dz}{v} \\
&= \mathbf{n}\frac{1}{r^2}\left|f^{(+)}(\mathbf{n},\mathbf{p})\right|^2\underbrace{\int_{-\infty}^{+\infty}\chi^2(0,0,z)dz}_{\Phi_{in}};
\end{aligned}$$

$$\vec{\Phi}_{em}=\mathbf{n}\Phi_{in}\frac{1}{r^2}\left|f^{(+)}(\mathbf{n},\mathbf{p})\right|^2 \tag{3.706}$$

We finally arrived in the position to calculate the probability that the detector to record the scattering process in unity of time in unity of its area $\Delta\mathbf{S}$, as enounced in the beginning of this section:

$$\begin{aligned}
\wp_{i\to f} &= \int_{\Delta S}\underbrace{\int_{-\infty}^{+\infty}\mathbf{j}_{em}dt}_{\vec{\Phi}_{em}}\,d\mathbf{S} = \int_{\Delta S}\vec{\Phi}_{em}d\mathbf{S} \\
&= \Phi_{in}\int_{\Delta S}\frac{1}{r^2}\left|f^{(+)}(\mathbf{n},\mathbf{p})\right|^2\underbrace{\mathbf{n}d\mathbf{S}}_{dS} \\
&= \Phi_{in}\int_{\Delta S}\left|f^{(+)}(\mathbf{n},\mathbf{p})\right|^2\frac{1}{r^2}\underbrace{dS}_{r^2\sin\theta d\theta d\varphi} \\
&= \Phi_{in}\int_{\Delta\Omega}\left|f^{(+)}(\mathbf{n},\mathbf{p})\right|^2\underbrace{\sin\theta d\theta d\varphi}_{d\Omega};
\end{aligned}$$

$$\wp_{i\to f} = \Phi_{in} \int_{\Delta\Omega} \left| f^{(+)}(\mathbf{n},\mathbf{p}) \right|^2 d\Omega \overset{!}{=} \Phi_{in} \Delta\sigma \tag{3.707}$$

From where results the searched crossing section,

$$\Delta\sigma = \int_{\Delta\Omega} \left| f^{(+)}(\mathbf{n},\mathbf{p}) \right|^2 d\Omega \overset{!}{=} \int_{\Delta\Omega} \left(\frac{d\sigma}{d\Omega} \right)_{(\mathbf{p},\mathbf{n})} d\Omega \tag{3.708}$$

as well as the effective differential cross section:

$$\left(\frac{d\sigma}{d\Omega} \right)_{(\mathbf{p},\mathbf{n})} = \left| f^{(+)}(\mathbf{n},\mathbf{p}) \right|^2 \tag{3.709}$$

as being directly determined by the scattering function that this way appears to be the key of the whole scattering observation. These results will be employed in the next section as well.

3.8.4 MORE ABOUT SCATTERING: BORN APPROXIMATION AND RESONANCE PROFILE

I. We saw in previous section that the effective differential cross section is independent on the incident flux of particle, meaning that one fascicle with low or higher density of particles produces the same observation on the apparatus detecting scattering phenomena. Yet, it depends on the scattering amplitude

$$f^{(+)}(\mathbf{n},\mathbf{P}) = -(2\pi)^2 \hbar m \left\langle P\mathbf{n} \middle| V(\mathbf{x}) \middle| \Psi_{\mathbf{P}}^{(+)} \right\rangle \tag{3.710}$$

in term of the rewriting the Dyson series of Section 3.6.4 here as the *Born series*:

$$\left| \Psi_{\mathbf{P}}^{(+)} \right\rangle = \left| \mathbf{P} \right\rangle + \sum_{n=1}^{\infty} \left[\frac{1}{E(\mathbf{P}) - \widehat{H}_0 + i\eta} V(\mathbf{x}) \right]^n \left| \mathbf{P} \right\rangle \tag{3.711}$$

of which limitation to the first order of the series consecrates the so-called *Born approximation*:

$$\left| \Psi_{\mathbf{P}}^{(+)B} \right\rangle = \left| \mathbf{P} \right\rangle + \frac{1}{E(\mathbf{P}) - \widehat{H}_0 + i\eta} V(\mathbf{x}) \left| \mathbf{P} \right\rangle \tag{3.712}$$

Within Born approximation, the scattering amplitude becomes:

$$f^{(+)B}(\mathbf{n},\mathbf{P}) = -(2\pi)^2 \hbar m \langle P\mathbf{n}|V(\mathbf{x})\hat{1}_{\{|\mathbf{x}\rangle\}}|\Psi_{\mathbf{P}}^{(+)B}\rangle$$

$$= -(2\pi)^2 \hbar m \int \langle P\mathbf{n}|\underbrace{V(\mathbf{x})|\mathbf{x}\rangle}_{|\mathbf{x}\rangle V(\mathbf{x})}\langle\mathbf{x}|\left(|\mathbf{P}\rangle + \frac{1}{E(\mathbf{P}) - \widehat{H}_0 + i\eta} V(\mathbf{x})\hat{1}_{\{|\mathbf{y}\rangle\}}|\mathbf{P}\rangle\right)d\mathbf{x}$$

$$= -(2\pi)^2 \hbar m \int \langle P\mathbf{n}|\mathbf{x}\rangle V(\mathbf{x})\left[\langle\mathbf{x}|\mathbf{P}\rangle + \int\langle\mathbf{x}|\frac{1}{E(\mathbf{P}) - \widehat{H}_0 + i\eta}\underbrace{V(\mathbf{x})|\mathbf{y}\rangle}_{|\mathbf{y}\rangle V(\mathbf{y})}\langle\mathbf{y}|\mathbf{P}\rangle d\mathbf{y}\right]d\mathbf{x}$$

$$= -(2\pi)^2 \hbar m \int \langle P\mathbf{n}|\mathbf{x}\rangle V(\mathbf{x})\left[\langle\mathbf{x}|\mathbf{P}\rangle + \int V(\mathbf{y})\underbrace{\langle\mathbf{x}|\frac{1}{E(\mathbf{P}) - \widehat{H}_0 + i\eta}|\mathbf{y}\rangle}_{g^{(+)}(\mathbf{x}-\mathbf{y})}\langle\mathbf{y}|\mathbf{P}\rangle d\mathbf{y}\right]d\mathbf{x}$$

$$= -(2\pi)^2 \hbar m \int \langle P\mathbf{n}|\mathbf{x}\rangle V(\mathbf{x})\left[\langle\mathbf{x}|\mathbf{P}\rangle + \int V(\mathbf{y}) g^{(+)}(\mathbf{x}-\mathbf{y})\langle\mathbf{y}|\mathbf{P}\rangle d\mathbf{y}\right]d\mathbf{x}$$

$$= -(2\pi)^2 \hbar m \int \langle P\mathbf{n}|\mathbf{x}\rangle V(\mathbf{x})\left[\langle\mathbf{x}|\mathbf{P}\rangle - \frac{1}{4\pi}\frac{2m}{\hbar^2}\int V(\mathbf{y})\frac{e^{\frac{i}{\hbar}P|\mathbf{x}-\mathbf{y}|}}{|\mathbf{x}-\mathbf{y}|}\langle\mathbf{y}|\mathbf{P}\rangle d\mathbf{y}\right]d\mathbf{x} \tag{3.713}$$

Now, the Born approximation is valid when the first term in parenthesis is much higher than the other that is equivalently of having the condition:

$$\left|\langle\mathbf{x}|\Psi_{\mathbf{P}}^{(+)(0)}\rangle\right| >> \left|\langle\mathbf{x}|\Psi_{\mathbf{P}}^{(+)(1)}\rangle\right| \tag{3.714}$$

or even more generally:

$$\frac{\left|\langle\mathbf{x}|\Psi_{\mathbf{P}}^{(+)(n+1)}\rangle\right|}{\left|\langle\mathbf{x}|\Psi_{\mathbf{P}}^{(+)(n)}\rangle\right|} << 1 \tag{3.715}$$

that is nothing than the D'Alembert criterion for the Born series convergence, meaning that the perturbation caused by scattering is nevertheless observable (or detectable).

Returning to our working scattering amplitude, the Born condition produces the constraint:

$$\left|\langle\mathbf{x}|\mathbf{P}\rangle\right| >> \left|\frac{1}{4\pi}\frac{2m}{\hbar^2}\int V(\mathbf{y})\frac{e^{\frac{i}{\hbar}P|\mathbf{x}-\mathbf{y}|}}{|\mathbf{x}-\mathbf{y}|}\langle\mathbf{y}|\mathbf{P}\rangle d\mathbf{y}\right| \tag{3.716}$$

which can be specialized on origin ($\mathbf{x}=0$) of interaction since for the most potentials of physical interest the maximum is located in their origin (i.e., the closer the particles the higher their interaction) to equivalently transforms:

$$\left|\langle 0|\mathbf{P}\rangle\right| >> \left|\frac{1}{4\pi}\frac{2m}{\hbar^2}\int V(\mathbf{y})\frac{e^{\frac{i}{\hbar}P|0-\mathbf{y}|}}{|0-\mathbf{y}|}\langle \mathbf{y}|\mathbf{P}\rangle d\mathbf{y}\right|$$

$$\Leftrightarrow \frac{1}{(2\pi\hbar)^{3/2}}\underbrace{e^{\frac{i}{\hbar}\mathbf{P}\cdot 0}}_{1} >> \left|\frac{1}{4\pi}\frac{2m}{\hbar^2}\int V(\mathbf{y})\frac{e^{\frac{i}{\hbar}P|\mathbf{y}|}}{|\mathbf{y}|}\frac{1}{(2\pi\hbar)^{3/2}}e^{\frac{i}{\hbar}\mathbf{P}\cdot\mathbf{y}}d\mathbf{y}\right|$$

$$\Leftrightarrow 1 >> \left|\frac{1}{4\pi}\frac{2m}{\hbar^2}\int V(\mathbf{y})\frac{e^{iK|\mathbf{y}|+\frac{i}{\hbar}\mathbf{P}\cdot\mathbf{y}}}{|\mathbf{y}|}d\mathbf{y}\right| \quad (3.717)$$

This condition may be further simplified for a central potential $V(r)$, so that the $\mathbf{y}$ vector, with modulus $|\mathbf{y}|=r$, is expressed in spherical coordinates so that $y_3\|\mathbf{P}$, that gives in the first instance

$$\frac{\mathbf{P}\cdot\mathbf{y}}{\hbar} = \underbrace{\frac{P}{\hbar}}_{K} r\cos\theta = Kr\cos\theta \quad (3.718)$$

and then for the entire Born condition

$$1 >> \left|\frac{1}{4\pi}\frac{2m}{\hbar^2}\int_0^{2\pi}d\varphi\int_0^{\pi}d\theta\int_0^{\infty}r^2\sin\theta\frac{V(r)}{r}e^{iKr(1+\cos\theta)}dr\right|$$

$$\Leftrightarrow 1 >> \left|\frac{m}{\hbar^2}\int_0^{\infty}rV(r)e^{iKr}\left(\int_0^{\theta}\sin\theta e^{iKr\cos\theta}d\theta\right)dr\right|$$

$$\Leftrightarrow 1 >> \left|\frac{m}{\hbar^2}\int_0^{\infty}rV(r)e^{iKr}\frac{e^{iKr}-e^{-iKr}}{iKr}dr\right|$$

$$\Leftrightarrow 1 >> \left|\frac{im}{\hbar^2 K}\int_0^{\infty}V(r)\left(1-e^{2iKr}\right)dr\right| \quad (3.719)$$

The final stage here regards the introduction of the so-called action radius r of the concerned potential through rewriting the last Born condition as:

$$\left|\int_0^\infty \underbrace{V(r)}_{energy}\left(1-e^{2iKr}\right)\underbrace{dr}_{lenght}\right| << \frac{\hbar^2 K}{m} \quad \Big| \times \frac{1}{r_0}$$

$$\Leftrightarrow \left|\frac{1}{r_0}\int_0^\infty V(r)\left(1-e^{2iKr}\right)dr\right| << 2Kr_0\Delta E \tag{3.720}$$

emphasizing the energy localization in the action radius of the potential

$$\Delta E = \frac{\hbar^2}{2mr_0^2} \tag{3.721}$$

that correlates with the sensibility of the observer.

2. Continuing on the same line of employing the observability of quantum elastic scattering, we may reinterpret the energy localization by reconsidering the scattering process in a more phenomenological way. The starting point is again the wave-function $\langle \mathbf{x} | \Psi^{(+)}(t) \rangle$ rewritten in the simplified form:

$$\Psi^{(+)}(\mathbf{x},t) \cong \chi e^{-\frac{i}{\hbar}Et}\left(e^{\frac{i}{\hbar}(\mathbf{p}\cdot\mathbf{n})r} + f^{(+)}(\mathbf{n},\mathbf{p})\frac{1}{r}e^{\frac{i}{\hbar}pr}\right) \equiv \Psi_{in}(\mathbf{x},t) + \Psi_{em}(\mathbf{x},t) \tag{3.722}$$

since the observation:

$$\chi(\mathbf{n}r - \mathbf{v}t) = \chi(0,0,r - vt) \tag{3.723}$$

Yet, this form may be slightly modified when for the incident plane wave one consider the averaged one over the angular integration, integrating over the angular angles (φ,θ) and then dividing the result to their solid angle 4π, while choosing the polar angle that one between the directions of incidence and scattering detection, $\angle(\mathbf{p},\mathbf{n}) = \theta$:

$$\tilde{\Psi}_{in}(\mathbf{x},t) = \chi e^{-\frac{i}{\hbar}Et}\frac{1}{4\pi}\int_\Omega e^{\frac{i}{\hbar}(\mathbf{p}\cdot\mathbf{n})r}d\Omega$$

$$= \chi e^{-\frac{i}{\hbar}Et} \frac{1}{4\pi} \int_0^{2\pi} d\varphi \int_0^{\pi} \sin\theta e^{\frac{i}{\hbar}(\mathbf{p}\cdot\mathbf{n})r} d\theta$$

$$= \chi e^{-\frac{i}{\hbar}Et} \frac{\hbar}{2} \frac{e^{i\frac{pr}{\hbar}} - e^{-i\frac{pr}{\hbar}}}{ipr} \tag{3.724}$$

The basic difference of this form respecting the previous incident one is that now the income wave displays a spherical form, in a similar way with the scattered one, besides containing both the forward and backward forms of propagation, being in this respect a generalized form to handle. However, now the so that the total (new) scattered wave-function looks like:

$$\tilde{\Psi}^{(+)}(\mathbf{x},t) = \tilde{\Psi}_{in}(\mathbf{x},t) + \Psi_{em}(\mathbf{x},t)$$

$$= \chi e^{-\frac{i}{\hbar}Et} \left[\frac{\hbar}{2ipr} \left(e^{i\frac{pr}{\hbar}} - e^{-i\frac{pr}{\hbar}} \right) + f^{(+)}(\mathbf{n},\mathbf{p}) \frac{1}{r} e^{\frac{i}{\hbar}pr} \right]$$

$$= \chi \frac{\hbar}{2ipr} e^{-\frac{i}{\hbar}Et} \left(e^{i\frac{pr}{\hbar}} - e^{-i\frac{pr}{\hbar}} + f^{(+)} \frac{2ip}{\hbar} e^{\frac{i}{\hbar}pr} \right)$$

$$= \chi \frac{\hbar}{2ipr} e^{-\frac{i}{\hbar}Et} \left[\left(1 + \frac{2i}{\hbar} p f^{(+)} \right) e^{i\frac{pr}{\hbar}} - e^{-i\frac{pr}{\hbar}} \right] \tag{3.725}$$

Over this form one may impose the condition that being about an elastic scattering process, the total forward and total backward wave amplitudes has to be equal that is to fulfill the constraint:

$$\left| 1 + \frac{2i}{\hbar} p f^{(+)} \right| = 1 \tag{3.726}$$

If this relationship is regarded as an equation for scattering amplitude, the general solution of it reads as:

$$\tilde{f}^{(+)} = \frac{\hbar}{2ip} \left(e^{2i\delta} - 1 \right) \tag{3.727}$$

where the introduced quantity δ plays the role of the *phasing difference* (between the incident and emergent waves in scattering process), being dependent in general by the momentum but is a function of energy in a more general framework, $\delta = \delta(E)$. Worth observing that each of its value with the form:

$$\delta(E_0) = \left(n_0 + \frac{1}{2}\right)\pi,\ n_0 \in \mathbf{Z} \tag{3.728}$$

the scattering amplitude become

$$\tilde{f}_0^{(+)} = -\frac{\hbar}{ip} \tag{3.729}$$

while the associate (observed) cross section takes its maximum value

$$\Delta\tilde{\sigma}_0 = \int_{\Omega} \left|\tilde{f}_0^{(+)}(E)\right|^2 d\Omega = \left|\tilde{f}_0^{(+)}(E)\right|^2 \underbrace{\int_{\Omega} d\Omega}_{4\pi} = 4\pi\left(\frac{\hbar}{p}\right)^2 \tag{3.730}$$

Such cases correspond with the called resonances, when the detection apparatus registers (though cross sections) moments (or de Broglie wave-numbers) close with those characterizing the income wave-function on the target. Thus, these cases practically detect the target states that are resonant with the incoming wave packet, the scattering process being the tool of detecting them. Naturally, the next step is to explore the scattering amplitude and the associate cross section behavior near such resonances. This can be achieved through rewriting the phase δ, in first instance, using the trigonometric identities:

$$\begin{aligned} \cot\delta &= \frac{\cos\delta}{\sin\delta} = \frac{i\left(e^{i\delta} + e^{-i\delta}\right)}{e^{i\delta} - e^{-i\delta}} = i\frac{1+e^{-2i\delta}}{1-e^{-2i\delta}} \\ \Leftrightarrow e^{2i\delta} &= \frac{\cot\delta + i}{\cot\delta - i} \\ \Rightarrow e^{2i\delta} - 1 &= \frac{2i}{\cot\delta - i} \end{aligned} \tag{3.731}$$

that rewrites the actual scattering amplitude as:

$$\tilde{f}^{(+)} = \frac{\hbar}{p}\frac{1}{\cot\delta(E) - i} \tag{3.732}$$

which can be further approximated by the first order expansion:

$$\cot\delta(E)_{E_0} \cong \underbrace{\cot\delta(E_0)}_{\cong 0} - \underbrace{\left[1+\cot^2\delta(E_0)\right]}_{\substack{\equiv 2/\Gamma \\ inverse \\ energy}}(E-E_0) \cong -\frac{2}{\Gamma}(E-E_0) \quad (3.733)$$

to become

$$\tilde{f}^{(+)} = \frac{\hbar}{p}\frac{1}{-\frac{2}{\Gamma}(E-E_0)-i} = -\frac{\hbar}{p}\frac{\Gamma/2}{E-E_0+i\Gamma/2} \quad (3.734)$$

With this, the observed crossing section around resonances looks like:

$$\Delta\tilde{\sigma} = \int_\Omega \left|\tilde{f}^{(+)}(E)\right|^2 d\Omega = 4\pi\left|\tilde{f}^{(+)}(E)\right|^2 = \underbrace{4\pi\left(\frac{\hbar}{p}\right)^2}_{\Delta\tilde{\sigma}_0}\frac{(\Gamma/2)^2}{(E-E_0)^2+(\Gamma/2)^2};$$

$$\Delta\tilde{\sigma}(E) = \Delta\tilde{\sigma}(E_0)\frac{(\Gamma/2)^2}{(E-E_0)^2+(\Gamma/2)^2} \quad (3.735)$$

which illustrates the profile of the recorded elastic scatterings near resonances (of the target), under the form known as the Breit-Wigner formula. Note that since the factor Γ was introduced like energy (for the target states, or more precisely as the width of the energy interval around the resonance E_0 – thus playing the role of localization energy interval) it can be further replaced by the *average life time* τ of those states through the Heisenberg spectroscopic relationship,

$$\frac{\hbar}{\Gamma} = \tau \quad (3.736)$$

with which occasion the Breit-Wigner formula rewrites as:

$$\Delta\tilde{\sigma}(\omega) = \Delta\tilde{\sigma}(\omega_0)\frac{(1/2\tau)^2}{(\omega-\omega_0)^2+(1/2\tau)^2} \quad (3.737)$$

in terms of the frequencies around the resonant one, corresponding with the scattered waves, according with the energy-frequency Planck celebrated formula.

Worth noting that the scattering Breit-Wigner profile, while being close to Gaussian one (specific for the normalized de Broglie wave-functions, see Section 1.3.3), see Figure 3.24, adds new type of curves that models the natural (quantum) phenomena since it is characteristic to light-atoms interaction as well as to a variety of nuclear scattering observation (e.g., proton scattering by nuclei, or of pion scattering by protons, etc.). In all cases this profiles indicates that in principle for any quantum system exists a "meta-stable" level that can be detectable/measured/observed through elastic scattering, being this a vary valuable insight of how quantum phenomena can be unveiled by their classical (scattering, lifetime, energy) manifestation.

3. Lastly, such quantum "resonance" curve may be recuperated from classical-to-quantum correspondence principle derived from *forced oscillations* as following. The starting point is the Newton second law of motion

$$m\partial_t^2 y = \sum F_{applied} \tag{3.738}$$

of an harmonic oscillator in the case its free motion under the elastic force

$$F_e = -m\omega_0^2 y \tag{3.739}$$

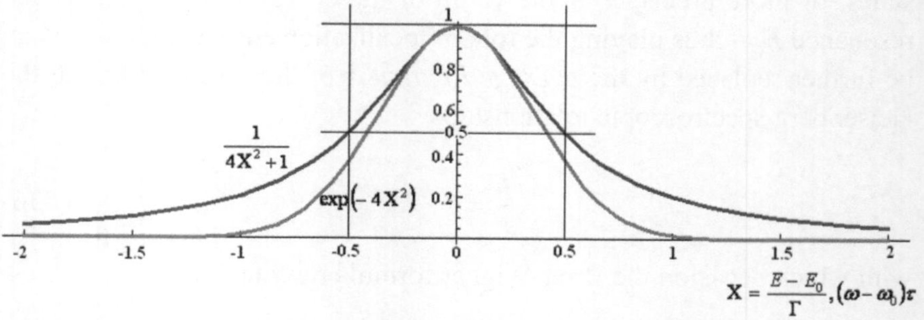

FIGURE 3.24 Comparison between the resonance (upper) and Gaussian (lower) curves: wile the last is fast decreasing to zero outside of the resonance (central) region, the first one presents the long (asymptotical) tail specific for (asymptotic) scatterings.

is amortized by the (inner) resistance force

$$F_r = -b\partial_t y, \quad b = ct. \tag{3.740}$$

while being as well under the external (perturbed) periodic (harmonic as well) force:

$$F_p = F_0 \exp(-i\omega_p t) \tag{3.741}$$

Thus the working classical equation looks like:

$$\partial_t^2 y + \gamma\partial_t y + \omega_0^2 y = \frac{F_0}{m}\exp(-i\omega_p t) \tag{3.742}$$

were the *amortization coefficient* $\gamma = b/m$ was introduced.

The general solution of this non-homogeneous equation is the sum of its homogeneous equation to which there is added the particular solution of the non-homogeneous equation that is usually taken as the free term form:

$$y = A_p \exp(-i\omega_p t) \tag{3.743}$$

Nevertheless, this solution has to fulfill the entirely above equation, leading with the new equation:

$$\left[(-i\omega_p)^2 + \gamma(-i\omega_p) + \omega_0^2\right] A_p \exp(-i\omega_p t) = \frac{F_0}{m}\exp(-i\omega_p t)$$
$$\Rightarrow A_p = \frac{F_0}{m}\frac{1}{\omega_0^2 - \omega_p^2 - i\gamma\omega_p} \tag{3.744}$$

For *resonance* effect to take place one has to have $\omega_p \cong \omega_0$ (beside the fact that in these circumstances the amortization coefficient is almost vanishing, $\gamma \to 0$) that produces the approximations:

$$\omega_0^2 - \omega_p^2 = (\omega_0 + \omega_p)(\omega_0 - \omega_p) \cong 2\omega_0(\omega_0 - \omega_p), \ \gamma\omega_p \cong \gamma\omega_0 \tag{3.745}$$

and the *resonance amplitude*:

$$A_{res} = -\frac{F_0}{2m\omega_0}\frac{1}{(\omega_p - \omega_0) + i\gamma} \tag{3.746}$$

in close correspondence with above scattering amplitude $\tilde{f}^{(+)}$ when applied the Planck quantification, $\omega = E / \hbar$, i.e., the classical-to-quantum correspondence as:

$$A_{res} = -\frac{F_0}{2mE_0}\hbar^2 \frac{1}{\left(E_p - E_0\right) + i\hbar\gamma} \tag{3.747}$$

Note that the maximum amplitude at resonance is given by:

$$A_{res}^{\max} = -\frac{F_0}{2im\gamma E_0}\hbar \leftrightarrow \tilde{f}_0^{(+)} = -\frac{\hbar}{ip} \tag{3.748}$$

so that the resonance amplitude rewrites as

$$A_{res} = iA_{res}^{\max} \frac{\hbar\gamma}{\left(E_p - E_0\right) + i\hbar\gamma} \leftrightarrow \tilde{f}^{(+)} = -\frac{\hbar}{p}\frac{\Gamma / 2}{E - E_0 + i\Gamma / 2} \tag{3.749}$$

while remarking the one-to-one correspondences (through reminding γ as being of ω in nature). From this point all is recovered equivalently since recognizing that the intensity of radiation emitted by an oscillator is given by the squared of the modulus of its elongation, $\left|A_{rez}\right|^2$, this way recovering the previously introduced quantum resonance (Breit-Wigner) curve.

Yet, being at the stage pointing on the classical counterpart of the quantum resonance curve, worth specifying that also in the electric circuits with resistance (R), inductance (L), and capacitance (C), the oscillating regime maintained by the applied tension $V(t)$ is governed by the dynamical (conservation) equation for the charge $Q(t)$:

$$\underbrace{L\partial_t^2 Q}_{V_L} + \underbrace{R\underbrace{\partial_t Q}_{i(t)}}_{V_R} + \underbrace{\frac{Q}{C}}_{V_C} = V(t) \tag{3.750}$$

Moreover, the resonance effect and its profile may be observed in Nature in various instances:

- The Earth atmospheric oscillation under the forced gravitation induced by the Moon, with a periodicity between 10.5–12.5 hours;
- Crystal ionic oscillations (e.g., in NaCl crystals) under the absorption of infrared radiation;

- Resonances of paramagnetic organic compounds that looses magnetic energy under external applied magnetic field;
- The absorption of gamma rays by nuclei (Mössbauer effect) proofing their inner oscillatory structure while predicting their dimension as well;
- Scatterings of elementary particles that while allowing different evolution reactions produce various resonances assimilated with new particles (or group of particles) as they are existing with the resonance energy or frequency;

and many others.

3.8.5 FROM YUKAWA POTENTIAL TO RUTHERFORD SCATTERING

Let's se whether one can employ the connection *structure-field interaction* in terms of an unifying or generalized potential of Nature; however, let's specify that, while interactions are intermediated by bosons (particles with integer number of spin, for example, the photons of light), the structure is quantified by fermions (particles with semi-integer number spin, for example, the electrons of atoms, nucleons of nuclei). Yet, their interaction is present and observed – therefore raising the question of their common potential roots. In this regard we will present in the following how the one boson quantum equation (Klein-Gordon) can be developed into a potential (Yukawa potential) that at its turn is to be attributed to influence fermions (protons for instance) until recovering the Coulombic central field and the "classical" Rutherford scattering formula. Over all, this section unfolds the observation of the boson-fermionic interaction from the Yukawa potential perspective.

1. The starting point it represents the Klein-Gordon equation of Section 2.2.3, Eq. (2.34); in the situation the wave-function solution is represented by the potential (its source) in a spherical (as a source) symmetry,

$$\psi_t(\mathbf{r}) = V(r) \tag{3.751}$$

the time-dependence term may be skipped, leaving with the working equation

$$\nabla^2 V(r) = \left(\frac{mc}{\hbar}\right)^2 V(r) \tag{3.752}$$

Thorough transforming the Cartesian Laplacian into the radial one through the recipe of the Section 3.3.1,

$$\nabla_r^2 \bullet = \frac{2}{r}\frac{d}{dr}\bullet + \frac{d^2}{dr^2}\bullet \tag{3.753}$$

the last equation rewrites:

$$\frac{d^2V(r)}{dr^2} + \frac{2}{r}\frac{dV(r)}{dr} - \left(\frac{mc}{\hbar}\right)^2 V(r) = 0 \tag{3.754}$$

whose solution is found through exploring the two extreme cases specific to the spherical potentials: firstly there is considered the asymptotic ($r \to \infty$) version

$$\frac{d^2V(r)}{dr^2} - a^2 V(r) = 0, \quad a = \frac{mc}{\hbar} \tag{3.755}$$

which though its associate secular equation, say

$$\lambda^2 - a^2 = 0 \tag{3.756}$$

it provides the physically solution

$$V_\infty(r) \cong \exp(-ar) \overset{r\to\infty}{\to} 0 \tag{3.757}$$

Then, considering the other extreme limit, namely the behavior in the origin, one observes that for $r \to 0$ the term with $2/r$ dominates the entire equation, thus suggesting the specific dependence:

$$V_0(r) \cong \frac{1}{r} \overset{r\to 0}{\to} \infty \tag{3.758}$$

Overall, combining the two potential forms in a continuous analytical formulation one establishes the so-called Yukawa potential

$$V(r) = \frac{A}{r}\exp(-ar) \tag{3.759}$$

which fulfills both above conditions, either in origin and at infinitum, while by summing up the expressions:

$$-a^2V(r) = -A\frac{a^2}{r}e^{-ar} \tag{3.760a}$$

$$\frac{2}{r}\partial_r V(r) = -A\frac{2}{r^3}e^{-ar} - A\frac{2a}{r^2}e^{-ar} \tag{3.760b}$$

$$\partial_r^2 V(r) = A\frac{2}{r^3}e^{-ar} + A\frac{2a}{r^2}e^{-ar} + A\frac{a^2}{r}e^{-ar} \tag{3.760c}$$

satisfies also the entire original equation.

With these the general working Yukawa potential looks like:

$$V_{Yuk}(r) = \frac{A}{r}\exp\left(-\frac{r}{r_0}\right) \tag{3.761}$$

where now we recognize from the previous notation

$$r_0 = \frac{1}{a} = \frac{\hbar}{mc} \tag{3.762}$$

as playing the role of the *action radius of the potential source* in scatterings processes, equivalent with the *Compton length* of the scattered particle (center). Remarkably, at this point the contact with scattering process may be done in order to determine the particle mass "m" from observing the cross section it produces against various incoming scattered waves. In this context the photon (as a boson) acts on an infinite radius $r_0^\gamma \to \infty$ since its zero rest mass $m_\gamma = 0$, while the working potential becomes Coulombic, $V_{Coulomb}(r) = A/r$, meaning either that this is the potential that scatters the photons (because the internal structure of photons is also mediated by such (electrostatic) potential – a picture in agreement with the electron-positron (matter-anti-matter) of light and also with the Dirac relativistic sea characteristic to Klein-Gordon framework, see Section 2.2.3) or this is the scattering potential between two electric charges (the sign and type of A depends on their type), one being the center of scattering and other scattered by it.

One also should note that this potential, known as Yukawa potential since it was firstly proposed by Hideki Yukawa (in 1935) while studying the mesonic interaction with nuclei, may also represent the electrostatic shielding potential of nuclei in electronic structure of atoms (generalized Coulomb), molecules (generalized Morse) or solid state (Born-Meyer) systems and will be closely studied with occasion unfolding specific discussions in the forthcoming dedicated volumes of this series. For the moment, we will continue the analysis of scattering studied on this type of potential, for preserving the actual general level of exposition.

2. Let's study, for testing the observability features of Yukawa potential, the possibility of applying for it the above deduced Born's approximation for scattering. That is we have to evaluate the inequality:

$$\left|\frac{1}{r_0}I(a)\right| << 2Kr_0\left(\frac{\hbar^2}{2mr_0^2}\right) \tag{3.763}$$

through evaluating the integral

$$I(a)=\int_0^\infty V_{Yuk}(r)\left(1-e^{2iKr}\right)dr=\int_0^\infty \frac{A}{r}\exp(-ar)\left(1-e^{2iKr}\right)dr \tag{3.764}$$

This is to be done by firstly taking its firs derivative respecting the parameter "a":

$$\begin{aligned}\frac{dI(a)}{da}&=-A\int_0^\infty e^{-ar}\left(1-e^{2iKr}\right)dr\\&=-A\int_0^\infty e^{-ar}dr+A\int_0^\infty e^{2iKr-ar}dr\\&=\frac{A}{a}e^{-ar}\Big|_0^\infty+\frac{A}{2iK-a}e^{(2iK-a)r}\Big|_0^\infty\\&=-\frac{A}{a}+\frac{A}{a-2iK};\end{aligned}$$

$$\Leftrightarrow dI(a) = -A\frac{da}{a} + A\frac{da}{a-2iK}$$
$$\Rightarrow I(a) = -A\ln a + A\ln(a-2iK) + \tilde{C} \tag{3.765}$$

The last relation may be further transformed in what regard the "unusual" logarithm of a complex number; for that one remember the basic complex relationships

$$z = |z|e^{i\theta} \Rightarrow \ln z = \ln|z| + i\theta$$
$$\Rightarrow \ln(x+iy) = \ln\sqrt{x^2+y^2} + i\arctan\frac{y}{x};$$
$$\ln(a-i2K) = \ln\sqrt{a^2+4K^2} + i\arctan\left(-\frac{2K}{a}\right) \tag{3.766}$$

to get, through employing $\arctan(-\bullet) = -\arctan(\bullet)$, the searched integral

$$I(a) = -A\ln a + A\ln\sqrt{a^2+4K^2} - iA\arctan\left(\frac{2K}{a}\right) + \tilde{C} \tag{3.767}$$

or in terms of action radius, $r_0 = 1/a$, the result:

$$I(r_0) = A\ln r_0 + A\ln\sqrt{\frac{1}{r_0^2}+4K^2} - iA\arctan(2Kr_0) + \tilde{C}$$
$$= A\ln\sqrt{1+4K^2r_0^2} - iA\arctan(2Kr_0) + \tilde{C}$$
$$= \frac{A}{2}\ln(1+4K^2r_0^2) - iA\arctan(2Kr_0) + \tilde{C} \tag{3.768}$$

Yet, under natural assumption that for zero action radius the scattering amplitude and above integral identically vanishes,

$$I(r_0 \to 0) \to 0 \Rightarrow \tilde{C} = 0 \tag{3.769}$$

the integration constant is determined and the Born approximation condition now reads:

$$\underbrace{\frac{A}{r_0}}_{energy}\underbrace{\left|\frac{1}{2}\ln(1+4K^2r_0^2) - i\arctan(2Kr_0)\right|}_{without\ \mathrm{dim}} << \underbrace{2Kr_0}_{no\ \mathrm{dim}}\underbrace{\left(\frac{\hbar^2}{2mr_0^2}\right)}_{energy} \tag{3.770}$$

which may be resumed as

$$\frac{A}{r_0} << \tilde{F}(2Kr_0)\frac{\hbar^2}{2mr_0^2} \tag{3.771}$$

with the *scattering control function*:

$$\tilde{F}(\rho) = \frac{\rho}{\sqrt{\frac{1}{4}\ln^2(1+\rho^2) + \arctan^2(\rho)}}, \quad \rho = 2Kr_0 \tag{3.772}$$

that can be at its turn considered for the two energetic extremes:

- For small energies one has $\rho << 1$ and the series expansion of $\tilde{F}$ function may be applied to yield in the second order restriction:

$$\tilde{F}_{<<}(\rho) \cong 1 + \frac{5}{24}\rho^2 + \vartheta^3(\rho) \tag{3.773}$$

while in the first order the Born condition simply gives

$$\frac{A}{r_0} << \frac{\hbar^2}{2mr_0^2} \tag{3.774}$$

from where follows the mass limitation:

$$m << \frac{\hbar^2}{2r_0 A} \tag{3.775}$$

Now, considering from Yukawa potential the Coulomb limit for scattering (by charged particles) on atomic centers, one deals with the potential:

$$|V_{Coulomb}(r)| = \frac{A_0}{r} \tag{3.776}$$

and therefore identifying the constant

$$A_0 = \frac{e^2}{4\pi\varepsilon_0} \cong 23.072 \cdot 10^{-29} \left\langle C^2 m / F \right\rangle \tag{3.777}$$

where one should note that a farad is the charge in coulombs a capacitor will accept for the potential across it to change 1 volt. Since a coulomb is 1 ampere second it has the equivalent representation until the SI units:

$$F = A \cdot s / V = J / V^2 = C / V = C^2 / J$$
$$= C^2 / (N \cdot m) = s^2 \cdot C^2 / (m^2 \cdot kg) = s^4 \cdot A^2 / (m^2 \cdot kg)$$

By replacing the action radius with its electronic working expression $r_0 = \hbar / (mc)$ the above condition translates as:

$$A_0 << \frac{c\hbar}{2} \tag{3.778}$$

a constraint fairly respected by the numerical inequality:

$$2.3072 \cdot 10^{-28} << 1.58 \cdot 10^{-26} \tag{3.779}$$

certifying therefore the validity of the Born approximation even for charges scattering with low energies by atomic electrons, as well as the consistency of the Yukawa potential when specialized to its Coulomb limit.

- For high energies one has $\rho >> 1$ and can neglect "1" respecting ρ and replacing arctan by $\pi / 2$ for great arguments, to yield for the control function

$$\tilde{F}_{>>}(\rho) = \frac{\rho}{\sqrt{\ln^2 \rho + \pi^2 / 4}} \tag{3.780}$$

that replaced in the actual Born condition provides an equation for appropriate wave-function K for the incoming (scattered) light used for structure investigation/observation:

$$\frac{A}{r_0} << \frac{2Kr_0}{\sqrt{\ln^2 (2Kr_0) + \pi^2 / 4}} \frac{\hbar^2}{2mr_0^2}$$
$$\Leftrightarrow A^2 << \frac{K^2}{\ln^2 (2Kr_0) + \pi^2 / 4} \frac{\hbar^4}{m^2} \tag{3.781}$$

while when considering electronic scattering, i.e., replacing the constants A_0 & r_0^e for Coulombic potential and electronic action radius, respectively, one gets the inequality:

$$\tilde{f}_{>>}(K) >> 0 \tag{3.782a}$$

$$\tilde{f}_{>>}(K) = \frac{\hbar^4}{m_0^2 A_0^2} K^2 - \ln^2\left(\frac{2\hbar}{m_0 c} K\right) - \frac{\pi^2}{4} \quad (3.782b)$$

whose validity is confirmed by representation given in Figure 3.25 for wave-lengths in the high energy regime, i.e.,

$$K = \frac{2\pi}{\lambda},\ \lambda < 10^{-10}\langle m\rangle \quad (3.783)$$

being of at least of the X-ray type.

The conclusion is that also in the higher energies level the Born approximation finely combines with the Yukawa potential to correctly model the photonic scatterings on Coulombic manifested interaction. Worth therefore to further compute, from the scattering amplitude, the differential and total cross section induced by the Yukawa (and then Coulombic) potentials – a matter in the next approached.

3. Having now the analytical and numerical confirmation that Born approximation is a reliable one for modeling scattering processes it will be

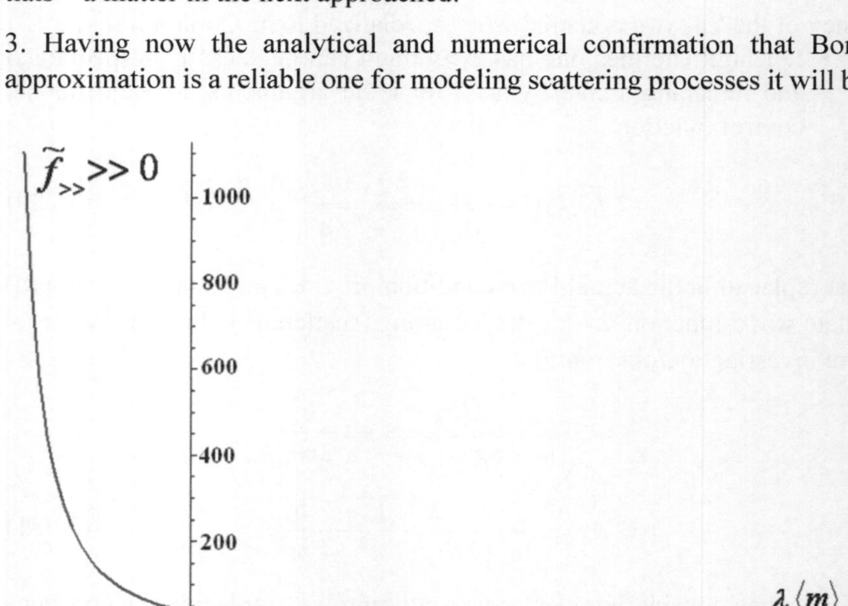

FIGURE 3.25 The graphical check of the reliability of the Born approximation for the Coulomb potential abstracted from Yukawa model for electronic scattering by higher photonic energies (lower wave-lengths).

adopted din what follows to actualize and compute the scattering amplitude successively as:

$$
\begin{aligned}
f^{(+)B}(\mathbf{n},\mathbf{P}) &\cong -(2\pi)^2 \hbar m \int \langle P\mathbf{n}|\mathbf{x}\rangle V(\mathbf{x}) \langle \mathbf{x}|\mathbf{P}\rangle d\mathbf{x} \\
&= -(2\pi)^2 \hbar m \int \frac{1}{(2\pi\hbar)^{3/2}} e^{-\frac{i}{\hbar}P\mathbf{n}\cdot\mathbf{x}} V(\mathbf{x}) \frac{1}{(2\pi\hbar)^{3/2}} e^{\frac{i}{\hbar}\mathbf{P}\cdot\mathbf{x}} d\mathbf{x} \\
&= -(2\pi)^2 \hbar m \frac{1}{(2\pi\hbar)^3} \int V(\mathbf{x}) \exp[-i\underbrace{(P\mathbf{n}-\mathbf{P})}_{\Pi}\cdot \mathbf{x}/\hbar] d\mathbf{x} \\
&= -(2\pi)^2 \hbar m \frac{1}{(2\pi\hbar)^3} \int_0^{2\pi} d\varphi \int_0^{\pi} d\theta \int_0^{\infty} dr\, r^2 \sin\theta V(r) e^{-\frac{i}{\hbar}\Pi r \cos\theta} \\
&= -\frac{m}{\hbar^2} \int_0^{\infty} dr\, r^2 V(r) \int_0^{\pi} d\theta \sin\theta e^{-\frac{i}{\hbar}\Pi r \cos\theta} \\
&= -\frac{m}{\hbar^2} \int_0^{\infty} dr\, r^2 V(r) \hbar \frac{e^{\frac{i}{\hbar}\Pi r} - e^{-\frac{i}{\hbar}\Pi r}}{i\Pi r}
\end{aligned}
\tag{3.784}
$$

releasing the working resumed form

$$f^{(+)B}(\mathbf{n},\mathbf{P}) = \frac{im}{\hbar\Pi}\left[J(\Pi) - J(-\Pi)\right] \tag{3.785}$$

$$J(\Pi) = \int_0^{\infty} r V(r) e^{\frac{i}{\hbar}\Pi r} dr \tag{3.786}$$

This scattering amplitude is to be in next employed to the Yukawa potential in the experimental conditions of elastic scattering that specifies the scattering angle $\tilde{\theta}$ and momentum vectorial relationships as

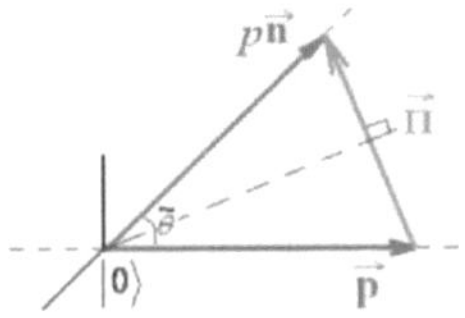

thus quoting that

$$\Pi = 2p\sin\frac{\tilde{\theta}}{2} \tag{3.787}$$

With this, one firstly yields:

$$J_{Yuk}(\Pi) = \int_0^{\infty} r V_{Yuk}(r) e^{\frac{i}{\hbar}\Pi r} dr$$

$$= A\int_0^{\infty} e^{-\frac{r}{r_0}} e^{\frac{i}{\hbar}\Pi r} dr = A\int_0^{\infty} e^{r\left(\frac{i}{\hbar}\Pi - \frac{1}{r_0}\right)} dr = \frac{A}{\frac{i}{\hbar}\Pi - \frac{1}{r_0}} e^{r\left(\frac{i}{\hbar}\Pi - \frac{1}{r_0}\right)}\Bigg|_0^{\infty}$$

$$= \frac{A}{\frac{i}{\hbar}\Pi - \frac{1}{r_0}}\left\{\underbrace{\lim_{r\to\infty} e^{-r/r_0}}_{0}\underbrace{\lim_{r\to\infty}\exp(ir\Pi/\hbar)}_{oscillant\ bounded} - \underbrace{\lim_{r\to 0} e^{r\left(\frac{i}{\hbar}\Pi - \frac{1}{r_0}\right)}}_{1}\right\};$$

$$J_{Yuk}(\Pi) = \frac{A}{\frac{1}{r_0} - \frac{i}{\hbar}\Pi};\ \Rightarrow J_{Yuk}(-\Pi) = \frac{A}{\frac{1}{r_0} + \frac{i}{\hbar}\Pi} \tag{3.788}$$

Then the scattering amplitude for Yukawa potential is obtained:

$$f_{Yuk}^{(+)B}(\mathbf{n},\mathbf{P}) = \frac{im}{\hbar\Pi}\left[J_{Yuk}(\Pi) - J_{Yuk}(-\Pi)\right]$$

$$= \frac{imA}{\hbar\Pi}\left\{\frac{\left(\frac{1}{r_0} - \frac{i}{\hbar}\Pi\right) - \left(\frac{1}{r_0} + \frac{i}{\hbar}\Pi\right)}{\frac{1}{r_0^2} + \frac{\Pi^2}{\hbar^2}}\right\} = -\frac{2mA}{\Pi^2 + \frac{\hbar^2}{r_0^2}};$$

$$f_{Yuk}^{(+)B}(\mathbf{n},\mathbf{P}) = -\frac{2mA}{4p^2\sin^2\frac{\tilde{\theta}}{2} + \frac{\hbar^2}{r_0^2}} \tag{3.789}$$

Now, the differential cross section for Yukawa potential reads:

$$\left(\frac{d\sigma}{d\Omega}\right)_{(\mathbf{p},\mathbf{n})-Yuk} = \left|f_{Yuk}^{(+)B}(\mathbf{n},\mathbf{P})\right|^2 = \frac{4m^2A^2}{\left(4p^2\sin^2\frac{\tilde{\theta}}{2} + \frac{\hbar^2}{r_0^2}\right)^2} \tag{3.790}$$

At this point the custom Coulombic limit, however slightly generalized to the interaction between two charged particles or centers,

$$\tilde{A}_0 = Z_1 Z_2 e_0^2 ,\ r_0 \to \infty \tag{3.791}$$

brings the radius action in the infinity limit (as specific to photon-electron scattering, in accordance with boson-fermion "production" from the Dirac sea, sea the discussion from the beginning of this section), so that the result recovers the celebrated *Rutherford scattering formula*:

$$\left(\frac{d\sigma}{d\Omega}\right)_{(\mathbf{p},\mathbf{n})-Rutherford} \cong \left(\frac{2mZ_1Z_2e_0^2}{4p^2\sin^2\dfrac{\tilde{\theta}}{2}}\right)^2 = \frac{Z_1^2Z_2^2e_0^4}{16E^2\sin^4\dfrac{\tilde{\theta}}{2}} \tag{3.792}$$

One formidable feature of this result is that although at the end of a long and somehow complex quantum theory exposition, thus being quantum in origins, it does not explicitly contain the quantum Planck constant. There seems that is this behavior that makes observable the quantum phenomena of Coulombic scatterings, which, nevertheless are manifestly with an infinite effective cross section

$$\sigma = \int d\Omega \left(\frac{d\sigma}{d\Omega}\right)_{(\mathbf{p},\mathbf{n})-Rutherford} = \frac{Z_1^2Z_2^2e_0^4}{16E^2}\int_0^{2\pi} d\varphi \int_0^{\pi} \frac{\sin\tilde{\theta}}{\sin^4\dfrac{\tilde{\theta}}{2}} d\tilde{\theta}$$

$$= \frac{Z_1^2Z_2^2e_0^4}{8E^2}\pi \int_0^{\pi} \frac{2\sin\dfrac{\tilde{\theta}}{2}\cos\dfrac{\tilde{\theta}}{2}}{\sin^4\dfrac{\tilde{\theta}}{2}} d\tilde{\theta} = \frac{Z_1^2Z_2^2e_0^4}{2E^2}\pi \int_0^{\pi} \sin^{-3}\frac{\tilde{\theta}}{2}\,\partial_{\tilde{\theta}}\left(\sin\frac{\tilde{\theta}}{2}\right) d\tilde{\theta}$$

$$= -\frac{Z_1^2Z_2^2e_0^4}{4E^2}\pi \left[\sin^{-2}\frac{\tilde{\theta}}{2}\right]_0^{\pi} = -\frac{Z_1^2Z_2^2e_0^4}{4E^2}\pi \left[1 - \frac{1}{\sin^2 0}\right] \to \infty \tag{3.793}$$

relaying on the infinite action radius that is characteristic to photons (bosons), which intermediates the electronic (fermions) interactions.

Finally, worth mentioning that the present scattering framework and Rutherford formulation may be generalized (in what is to be called Mott scattering formula) through fully involving the Dirac sea of electron-positron production, within Dirac theory, thus by means of quantum relativity approach. This will be nevertheless presented with other occasion.

3.9 CONCLUSION

The main lessons to be kept for the further theoretical and practical investigations of the quantum mechanics postulates and basic applications that are presented in the present chapter pertain to the following:

- identifying the fundamental quantum mechanical paradox: the continuity of the wave-function associated with the quantification of eigen-energy/spectrum;
- employing wave-function continuity towards modeling quantum tunneling in general and (alpha) nuclei disintegration as a fundamental application;
- writing the eigen-spectra and eigen-functions of atomic Hydrogenic system, molecular vibration as well as for solid-state free electronic states;
- dealing with semi-classical treatment of quantum basic systems: H-atoms, ω-molecular oscillations, and polynomial potential for electrons in solid states;
- characterizing the quantum systems at equilibrium by variational principle of eigen-energy as averaged Hamiltonian over the optimized (parameter) eigen-functions;
- understanding the substance stability by variational principle as applied to the main fundamental forms of evolution in closed aggregation, i.e., the rotational symmetry in atoms, vibrational motion in molecular, translation in solid state;
- describing the electrons motion in solid state as being specific to excited state rather than to the ground state (the so-called solid state paradox) from where also the phenomenological understanding of conductibility;
- learning the equivalence between the Schrodinger and Heisenberg quantum pictures as relating with the wave-functions and operator evolutions, respectively (in metaphorical analogy with a fisherman moving on a lake or the lake moving around the fixed fisherman);
- treating the quantum evolution by Green function as a quantum amplitude of its first cause;

- solving the free quantum evolution for its Green function as referential for any further observation and measurement by perturbation action(s);
- formulating the general algorithm for perturbing quantum states in various orders while noting their intra-related behavior in combining eigen -functions and -energies;
- interpreting the substance reality by perturbations: atoms with nuclear isotopic corrections, harmonic oscillator in higher orders of oscillation spectra, electrons in quasi-free in solid states;
- connecting Dirac bra-ket formalism with the Dirac creation-annihilation formalism in spectra quantification, with specific application to harmonic oscillator, as the paradigmatic system for quantum fluctuation themselves;
- developing a unitary view on measuring the quantum phenomena as based on conservation laws described by associate quantum dynamical operators for translation, and rotation;
- finding applications for observing quantum phenomena: from particle transport (e.g., electronic current by dissipative media thus producing the effect of electrical resistance) to the particle scattering and interaction resonance as the quantum counterpart of scattering phenomena in classical mechanics, here generalized to quantum approach by means of the Born approximation and Dyson expansion series of scattering amplitudes.

KEYWORDS

- **Born approximation**
- **Green function**
- **Heisenberg picture**
- **hydrogenic states**
- **interaction picture**
- **quantum electric resistance**
- **quantum ground level**
- **quantum mechanics postulates**
- **quantum resonance**

- **quantum scattering**
- **quantum transition**
- **quantum tunneling**
- **Rutherford Scattering**
- **solid state free electronic states**
- **Unitary transformation**
- **vibrational states**
- **Yukawa potential**

REFERENCES

AUTHOR'S MAIN REFERENCES

Putz, M. V. (2016). *Quantum Nanochemistry. A Fully Integrated Approach: Vol. II. Quantum Atoms and Periodicity*. Apple Academic Press & CRC Press, Toronto-New Jersey, Canada-USA.

Putz, M. V., Lazea, M., Chiriac, A. (2010). *Introduction in Physical Chemistry. The Structure and Properties of Atoms and Molecules* (in Romanian), Mirton Publishing House, Timişoara.

Putz, M. V. (2006). *The Structure of Quantum Nanosystems* (in Romanian), West University of Timişoara Publishing House, Timişoara.

SPECIFIC REFERENCES

Gamow, G. (1928). Zur Quantentheorie des Atomkernes, *Zeitschrift für Physik*, 51, 204–212.

Gurney, R. W., Condon, E. U. (1928). Wave mechanics and radioactive disintegration. *Nature*, 122, 439.

HyperPhysics (2010). http://hyperphysics.phy-astr.gsu.edu/Hbase/hframe.html

Rohlf, J. W. (1994). *Modern Physics from A to Z0*, Wiley, NY.

Yukawa, H. (1935). On the interaction of elementary particles, I. *Proc. Phys. ·Math. Soc. Japan*, 17, 48–57.

FURTHER READINGS

Byron, F. W., Fuller, R. W. (1992). *Mathematics of Classical and Quantum Physics*, Courier Dover Publications, New York.

Das, A., Ferbel, T. (1994). *Introduction to Nuclear and Particle Physics*, Wiley, NY.

Dicke, R. H., Wittke, J. P. (1960). *Introduction to Quantum Mechanics*, Addison Wesley, Boston.

Dirac, P. A. M. (1925). The fundamental equations of quantum mechanics. *Proceedings of the Royal Society A: Mathematical, Physical and Engineering Sciences* 109 (752), 642–653.

Dirac, P. A. M. (1944). *The Principles of Quantum Mechanics*; Oxford University Press: Oxford, UK.

Eisberg, R., Resnick, R. (1985). *Quantum Physics*, 2nd Ed., Wiley, New Jersey.

Greene, B. (1999). *The Elegant Universe*, W. W. Norton, NY.

Haken, H., Wolf, H. C. (1996). *The Physics of Atoms and Quanta*, 5th Ed., Springer-Verlag, Berlin-Heidelberg.

Jauch, J. M. (1968). *Foundations of Quantum Mechanics*, Addison-Wesley Publ. Cy., Reading, Massachusetts.

Krane, K. (1987). *Introductory Nuclear Physics*, Wiley, NY.

Levine, I. N. (1991). *Quantum Chemistry*, Prentice Hall, Englewood Cliffs, NJ, 4th edition.

Löwdin, P.-O., Ed. (1999). *Advances in Quantum Chemistry*, Vol. 34. Academic Press, New York.

Mackey, G. (1963). *Mathematical Foundations of Quantum Mechanics*, W. A. Benjamin, New York (2004 paperback reprint by Dover, New York).

Mandl, F. (1992). *Quantum Mechanics*, John Wiley & Sons, Chichester.

McQuarrie, D. A. (1983). *Quantum Chemistry*, University Science Books, Mill Valey, CA.

Merzbacher, E. (1998). *Quantum Mechanics*, 3rd Ed., Wiley, NY.

Pauling, L., Wilson, E. B. (1935). *Introduction to Quantum Mechanics*, McGraw-Hill, New York.

Schaefer, H. F., Ed. (1977). *Methods of Electronic Structure Theory*, Plenum Press, New York.

Steinfeld, J. I. (1989). *Molecules and Radiation: An Introduction to Modern Molecular Spectroscopy*, MIT Press, Cambridge, MA, Second Edition.

Szabo, A., Ostlund, N. S. (1989). *Modern Quantum Chemistry: Introduction to Advanced Electronic Structure Theory*, McGraw-Hill, New York.

von Neumann, J. (1932). *Mathematical Foundations of Quantum Mechanics*, Princeton University Press, New Jersey.

Weyl, H. (1950). *The Theory of Groups and Quantum Mechanics*, Dover Publications, New York.

CHAPTER 4

QUANTUM MECHANICS FOR QUANTUM CHEMISTRY

CONTENTS

ABSTRACT

Basing on the first principles of Quantum mechanics as exposed in the previous chapters and sections, special chapters of quantum theory are here unfolded in order to further extend and caching the quantum information from free to observed evolution within the matter systems with constraints (boundaries). As such, the Feynman path integral formalism is firstly exposed and then applied to atomic, quantum barrier and quantum harmonically vibration, followed by density matrix approach, opening the Hartree-Fock and Density Functional pictures of many-electronic systems, with a worthy perspective of electronic occupancies via Koopmans theorem, while ending with a further generalization of the Heisenberg observability and of its first application to mesosystems.

4.1 INTRODUCTION

Not necessarily in an historical order but rather as a phenomenological classification, one should learn that the actual quantum chemistry originates in five levels of quantum approximations imposed on the many-electronic-many-nuclei systems, either in isolate or interacting state. They are summarized below along mentioning the current limitations, controversies and prospects.

1. *The Born-Oppenheimer approximation* (Born & Oppenheimer, 1927), while intended in producing a simplification of the electronic calculus for frozen nuclei approximation, breaks down actually when, for instance, computing the magnetic dipole moment and its derivative with respect to the nuclear velocities

or momenta, for assessing the molecular properties of surfaces (Buckingham et al., 1987).

2. *The single Slater determinant representation of the ground electronic state* (Slater, 1929), which nicely solved the exchange behavior of electrons by incorporating the Pauli repulsion in antisymmetric determinants (Pauli, 1940), was conceptually extended to the configuration interaction by the seminal works of Lowdin (Löwdin, 1955), while noting afterwards further generalization by the fruitful notion of the so-called complete active space (CAS) (Roos et al., 1980) that, when combined with other quantum chemical methods such as the self consistent field (SCF) and density functional theory (DFT) – see below, become very productive in accounting for *all* electronic states which contribute to the *reactive* space, be it which electronic states of species (reactants, intermediates and products) are involved in chemical reactions, thermally or photo-induced ones (Roos et al., 1982; Roos & Malmqvist, 2004).
3. *Simple Hückel* (Hückel, 1931) *and molecular orbitals' theories* (Parr et al., 1950; Roothaan, 1951, 1958; Purvis & Bartlett, 1982; Roothaan & Detrich, 1983; Pople et al., 1987; Curtiss et al., 1998; Ohlinger et al., 2009), nevertheless viewed as the next natural step over the paradigmatic Heitler-London theory of homopolar chemical bonding (Heitler & London, 1927), have been unlocking the door for self-consistent field Hartree-Fock-Slater theories (Pople & Nesbet, 1954; Roothaan, 1960; Corongiu, 2007; Glaesemann & Schmidt, 2010) and of associate semi-empirical formulations (Pariser & Parr, 1953; Pople, 1953) in treating a plethora of chemical system and phenomena on the base of their internal symmetry, while remarkably agreeing (and sometimes predicting) the observed spectra and reactivity, among which the pericyclic reactions (Beaudry et al., 2005; Hickenboth et al., 2007) and the Woodward-Hoffman rules (Woodward & Hoffmann, 1965; Hoffmann & Woodward, 1968) are eminent examples; yet, this direction let with the so-called quantum correlation problem (i.e., modeling the electronic movement in the dynamical field of the other electrons present in the system) that remains little tractable within the Slater (or even

with configurational interaction) framework; the solution in getting accurate correlations arrives with the advent of Density Functional Theory (DFT) and with the price of modifying the overall wave-function of the system and of its spectra.

4. *Thomas-Fermi theory* (Balàzs, 1967; Fermi, 1927; Lieb & Simon, 1977; Teller, 1962; Thomas, 1927) *along Walter Kohn's developments* (Hohenberg & Kohn, 1964; Kohn & Sham, 1965) *while* merging in the celebrated Density Functional Theory (Kohn et al., 1996) have the merit in being conceptually exact, i.e., performing the ab initio analysis of the electronic spectra relying only upon the universal constants of electronic charge, mass, Planck constant, and of their combinations in bare and effective potential, while providing an approximate set of orbitals (called as Kohn-Sham orbitals); they eventually correctly resemble the observed electronic density of the system along the measured energies (with correlation effects included), yet being with less significance even than the classical wave-function concept; actually, the current DFT uses as input the so-called basis function just like a mathematical tool, that can be adapted or optimized depending on the accuracy needed in relation with optimized effective potential (Bokhan & Bartlett, 2006), adding dispersion effects, etc. After all, the computationally implementation of DFT becomes so parameterized procedure that makes from it a sort of semi-empirical based DFT (Derosa, 2009), that can be nevertheless extended to include time-dependency excited states effects (Besley et al., 2009; Burke et al., 2005; Runge & Gross, 1984; March et al., 1999; Ploetner et al., 2010), as well as modeling the actual hot topic of Bose-Einstein condensates (Putz, 2011b). Moreover, it is worth saying that the great merit and paradox of DFT, is that the theory provides the recipe to compute two-body interactions (as exchange and correlations) by approximations to single body (density) behavior; from the physical point of view the picture is flawed, yet it turned out that the approximations works very well – markings therefore a landmark in quantum chemistry achievement; in passing this was perhaps also the reason the theory was Nobel awarded in Chemistry (in 1998) and not in Physics since there is no new physics inside but useful

reformulations (thus a new theory) for Chemistry. Recent works on DFT also try to further lighten on the DFT limits, i.e., formulating various approximations for exchange-correlation functionals (Putz, 2008) in more or less agreement with the fundamental theorems and limits at asymptotic and nuclei ranges (Capelle, 2006), until attempting to formulate expectation values of various physical observables based only on density similar to those based on the expectation value of quantum mechanical wave function (Bartlett & Musial, 2007), leads with the believe that true *ab initio DFT* is far for being fully engaged (Bartlett et al., 2004; Bartlett et al., 2005), while only *semi-empirical DFT* seems to prevail in a way or other when is about computational implementation. On the other side, still, the conceptual DFT (Geerlings et al., 2003), is of the first importance in formulating the chemical reactivity and its indices that help in understanding and modeling the chemical systems to a large extent (De Proft & Geerlings, 2001; Chermette, 1999).

5. *The solvent effects* seem to need almost always be taken into account when using quantum chemical treatment for describing chemical systems' reactivity. The environment interaction, and sometimes strongly interacting solvent molecules (e.g., water molecules in the case of biomolecules: amino acids, peptides, nucleic acids and their complexes) need to be considered in any modeling study of open chemical systems in order to fully understand and interpret the experimental results such as the vibrational, NMR and electronic spectra, and the chiral analogues (Tiwari et al., 2008). Overall, the interaction of the system with environment stands also in the foreground of the quantum theory when always predicting an additional quantum fluctuation upon the concerned system due to its coupling with the media/observer/solvent (Corni et al., 2003; Fung et al., 2006); it can be nevertheless implemented by counting for additional reactions and stability of the investigated chemical systems, while having also at side the quantum statistical tools for treating the macro-canonical samples in a correct physical way; worth saying that at this point DFT is well equipped from its basic definition of density - associated with the total number of electrons in the system - that can be then easily extended to

include also those effects coming from the environment (Nandini & Sathyanarayana, 2003).

6. *The density matrix theory*, the ancestor of density functional theory, provides the immediate framework for Path Integral (PI) development, allowing the canonical density be extended for the many-electronic systems through the density functional closure relationship. Yet, the use of path integral formalism for electronic density prescription presents several advantages: assures the inner quantum mechanical description of the system by parameterized paths; averages the quantum fluctuations; behaves as the propagator for time-space evolution of quantum information; resembles Schrödinger equation; allows quantum statistical description of the system through partition function computing. In this framework, four levels of path integral formalism can be approached: (1i) the Feynman quantum mechanical (present Chapter); (2i) the semiclassical, (3i) the Feynman-Kleinert effective classical, and the (4i) Fokker-Planck non-equilibrium ones (for the last three levels see Volume II of the present five-volume book). They lead with the practical specializations for quantum free and harmonic motions, for statistical high and low temperature limits, the smearing justification for the Bohr's quantum stability postulate with the paradigmatic Hydrogen atomic excursion, along the quantum chemical calculation of semiclassical electronegativity and hardness, of chemical action and Mulliken electronegativity, as well as by the Markovian generalizations of Becke-Edgecombe electronic focalization functions – all advocate for the reliability of assuming PI formalism of quantum mechanics as a versatile one, suited for analytically and/or computationally modeling of a variety of fundamental physical and chemical reactivity concepts characterizing the (density driving) many-electronic systems.

Accordingly, the present chapter combines these issues in a first systematical inside look on quantum chemistry by quantum mechanics, so planting the "seeds" for the next applications on atoms, molecules, nanostructures and bio-chemical interactions (see the next Volumes II–V of the present five-volume book).

4.2 FEYNMAN'S PATH INTEGRAL QUANTUM FORMALISM

4.2.1 CONSTRUCTION OF PATH INTEGRAL

One starts considering the slicing for the time interval $[t_b, t_a]$

$$t_b = t_{n+1} > t_n > \ldots > t_2 > t_1 > t_0 = t_a \tag{4.1}$$

with the spatial ending points recalled as $x' = x_b, x = x_a$ for the quantum propagator (Green function) of Chapters 2 and 3, see for instance Eq. (3.310) as the actual space-time evolution amplitude

$$(x_b t_b; x_a t_a) = \langle x_b | \exp\left(-\frac{i}{\hbar}(t_b - t_a)\widehat{H}\right) | x_a \rangle \tag{4.2}$$

may be firstly rewritten in terms of associate evolution operator

$$\widehat{U}(t_b - t_a) = \exp\left(-\frac{i}{\hbar}(t_b - t_a)\widehat{H}\right) \tag{4.3}$$

to successively become

$$\begin{aligned}(x_b t_b; x_a t_a) &= \langle x_b | \widehat{U}(t_b - t_a) | x_a \rangle \\ &= \langle x_b | \widehat{U}(t_b, t_n)\widehat{U}(t_n, t_{n-1}) \ldots \widehat{U}(t_j, t_{j-1}) \ldots \widehat{U}(t_2, t_1)\widehat{U}(t_1, t_a) | x_a \rangle \\ &= \prod_{j=1}^{n}\left[\int_{-\infty}^{+\infty} dx_j\right] \prod_{j=1}^{n+1} (x_j t_j; x_{j-1} t_{j-1})\end{aligned} \tag{4.4}$$

when n-times the complete eigen-coordinate set

$$\hat{1} = \int_{-\infty}^{+\infty} |x_j\rangle\langle x_j| dx_j,\ j = \overline{1, n} \tag{4.5}$$

was introduced for each pair of events, with the elementary propagator between them:

$$(x_j t_j; x_{j-1} t_{j-1}) = \langle x_j | \exp\left(-\frac{i}{\hbar}(t_j - t_{j-1})\widehat{H}\right) | x_{j-1} \rangle = \langle x_j | \exp\left(-\frac{i}{\hbar}\varepsilon\widehat{H}\right) | x_{j-1} \rangle \tag{4.6}$$

where the elementary time interval was set as

$$\varepsilon = t_j - t_{j-1} = \frac{t_b - t_a}{n+1} > 0 \tag{4.7}$$

Now, the elementary quantum evolution amplitude (4.7) is to be evaluated, firstly by reconsidering the eigen-coordinate unitary operator, in the working form

$$\hat{1}_x = \int_{-\infty}^{+\infty} |x\rangle\langle x| dx \tag{4.8}$$

to separate the operatorial Hamiltonian contributions to the kinetic and potential ones,

$$\widehat{H} = \widehat{T} + \widehat{V} \tag{4.9}$$

as:

$$\begin{aligned}(x_j t_j; x_{j-1} t_{j-1}) &\cong \langle x_j | e^{-\frac{i}{\hbar}\varepsilon \widehat{V}(\hat{x}, t_j)} \hat{1}_x e^{-\frac{i}{\hbar}\varepsilon \widehat{T}(\hat{p}, t_j)} | x_{j-1} \rangle \\ &= \int_{-\infty}^{+\infty} \langle x_j | e^{-\frac{i}{\hbar}\varepsilon \widehat{V}(\hat{x}, t_j)} | x \rangle \langle x | e^{-\frac{i}{\hbar}\varepsilon \widehat{T}(\hat{p}, t_j)} | x_{j-1} \rangle dx\end{aligned} \tag{4.10}$$

where we have used the first order limitation of the Baker-Hausdorff formula, see Eq. (2.347)

$$e^{-\frac{i}{\hbar}\varepsilon\left[\widehat{V}(\hat{x}, t_j) + \widehat{T}(\hat{p}, t_j)\right]} = e^{-\frac{i}{\hbar}\varepsilon \widehat{V}(\hat{x}, t_j)} e^{-\frac{i}{\hbar}\varepsilon \widehat{T}(\hat{p}, t_j)} + \underbrace{O(\varepsilon^2)}_{\cong 0} \tag{4.11}$$

that is we assumed the second order of elementary time intervals be vanishing

$$\varepsilon^2 \cong 0 \tag{4.12}$$

Next, each obtained working energetic contribution are evaluated separated as: for kinetic contribution the inserting of the momentum complete eigen-set

$$\hat{1}_{p_j} = \int_{-\infty}^{+\infty} |p_j\rangle\langle p_j| dp_j \tag{4.13}$$

yields:

$$\left\langle x\left|e^{-\frac{i}{\hbar}\varepsilon\hat{T}(\hat{p},t_j)}\right|x_{j-1}\right\rangle=\int_{-\infty}^{+\infty}\left\langle x\left|e^{-\frac{i}{\hbar}\varepsilon\hat{T}(\hat{p},t_j)}\right|p_j\right\rangle\left\langle p_j|x_{j-1}\right\rangle dp_j$$
$$=\int_{-\infty}^{+\infty}e^{-\frac{i}{\hbar}\varepsilon T(p_j,t_j)}\underbrace{\left\langle x|p_j\right\rangle}_{\frac{1}{\sqrt{2\pi\hbar}}e^{\frac{i}{\hbar}p_jx}}\underbrace{\left\langle p_j|x_{j-1}\right\rangle}_{\frac{1}{\sqrt{2\pi\hbar}}e^{-\frac{i}{\hbar}p_jx_{j-1}}}dp_j$$
$$=\int_{-\infty}^{+\infty}e^{\frac{i}{\hbar}p_j(x-x_{j-1})}e^{-\frac{i}{\hbar}\varepsilon T(p_j,t_j)}\frac{dp_j}{2\pi\hbar} \tag{4.14}$$

While for potential elementary amplitude we get:

$$\left\langle x_j\left|e^{-\frac{i}{\hbar}\varepsilon\hat{V}(\hat{x},t_j)}\right|x\right\rangle=e^{-\frac{i}{\hbar}\varepsilon V(x,t_j)}\underbrace{\left\langle x_j|x\right\rangle}_{\delta(x_j-x)}=\delta(x_j-x)e^{-\frac{i}{\hbar}\varepsilon V(x,t_j)} \tag{4.15}$$

With relations (4.14) and (4.15) back in Eq. (4.11) the elementary propagator takes the form:

$$\left(x_jt_j;x_{j-1}t_{j-1}\right)=\int_{-\infty}^{+\infty}\left(\int_{-\infty}^{+\infty}e^{\frac{i}{\hbar}p_j(x-x_{j-1})}e^{-\frac{i}{\hbar}\varepsilon T(p_j,t_j)}\frac{dp_j}{2\pi\hbar}\right)\left(\delta(x_j-x)e^{-\frac{i}{\hbar}\varepsilon V(x,t_j)}\right)dx$$
$$=\int_{-\infty}^{+\infty}\frac{dp_j}{2\pi\hbar}\exp\left[\frac{ip_j(x_j-x_{j-1})}{\hbar}-i\varepsilon\frac{T(p_j,t_j)+V(x_j,t_j)}{\hbar}\right]$$
$$=\int_{-\infty}^{+\infty}\frac{dp_j}{2\pi\hbar}\exp\left\{\frac{i}{\hbar}\left[p_j(x_j-x_{j-1})-\varepsilon H(x_j,p_j,t_j)\right]\right\} \tag{4.16}$$

Replacing the elementary quantum amplitude (4.16) back in the global one given by Eq. (4.4), it assumes the form:

$$\left(x_bt_b;x_at_a\right)\cong\prod_{j=1}^{n}\left[\int_{-\infty}^{+\infty}dx_j\right]\prod_{j=1}^{n+1}\int_{-\infty}^{+\infty}\frac{dp_j}{2\pi\hbar}\exp\left\{\varepsilon\frac{i}{\hbar}\left[\begin{array}{l}p_j\dfrac{x_j-x_{j-1}}{\varepsilon}\\-H(x_j,p_j,t_j)\end{array}\right]\right\}$$
$$=\left(\prod_{j=1}^{n}\left[\int_{-\infty}^{+\infty}dx_j\right]\right)\left(\prod_{j=1}^{n+1}\left[\int_{-\infty}^{+\infty}\frac{dp_j}{2\pi\hbar}\right]\right)\exp\left\{\frac{i}{\hbar}\varepsilon\sum_{j=1}^{n+1}\left[\begin{array}{l}p_j\dfrac{x_j-x_{j-1}}{\varepsilon}\\-H(x_j,p_j,t_j)\end{array}\right]\right\} \tag{4.17}$$

which for in infinitesimal temporal partition,

$$n \to \infty; \varepsilon \to 0 \tag{4.18}$$

the quantum propagator behaves like the *Feynman path integral* (Dirac, 1933; Feynman, 1948; Wiener, 1923; Infeld & Hull, 1951; Schulmann, 1968; Abarbanel & Itzykson, 1969; Campbell et al., 1975; Laidlaw & DeWitt-Morette, 1971):

$$\left(x_b t_b; x_a t_a\right) \equiv \int_{x(t_a)=x_a}^{x(t_b)=x_b} \mathrm{D}x\mathrm{D}p \exp\left\{\frac{i}{\hbar} S[x,p,t]\right\} \tag{4.19}$$

where we have considered the limiting notations for the path integral measure:

$$\lim_{n\to\infty}\left(\prod_{j=1}^{n}\left[\int_{-\infty}^{+\infty} dx_j\right]\right)\left(\prod_{j=1}^{n+1}\left[\int_{-\infty}^{+\infty}\frac{dp_j}{2\pi\hbar}\right]\right) \equiv \int_{x(t_a)=x_a}^{x(t_b)=x_b} \mathrm{D}x\mathrm{D}p \tag{4.20}$$

and for the involved action:

$$\lim_{\substack{n\to\infty \\ \varepsilon\to 0}} \varepsilon \sum_{j=1}^{n+1}\left[p_j \frac{x_j - x_{j-1}}{\varepsilon} - H\left(x_j, p_j, t_j\right)\right]$$
$$\overset{(4.7)}{\equiv} \int_{t_a}^{t_b}\left[p(t)\dot{x}(t) - H\left(x(t), p(t), t\right)\right]dt = \int_{t_a}^{t_b} L\left(x(t), p(t), t\right)dt = S[x,p,t] \tag{4.21}$$

Note that the results (4.19)–(4.21) account for quantum information for the quantum evolution of a system throughout accounting all histories (possibilities for linking two events in time-space) for a quantum evolution (Peak & Inomata, 1969; Gerry & Singh, 1979; Kleinert, 1989):

$$\left(x_b t_b; x_a t_a\right) = \sum_{\substack{ALL \\ HISTORIES}} \exp\left\{\frac{i}{\hbar} S[x,p,t]\right\} \tag{4.22}$$

thus being suitable to be implemented in the N-particle functional scheme once it is analytically computed (see the Section 4.3).

However, for achieving such goal, a more practical form of the Feynman integral may be obtained once the Hamiltonian is implemented as

$$H = \frac{p^2}{2m} + V(x,t) \tag{4.23}$$

leading with the action (4.21) unfolding

$$\begin{aligned} S[x,p,t] &= \sum_{j=1}^{n+1}\left[p_j\left(x_j - x_{j-1}\right) - \varepsilon\frac{p_j^2}{2m} - \varepsilon V(x_j,t_j)\right] \\ &= \sum_{j=1}^{n+1}\left[-\frac{\varepsilon}{2m}\left(p_j - \frac{x_j - x_{j-1}}{\varepsilon}m\right)^2 + \varepsilon\frac{m}{2}\left(\frac{x_j - x_{j-1}}{\varepsilon}\right)^2 - \varepsilon V(x_j,t_j)\right] \end{aligned} \tag{4.24}$$

allowing the momentum integrals in Eq. (4.20) to be solved out as

$$\int_{-\infty}^{+\infty} \frac{dp_j}{2\pi\hbar}\exp\left\{-\frac{i}{\hbar}\frac{\varepsilon}{2m}\left(p_j - \frac{x_j - x_{j-1}}{\varepsilon}m\right)^2\right\} = \sqrt{\frac{m}{2\pi\hbar i\varepsilon}} \tag{4.25}$$

through formally applying the Poisson formula (see Appendix A.1.2):

$$\int_{-\infty}^{+\infty} \exp\left(-ay^2\right)dy = \sqrt{\frac{\pi}{a}} \tag{4.26}$$

The remaining quantum evolution amplitude reads as the spatial path integral only:

$$\begin{aligned} (x_b t_b; x_a t_a) &= \left(\prod_{j=1}^{n}\left[\int_{-\infty}^{+\infty} dx_j\right]\right)\left(\prod_{j=1}^{n+1}\left[\sqrt{\frac{m}{2\pi\hbar i\varepsilon}}\right]\right)\exp\left\{\frac{i}{\hbar}\sum_{j=1}^{n+1}\left[\varepsilon\frac{m}{2}\left(\frac{x_j - x_{j-1}}{\varepsilon}\right)^2 - \varepsilon V(x_j,t_j)\right]\right\} \\ &= \frac{1}{\sqrt{2\pi\hbar i\varepsilon / m}}\prod_{j=1}^{n}\left[\int_{-\infty}^{+\infty}\frac{dx_j}{\sqrt{2\pi\hbar i\varepsilon / m}}\right]\exp\left\{\frac{i}{\hbar}\varepsilon\sum_{j=1}^{n+1}\left[\frac{m}{2}\left(\frac{x_j - x_{j-1}}{\varepsilon}\right)^2 - V(x_j,t_j)\right]\right\} \\ &\underset{\varepsilon\to 0}{\overset{n\to\infty}{\to}} \int_{x(t_a)=x_a}^{x(t_b)=x_b} \mathrm{D}'x \exp\left\{\frac{i}{\hbar}S[x,\dot{x},t]\right\} \end{aligned} \tag{4.27}$$

with the actual modified measure of integration:

$$\int_{x(t_a)=x_a}^{x(t_b)=x_b} \mathsf{D}'x \equiv \lim_{\substack{n\to\infty \\ \varepsilon\to 0}} \left(\frac{1}{\sqrt{2\pi\hbar i\varepsilon / m}} \prod_{j=1}^{n} \left[\int_{-\infty}^{+\infty} \frac{dx_j}{\sqrt{2\pi\hbar i\varepsilon / m}} \right] \right) \tag{4.28}$$

and the working action:

$$S[x,\dot{x},t] = \int_{t_a}^{t_b} L(x,\dot{x},t)\,dt = \int_{t_a}^{t_b} \left[\frac{m\dot{x}^2}{2} - V(x,t) \right] dt \tag{4.29}$$

Note that when the *partition function* is considered, another space coordinate is to be taken over the path integral (4.27), namely:

$$Z(t_b;t_a) = \int_{-\infty}^{+\infty} (xt_b;xt_a)\,dx = \int_{x(t_a)=x(t_b)} \mathsf{D}''x \exp\left\{ \frac{i}{\hbar} S[x,\dot{x},t] \right\} \tag{4.30}$$

while the new integration measure

$$\int_{x(t_a)=x(t_b)} \mathsf{D}''x \equiv \lim_{n\to\infty} \left(\prod_{j=1}^{n+1} \left[\int_{-\infty}^{+\infty} \frac{dx_j}{\sqrt{2\pi\hbar i\varepsilon / m}} \right] \right) \tag{4.31}$$

Ones of the main advantages dealing with path integrals relays on following:

- Attractive conceptual representation of dynamical quantum processes without operatorial excursion;
- Allows for quantum fluctuation description in analogy with thermal analogies,

$$(x_b t_b; x_a t_a)_{QM} = \int_{x(t_a)=x_a}^{x(t_b)=x_b} \mathsf{D}'x(t) \exp\left\{ \frac{i}{\hbar} \int_{t_a}^{t_b} \left[\frac{m\dot{x}^2}{2} - V(x(t),t) \right] dt \right\} \tag{4.32}$$

through changing the temporal intervals with the thermodynamic temperature (T) by means of Wick transformation

$$\begin{cases} t := -i\tau,\ \tau = \hbar\beta, \beta = \dfrac{1}{k_B T} \\ dt = -i d\tau \\ \dfrac{d}{dt} = i\dfrac{d}{d\tau} \end{cases} \tag{4.33}$$

i.e. transforming quantum mechanical (QM) into quantum statistical (QS) propagators:

$$(x_b \hbar\beta; x_a 0)_{QS} = \int_{x(0)=x_a}^{x(\hbar\beta)=x_b} \mathrm{D}'x(\tau)\exp\left\{-\frac{1}{\hbar}\int_0^{\hbar\beta}\begin{bmatrix}\frac{m}{2}\dot{x}^2(\tau)\\ +V(x(\tau),\tau)\end{bmatrix}d\tau\right\} \quad (4.34)$$

from where immediately writing also the associate QS-partition function:

$$Z_{QS}(\beta) = \int_{-\infty}^{+\infty}(x_b \hbar\beta; x_a 0)_{QS}\, dx$$
$$= \int_{x(0)=x(\hbar\beta)} \mathrm{D}''x \exp\left\{-\frac{1}{\hbar}\int_0^{\hbar\beta} L^+(x,\dot{x},\tau)d\tau\right\} \quad (4.35)$$

both QS object having the effect of transforming the *canonical Lagrangian* of action into the so-called *Euclidian* one

$$L^+(x,\dot{x},\tau) = \frac{m}{2}\dot{x}^2(\tau) + V(x(\tau),\tau) \quad (4.36)$$

analogous with the fact the Euclidian metric has all its diagonal terms as with positive sign.

Yet, the connection of the path integrals of propagators with the Schrödinger quantum formalism is to be revealed, and in next addressed.

4.2.2 PROPAGATOR'S EQUATION

There are two ways for showing the propagator path integral links with Schrödinger equation.

Firstly, there is by employing one of the above path integral, say that of Eq. (4.27) with Eq. (4.29),

$$(x_b t_b; x_a t_a) = \int_{x(t_a)=x_a}^{x(t_b)=x_b} \mathrm{D}'x \exp\left\{\frac{i}{\hbar}S[x,\dot{x},t]\right\} \quad (4.37)$$

to perform the derivative:

$$\frac{\partial}{\partial x_b}(x_b,t_b;x_a,t_a)=\frac{(x_b+\delta x(t_b),t_b;x_a,t_a)-(x_b,t_b;x_a,t_a)}{\delta x(t_b)}$$

$$=\frac{1}{\delta x(t_b)}\left[\begin{array}{l}\displaystyle\int_{x(t_a)=x_a}^{x(t_b)=x_b}\mathrm{D'}x\exp\left\{\begin{array}{l}\frac{i}{\hbar}S[x(t)\\+\delta x(t)]\end{array}\right\}\\ \displaystyle-\int_{x(t_a)=x_a}^{x(t_b)=x_b}\mathrm{D'}x\exp\left\{\frac{i}{\hbar}S[x(t)]\right\}\end{array}\right]$$

$$=\frac{1}{\delta x(t_b)}\delta S\frac{\displaystyle\int_{x(t_a)=x_a}^{x(t_b)=x_b}\mathrm{D'}x\exp\left\{\begin{array}{l}\frac{i}{\hbar}S[x(t)\\+\delta x(t)]\end{array}\right\}-\int_{x(t_a)=x_a}^{x(t_b)=x_b}\mathrm{D'}x\exp\left\{\frac{i}{\hbar}S[x(t)]\right\}}{\delta S}$$

$$=\frac{1}{\delta x(t_b)}\delta S\frac{\delta(x_b,t_b;x_a,t_a)}{\delta S}$$

$$=\frac{1}{\delta x(t_b)}\delta S\frac{i}{\hbar}(x_b,t_b;x_a,t_a)$$

$$=\frac{1}{\delta x(t_b)}\left[\int_{t_a}^{t_b}\left(\frac{\partial L}{\partial x}\delta x+\frac{\partial L}{\partial \dot{x}}\delta\dot{x}\right)dt\right]\frac{i}{\hbar}(x_b,t_b;x_a,t_a)$$

$$=\frac{1}{\delta x(t_b)}\left[\int_{t_a}^{t_b}\left(\frac{\partial L}{\partial x}\delta x+\frac{\partial L}{\partial \dot{x}}\frac{d}{dt}\delta x\right)dt\right]\frac{i}{\hbar}(x_b,t_b;x_a,t_a)$$

$$=\frac{1}{\delta x(t_b)}\left[\begin{array}{l}\left.\frac{\partial L}{\partial \dot{x}}\frac{\partial}{\partial t}\delta x\right|_{t_a}^{t_b}\\ \displaystyle+\int_{t_a}^{t_b}\underbrace{\left(\frac{\partial L}{\partial x}-\frac{d}{dt}\frac{\partial L}{\partial \dot{x}}\right)}_{\substack{=0\\ \text{Euler-Lagrange equation}}}\delta x dt\end{array}\right]\frac{i}{\hbar}(x_b,t_b;x_a,t_a)$$

$$=\frac{1}{\delta x(t_b)}\left[\frac{\partial L}{\partial \dot{x}}(t_b)\frac{\partial}{\partial t}\underbrace{\delta x(t_b)}_{\neq 0}-\frac{\partial L}{\partial \dot{x}}(t_a)\frac{\partial}{\partial t}\underbrace{\delta x(t_a)}_{=0}\right]\frac{i}{\hbar}(x_b,t_b;x_a,t_a)$$

$$=\underbrace{\frac{\partial L}{\partial \dot{x}}(t_b)}_{p(t_b)}\frac{i}{\hbar}(x_b,t_b;x_a,t_a);$$

$$\frac{\partial}{\partial x_b}(x_b,t_b;x_a,t_a)=\left[\frac{i}{\hbar}p(t_b)\right](x_b,t_b;x_a,t_a) \tag{4.38}$$

Similarly for the second derivative we have:

$$\frac{\partial}{\partial x_b^2}(x_b,t_b;x_a,t_a)=\left[\frac{i}{\hbar}p(t_b)\right]^2(x_b,t_b;x_a,t_a)=-\frac{p^2(t_b)}{\hbar^2}(x_b,t_b;x_a,t_a) \tag{4.39}$$

while for time derivative we obtain following the same formal steps as before for coordinate derivatives:

$$\frac{\partial}{\partial t_b}(x_b,t_b;x_a,t_a)=\frac{\delta S}{\delta t_b}\frac{i}{\hbar}(x_b,t_b;x_a,t_a)=-\frac{i}{\hbar}H(t_b)(x_b,t_b;x_a,t_a) \tag{4.40}$$

by recalling the Hamilton-Jacobi equation of motion in the form

$$\frac{\delta S}{\delta t_b}=-H(t_b) \tag{4.41}$$

Now, there is immediate that for a Hamiltonian of the form (4.23) one gets through multiplying both its side with the propagator (4.37) and then considering for the square momentum and Hamiltonian terms the relations (4.39) and (4.41), respectively, one leaves with the Schrödinger type equation for the path integral:

$$i\hbar\frac{\partial}{\partial t_b}(x_b,t_b;x_a,t_a)=\left[-\frac{\hbar^2}{2m}\frac{\partial}{\partial x_b^2}+V(x_b)\right](x_b,t_b;x_a,t_a) \tag{4.42}$$

Remarkably, besides establishing the link with the Schrödinger picture, Eq. (4.42) tells something more important, namely that the wave function itself, i.e., $\Psi(x_b,t_b)$ that usually fulfills equations like Eq. (4.42) may be replaced (and generalized as well) by the quantum propagator $(x_b,t_b;x_a,t_a)$ with the crucial consequence in that the propagator is providing the *N*-electronic density in the direct and elegant manner as

$$\rho_N\left(x;t_b-t_a\right)=\frac{N}{Z\left(t_b;t_a\right)}\left(x_b,t_b;x_a,t_a\right)\Big|_{x_b=x_b=x} \tag{4.43}$$

with partition function given as in (4.30), assuring for the correct *N*-representability (as is fundamental in density functional theory (DFT)) constraint:

$$\int \rho_N\left(x;t_b-t_a\right)dx=N \tag{4.44}$$

thus nicely replacing the complicated many-body wave function calculations.

Nevertheless, the path integral formalism is able to provide also *the exact* Schrödinger equation for the wave function, as will be shown in the sequel.

4.2.3 RECOVERING WAVE FUNCTION'S EQUATION

The starting point is the manifested *equivalence* between the path integral propagator and the Green function, with the role in transforming one wave-function registered on one space-time event into other one, either in the future or past quantum evolution. Here, we consider only retarded phenomena,

$$\left(x_2,t_2;x_1,t_1\right)=iG^{+}\left(x_2,t_2;x_1,t_1\right) \tag{4.45}$$

in accordance with the very beginning path integral construction, see the grid (4.1) and the relation (4.2), and consider the so-called quantum Huygens principle of wave-packet propagation (Greiner & Reinhardt, 1994):

$$\psi\left(x_2,t_2\right)=\int\left(x_2,t_2;x_1,t_1\right)\psi\left(x_1,t_1\right)dx_1\ ,\ t_2>t_1 \tag{4.46}$$

Yet, we will employ Eq. (4.46) for an *elementary* propagator, for a quantum evolution as presented in Figure 4.1, thus becoming like:

$$\psi\left(x,t+\varepsilon\right)=A\int\exp\left[\frac{i}{\hbar}\varepsilon L\left(\frac{x+x_0}{2},\frac{x-x_0}{2},t+\frac{\varepsilon}{2}\right)\right]\psi\left(x-\xi,t\right)dx_0 \tag{4.47}$$

where *A* plays the role of the normalization constant in Eq. (4.47) to assure the convergence of the wave function result. Equation (4.47) may be still transformed through employing the geometrical relation:

$$x=x_0+\xi \tag{4.48}$$

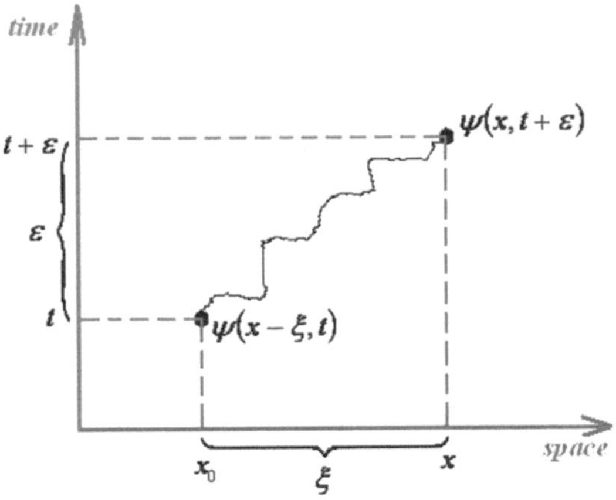

FIGURE 4.1 Depiction of the space-time elementary retarded path connecting two events characterized by their dynamic wave-functions.

to compute the space and velocity averages:

$$\frac{x+x_0}{2}=\frac{2x-\xi}{2}=x-\frac{\xi}{2} \tag{4.49}$$

$$\frac{x-x_0}{\varepsilon}=\frac{\xi}{\varepsilon} \tag{4.50}$$

respectively, while changing the variable

$$dx_0=-d\xi \tag{4.51}$$

to the actual form:

$$\begin{aligned}\psi(x,t+\varepsilon)&=\tilde{A}\int\exp\left\{\frac{i}{\hbar}\varepsilon\left[\begin{matrix}\frac{m}{2}\frac{\xi^2}{\varepsilon^2}\\ -V\left(x-\frac{\xi}{2},t+\frac{\varepsilon}{2}\right)\end{matrix}\right]\right\}\psi(x-\xi,t)d\xi\\ &=\tilde{A}\int\exp\left[\frac{im}{2\hbar\varepsilon}\xi^2\right]\exp\left[-\frac{i}{\hbar}\varepsilon V\left(\begin{matrix}x-\frac{\xi}{2},t\\ +\frac{\varepsilon}{2}\end{matrix}\right)\right]\psi(x-\xi,t)d\xi\end{aligned} \tag{4.52}$$

where Lagrangian was considered with its canonical form, as in Eq. (4.29), and the new constant factor was considered assimilating the minus sign of (4.52).

Next, since noticing the square dependence of ξ in Eq. (4.52) there will be assumed the series expansion in coordinate (ξ) and time (ε) elementary steps restrained to the second and first order, respectively, being the time interval cut-off in accordance with the general (4.12) prescription. Thus we have:

$$\psi(x-\xi,t)\cong\psi(x,t)-\xi\left[\frac{\partial}{\partial x}\psi(x,t)\right]_{\xi\to 0}+\frac{\xi^2}{2}\left[\frac{\partial^2}{\partial x^2}\psi(x,t)\right]_{\xi\to 0} \tag{4.53}$$

$$\psi(x,t+\varepsilon)\cong\psi(x,t)+\varepsilon\left[\frac{\partial}{\partial t}\psi(x,t)\right]_{\varepsilon\to 0} \tag{4.54}$$

$$\exp\left[-\frac{i}{\hbar}\varepsilon V\left(x-\frac{\xi}{2},t+\frac{\varepsilon}{2}\right)\right]\cong 1-\frac{i}{\hbar}\varepsilon V(x,t) \tag{4.55}$$

and the form Eq. (4.52) successively rearranges:

$$\begin{aligned}
&\psi(x,t)+\varepsilon\left[\frac{\partial}{\partial t}\psi(x,t)\right]\\
&=\tilde{A}\int e^{-\frac{m}{2i\hbar\varepsilon}\xi^2}\left[1-\frac{i}{\hbar}\varepsilon V(x,t)\right]\left\{\begin{array}{l}\psi(x,t)\\ -\xi\left[\frac{\partial}{\partial x}\psi(x,t)\right]\\ +\frac{\xi^2}{2}\left[\frac{\partial^2}{\partial x^2}\psi(x,t)\right]\end{array}\right\}d\xi\\
&=\tilde{A}\psi(x,t)\int e^{-\frac{m}{2i\hbar\varepsilon}\xi^2}d\xi-\tilde{A}\left[\frac{\partial}{\partial x}\psi(x,t)\right]\int\xi e^{-\frac{m}{2i\hbar\varepsilon}\xi^2}d\xi\\
&\qquad+\tilde{A}\frac{1}{2}\left[\frac{\partial^2}{\partial x^2}\psi(x,t)\right]\int\xi^2 e^{-\frac{m}{2i\hbar\varepsilon}\xi^2}d\xi\\
&-\tilde{A}\frac{i}{\hbar}\varepsilon V(x,t)\psi(x,t)\int e^{-\frac{m}{2i\hbar\varepsilon}\xi^2}d\xi\\
&+\tilde{A}\frac{i}{\hbar}\varepsilon V(x,t)\left[\frac{\partial}{\partial x}\psi(x,t)\right]\int\xi e^{-\frac{m}{2i\hbar\varepsilon}\xi^2}d\xi
\end{aligned} \tag{4.56}$$

where we have neglected the mixed orders producing a total order beyond maximum two, for example, $\varepsilon\xi^2\cong 0$, and were we arranged the exponentials under integrals such that be of Gaussian type (i.e., employing the identity $-i = 1/i$). Now, the integrals appearing on Eq. (4.56) are of Poisson type of various orders, and solves for notation

$$\frac{m}{2\hbar\varepsilon i} \equiv a \tag{4.57}$$

as:

$$\int e^{-\frac{m}{2i\hbar\varepsilon}\xi^2} d\xi \to \int_{-\infty}^{+\infty} \exp\left(-a\xi^2\right) d\xi = \sqrt{\frac{\pi}{a}} = \sqrt{\frac{2\pi\hbar\varepsilon i}{m}} \tag{4.58}$$

$$\int \xi e^{-\frac{m}{2i\hbar\varepsilon}\xi^2} d\xi \to \int_{-\infty}^{+\infty} \xi \exp\left(-a\xi^2\right) d\xi = 0 \tag{4.59}$$

$$\int \xi^2 e^{-\frac{m}{2i\hbar\varepsilon}\xi^2} d\xi \to \int_{-\infty}^{+\infty} \xi^2 \exp\left(-a\xi^2\right) d\xi = \frac{1}{2a}\sqrt{\frac{\pi}{a}} = \frac{\hbar\varepsilon i}{m}\sqrt{\frac{2\pi\hbar\varepsilon i}{m}} \tag{4.60}$$

With these the expression (4.56) simplifies to:

$$\psi(x,t) + \varepsilon\left[\frac{\partial}{\partial t}\psi(x,t)\right] = \tilde{A}\sqrt{\frac{2\pi\hbar\varepsilon i}{m}}\begin{bmatrix} 1 + \frac{1}{2}\frac{\hbar\varepsilon i}{m}\frac{\partial^2}{\partial x^2} \\ -\frac{i}{\hbar}\varepsilon V(x,t) \end{bmatrix}\psi(x,t) \tag{4.61}$$

which in the limit $\varepsilon \to 0$, common for path integrals, leaves with identity:

$$\psi(x,t) = \lim_{\varepsilon \to 0}\left(\tilde{A}\sqrt{\frac{2\pi\hbar\varepsilon i}{m}}\right)\psi(x,t) \tag{4.62}$$

from where the convergence constant of path integral (4.52) is found

$$\tilde{A}(\varepsilon) = \sqrt{\frac{m}{2\pi\hbar\varepsilon i}} \tag{4.63}$$

with identical form as previously, see Eq. (4.25), thus confirming the consistency of the present approach. Nevertheless, with the constant (4.63) back in Eq. (4.61) we get the equivalent forms:

$$\psi(x,t) + \varepsilon\left[\frac{\partial}{\partial t}\psi(x,t)\right] = \psi(x,t) + \frac{1}{2}\frac{\hbar\varepsilon i}{m}\frac{\partial^2}{\partial x^2}\psi(x,t) - \frac{i}{\hbar}\varepsilon V(x,t)\psi(x,t)$$

$$\Leftrightarrow \frac{\partial}{\partial t}\psi(x,t) = \frac{1}{2}\frac{\hbar i}{m}\frac{\partial^2}{\partial x^2}\psi(x,t) - \frac{i}{\hbar}V(x,t)\psi(x,t)$$

$$\Leftrightarrow i\hbar\frac{\partial}{\partial t}\psi(x,t)=\left[-\frac{1}{2}\frac{\hbar^2}{m}\frac{\partial^2}{\partial x^2}+V(x,t)\right]\psi(x,t) \tag{4.64}$$

being this last one identically recovering the Schrödinger wave function equation.

There was therefore thoroughly proofed that the Feynman path integral is may be reduced to the quantum wave-packet motion while carrying also the information that connects coupled events across the paths' evolution, being in this a general approach of quantum mechanics and statistics.

Next section(s) will deal with presenting practical application/calculation of the path integrals for fundamental quantum systems, from free and harmonic oscillator motion to Bohr and quantum barrier too.

4.3 PATH INTEGRALS FOR BASIC MATTER'S STRUCTURES

4.3.1 *GENERAL PATH INTEGRAL'S PROPERTIES*

There are three fundamental properties most useful for path integral calculations (Dittrich & Reuter, 1994).

1. Firstly, one may combine the two above Schrödinger type information about path integrals: the fact that propagator itself $(x_b,t_b;x_a,t_a)$ obeys the Schrödinger equation, see Eq. (4.42), thus behaving as a sort of wave-function and the fact that Schrödinger equation of the wave-function is recovered by the quantum Huygens principle of wave-packet propagation, see Eq. (4.46). Thus it makes sense to rewrite Eq. (4.46) with the propagator instead of wave-function obtaining the so-called *group property for propagators*:

$$(x_3,t_3;x_1,t_1)=\int(x_3,t_3;x_2,t_2)(x_2,t_2;x_1,t_1)dx_2,\ t_3>t_2>t_1 \tag{4.65}$$

which, nevertheless, may be recursively applied until covering the entire time slicing of the interval $[t_a,t_b]$ as given in (4.1):

$$(x_b,t_b;x_a,t_a)=\int(x_b,t_b;x_n,t_n)(x_n,t_n;x_{n-1},t_{n-1})\cdots(x_1,t_1;x_a,t_a)\prod_{j=1}^{n}dx_j \tag{4.66}$$

while remarking the absence of time intermediate integration.

2. Secondly, from the Huygens principle Eq. (4.46) there is abstracted also the limiting delta Dirac-function for a propagator connecting two space events simultaneously:

$$(x,t_1;x_1,t_1)=\delta(x-x_1) \tag{4.67}$$

as immediately is checked out:

$$\psi(x,t_1)=\int(x,t_1;x_1,t_1)\psi(x_1,t_1)dx_1=\int\delta(x-x_1)\psi(x_1,t_1)dx_1 \tag{4.68}$$

This property is often used as the analytical check once a path integral propagator is calculated for a given system.

3. Thirdly, and perhaps most practically, one would like to be able to solve the path integrals, say with canonical Lagrangian form (4.32), in more direct way than to consider all multiple integrals involved in the measure (4.28).

Hopefully, this is possible working out the quantum fluctuations along the classical path connecting two space-time events. In other words, this is to disturb the classical path $x_{cl}(t)$ by the quantum fluctuations $\delta x(t)$ to obtain the quantum evolution path:

$$x(t)=x_{cl}(t)+\delta x(t) \tag{4.69}$$

and its first time derivation:

$$\dot{x}(t)=\dot{x}_{cl}(t)+\delta\dot{x}(t) \tag{4.70}$$

Very important, note that the quantum fluctuation vanishes at the endpoints of the evolution path since "meeting" with the classical (observed) path, see Figure 4.2:

$$\delta x(t_a)=0=\delta x(t_b) \tag{4.71}$$

being these known as the *Dirichlet boundary conditions*.

With these the purpose is to separate the classical by the quantum fluctuation contributions also in the path integral propagator. Fortunately this is possible for enough large class of potentials, more precisely for quadratic Lagrangian's of general type:

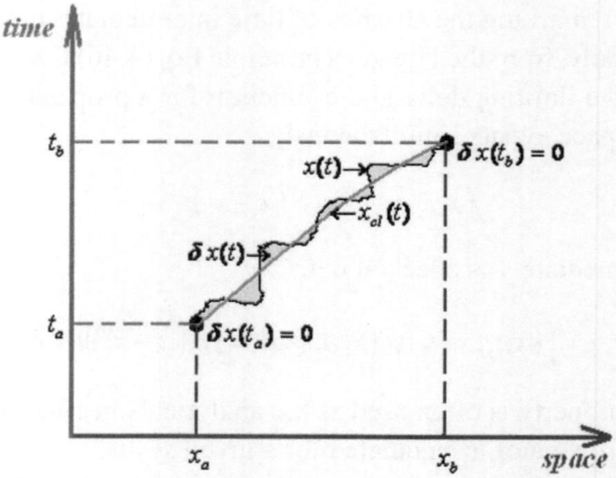

FIGURE 4.2 Illustration of the quantum fluctuations $\delta x(t)$ around the classical path $x_{cl}(t)$ producing the space-time evolution of the Figure 4.1.

$$L(x,\dot{x},t) = \alpha(t)x^2 + \beta(t)x\dot{x} + \gamma(t)\dot{x}^2 + \lambda(t)x + \chi(t)\dot{x} + \sigma(t) \tag{4.72}$$

Expanding the path integral action (4.29) around the classical path requires the expansion of its associate Lagrangian (4.72); we get accordingly:

$$L(x,\dot{x},t) = L(x_{cl},\dot{x}_{cl},t) + \left[\frac{\partial L}{\partial x}\delta x + \frac{\partial L}{\partial \dot{x}}\delta \dot{x}\right]_{x_{cl},\dot{x}_{cl}} + \frac{1}{2}\left[\begin{array}{l}\frac{\partial^2 L}{\partial x^2}\delta x + 2\frac{\partial^2 L}{\partial x \partial \dot{x}}\delta x \delta \dot{x} \\ + \frac{\partial^2 L}{\partial \dot{x}^2}\delta \dot{x}^2\end{array}\right]_{x_{cl},\dot{x}_{cl}} \tag{4.73}$$

$$\begin{aligned} S[x,\dot{x},t] &= \int_{t_a}^{t_b} L(x,\dot{x},t)\,dt \\ &= \int_{t_a}^{t_b} L(x_{cl},\dot{x}_{cl},t)\,dt + \int_{t_a}^{t_b}\left[\frac{\partial L}{\partial x}\delta x + \frac{\partial L}{\partial \dot{x}}\delta \dot{x}\right]_{x_{cl},\dot{x}_{cl}} dt \\ &+ \int_{t_a}^{t_b}\left[\alpha(t)\delta x + \beta(t)\delta x \delta \dot{x} + \gamma(t)\delta \dot{x}^2\right]dt \end{aligned}$$

$$= S_{cl}[x_{cl}, \dot{x}_{cl}, t] + \underbrace{\left.\frac{\partial L}{\partial \dot{x}}\delta x\right|_{t_a}^{t_b}}_{\substack{=0 \\ \delta x(t_a)=0=\delta x(t_b)}} + \int_{t_a}^{t_b} \underbrace{\left(\frac{\partial L}{\partial x} - \frac{d}{dt}\frac{\partial L}{\partial \dot{x}}\right)}_{\substack{=0 \\ Euler-Lagrange}} \delta x dt$$

$$+\int_{t_a}^{t_b}\left[\alpha(t)\delta x + \beta(t) y\delta\dot{x} + \gamma(t)\delta\dot{x}^2\right]dt$$

$$= S_{cl}[x_{cl}, \dot{x}_{cl}, t] + \int_{t_a}^{t_b}\left[\alpha(t)\delta x + \beta(t) y\delta\dot{x} + \gamma(t)\delta\dot{x}^2\right]dt \tag{4.74}$$

With action (4.74) one observes it practically *separates* into the classical and quantum fluctuation contributions; this has two major consequences:

- The classical action goes outside of the path integration simply becoming the multiplication factor $\exp\left[(i/\hbar)S_{cl}\right]$;
- The remaining contribution since depending only on quantum fluctuation $\delta x(t)$ allow in the changing of integration measure:

$$\int_{x(t_a)=x_a}^{x(t_b)=x_b} \mathrm{D}'x(t) \to \int_{\delta x(t_a)=0}^{\delta x(t_b)=0} \underbrace{\left|\frac{\delta x(t)}{\delta(\delta x(t))}\right|}_{\substack{=1 \\ eq.\ (4.69)}} \mathrm{D}'\delta x(t) = \int_{\delta x(t_a)=0}^{\delta x(t_b)=0} \mathrm{D}'\delta x(t) \tag{4.75}$$

In these circumstances the path integral propagator factorizes as:

$$(x_b, t_b; x_a, t_a) = \underbrace{\exp\left\{\frac{i}{\hbar}S_{cl}[x_{cl}, \dot{x}_{cl}, t]\right\}}_{\substack{classical \\ contribution}} \underbrace{\int_{\delta x(t_a)=0}^{\delta x(t_b)=0} \mathrm{D}'\delta x(t) \exp\left\{\frac{i}{\hbar}\int_{t_a}^{t_b}\begin{bmatrix}\alpha(t)\delta x \\ +\beta(t) y\delta\dot{x} \\ +\gamma(t)\delta\dot{x}^2\end{bmatrix}dt\right\}}_{quantum\ \ fluctuations} \tag{4.76}$$

Few conceptual comments are now compulsory based on the path integral form (4.76):

- There is clear that since the quantum fluctuation term does not depend on ending space coordinates but only on their time coordinates, so that in the end will depend only on the time difference $(t_b - t_a)$ since by means of energy conservation all the quantum fluctuation is a time-translation invariant, see for instance the Hamilton-Jacobi Eq. (4.41); therefore it may be further resumed as the *fluctuation factor*:

$$F(t_b - t_a) \equiv \int_{\delta x(t_a)=0}^{\delta x(t_b)=0} \mathcal{D}\delta x(t) \exp\left\{ \frac{i}{\hbar} \int_{t_a}^{t_b} \begin{bmatrix} \alpha(t)\delta x + \beta(t) y \delta \dot{x} \\ +\gamma(t)\delta \dot{x}^2 \end{bmatrix} dt \right\} \tag{4.77}$$

- Looking at the terms appearing in the whole Lagrangian (4.72) and to those present on the factor (4.77) it seems that once the factor (4.77) is known for a given Lagrangian, say L, then the same is characterizing also the modified one with the terms that are not present in the forms (4.77):

$$\tilde{L} = L + \lambda(t)x + \chi(t)\dot{x} + \sigma(t) \tag{4.78}$$

- The resulting working path integral of the propagator now simply reading as:

$$(x_b, t_b; x_a, t_a) = F(t_b - t_a) \exp\left\{ \frac{i}{\hbar} S_{cl}[x_{cl}, \dot{x}_{cl}, t] \right\} \tag{4.79}$$

gives intuitive inside of what path integral formalism of quantum mechanics really does: corrects the classical paths by the *quantum fluctuations viewed as amplitude of the (semi) classical wave*.

Next, the big challenge is to compute the above fluctuation factor (4.79); here there are two possibilities of approach. One is considering the fluctuations as a Fourier series expansion so that directly (although through enough involving procedure) solving the multiple integrals appearing in Eq. (4.77). Yet, this route was that originally proposed by Feynman in his quantum mechanically devoted monograph (Feynman & Hibbs, 1965), while recently refined in an extended textbook (Kleinert, 2004).

The second way is trickier, although with its limitation, but avoids performing the direct integration prescribed by Eq. (4.77), while instructive since computing the quantum fluctuation again in terms of classical path action (Dittrich & Reuter, 1994), however through employing the present first two propagator properties, the group property (4.65) and the delta-Dirac limit (4.67), to the quantum wave (4.79).

As such, combining the stipulated propagator properties, one starts equivalently writing

$$\begin{aligned}\delta(x_b - x_a) &= (x_b, t; x_a, t)\\ &= \int (x_b, t; x, 0)(x, 0; x_a, t)\,dx = \int (x_b, t; x, 0)(x_a, t; x, 0)^* \,dx \end{aligned} \tag{4.80}$$

where the last identity follows by employing the identity between the retarded (+) and advanced (–) Green functions (Greiner & Reinhardt, 1994)

$$G^+(x_b, t_b; x_a, t_a) = \left[G^-(x_a, t_a; x_b, t_b)\right]^* \tag{4.81}$$

combined with the propagator-Green function relationship (4.45), here supplemented with the advanced propagator version:

$$(x_a, t_a; x_b, t_b) = -iG^-(x_a, t_a; x_b, t_b) \tag{4.82}$$

Now, the propagators from Eq. (4.80) may be written immediately from the general form (4.79):

$$(x_b, t; x, 0) = F(t)\exp\left\{\frac{i}{\hbar} S_{cl}(x_b, t; x, 0)\right\} \tag{4.83}$$

$$(x_a, t; x, 0)^* = F^*(t)\exp\left\{-\frac{i}{\hbar} S_{cl}(x_a, t; x, 0)\right\} \tag{4.84}$$

which help in rewrite (4.80) as:

$$\delta(x_b - x_a) = \int dx |F(t)|^2 \exp\left\{\frac{i}{\hbar}\left[S_{cl}(x_b,t;x,0) - S_{cl}(x_a,t;x,0)\right]\right\}$$

$$\underset{\Delta x = x_b - x_a}{\overset{x_b = x_a + \Delta x}{=}} |F(t)|^2 \int dx \exp\left\{\frac{i}{\hbar}\begin{bmatrix} S_{cl}(x_a + \Delta x,t;x,0) \\ -S_{cl}(x_a,t;x,0) \end{bmatrix}\right\}$$

$$= |F(t)|^2 \int dx \exp\left\{\frac{i}{\hbar}\frac{\partial S_{cl}(x_a,t;x,0)}{\partial x_a}\Delta x\right\}$$

$$= |F(t)|^2 \int dx \exp\left\{\frac{i}{\hbar}\frac{\partial S_{cl}(x_a,t;x,0)}{\partial x_a}(x_b - x_a)\right\} \tag{4.85}$$

Next, assuming the notation:

$$s(x) \equiv \frac{\partial S_{cl}(x_a,t;x,0)}{\partial x_a} \tag{4.86}$$

if its derivative $ds(x)/dx$ is independent of x it goes out the integral if the changing in variable may apply on Eq. (4.85) leaving with the identity:

$$\delta(x_b - x_a) = \frac{2\pi\hbar |F(t)|^2}{|ds/dx|} \underbrace{\int \frac{ds}{2\pi\hbar} \exp\left\{\frac{i}{\hbar}(x_b - x_a)s\right\}}_{\delta(x_b - x_a)} \tag{4.87}$$

from where the quantum fluctuation factor follows at once with the analytical general form:

$$F(t) = \sqrt{\frac{1}{2\pi\hbar}\left|\frac{ds}{dx}\right|} \underset{eq.(4.86)}{=} \frac{1}{\sqrt{2\pi\hbar}}\left|\frac{\partial^2 S_{cl}(x_a,t;x,0)}{\partial x \partial x_a}\right|^{1/2} \tag{4.88}$$

With expression (4.88) the propagator (4.79) is fully expressed in terms of classical action as:

$$(x_b,t_b;x_a,t_a) = \frac{1}{\sqrt{2\pi\hbar}}\left|\frac{\partial^2 S_{cl}(x_a,t;x,0)}{\partial x \partial x_a}\right|^{1/2}$$

$$\exp\left\{\frac{i}{\hbar} S_{cl}(x_b,t_b;x_a,t_a)\right\}, \; t = t_b - t_a \tag{4.89}$$

or in the more appealing form:

$$(x_b,t_b;x_a,t_a)=\frac{1}{\sqrt{2\pi\hbar}}\left|\frac{\partial^2 S_{cl}(x_b,t_b;x_a,t_a)}{\partial x_b \partial x_a}\right|^{1/2}\exp\left\{\frac{i}{\hbar}S_{cl}(x_b,t_b;x_a,t_a)\right\} \quad (4.90)$$

usually referred to as the Van Vleck-Pauli-Morette formula, emphasizing on the importance of solving the classical problem for a given canonical Lagrangian (Laidlaw & DeWitt-Morette, 1971; Peak & Inomata, 1969).

However, the path integral solution (4.90) has to be used with two amendments:

- the procedure is valid only when the quantity (4.86), here rewritten in the spirit of Eq. (4.90) as $\partial S_{cl}(x_b,t_b;x_a,t_a)/\partial x_a$, performed respecting one end-point coordinate remains linear in the other space (end-point) coordinate x_b, so that the identity (4.87) holds; this is true for the quadratic Lagrangian's of type (4.72) but not when higher orders are involved, when the previously stipulated Fourier analysis has to be undertaken (one such case will be in foregoing sections presented).
- In the case the formula (4.90) is applicable, i.e., when previous condition applies, the obtained result has to be still verified to recover the delta-Dirac function in the limit:

$$\lim_{t_b\to t_a}(x_b,t_b;x_a,t_a)=\delta(x_b-x_a) \quad (4.91)$$

in accordance with the implemented recipe, see Eq. (4.80); usually this step is providing additional phase correction to the solution (4.90).

The present algorithm is in next exemplified on two paradigmatic quantum problems: the free motion and the motion under harmonic oscillator influence. In each case the set of the classical action will almost solve the entire path integral problem.

4.3.2 PATH INTEGRAL FOR THE FREE PARTICLE

Given a free particle with the Lagrangian

$$L_{(0)}(x,\dot{x},t)=\frac{m}{2}\dot{x}^2 \quad (4.92)$$

it leaves by means of Euler-Lagrange equation

$$\frac{d}{dt}\left(\frac{\partial L}{\partial \dot{x}}\right)=\frac{\partial L}{\partial x} \tag{4.93}$$

to the classical (Newtonian) motion:

$$\ddot{x}_{cl}(t)=0 \tag{4.94}$$

with the obvious solution

$$x_{cl}(t)=x_a+\frac{x_b-x_a}{t_b-t_a}(t-t_a) \tag{4.95}$$

fulfilling the boundary conditions:

$$x_{cl}(t_a)=x_a;\, x_{cl}(t_b)=x_b \tag{4.96}$$

being these endpoints the states where the system is observable, i.e., when quantum fluctuations vanishes, see Eq. (4.71) and Figure 4.2.

Replacing solution (4.95) back in Lagrangian (4.92) the classical action is immediately found:

$$S_{(0)cl}(x_b,t_b;x_a,t_a)=\int_{t_a}^{t_b} L_{(0)}(x_{cl},\dot{x}_{cl},t)\,dt=\frac{m}{2}\int_{t_a}^{t_b}\dot{x}_{cl}^2(t)dt=\frac{m}{2}\frac{(x_b-x_a)^2}{t_b-t_a} \tag{4.97}$$

Next, the quantity (4.86) is firstly evaluated in the spirit of Eq. (4.90) as

$$s_{(0)}(x)=\frac{\partial S_{(0)cl}(x_b,t_b;x_a,t_a)}{\partial x_a}=\frac{m}{t_b-t_a}(x_b-x_a) \tag{4.98}$$

and recognized as linear in the other end-point space coordinate x_b. Thus, the formula (4.90) may be applied, with the actual yield:

$$(x_b,t_b;x_a,t_a)_{(0)}=\sqrt{\frac{m}{2\pi\hbar(t_b-t_a)}}\exp\left\{\frac{i}{\hbar}\frac{m}{2}\frac{(x_b-x_a)^2}{t_b-t_a}\right\} \tag{4.99}$$

Finally, the result of Eq. (4.99) has to be arranged so that to satisfy the limit (4.91) as well. For that we use the delta-Dirac representation:

$$\delta(x_b - x_a) = \frac{1}{\sqrt{\pi}} \lim_{T \to 0} \left\{ \frac{1}{\sqrt{T}} \exp\left[-\frac{(x_b - x_a)^2}{T} \right] \right\} \tag{4.100}$$

Comparison between Eqs. (4.99) and (4.100) leads with identification:

$$\frac{1}{T} = -\frac{im}{2\hbar(t_b - t_a)} = \frac{m}{2i\hbar(t_b - t_a)} \tag{4.101}$$

thus correcting the factor of Eq. (4.99) so that to have the correct limiting path integral solution:

$$(x_b, t_b; x_a, t_a)_{(0)} = \sqrt{\frac{m}{2\pi i\hbar(t_b - t_a)}} \exp\left\{ \frac{i}{\hbar} \frac{m}{2} \frac{(x_b - x_a)^2}{t_b - t_a} \right\} \tag{4.102}$$

Remarkably, this solution is indeed identical with the Green function of the free particle, until the complex factor of Eq. (4.45), this way confirming the reliability of the path integral approach. Moreover, beside of its foreground character in quantum mechanics, the present path integral of the free particle can be further used in the paradigmatic vibrational motion by using the basic rules in using path integrals propagator for density computations:

- The reliable application of the density computation upon the partition function algorithm, see Eqs. (4.43) and (4.44), prescribes the transformation of the obtained quantum result in the quantum statistical counterpart by means of Wick transformation (4.33) supplemented by the trigonometric to hyperbolic functions conversions;
- In computation of the path integral propagator the workable *Van Vleck-Pauli-Morette formula* looks like

$$(x_b, t_b; x_a, t_a) = \sqrt{\frac{1}{2\pi i\hbar} \left| \frac{\partial^2 S_{cl}(x_b, t_b; x_a, t_a)}{\partial x_b \partial x_a} \right|} \exp\left\{ \frac{i}{\hbar} S_{cl}(x_b, t_b; x_a, t_a) \right\} \tag{4.103}$$

with the complex factor "*i*" included, as confirmed by both the free and harmonic oscillator quantum motions, for being used for classical actions linear upon derivation respecting one of the end-point space coordinates in the other one; yet the formula (4.103) should be always checked for fulfilling the limiting (4.91) delta-Dirac function for simultaneous events for any applied potential.

Beside of these working rules, regaining the energy quantification of free electrons in solid state (motion within the infinite high box) as well as the obligatory Bohr quantification for the continuous deformation of the path on the circle (Kleinert, 2004; Dittrich & Reuter, 1994), will complete the 3-fold fundamental types of quantum evolution (i.e., covering the rotation, vibration and translations for motion in atomic quantified circles, in molecule and along the solid state bands, respectively) as loaded and reloaded from various perspectives through this volume, see the preceding chapters.

4.3.3 PATH INTEGRAL FOR MOTION AS THE HARMONIC OSCILLATOR

The characteristic Lagrangian of the harmonic oscillator,

$$L_{(\omega)}\left(x,\dot{x},t\right)=\frac{m}{2}\dot{x}^2-\frac{m}{2}\omega^2x^2 \tag{4.104}$$

provides, when considered in the Euler-Lagrange equation (4.93), the classical equation of motion:

$$\ddot{x}_{cl}(t)+\omega^2x_{cl}(t)=0 \tag{4.105}$$

with the well known solution

$$x_{cl}(t)=C\sin\left(\omega t+\varphi\right) \tag{4.106}$$

specialized on the end-point events of motion as:

$$x_a=x_{cl}\left(t_a\right)=C\sin\left(\omega t_a+\varphi\right) \tag{4.107}$$

$$x_b = x_{cl}(t_b) = C\sin(\omega t_b + \varphi) \tag{4.108}$$

In the same way as done for free motion, see solution (4.95), worth rewritten the actual classical solution (4.106) in terms of relations (4.107) and (4.108), for instance as

$$\begin{aligned} x_{cl}(t) &= C\sin\left[\omega(t-t_a)+(\omega t_a+\varphi)\right] \\ &= \underbrace{C\cos(\omega t_a+\varphi)}_{\dot{x}_a/\omega}\sin\left[\omega(t-t_a)\right] \\ &\quad + \underbrace{C\sin(\omega t_a+\varphi)}_{x_a}\cos\left[\omega(t-t_a)\right]; \end{aligned}$$

$$x_{cl}(t) = \frac{1}{\omega}\dot{x}_a\sin\left[\omega(t-t_a)\right] + x_a\cos\left[\omega(t-t_a)\right] \tag{4.109}$$

or similarly as:

$$x_{cl}(t) = \frac{1}{\omega}\dot{x}_b\sin\left[\omega(t-t_b)\right] + x_b\cos\left[\omega(t-t_b)\right] \tag{4.110}$$

On the other side the classical action of the Lagrangian (4.104) looks like:

$$\begin{aligned} S_{(\omega)cl}(x_b,t_b;x_a,t_a) &= \int_{t_a}^{t_b} L_{(\omega)}(x_{cl},\dot{x}_{cl},t)\,dt \\ &= \frac{m}{2}\int_{t_a}^{t_b}\left[\left(\frac{dx_{cl}}{dt}\right)\left(\frac{dx_{cl}}{dt}\right) - \omega^2 x_{cl}^2\right]dt \\ &= \frac{m}{2}\int_{t_a}^{t_b}\left[\frac{d}{dt}(x_{cl}\dot{x}_{cl}) - x_{cl}\ddot{x}_{cl} - \omega^2 x_{cl}^2\right]dt \\ &= \left[\frac{m}{2}x_{cl}(t)\dot{x}_{cl}(t)\right]_{t_a}^{t_b} - \frac{m}{2}\int_{t_a}^{t_b}\underbrace{\left(\ddot{x}_{cl}+\omega^2 x_{cl}\right)}_{\substack{=0 \\ eq.\,(4.105)}}x_{cl}\,dt \\ &= \frac{m}{2}\left[x_b(t_b)\dot{x}_b(t_b) - x_a(t_a)\dot{x}_a(t_a)\right] \end{aligned} \tag{4.111}$$

Now, in order having classical action in terms of only space-time coordinate of the ending points, one has to replace the end-point velocities in Eq. (4.111) with the aid of relations (4.109) and (4.110) in which the current time is taken as the $t = t_b$ and $t = t_a$, respectively; thus we firstly get:

$$\dot{x}_a = \frac{\omega}{\sin\left[\omega(t_b - t_a)\right]}\left\{x_b - x_a \cos\left[\omega(t_b - t_a)\right]\right\} \tag{4.112}$$

$$\dot{x}_b = \frac{\omega}{\sin\left[\omega(t_b - t_a)\right]}\left\{-x_a + x_b \cos\left[\omega(t_b - t_a)\right]\right\} \tag{4.113}$$

then we form the needed products:

$$x_b\dot{x}_b = \frac{\omega}{\sin\left[\omega(t_b - t_a)\right]}\left\{x_b^2 \cos\left[\omega(t_b - t_a)\right] - x_a x_b\right\} \tag{4.114}$$

$$x_a\dot{x}_a = \frac{\omega}{\sin\left[\omega(t_b - t_a)\right]}\left\{-x_a^2 \cos\left[\omega(t_b - t_a)\right] + x_a x_b\right\} \tag{4.115}$$

to finally replace them in expression (4.111) to provide the computed classical action:

$$S_{(\omega)cl}\left(x_b, t_b; x_a, t_a\right) = \frac{m\omega}{2\sin\left[\omega(t_b - t_a)\right]}\left\{\begin{array}{l}\left(x_a^2 + x_b^2\right) \\ \cos\left[\omega(t_b - t_a)\right] - 2x_a x_b\end{array}\right\} \tag{4.116}$$

Note that the correctness of Eq. (4.116) may be also checked by imposing the limit $\omega \to 0$ in which case the previous free motion has to be recovered; indeed by employing the consecrated limit

$$\lim_{y \to 0} \frac{\sin y}{y} = 1 \tag{4.117}$$

one immediately get:

$$\lim_{\omega \to 0} S_{(\omega)cl}\left(x_b, t_b; x_a, t_a\right)$$

$$=\frac{m}{2(t_b-t_a)}\underbrace{\lim_{\omega\to 0}\frac{\omega(t_b-t_a)}{\sin[\omega(t_b-t_a)]}}_{=1}\underbrace{\lim_{\omega\to 0}\begin{Bmatrix}(x_a^2+x_b^2)\cos[\omega(t_b-t_a)]\\-2x_a x_b\end{Bmatrix}}_{(x_b-x_a)^2}$$

$$=\frac{m}{2}\frac{(x_b-x_a)^2}{t_b-t_a}$$

$$=S_{(0)cl}(x_b,t_b;x_a,t_a) \tag{4.118}$$

Such kind of check is most useful and has to hold also for the quantum propagator as a whole. Going to determine it one has to reconsider the classical action (4.116) so that the quantity (4.86) is directly evaluated in the spirit of Eq. (4.90) as:

$$\frac{\partial S_{(\omega)cl}(x_b,t_b;x_a,t_a)}{\partial x_a}=\frac{m\omega}{\sin[\omega(t_b-t_a)]}\{x_a\cos[\omega(t_b-t_a)]-x_b\} \tag{4.119}$$

thus again encountering it as being linear in the other end-point coordinate x_b, being this the fortunate situation in which the previous section algorithm for path integral computation may be applied though the expression (4.90), here manifested with the harmonic oscillator result:

$$(x_b,t_b;x_a,t_a)=\sqrt{\frac{m\omega}{2\pi\hbar\sin[\omega(t_b-t_a)]}}$$
$$\exp\left\{\frac{i}{\hbar}\frac{m\omega}{2\sin[\omega(t_b-t_a)]}\begin{Bmatrix}(x_a^2+x_b^2)\\\cos[\omega(t_b-t_a)]-2x_a x_b\end{Bmatrix}\right\} \tag{4.120}$$

Yet, as above was the case for the classical action itself, also the pre-exponential quantum fluctuation factor of Eq. (4.120) has to overlap with that appearing in the path integral of free motion of Eq. (4.102) under the limit $\omega\to 0$:

$$\lim_{\omega\to 0}\sqrt{\frac{m\omega}{2\pi\hbar\sin[\omega(t_b-t_a)]}}\underset{eq.(4.117)}{=}\sqrt{\frac{m}{2\pi\hbar(t_b-t_a)}} \tag{4.121}$$

Thus we have to have the propagator (4.120) with the exponential pre-factor gaining an appropriate placed complex factor "*i*:

$$(x_b,t_b;x_a,t_a)_{(\omega)}=\sqrt{\frac{m\omega}{2\pi i\hbar\sin\left[\omega(t_b-t_a)\right]}}\exp\left\{\begin{array}{l}\frac{i}{\hbar}\frac{m\omega}{2\sin\left[\omega(t_b-t_a)\right]}\\ \times\left\{\left(x_a^2+x_b^2\right)\cos\left[\omega(t_b-t_a)\right]-2x_ax_b\right\}\end{array}\right\} \tag{4.122}$$

This is the searched propagator of the (electronic) motion under the harmonic oscillating potential, computed by means of path integral; it provides the canonical density to be implemented in the DFT algorithm (4.43) and (4.44):

$$\rho_{\otimes}(x,t_b-t_a)=(x,t_b;x,t_a)_{(\omega)}=\sqrt{\frac{m\omega}{2\pi i\hbar\sin\left[\omega(t_b-t_a)\right]}}\times\exp\left\{\frac{i}{\hbar}\frac{m\omega\left(\cos\left[\omega(t_b-t_a)\right]-1\right)}{\sin\left[\omega(t_b-t_a)\right]}x^2\right\} \tag{4.123}$$

Yet, for practical implementations, the passage from quantum mechanics (QM) to quantum statistics (QS) is to be considered based on the Wick transformation (4.33) here rewritten as:

$$(t_b-t_a)\to -i\hbar\beta\equiv -i(\tau_b-\tau_a) \tag{4.124}$$

providing the Euler based trigonometric to hyperbolic function conversions (by analytic continuations):

$$\sin\omega(t_b-t_a)=\frac{1}{i}\frac{e^{i\omega(t_b-t_a)}-e^{-i\omega(t_b-t_a)}}{2}\to\frac{1}{i}\frac{e^{\omega(\tau_b-\tau_a)}-e^{-i\omega(\tau_b-\tau_a)}}{2}$$
$$=\frac{1}{i}\sinh\omega(\tau_b-\tau_a) \tag{4.125}$$

$$\cos\omega(t_b - t_a) = \frac{e^{i\omega(t_b - t_a)} + e^{-i\omega(t_b - t_a)}}{2} \to \frac{e^{\omega(\tau_b - \tau_a)} + e^{-i\omega(\tau_b - \tau_a)}}{2}$$
$$= \cosh\omega(\tau_b - \tau_a) \tag{4.126}$$

allowing for density (4.123) the counterpart formulation:

$$\rho_{\otimes}(x, \tau_b - \tau_a) = \sqrt{\frac{m\omega}{2\pi\hbar\sinh\left[\omega(\tau_b - \tau_a)\right]}}$$
$$\exp\left\{-\frac{1}{\hbar}\frac{m\omega\left(\cosh\left[\omega(\tau_b - \tau_a)\right] - 1\right)}{\sinh\left[\omega(\tau_b - \tau_a)\right]}x^2\right\} \tag{4.127}$$

The uni-particle (electronic) density (4.127) is then used for computing the harmonic oscillator partition function:

$$Z_{\omega} = \int_{-\infty}^{+\infty} \rho_{\otimes}(x, \tau_b - \tau_a)dx$$
$$= \sqrt{\frac{m\omega}{2\pi\hbar\sinh\left[\omega(\tau_b - \tau_a)\right]}}$$
$$\int_{-\infty}^{-\infty} \exp\left\{-\frac{1}{\hbar}\frac{m\omega\left(\cosh\left[\omega(\tau_b - \tau_a)\right] - 1\right)}{\sinh\left[\omega(\tau_b - \tau_a)\right]}x^2\right\}dx$$
$$= \sqrt{\frac{m\omega}{2\pi\hbar\sinh\left[\omega(\tau_b - \tau_a)\right]}}\sqrt{\frac{\pi\hbar\sinh\left[\omega(\tau_b - \tau_a)\right]}{m\omega\left(\cosh\left[\omega(\tau_b - \tau_a)\right] - 1\right)}}$$
$$= \sqrt{\frac{1}{2\left(\cosh\left[\omega(\tau_b - \tau_a)\right] - 1\right)}} \tag{4.128}$$

Now using the "double angle" formula:

$$\cosh 2y = \cosh^2 y + \sinh^2 y = 2\cosh^2 y - 1 = 2\sinh^2 y + 1 \tag{4.129}$$

partition function (4.128) further becomes:

$$Z_\omega = \frac{1}{2\sinh\left[\dfrac{\omega(\tau_b - \tau_a)}{2}\right]} \underset{eq.(4.124)}{=} \frac{1}{2\sinh\left[\dfrac{\omega\hbar\beta}{2}\right]} \tag{4.130}$$

Remarkably, the result (4.130) recovers also the energy quantification of the quantum motion under the harmonic oscillator influence, through the successive transformations:

$$\begin{aligned}
Z_\omega &= \frac{1}{\exp(\omega\hbar\beta/2) - \exp(-\omega\hbar\beta/2)} \\
&= \exp(-\omega\hbar\beta/2)\frac{1}{1-\exp(-\omega\hbar\beta)} \\
&= \exp(-\omega\hbar\beta/2)\sum_{n=0}^{\infty}\left[\exp(-\omega\hbar\beta)\right]^n \\
&= \exp(-\omega\hbar\beta/2)\sum_{n=0}^{\infty}\left[\exp(-n\omega\hbar\beta)\right] \\
&= \sum_{n=0}^{\infty}\exp\left[-\beta\hbar\omega\left(n+\frac{1}{2}\right)\right]
\end{aligned} \tag{4.131}$$

Comparing the expression (4.131) with the canonical formulation of the partition function

$$Z \equiv \sum_{n=0}^{\infty}\exp[-\beta E_n] \tag{4.132}$$

there follows immediately the harmonic oscillator energy quantification:

$$E_n(\omega) = \hbar\omega\left(n+\frac{1}{2}\right) \tag{4.133}$$

in perfect agreement with the consecrated expression, see Eqs. (2.525), (3.149), (3.231), and (3.280).

4.3.4 PATH INTEGRAL REPRESENTATION FOR THE BOHR'S ATOM

Consider a particle in a circular closed path motion (i.e., an orbit continuously deformed into a circle), with the specific parameters:

- the fixed (not necessarily observed) end-points $\varphi = 0$ & $\varphi = 2\pi$;
- the circular length parameter $s_{a...b}$ along the circle length L.

This circular motion, being along a simple connected line (the circle) can be projected on the free-particle (line) motion (on real space $\Re$) over which specific constraints are imposed to regain the circular path and motion towards the present path integral model of the Bohr's atom. Therefore, one re-considers the free-motion propagator of Eq. (4.102)

$$\left(x_b^{(n)}, t_b; x_a, t_a\right)_{(0,\Re)} = \sqrt{\frac{m}{2\pi i\hbar\left(t_b - t_a\right)}} \exp\left\{\frac{i}{\hbar}\frac{m}{2}\frac{\left(x_b^{(n)} - x_a\right)^2}{t_b - t_a}\right\} \tag{4.134}$$

by assuming that while going from x_a to $x_b^{(n)}$ the particle covered the entire circular orbit as much as n-times the so that adding nL to the initial circular parameter on ring, that writes:

$$\begin{cases} x_b^{(n)} - x_a = s_b - s_a + nL \equiv \Delta s + nL \\ t_b - t_a \equiv \Delta t \end{cases} \tag{4.135}$$

Such that the evolution amplitude (the propagator) (4.134) acquires the atomic circular orbit information under the working form

$$\left(s_b, t_b; s_a, t_a\right)_L = \sqrt{\frac{m}{2\pi i\hbar\Delta t}} \exp\left\{\frac{i}{\hbar}\frac{m}{2}\frac{\left(\Delta s + nL\right)^2}{\Delta t}\right\} \tag{4.136}$$

At this point one would seek to unfold this information in a sum of orbit paths in order to correctly describe the particle motion within atomic circular orbits; to this aim, one would benefit from turning back the exponential form of Eq. (4.136) into a path integral; this can be achieved in two steps:

- Firstly, the Poisson formula (4.26) is employed to provide the extended useful integral formulation

$$\int_{-\infty}^{+\infty} e^{-\left(ay^2+2by+c\right)}dy = e^{\frac{b^2-ac}{a}}\int_{-\infty}^{+\infty} e^{-\frac{(ay+b)^2}{a}}dy = \frac{1}{\sqrt{a}}e^{\frac{b^2-ac}{a}}\int_{-\infty}^{+\infty} e^{-x^2}dx = \sqrt{\frac{\pi}{a}}e^{\frac{b^2-ac}{a}} \quad (4.137)$$

which is adapted for the present purpose ($c = 0$) to the identity

$$\frac{1}{\sqrt{a\pi}}e^{\frac{b^2}{a}} = \frac{1}{\pi}\int_{-\infty}^{+\infty} e^{-\left(ap^2+2bp\right)}dp \quad (4.138)$$

of which l.h.s. may be recognized in Eq. (4.136) rearranged as

$$\left(s_b,t_b;s_a,t_a\right)_L = \sqrt{\frac{m}{8\hbar}}\frac{1}{\sqrt{\pi}}\sqrt{\frac{4}{i\Delta t}}\exp\left\{\frac{4}{i\Delta t}\left[\frac{1}{2i}\sqrt{\frac{m}{2\hbar}}\left(\Delta s+nL\right)\right]^2\right\} \quad (4.139)$$

This way the circular propagator may be considered as the path integral

$$G\left(\Delta s,\Delta t\right)_L = \frac{1}{\pi}\sqrt{\frac{m}{8\hbar}}\int \exp\left\{-i\frac{\Delta t}{4}p^2 + ip\sqrt{\frac{m}{2\hbar}}\left(\Delta s+nL\right)\right\}dp \quad (4.140)$$

Which can be further rewritten by changing the integrand in order to achieve the non-dimensional quantities under exponential, by the transformation

$$p \to p\sqrt{\frac{2}{m\hbar}} \quad (4.141)$$

leading with the working propagator

$$G\left(\Delta s,\Delta t\right)_L = \frac{1}{2\pi\hbar}\int \exp\left\{-i\frac{\Delta t}{2m\hbar}p^2 + \frac{i}{\hbar}p\left(\Delta s+nL\right)\right\}dp \quad (4.142)$$

- The second step is considering in Eq. (4.142) of the complete orbits as factorization contribution

$$G(\Delta s, \Delta t)_L = \frac{1}{2\pi\hbar}\int\left[e^{\frac{i}{\hbar}pnL}\right]e^{-i\frac{\Delta t}{2m\hbar}p^2+\frac{i}{\hbar}p\Delta s}dp \tag{4.143}$$

And to consider on this last expression all possible paths (n-th order) by which the final circular position on orbit is reached; equivalently, this means summing up over all such possibilities (like in counting states within the partition functions) to yield

$$G(\Delta s, \Delta t)_L = \frac{1}{2\pi\hbar}\int\left[\sum_{n=-\infty}^{+\infty} e^{\frac{i}{\hbar}pnL}\right]e^{-i\frac{\Delta t}{2m\hbar}p^2+\frac{i}{\hbar}p\Delta s}dp \tag{4.144}$$

Now we are in position to recognize the Poisson summation (the comb) formula (see Appendix A.1.2) linking the exponential fluctuations with the delta Dirac point-contributions

$$\sum_{n=-\infty}^{+\infty} e^{i2\pi nx} = \sum_{n=-\infty}^{+\infty} \delta(x-n) \tag{4.145}$$

Specifically, in the case of Eq. (4.144), one has in the first instance the transformation

$$\sum_{n=-\infty}^{+\infty} e^{\frac{i}{\hbar}pnL} = \sum_{n=-\infty}^{+\infty} e^{i2\pi n\frac{pL}{2\pi\hbar}} = \sum_{n=-\infty}^{+\infty} \delta\left(\frac{pL}{2\pi\hbar}-n\right) = \frac{2\pi\hbar}{L}\sum_{n=-\infty}^{+\infty} \delta\left(p-\frac{2\pi\hbar}{L}n\right) \tag{4.146}$$

where the Dirac factorization property was considered

$$\delta\left[a(X-Y)\right] = \frac{1}{|a|}\delta(X-Y) \tag{4.147}$$

One has therefore the actual propagator of the Bohr's atom

$$G(\Delta s, \Delta t)_L = \frac{1}{L}\int\sum_{n=-\infty}^{+\infty} \delta\left(p-\frac{2\pi\hbar}{L}n\right)e^{-i\frac{\Delta t}{2m\hbar}p^2+\frac{i}{\hbar}p\Delta s}dp \tag{4.148}$$

Which allows further solution my means of the filtration property of delta-Dirac function, see Eq. (2.11), to get the almost final result, namely:

$$G(\Delta s,\Delta t)_L = \frac{1}{L}\sum_{n=-\infty}^{+\infty} \exp\left\{-i\frac{\Delta t}{2m\hbar}\left(\frac{2\pi\hbar}{L}n\right)^2 + i2\pi n\frac{\Delta s}{L}\right\} \quad (4.149)$$

The final results will be displayed in terms of the natural parameter for the circular motion, i.e., by the angle φ: for a radius R of the given orbit, one readily can complete the list of Eq. (4.135) with the actual one including the inertia momentum (I):

$$\begin{cases} L = 2\pi R \\ \Delta s = R\left(\varphi_b^{(n)} - \varphi_a\right) \overset{!}{=} R\varphi \\ p = mR\dot{\varphi} \\ t_b - t_a \equiv \Delta t \\ I = mR^2 \end{cases} \quad (4.150)$$

Now we arrive to the quantification conclusions:

- The delta Dirac function in Eq. (4.146) gives the momentum quantification

$$p = \frac{2\pi\hbar}{L}n \quad (4.151)$$

Which is quite equivalent with the phenomenological deduction in Eq. (1.79) when considering also length-radius connection from (4.150); moreover, with the present approach also the kinetic momentum is found quantified since:

$$n\hbar = \frac{pL}{2\pi} = \frac{(mR\dot{\varphi})(2\pi R)}{2\pi} = mR^2\dot{\varphi} = L_z \quad (4.152)$$

thus completing the Bohr's quantification with the so-called "first integral" of motion, characteristic to the orbital motion itself, apart of the de Broglie closure quantification on (whatever) orbit (4.151).

Nevertheless, the total energy on orbit is directly quantified by the kinetic energy information absorbing all potential information (by circular motion and closing paths), i.e., directly writing as:

$$E_n = \frac{p^2}{2m} = \frac{\hbar^2}{2mR^2}n^2 = \frac{\hbar^2}{2I}n^2 \tag{4.153}$$

certainly in accordance with the phenomenological expression obtained from optimum quantities in Eq. (1.82), apart of the minus sign (indicating the binding energy nature).

Eventually, the atomic Bohr's propagator (4.149) finally looks like

$$G(R\varphi, \Delta t) = \frac{1}{2\pi R}\sum_{n=-\infty}^{+\infty} \exp\left\{-i\frac{\Delta t}{2I}\hbar n^2 + in\varphi\right\} \tag{4.154}$$

Or under the non-dimensional form:

$$G(\varphi, \Delta t) = RG(R\varphi, \Delta t) = \frac{1}{2\pi}\sum_{n=-\infty}^{+\infty} \exp\left\{-i\frac{\Delta t}{2I}\hbar n^2 + in\varphi\right\} \tag{4.155}$$

such that the evolution amplitude of the propagator is shaped as a trigonometric normalized output itself. Worth noting that accommodating the Bohr's atom and quantification within path integral formalism gives it strength and elegancy, while offering enough consistency and richness in quantum information in order to have preeminence in treating complex quantum structure and interactions.

4.3.5 PATH INTEGRAL FOR MOTION IN THE QUANTUM WELL

The particle in a quantum well of potential

$$V(x) = \begin{cases} 0 & \ldots\ 0 < x < L \\ \infty & \ldots\ x \le 0\ \&\ x \ge L \end{cases} \tag{4.156}$$

may be described as a combination between:

- the free particle propagation due to the translational movement inside the walls
- the Bohr's atom movement due to the closing paths by bouncing off the walls

However, there are also two characteristics, due to the collisions and the turning points along the path on the walls, namely:

- starting the initial point (x_a, t_a) the final point may be reached either from the direct path (x_b, t_b) as well as from the opposite direction of path $(-x_b, t_b)$; therefore, the propagator of the particle inside the quantum well should be regarded as a superposition of two free particle contributions

$$(x_b, t_b; x_a, t_a)_{well} = \sqrt{\frac{m}{2\pi i\hbar(t_b - t_a)}} \left\{ \begin{array}{l} \exp\left[\dfrac{i}{\hbar}\dfrac{m}{2}\dfrac{(x_b - x_a)^2}{t_b - t_a}\right] \\ -\exp\left[\dfrac{i}{\hbar}\dfrac{m}{2}\dfrac{(x_b - x_a)^2}{t_b - t_a}\right] \end{array} \right\} \quad (4.157)$$

- consequently the space-time information in Eq. (4.135) is now modified such that, due to the wall turning points, the traveled space is *doubled* by the forward and backward movements such that the for the r-th trip between walls we will have

$$\begin{cases} x_b - x_a = x_f - x_i + 2rL \\ t_b - t_a \equiv \Delta t \end{cases} \quad (4.158)$$

Combining these information, as previously for the Bohr's atom, with the sum over infinite histories with the same scenario and output, while performing for each of the exponentials of Eq. (4.157) the same analytical transformation by means of Eq. (4.138) as previously done for atomic circular motion, we adapt the result (4.142) to the present quantum well situation to firstly get

$$G\left(x_f;x_i,\Delta t\right)_L = \frac{1}{2\pi\hbar}\sum_{r=-\infty}^{\infty}\int\left[\begin{array}{l}\exp\left\{-i\frac{\Delta t}{2m\hbar}p^2+\frac{i}{\hbar}p\left(x_f-x_i+2rL\right)\right\}\\ -\exp\left\{-i\frac{\Delta t}{2m\hbar}p^2+\frac{i}{\hbar}p\left(-x_f-x_i+2rL\right)\right\}\end{array}\right]dp \tag{4.159}$$

from where one may separate the final point contribution and apply the sum contribution over paths on the specific changing phase (by $2Lp/\hbar$) term

$$G\left(x_f;x_i,\Delta t\right)_L = \frac{1}{2\pi\hbar}\int e^{-i\frac{\Delta t}{2m\hbar}p^2}e^{-\frac{i}{\hbar}px_i}\left\{\sum_{r=-\infty}^{+\infty}e^{2\frac{i}{\hbar}rLp}\right\}2i\underbrace{\left\{\frac{e^{\frac{i}{\hbar}px_f}-e^{-\frac{i}{\hbar}px_f}}{2i}\right\}}_{\sin\left[px_f/\hbar\right]}dp \tag{4.160}$$

while now, one recognizes, as before, the Poisson –comb function on which the similar transformation as in Eq. (4.146) holds

$$\begin{aligned}\sum_{r=-\infty}^{+\infty}e^{2\frac{i}{\hbar}prL} &= \sum_{r=-\infty}^{+\infty}e^{i2\pi r\frac{pL}{\pi\hbar}} = \sum_{n=-\infty}^{+\infty}\delta\left(\frac{pL}{\pi\hbar}-n\right)\\ &= \frac{\pi\hbar}{L}\sum_{n=-\infty}^{+\infty}\delta\left(p-\frac{\pi\hbar}{L}n\right)\end{aligned} \tag{4.161}$$

This way, the propagator of the particle within the quantum well further writes from Eq. (4.160) under delta-Dirac form

$$G\left(x_f;x_i,\Delta t\right)_L = \frac{i}{L}\int e^{-i\frac{\Delta t}{2m\hbar}p^2}e^{-\frac{i}{\hbar}px_i}\sin\left[\frac{px_f}{\hbar}\right]\sum_{n=-\infty}^{+\infty}\delta\left(p-\frac{\pi\hbar}{L}n\right)dp \tag{4.162}$$

Yielding upon performing the integration by the filtration property of delta-Dirac function, see Eq. (2.11), to leave with the Green function result

$$\begin{aligned}G\left(x_f;x_i,\Delta t\right)_L &= \frac{i}{L}\sum_{n=-\infty}^{+\infty}\exp\left\{-\frac{i}{\hbar}E_n\Delta t\right\}\\ &\times\exp\left\{-ik_nx_i\right\}\sin\left[k_nx_f\right]\end{aligned} \tag{4.163}$$

Where we have recorded the energy and wave vector quantifications, respectively

$$\begin{cases} E_n = \dfrac{p_n^2}{2m} = \dfrac{1}{2m}\dfrac{\pi^2\hbar^2}{L^2}n^2 \\ k_n = \dfrac{p_n}{\hbar} = \dfrac{\pi n}{L} \end{cases} \tag{4.164}$$

as driven by the delta-Dirac appearance in Eq. (4.161)

$$p_n = \frac{\pi n}{L}\hbar \tag{4.165}$$

Worth remarking that the energy quantification (4.164) for the particle trapped in the quantum wall exactly matches the earlier results (3.190a), (3.258), and (3.296), thus affirming once more the correctness of the path integral approach, while being richer in history of quantum paths' contribution.

Finally, one notes that with the quantifications in Eq. (4.164) we have also the level-properties:

$$\begin{cases} E_{-n} = E_n \\ E_{n=0} = 0 \\ k_{-n} = -k_n \end{cases} \tag{4.166}$$

With which the propagator (4.163) may be further evaluated by playing with the sum over states, i.e., by excluding the zero-state (see also the ground state solid-state paradox exposed in Section 3.5.3), while considering the remaining terms grouped under two anti-symmetrical sums:

$$\sum_{n=-\infty}^{+\infty} \bullet = \sum_{n=-\infty}^{-1} \bullet + 0 + \sum_{n=+1}^{+\infty} \bullet = \sum_{n=-1}^{+\infty}\left([\bullet]_{+n} - [\bullet]_{-n}\right) \tag{4.167}$$

to finally yield

$$G\left(x_f; x_i, \Delta t\right)_L = \frac{i}{L}\sum_{n=1}^{+\infty} \exp\left\{-\frac{i}{\hbar}E_n \Delta t\right\} 2i \underbrace{\frac{\exp\{-ik_n x_i\} - \exp\{ik_n x_i\}}{2i}}_{-\sin[px_i/\hbar]} \sin\left[k_n x_f\right]$$

$$= \frac{2}{L} \sum_{n=1}^{+\infty} \exp\left\{ -\frac{i}{\hbar} E_n \Delta t \right\} \sin[k_n x_i] \sin[k_n x_f] \tag{4.168}$$

This way completing the earlier wave-function information, see for instance Eq. (3.296) or Eq. (3.596), with the actual evolution amplitude (4.168) of the quantum propagator (Green function) for electrons in the valence band of solids.

The following approach will show how the already proved quite reliable approach of path integrals is naturally needed within the Dirac formalism of quantum mechanics applied on many-particle systems, specific to chemical structures formed by many-electrons in valence state, by means of the celebrated density matrix formalism – from where there is just a step to the "observable" density functional theory of many-body systems.

4.4 DENSITY MATRIX APPROACH LINKING PATH INTEGRAL FORMALISM

4.4.1 *ON MONO-, MANY-, AND REDUCED-ELECTRONIC DENSITY MATRICES*

Given a spectral representation $\{|n\rangle\}_{n\in N}$ for a set of quantum mono-electronic states,

$$|\varphi_k\rangle = \sum_n c_{kn} |n\rangle \tag{4.169}$$

one may employ its closure relation

$$\hat{1} = \sum_n |n\rangle\langle n| \tag{4.170}$$

to generally express the average of an observable (i.e., the operator $\hat{A}$) on a selected state as:

$$\left\langle \hat{A} \right\rangle_k = \frac{\langle \varphi_k | \hat{A} | \varphi_k \rangle}{\langle \varphi_k | \varphi_k \rangle} = \frac{\sum_{n,n'} \langle \varphi_k | n' \rangle \langle n' | \hat{A} | n \rangle \langle n | \varphi_k \rangle}{\sum_n \langle \varphi_k | n \rangle \langle n | \varphi_k \rangle} = \frac{\sum_{n,n'} c_{kn} c_{kn'}^* \langle n' | \hat{A} | n \rangle}{\sum_n |c_{kn}|^2} \tag{4.171}$$

while for the observable average over the entire sample the individual weight w_k should counted to provide the statistical result:

$$\left\langle \hat{A} \right\rangle = \frac{\sum_k w_k \left\langle \hat{A} \right\rangle_k}{\sum_k w_k} \tag{4.172}$$

When rewrite the global average in similar formal way as the selected k-average, actually in terms of it:

$$\left\langle \hat{A} \right\rangle = \frac{\sum_{n,n'} \langle n|\hat{\rho}|n'\rangle \langle n'|\hat{A}|n\rangle}{\sum_k w_k} \tag{4.173}$$

we introduced in fact the *density matrix elements*:

$$\langle n|\hat{\rho}|n'\rangle = \sum_k w_k \frac{c_{kn} c_{kn'}^*}{\sum_n |c_{kn}|^2} \tag{4.174}$$

which provides the *density operator*:

$$\hat{\rho} = \sum_{n,n'} |n\rangle\langle n|\hat{\rho}|n'\rangle\langle n'| = \sum_k \frac{w_k}{\sum_n |c_{kn}|^2} \left(\sum_n c_{kn} |n\rangle \right) \left(\sum_{n'} \langle n'| c_{kn'}^* \right)$$

$$= \sum_k \frac{w_k}{\sum_n |c_{kn}|^2} |\varphi_k\rangle\langle\varphi_k| \tag{4.175}$$

with the sum of diagonal matrix elements (the "trace" function)

$$\mathrm{Tr}\hat{\rho} = \sum_n \langle n|\hat{\rho}|n\rangle = \sum_k w_k \frac{\sum_n |c_{kn}|^2}{\sum_n |c_{kn}|^2} = \sum_k w_k \tag{4.176}$$

while the searched operatorial average now becomes:

$$\left\langle \hat{A} \right\rangle = \frac{\sum_{n,n'} \langle n|\hat{\rho}|n'\rangle \langle n'|\hat{A}|n\rangle}{\sum_k w_k} = \frac{\sum_n \langle n|\hat{\rho}\hat{A}|n\rangle}{\sum_n \langle n|\hat{\rho}|n\rangle} = \frac{\mathrm{Tr}\left(\hat{\rho}\hat{A}\right)}{\mathrm{Tr}\hat{\rho}} \tag{4.177}$$

Note that in above deductions the double (independent) averages technique was adopted, exploiting therefore the associate sums inter-conversions to produce the simplified results (Park et al., 1980; Blanchard, 1982; Snygg, 1982). Yet, this technique is equivalent with quantum mechanically factorization of the entire Hilbert space into sub-spaces, or at the limit into the subspace of interest (that selected to be measured, for instance) and the rest of the space being thus this approach equivalent with a system-bath sample; this note is useful for latter better understanding of the stochastic phenomena that underlay to open quantum systems, being this the physical foundation for chemical reactivity.

Next, in the case the concerned quantum states are *eigen-states*, they fulfill the normalization constraint:

$$\delta_{kk'} = \langle \varphi_k | \varphi_{k'} \rangle = \sum_n c_{kn}^* c_{k'n} \Rightarrow \sum_n |c_{kn}|^2 = 1 \tag{4.178}$$

on which base the above density operator now reads with the eigen-equation

$$\hat{\rho} | \varphi_k \rangle = \sum_{k'} w_{k'} | \varphi_{k'} \rangle \underbrace{\langle \varphi_{k'} | \varphi_k \rangle}_{\delta_{k'k}} = w_k | \varphi_k \rangle \tag{4.179}$$

leading with the eigen-values (as the diagonal elements) just the weights

$$\langle \varphi_k | \hat{\rho} | \varphi_k \rangle = w_k \tag{4.180}$$

as the observed values of the averaged density operator. Thus they have to naturally fulfill the closure probability relationship over the entire sample,

$$\sum_k w_k = 1 \tag{4.181}$$

from where the "normalization of density operator" through its above Trace property of Eq. (4.176):

$$\mathrm{Tr}\hat{\rho} = 1 \tag{4.182}$$

Moreover, in these eigen-conditions, the operatorial average further reads from Eq. (4.177):

$$\langle \hat{A} \rangle = \mathrm{Tr}(\hat{\rho}\hat{A}) \tag{4.183}$$

Now, there appears with better clarity the major role the density operator plays in quantum measurements, since convolutes with given operator to produce its (averaged) measured value on the prepared eigen-states. Nevertheless, when the so-called *pure states* are employed or prepared, the precedent distinction between the subsystem and system vanishes, and the density operator takes the pure quantum mechanical form of an elementary projector:

$$\widehat{\rho} = |\varphi\rangle\langle\varphi| \equiv \widehat{\Lambda} \tag{4.184}$$

This is a very useful expression for considering it associated with the mono-density operators when the many-fermionic systems are treated, although similar procedure applies for mixed (sample) states as well. There is immediate to see that for N formally independent partitions the Hilbert space corresponding to the N-mono-particle densities on pure states, we individually have, see Eqs. (4.176), (4.181), (4.182) and (4.184),

$$\widehat{\Lambda}_i = |\varphi_i\rangle\langle\varphi_i|, \mathrm{Tr}\widehat{\Lambda}_i = 1,\ i = \overline{1,N} \tag{4.185}$$

producing the total operator – projector constructed by their sum

$$\widehat{\Lambda}_N = \sum_{i=1}^{N} \widehat{\Lambda}_i \tag{4.186}$$

is correctly normalized to the total number of particle:

$$\mathrm{Tr}\widehat{\Lambda}_N = \mathrm{Tr}\left(\sum_{i=1}^{N} \widehat{\Lambda}_i\right) = \sum_{i=1}^{N} \underbrace{\mathrm{Tr}\widehat{\Lambda}_i}_{1} = N \tag{4.187}$$

Yet, the anti-symmetric restriction the N-fermionic state may be accounted from the mono-electronic states through considering Slater permutated (P_α) products (Putz & Chiriac, 2008; Thouless, 1972):

$$|\Phi_N\rangle = \frac{1}{\sqrt{N!}} \sum_{P_\alpha} (-1)^{P_\alpha} P_\alpha \left[\prod_{i=1}^{N} |\varphi_i\rangle\right] \tag{4.188}$$

for constructing the N-electronic density operator:

$$\hat{\rho}^{(N)} = |\Phi_N\rangle\langle\Phi_N| \tag{4.189}$$

with which help the $N \times N$ density matrix writes as (in coordinate representation):

$$\begin{aligned}\rho^{(N)}(x'_1...x'_N; x_1...x_N) &= \langle x'_1...x'_N|\hat{\rho}^{(N)}|x_1...x_N\rangle \\ &= \langle x'_1...x'_N|\Phi_N\rangle\langle\Phi_N|x_1...x_N\rangle \\ &= \frac{1}{N!}\sum_{P'_\alpha}(-1)^{P'_\alpha}P'_\alpha\left[\prod_{i=1}^{N}\langle x'_i|\varphi_i\rangle\right] \\ &\times\sum_{P_\alpha}(-1)^{P_\alpha}P_\alpha\left[\prod_{i=1}^{N}\langle\varphi_i|x_i\rangle\right] \\ &= \frac{1}{N!}\sum_{P'_\alpha}(-1)^{P'_\alpha}P'_\alpha\left[\prod_{i=1}^{N}\varphi_i(x'_i)\right] \\ &\times\sum_{P_\alpha}(-1)^{P_\alpha}P_\alpha\left[\prod_{i=1}^{N}\varphi_i^*(x_i)\right]\end{aligned} \tag{4.190}$$

However, in practice, due to the fact the multi-particle operators have properties associate with number of systemic properties less than the total number of particle, say of order $p < N$, worth working with the *p-order reduced density matrix* introduced as:

$$\rho^{(p)}(x'_1...x'_p; x_1...x_p) = \binom{N}{p}\int\Phi_N^*(x_1...x_N)\Phi_N(x'_1...x'_N)\prod_{j=p+1}^{N}dx_j \tag{4.191}$$

with the following useful properties (Blum, 1981):

- Normalization:

$$\int\rho^{(p)}(x'_1...x'_p; x_1...x_p)\prod_{j=1}^{p}dx_j = \binom{N}{p} \tag{4.192}$$

- Recursion:

$$\int\rho^{(p)}(x'_1...x'_p; x_1...x_p)dx_p = \frac{N+1-p}{p}\rho^{(p-1)}(x'_1...x'_{p-1}; x_1...x_{p-1}) \tag{4.193}$$

- First order Löwdin reduction:

$$\rho^{(p)}\left(x'_1...x'_p;x_1...x_p\right)=\frac{1}{p!}\det\left[\rho^{(1)}\left(x'_k;x_k\right)\right]$$

$$=\frac{1}{p!}\begin{vmatrix}\rho^{(1)}\left(x'_1;x_1\right) & \rho^{(1)}\left(x'_1;x_2\right) & \cdots & \rho^{(1)}\left(x'_1;x_p\right)\\ \rho^{(1)}\left(x'_2;x_1\right) & \rho^{(1)}\left(x'_2;x_2\right) & \cdots & \rho^{(1)}\left(x'_2;x_p\right)\\ \vdots & \vdots & & \vdots\\ \rho^{(1)}\left(x'_p;x_1\right) & \rho^{(1)}\left(x'_p;x_2\right) & \cdots & \rho^{(1)}\left(x'_p;x_p\right)\end{vmatrix} \tag{4.194}$$

where the first order density matrix casts, abstracted from general definition:

$$\rho^{(1)}\left(x'_1;x_1\right)=N\int\Phi_N^*\left(x_1...x_N\right)\Phi_N\left(x'_1...x'_N\right)\prod_{j=2}^{N}dx_j \tag{4.195}$$

By these there is already noted the major importance the first order density plays in computing the higher order reduced density matrices that on their turn enters the operatorial averages, for instance:

$$\left\langle\hat{A}\right\rangle=\sum_{p=1}^{N}\mathrm{Tr}^{(p)}\left[\hat{A}\left(x_1...x_p\right)\hat{\rho}^{(p)}\right] \tag{4.196}$$

A special reference worth be made in regard of the free-relativist treatment of many-electronic atoms, ions, bi- or poly- atomic molecules, governed by the working Hamiltonian:

$$\hat{H}=e^2\sum_{G<H}\frac{Z_G Z_H}{R_{GH}}+\sum_{i=1}^{N}\frac{p_i^2}{2m}-e^2\sum_{i,G}\frac{Z_G}{r_{iG}}+e^2\sum_{i<j}\frac{1}{r_{ij}} \tag{4.197}$$

those terms are represented the inter-nuclear repulsion (only for molecules), free electronic motion, electron-nuclei Coulombic attraction, and inter-electronic Coulombian repulsion, respectively. For it, the average value is computed through considering electronic density of the first or second order only there where the electronic influence is present while the degree of matrix density is fixed by the type of electronic interaction:

$$\begin{aligned}\left\langle \widehat{H} \right\rangle &= e^2 \sum_{G<H} \frac{Z_G Z_H}{R_{GH}} \\ &+ \frac{1}{2m} \int p_1^2 \rho^{(1)}\left(x'_1;x_1\right)\Big|_{x'_1=x_1} dx_1 \\ &- e^2 \sum_G Z_G \int \frac{\rho^{(1)}\left(x'_1;x_1\right)}{r_{1G}}\Big|_{x'_1=x_1} dx_1 \\ &+ e^2 \iint \frac{\rho^{(2)}\left(x'_1,x'_2;x_1,x_2\right)}{r_{12}}\Bigg|_{\substack{x'_1=x_1 \\ x'_2=x_2}} dx_1 dx_2 \end{aligned} \quad (4.198)$$

There is obvious that even the second order reduced matrix has appeared,

$$\rho^{(2)}\left(x'_1,x'_2;x_1,x_2\right) = \binom{N}{2} \int \Phi_N^*\left(x_1...x_N\right)\Phi_N\left(x'_1...x'_N\right)\prod_{j=3}^{N} dx_j \quad (4.199)$$

it may be further reduced to the first one through the above determinant rule:

$$\begin{aligned}\rho^{(2)}\left(x'_1,x'_2;x_1,x_2\right) &= \frac{1}{2!}\begin{vmatrix} \rho^{(1)}\left(x'_1;x_1\right) & \rho^{(1)}\left(x'_1;x_2\right) \\ \rho^{(1)}\left(x'_2;x_1\right) & \rho^{(1)}\left(x'_2;x_2\right) \end{vmatrix} \\ &= \frac{1}{2}\begin{bmatrix} \rho^{(1)}\left(x'_1;x_1\right)\rho^{(1)}\left(x'_2;x_2\right) \\ -\rho^{(1)}\left(x'_1;x_2\right)\rho^{(1)}\left(x'_2;x_1\right) \end{bmatrix} \end{aligned} \quad (4.200)$$

emphasizing therefore on the importance of the first order reduced matrix knowledge.

The astonishing physical meaning behind this formalism relays in the fact that any multi-particle interaction (two-particle interaction included) may be reduced to the single particle behavior; in other terms, vice-versa, the appropriate perturbation (including strong-coupling) of the single particle evolution caries the equivalent information as that characterizing the whole many-body system.

In fact in this resides the power of the density matrix formalism: reducing a many-body problem to the single particle density matrix, abstracted from the single Slater determinant of Eq. (4.190) called also as *Fock-Dirac matrix*

$$\rho_{FD}^{(1)}(x'_1;x_1)=\sum_{i=1}^{N}\varphi_i^*(x_1)\varphi_i(x'_1) \tag{4.201}$$

and the associate operator

$$\widehat{\rho}_{FD}^{(1)}=\sum_{i=1}^{N}|\varphi_i\rangle\langle\varphi_i| \tag{4.202}$$

that is considerably simplifying the quantum problem to be solved. Let's illustrate this by firstly quoting that Fock-Dirac density operator of Eq. (4.202) has two fundamental properties, namely:

- The idem potency:

$$\widehat{\rho}_{FD}^{(1)}\widehat{\rho}_{FD}^{(1)}=\sum_{i,j=1}^{N}|\varphi_i\rangle\underbrace{\langle\varphi_i|\varphi_j\rangle}_{\delta_{ij}}\langle\varphi_j|=\sum_{i=1}^{N}|\varphi_i\rangle\langle\varphi_i|=\widehat{\rho}_{FD}^{(1)} \tag{4.203}$$

- The normal additivity, see Eqs. (4.187):

$$\mathrm{Tr}\,\widehat{\rho}_{FD}^{(1)}=\mathrm{Tr}\left(\sum_{i=1}^{N}|\varphi_i\rangle\langle\varphi_i|\right)=\mathrm{Tr}\left(\sum_{i=1}^{N}\widehat{\Lambda}_i\right)=N \tag{4.204}$$

while having the corresponding coordinate integral representations:

- Kernel multiplicity:

$$\int\rho_{FD}^{(1)}(x'_1;x''_1)\rho_{FD}^{(1)}(x''_1;x_1)dx''_1=\rho_{FD}^{(1)}(x'_1;x_1) \tag{4.205}$$

- Many-body normalization:

$$\int\rho_{FD}^{(1)}(x_1;x_1)dx_1=N \tag{4.206}$$

Remarkably, the last two identities may serve as the constraints when minimizing the above Hamiltonian average, here appropriately rewritten employing Eqs. (4.198) and (4.200) and where all external applied potential was resumed under $V(x_1)$ under the actual so-called *Hartree-Fock trial density matrix energy functional*

$$E_{HF}\left[\rho_{FD}^{(1)}\right]=\int\left[-\frac{\hbar^2}{2m}\nabla_1^2+V(x_1)\right]\rho_{FD}^{(1)}\left(x'_1;x_1\right)\Big|_{x'_1=x_1}dx_1$$
$$+\frac{e^2}{2}\iint\frac{1}{r_{12}}\begin{bmatrix}\rho_{FD}^{(1)}\left(x_1;x_1\right)\rho_{FD}^{(1)}\left(x_2;x_2\right)\\-\rho_{FD}^{(1)}\left(x_1;x_2\right)\rho_{FD}^{(1)}\left(x_2;x_1\right)\end{bmatrix}dx_1dx_2 \quad (4.207)$$

with the (Lagrange) variational principle:

$$\delta\begin{Bmatrix}E_{HF}\left[\rho_{FD}^{(1)}\right]\\-\iint\alpha\left(x'_1;x_1\right)\begin{bmatrix}\int\rho_{FD}^{(1)}\left(x'_1;x''_1\right)\rho_{FD}^{(1)}\left(x''_1;x_1\right)dx''_1\\-\rho_{FD}^{(1)}\left(x'_1;x_1\right)\end{bmatrix}dx'_1\,dx_1\\-\beta\left[\iint\delta\left(x'_1-x_1\right)\rho_{FD}^{(1)}\left(x'_1;x_1\right)dx'_1\,dx_1-N\right]\end{Bmatrix}=0 \quad (4.208)$$

By the functional derivative respecting the Fock-Dirac electron density one gets:

$$\frac{\delta E_{HF}\left[\rho_{FD}^{(1)}\right]}{\delta\rho_{FD}^{(1)}\left(x'_1;x_1\right)}-\int\rho_{FD}^{(1)}\left(x'_1;\bar{x}\right)\alpha\left(\bar{x};x_1\right)d\bar{x}$$
$$-\int\alpha\left(x'_1;\bar{\bar{x}}\right)\rho_{FD}^{(1)}\left(\bar{\bar{x}};x_1\right)d\bar{\bar{x}}+\alpha\left(x'_1;x_1\right)-\beta\delta\left(x'_1-x_1\right)=0 \quad (4.209)$$

which eventually transcribes at the operatorial level:

$$\hat{F}-\hat{\rho}_{FD}^{(1)}\hat{\alpha}-\hat{\alpha}\hat{\rho}_{FD}^{(1)}+\hat{\alpha}-\beta\hat{1}_\delta=0 \quad (4.210)$$

with $\hat{1}_\delta$ staying for the operator of the delta-Dirac matrix $\delta\left(x'_1-x_1\right)$, while $\hat{F}$ being the Fock operator corresponding to the coordinate matrix representation (Parr & Yang, 1989):

$$F\left(x'_1;x_1\right)=\frac{\delta E_{HF}\left[\rho_{FD}^{(1)}\right]}{\delta\rho_{FD}^{(1)}\left(x'_1;x_1\right)}$$
$$=\left[-\frac{\hbar^2}{2m}\nabla_1^2+V(x_1)\right]\delta\left(x'_1-x_1\right)$$
$$\underbrace{+\delta\left(x'_1-x_1\right)e^2\int\frac{1}{r_{12}}\rho_{FD}^{(1)}\left(x_2;x_2\right)dx_2-\frac{e^2}{r_{1'1}}\rho_{FD}^{(1)}\left(x'_1;x_1\right)}_{EXCHANGE\ \ CONTRIBUTION} \quad (4.211)$$

Equation (4.210) is most informative since, basing on the idem potency property of Eq. (4.203), through multiplying it on the right with Fock-Dirac density operator,

$$\hat{F}\hat{\rho}_{FD}^{(1)}-\hat{\rho}_{FD}^{(1)}\hat{\alpha}\hat{\rho}_{FD}^{(1)}-\hat{\alpha}\underbrace{\left(\hat{\rho}_{FD}^{(1)}\right)^2}_{\hat{\rho}_{FD}^{(1)}}+\hat{\alpha}\hat{\rho}_{FD}^{(1)}-\beta\hat{1}_\delta\hat{\rho}_{FD}^{(1)}=0 \quad (4.212)$$

and then with the same on left side,

$$\hat{\rho}_{FD}^{(1)}\hat{F}-\underbrace{\left(\hat{\rho}_{FD}^{(1)}\right)^2}_{\hat{\rho}_{FD}^{(1)}}\hat{\alpha}-\hat{\rho}_{FD}^{(1)}\hat{\alpha}\hat{\rho}_{FD}^{(1)}+\hat{\rho}_{FD}^{(1)}\hat{\alpha}-\beta\hat{\rho}_{FD}^{(1)}\hat{1}_\delta=0 \quad (4.213)$$

and subtracting the results, it yields:

$$\hat{F}\hat{\rho}_{FD}^{(1)}-\hat{\rho}_{FD}^{(1)}\hat{F}=0 \quad (4.214)$$

that is equivalently of saying that Fock energy operator commutes with the Fock-Dirac density operator,

$$\left[\hat{F},\hat{\rho}_{FD}^{(1)}\right]=0 \quad (4.215)$$

meaning that they both admit the same set of eigen-functions. This is nevertheless the gate for obtaining the density (matrix) functional energy expressions by means of finding the density (matrix) eigen-solutions only.

Yet, condition (4.215) is indeed a workable (reduced) condition raised from optimization of the averaged Hamiltonian of a many-electronic system, since the more general one referring to the whole Hamiltonian, known as the *Liouville or Neumann equation*, is obtained employing the temporal Schrödinger equation:

$$i\hbar\frac{\partial}{\partial t}|\varphi_i\rangle = \widehat{H}|\varphi_i\rangle \tag{4.216}$$

to the evolution equation of Fock-Dirac density operator evolution:

$$i\hbar\frac{\partial}{\partial t}\hat{\rho}^{(1)}_{FD} = \sum_{i=1}^{N} i\hbar\frac{\partial}{\partial t}\left(|\varphi_i\rangle\langle\varphi_i|\right) = \sum_{i=1}^{N}\underbrace{i\hbar\left(\frac{\partial}{\partial t}|\varphi_i\rangle\right)}_{\widehat{H}|\varphi_i\rangle}\langle\varphi_i|$$
$$+\sum_{i=1}^{N}|\varphi_i\rangle\underbrace{i\hbar\left(\frac{\partial}{\partial t}\langle\varphi_i|\right)}_{\langle\phi_i|\widehat{H}}$$
$$=\widehat{H}\left(\sum_{i=1}^{N}|\varphi_i\rangle\langle\varphi_i|\right)-\left(\sum_{i=1}^{N}|\varphi_i\rangle\langle\varphi_i|\right)\widehat{H} = \left[\widehat{H},\hat{\rho}^{(1)}_{FD}\right] \tag{4.217}$$

Lastly, note that all above properties may be rewritten since considering the *mixed p-order reduced matrix* with the form

$$\rho^{(p)}\left(x'_1...x'_p;x_1...x_p\right) = \sum_k w_k\rho_k^{(p)}\left(x'_1...x'_p;x_1...x_p\right) \tag{4.218}$$

as a natural extension of that characterizing the pure states. However, the sample statistical effects may be better considered by further expressing the electronic density operator and its matrix, equation and properties for systems in thermodynamic equilibrium (with environment), a mater in next section addressed.

4.4.2 CANONICAL DENSITY, BLOCH EQUATION, AND THE NEED OF PATH INTEGRAL

For a quantum system obeying the N-mono-electronic eigen-equations

$$\widehat{H}|\varphi_k\rangle = E_k|\varphi_k\rangle \tag{4.219}$$

the probability of finding one particle in the state $|\varphi_k\rangle$ at thermodynamical equilibrium with others, while the state + rest of states is considered a closed supra-system with no mass or charge transfer allowed, is given by the canonical distribution (Isihara, 1980):

$$w_k = \frac{1}{Z(\beta)} \exp(-\beta E_k) \tag{4.220}$$

providing the *mixed Fock-Dirac density* with the form:

$$\begin{aligned}
\hat{\rho}_N(\beta) &= \sum_{k=1}^{N} \frac{1}{Z(\beta)} \underbrace{\exp(-\beta E_k)|\varphi_k\rangle}_{\exp(-\beta\hat{H})|\varphi_k\rangle}\langle\varphi_k| \\
&= \sum_{k=1}^{N} \frac{1}{Z(\beta)} \exp\left(-\beta\hat{H}\right)|\varphi_k\rangle\langle\varphi_k| \\
&= \frac{1}{Z(\beta)} \exp\left(-\beta\hat{H}\right) \sum_{k=1}^{N} |\varphi_k\rangle\langle\varphi_k| \\
&= \frac{1}{Z(\beta)} \exp\left(-\beta\hat{H}\right) \sum_{k=1}^{N} \left(\sum_n c_{kn}|n\rangle\right)\left(\sum_n \langle n|c_{kn}^*\right) \\
&= \frac{1}{Z(\beta)} \exp\left(-\beta\hat{H}\right) \sum_{k=1}^{N} \underbrace{\left(\sum_n |c_{kn}|^2\right)}_{1} \underbrace{\left(\sum_n |n\rangle\langle n|\right)}_{1} \\
&= \frac{N}{Z(\beta)} \exp\left(-\beta\hat{H}\right)
\end{aligned} \tag{4.221}$$

This is a very interesting and important result motivating the quantum statistical approach of determining the density of states since it corresponds to the N-sample particle throughout simple *N*-multiplication. Note that Eq. (4.221) is very suited for handling since its normalization factor, the partition function $Z(\beta)$, obeys the consecrated expression

$$Z(\beta) = \mathrm{Tr}\left[\exp\left(-\beta\hat{H}\right)\right] = \int \langle x|e^{-\beta\hat{H}}|x\rangle dx \tag{4.222}$$

which is reflecting in density normalization

$$N[\rho] = \int \rho(x_1) dx_1 \tag{4.223}$$

being of paramount importance in density functional theory, the same as Eq. (4.206), because it opens the doors of observable quantities through electronic density rather than by means of wave function.

The recognized importance of partition function, in computing the internal energy as the average of the Hamiltonian of the system

$$U_N := \langle \hat{H} \rangle = \mathrm{Tr}\left[\hat{\rho}_N(\beta)\hat{H}\right] = \frac{N}{Z(\beta)}\mathrm{Tr}\left[\hat{H}\exp\left(-\beta\hat{H}\right)\right]$$
$$= -N\frac{\partial}{\partial\beta}\ln Z(\beta) \tag{4.224}$$

or to evaluate the free energy of the system:

$$F_N = -N\frac{1}{\beta}\ln Z(\beta) \tag{4.225}$$

is thus transferred to the knowledge of the closed evolution amplitude $\langle x|e^{-\beta\hat{H}}|x\rangle$, that at its turn is based on the *genuine* (not-normalized) *density operator*:

$$\hat{\rho}_{\otimes}(\beta) = \exp\left(-\beta\hat{H}\right) \tag{4.226}$$

sometimes called also like *canonic density operator*.

The great importance of density operator of Eq. (4.226) is immediately visualized in three ways

- It identifies the evolution operator

$$\hat{U}(t_b, t_a) = \exp\left[-\frac{i}{\hbar}\hat{H}(t_b - t_a)\right] \tag{4.227}$$

  on the ground of Wick equivalence relationship of Eqs. (4.33) or (4.124), which allows the transformation of the Schrödinger into Heisenberg or Interaction pictures for better describing the quantum interactions;
- It produces the so-called *Bloch equation* (Bloch, 1932) by taking its β – derivative,

$$-\frac{\partial \hat{\rho}_{\otimes}(\beta)}{\partial \beta}=\hat{H}\hat{\rho}_{\otimes}(\beta) \tag{4.228}$$

that identifies with the Schrödinger equation for genuine density operator

$$i\hbar\frac{\partial}{\partial t}\hat{\rho}_{\otimes}(\beta)=\hat{H}\hat{\rho}_{\otimes}(\beta) \tag{4.229}$$

through the same Wick transformation given by Eqs. (4.33) or (4.124), thus providing the quantum-mechanically to quantum-statistical equivalence.

- Fulfills the (short times, higher temperature) so-called Markovian limiting condition:

$$\lim_{\beta\to 0}\hat{\rho}_{\otimes}(\beta)=\hat{1} \tag{4.230}$$

a very useful constraint for developing either the perturbation or the variational formalism respecting electronic density and/or partition function, see bellow.

In the frame of coordinate representation the Bloch problem, i.e., differential equation and its initial (Cauchy) condition, looks like:

$$\begin{cases}-\dfrac{\partial}{\partial \beta}\rho_{\otimes}(x';x;\beta)=\hat{H}\rho_{\otimes}(x';x;\beta)\\ \lim\limits_{\beta\to 0}\rho_{\otimes}(x';x;\beta)=\delta(x'-x)\end{cases} \tag{4.231}$$

Solution of this system is a great task in general case, unless the perturbation method is undertaken for writing the Hamiltonian a sum of a free and small interaction components,

$$\hat{H}=\hat{H}_0+\hat{H}_1 \tag{4.232}$$

for which the free Hamiltonian solution is completely known, say

$$\hat{\rho}_0(\beta)=\exp\left(-\beta\hat{H}_0\right) \tag{4.233}$$

In these conditions, one may firstly write:

$$\frac{\partial}{\partial\beta}\left(e^{\beta\hat{H}_0}\hat{\rho}_\otimes\right)=\hat{H}_0 e^{\beta\hat{H}_0}\hat{\rho}_\otimes+e^{\beta\hat{H}_0}\underbrace{\frac{\partial\hat{\rho}_\otimes}{\partial\beta}}_{-\hat{H}\hat{\rho}_\otimes}=\hat{H}_0 e^{\beta\hat{H}_0}\hat{\rho}_\otimes$$

$$-\left(\hat{H}_0+\hat{H}_1\right)e^{\beta\hat{H}_0}\hat{\rho}_\otimes=-e^{\beta\hat{H}_0}\hat{H}_1\hat{\rho}_\otimes \tag{4.234}$$

where the inter-Hamiltonian components were considered to freely commute as per whish; then, the Eq. (4.234) is integrated on the realm $[0,\beta]$ to get:

$$e^{\beta\hat{H}_0}\hat{\rho}_\otimes(\beta)-\hat{1}=-\int_0^\beta e^{\beta'\hat{H}_0}\hat{H}_1(\beta')\hat{\rho}_\otimes(\beta')d\beta' \tag{4.235}$$

rearranged under the perturbative fashion:

$$\hat{\rho}_\otimes(\beta)=\hat{\rho}_0(\beta)-\int_0^\beta\hat{\rho}_0(\beta-\beta')\hat{H}_1(\beta')\hat{\rho}_\otimes(\beta')d\beta' \tag{4.236}$$

in the form reminding by the *Lippmann-Schwinger equation* for the perturbed dynamical wave-function (Messiah, 1961), with $\hat{\rho}_0(\beta-\beta')$ playing the role of the retarded Green function $G_0(t_b-t_a)$ (Feynman, 1972). Yet, expression (4.236) may be more generalized for the *p*-order approximation throughout choosing various *p*-paths of spanning the statistical realm $[0,\beta]$ by intermediate sub-intervals:

$$\beta=\beta_{n+1}>\beta_n>...>\beta_2>\beta_1>\beta_0=0 \tag{4.237}$$

leading wit the expansion:

$$\hat{\rho}_\otimes(\beta)=\hat{\rho}_0(\beta)+\sum_{l=1}^{n}(-1)^l\int_0^\beta d\beta_l\int_0^{\beta_l}d\beta_{l-1}...\int_0^{\beta_2}d\beta_1$$

$$\times\hat{\rho}_0(\beta-\beta_l)\left[\hat{H}_1(\beta_l)\hat{\rho}_0(\beta_l-\beta_{l-1})\right]$$

$$\cdots\left[\hat{H}_1(\beta_2)\hat{\rho}_0(\beta_2-\beta_1)\right]\hat{H}_1(\beta_1)\hat{\rho}_0(\beta_1) \tag{4.238}$$

or in coordinate representation:

$$\rho_{\otimes}(x';x;\beta)=\rho_0(x';x;\beta)+\sum_{l=1}^{n}(-1)^l\left(\int_0^{\beta}d\beta_l\int_0^{\beta_l}d\beta_{l-1}...\int_0^{\beta_2}d\beta_1\right)$$

$$\times\left(\int_{-\infty}^{+\infty}\cdots\int_{-\infty}^{+\infty}\prod_{j=1}^{l}dx_j\right)$$

$$\times\rho_0(x',x_l;\beta-\beta_l)\left[\widehat{H}_1(\beta_l)\rho_0(x_l;x_{l-1};\beta_l-\beta_{l-1})\right]$$

$$\cdots\left[\widehat{H}_1(\beta_2)\rho_0(x_2;x_1;\beta_2-\beta_1)\right]\widehat{H}_1(\beta_1)\rho_0(x_1;x;\beta_1) \tag{4.239}$$

for a parallel space discrimination of the spatial interval $[x',x]$ through the subdivisions:

$$x'=x_{n+1}>x_n>...>x_2>x_1>x_0=x \tag{4.240}$$

Such slicing procedure in solving the Bloch equation (4.231) for canonic density solution (4.239) seems an elegant way of avoiding the self-consistent equation (4.236). Therefore, it may further employed through reconsidering the problem (4.231) in a slightly modified variant, namely within the temporal approach

$$\begin{cases}-\hbar\dfrac{\partial}{\partial u}\rho_{\otimes}(x';x;u)=H(x')\rho_{\otimes}(x';x;u)\\ \rho_{\otimes}(x';x;u=0)=1\end{cases} \tag{4.241}$$

where the variable $u=\hbar\beta$ was considered for the time dimension.

Now, in the first instance the new problem has the *formal* total solution

$$\rho_{\otimes}(x';x;u)=\exp\left[-\frac{1}{\hbar}H(x')u\right] \tag{4.242}$$

that being of exponential type allows for direct slicing through factorization. That is, when considering the space division given by coordinate cuts of Eq. (4.240), and assuming the times flows equally on each sub-interval in quota of ε, $u=(n+1)\varepsilon$, the density solution (4.242) may be written as a product of intermediary solutions:

$$\begin{aligned}\rho_{\otimes}(x';x;u) &= \prod_{j=0}^{n+1}\exp\left[-\frac{1}{\hbar}H(x_j)\varepsilon\right]\\ &= \int\cdots\int\rho_{\otimes}(x';x_l;\varepsilon)\rho_{\otimes}(x_l;x_{l-1};\varepsilon)\cdots\rho_{\otimes}(x_1;x;\varepsilon)\prod_{j=1}^{n}dx_j\\ &\underset{\substack{\varepsilon\to 0\\(n+1)\varepsilon=u}}{\overset{n\to\infty}{\to}} \int\cdots\int\Lambda[x(u)]\mathrm{D}x(u)\end{aligned} \tag{4.243}$$

where we introduced the chained covariant density product:

$$\Lambda[x(u)] = \lim_{\substack{n\to\infty\\ \varepsilon\to 0\\ (n+1)\varepsilon=u}} \rho_{\otimes}(x';x_l;\varepsilon)\rho_{\otimes}(x_l;x_{l-1};\varepsilon)\cdots\rho_{\otimes}(x_1;x;\varepsilon) \tag{4.244}$$

and the extended integration metric:

$$\mathrm{D}x(u) = \lim_{n\to\infty}\prod_{j=1}^{n}dx_j \tag{4.245}$$

The general canonic solution (4.243) is called as the path integral solution for the Bloch equation (4.241), being therefore as a necessity when looking to general solutions for a given Hamiltonian. It gives general solution for electronic density (4.226) since accounting for all path connecting two end-points either in space and time (or temperatures) through in principle an infinite intermediary points; this way the resulted path integral comprises all quantum information contained by the particle' evolution between two states in thermodynamical equilibrium with environment (the other mono-particle states). However, once the canonical density evaluated through computed its path integral the associate mixed density matrix may be immediately written employing the operatorial form (4.221) to actual spatial representation

$$\rho_N(x';x;u) = \frac{N}{Z(u)}\rho_{\otimes}(x';x;u) \tag{4.246}$$

with the path integral based partition function written in accordance with Eq. (4.222):

$$Z(u) = \int \rho_{\otimes}(x;x;u)\,dx \tag{4.247}$$

while preserving the general DFT normalization condition:

$$\int \rho_N(x;x;u)\,dx = N \tag{4.248}$$

This way the general algorithm linking path integral to density matrix to electronic density employed by DFT for computing various density functionals (energies, reactivity indices) for characterizing chemical structure and reactivity was established, while emphasizing the basic role the path integral evaluation has in analytical evaluations towards a conceptual understanding of many-electronic quantum systems in their dynamics and interaction.

Being thus established the role and usefulness of path integral in density functional theory the next section will give more insight in appropriately defining (constructing) quantum chemical modern theories as are Hartree-Fock and density functional formalisms, being nowadays employed in various computational and conceptual schemes and applications for large classes of physico-chemical systems.

4.5 ROOTS OF SELF-CONSISTENT METHODS IN QUANTUM CHEMISTRY

Very often, the famous words of Dirac, i.e., "The underlying physical laws necessary for the mathematical theory of a large part of physics and the whole of chemistry are thus completely known", are quoted by theorists in physics when they like to underline that chemistry is in principle solved by the basics of quantum mechanics so that some more interesting problems should be solved. Despite this, from 1929 nowadays, quantum physics of atoms and molecules largely turns into quantum chemistry, an interdisciplinary discipline that still struggles with the elucidation of the actual behavior of electrons in nano- and bio- systems. While the total success is still not in sight, the achievements in the arsenal of concepts, principles, and implementation was considerable and already enters goes into the arsenal of humankind hall-of-fame giving thus hope for a shining dawn in the poly-electronic interaction arena (Preuss, 1969; Kryachko & Ludeña, 1987; Atkins & Friedman, 1997; Christofferson, 1989; Szabo & Ostlund, 1996). However, when questing for the underlying principles of the chemical bond, the first compulsory level of expertise may be called as the intensive level of analysis in which the main ingredients of a many-electronic-many-nuclear problem has to be clarified. These are subjected in the below following sections (Putz & Chiriac, 2008).

4.5.1 MOLECULAR ORBITAL APPROACH

The basic starting point is the consecrated time-independent Schrödinger equation

$$\hat{H}\Psi = E\Psi \tag{4.249}$$

with non-relativistic Hamiltonian

$$\begin{aligned}\hat{H} &= \hat{T}_e + \hat{T}_n + \hat{V}_{ee} + \hat{V}_{en} + \hat{V}_{nn} \\ &= -\frac{1}{2}\sum_i \nabla_i^2 - \frac{1}{2}\sum_A \frac{\nabla_A^2}{M_A} - \sum_{i,A}\frac{Z_A}{r_{iA}} + \sum_{i>j}\frac{1}{r_{ij}} \\ &\quad + \sum_{A>B}\frac{Z_A Z_B}{R_{AB}},\ i,j=\overline{1,N};\ A,B=\overline{1,M}\end{aligned} \tag{4.250}$$

accounting for the electron kinetic, nuclear kinetic, electron-electron repulsion, electron-nuclear attraction and nuclear-nuclear repulsion energetic terms, respectively.

Of course, as it is, Eq. (4.249) cannot be solved exactly, in its most general way. The approximations have to be implemented in such a way as to include the specific reality of the dynamic electronic-nuclear system. In this respect, considering an approximation is not viewed as a limitation here, but rather as a sort of rescaling of the concerned issue. Epistemologically, it is equivalent with Descartes' scholastic methodology of reducing a problem to smaller problems through the method of analysis.

Such a procedure has been long verified in mathematical-physics with impressive practical applications, for example, the integral-differential recipes, and with be thus safely implemented also here without loss in generality of the basic problem.

In quantum chemistry the specific method was consecrated as Born-Oppenheimer approximation that separate the electronic-nuclear system and problem in two smaller parametrically linked subsystems associated with an electronic motion, defined by equations

$$\begin{cases} \hat{H}_e \Psi_e = E_e \Psi_e \\ \hat{H}_e = \hat{T}_e + \hat{V}_{ee} + \hat{V}_{en} \end{cases} \Rightarrow \Psi_e(\{r_i\},\{R_A\});\ E_e(\{R_A\}) \tag{4.251}$$

and the corresponding nuclear motion

$$\begin{cases} \hat{H}_n \Psi_n = E \Psi_n \\ \hat{H}_n = \hat{T}_n + E_e(\{R_A\}) + \hat{V}_{nn} \end{cases} \tag{4.252}$$

It is worth noting that this phenomenological separation of electronic and nuclear problems may be possible due to the impressive difference in their mass that practically fixes the nuclei as the reference system in which the electronic system evolves. This is, nevertheless, only the first and most straight (however appropriate) approximation considered upon a many-body (electrons and nuclei) problem.

As a consequence, two stages of the overall solution can be given. One is obtained when solving the electronic problem only, therefore producing the so-called *single-point* calculation, i.e., the clamped nuclei remaining in a single inter-position.

The next stage is when replacing the electronic coordinates by their average values, since they move much faster than the nuclei, solving the nuclear Schrödinger equation (4.252) thus furnishing the vibration, rotation and translation solutions of a molecule. This way, the so-called *potential-energy surface* solution has been provided since $E_e(\{R_A\}) + \hat{V}_{nn}$ constitutes the potential for the nuclear motion as a whole.

While, molecular mechanics methods fairly provides nuclear solution of motion the electronic problem remains as the main, first cut, challenge to be addressed also because its elucidation leaves the sign also on the electronic pairing problem, the cornerstone concept in chemical bonding nature.

Thus, focusing only on the electronic Schrödinger equation (4.251), it can further be seen as a composite Hamiltonian, namely

$$\hat{H}_e = \hat{H}_e^I + \hat{H}_e^{II} \tag{4.253}$$

in terms of the electron solely and external electron energies

$$\hat{H}_e^I = \hat{T}_e + \hat{V}_{en} \tag{4.254}$$

for the kinetic and nuclear potential, respectively, on the one hand, and

$$\hat{H}_e^{II} = \hat{V}_{ee} \tag{4.255}$$

separating the electron-electron contribution, that already feel that has to have a specific behavior, both at classical and quantum levels of manifestations, at other hand.

Now, moving on to the specific electronic wave functions, let us consider the spin-orbitals

$$\chi_i^\sigma(1) = \phi_i(1)\sigma(1) \tag{4.256}$$

with their intrinsic ortho-normalized conditions fulfilled,

$$\left\langle \chi_i^\sigma \middle| \chi_j^\rho \right\rangle = \int \chi_i^{\sigma*}(1)\chi_j^\rho(1)d\tau_1 = \delta_{ij}\delta_{\sigma\rho} \tag{4.257}$$

as being *one-electron functions* or *molecular orbitals* MO, each as a product of a spatial orbital $\phi_i(1)$ and a spin function $\sigma(1) = \alpha, \beta$.

In these conditions, for a system with N electrons, the trial wave function (equivalent with the so-called Slater determinant) takes the so-called *trial Hartree-Fock (HF) form*:

$$\Psi_e^{HF} = \sqrt{N!}\,\hat{\wp}\,\Psi_e^H \tag{4.258}$$

with the Hartree wave function as simple product of spin-orbitals (the so-called *orbital approximation*)

$$\Psi_e^H = \prod_{\substack{i=1 \\ \sigma=\alpha,\beta}}^{N} \chi_i^\sigma(1) \tag{4.259}$$

and the antisymmetrization operator

$$\hat{\wp} = \frac{1}{N!}\sum_P (-1)^P P \tag{4.260}$$

having Hermitian and commutation properties:

$$\hat{\wp}^2 = \hat{\wp} \tag{4.261}$$

$$\left[\hat{H}_e^I, \hat{\wp}\right] = \left[\hat{H}_e^{II}, \hat{\wp}\right] = 0 \tag{4.262}$$

respectively.

This way, we formally succeed to further separate the many-electronic problem in as many one-electronic problems as electrons are considered in the molecular system.

From now on, basically, one can solve the many-electronic equation by manipulating the one-electronic properties of the system. How this can best be performed, at what cost and under what conditions, will be in next addressed.

4.5.2 ROOTHAAN APPROACH

Following the Dirac's quote, once the Schrodinger equation (4.249) was established "The underlying physical laws necessary for the mathematical theory of a large part of physics and the whole of chemistry are thus completely known" (Dirac, 1929). Unfortunately, the molecular spectra based on the eigen-problem Equation (4.249) is neither directly nor completely solved without specific atoms-in-molecule and/or symmetry constraints and approximation. As such at the mono-electronic level of approximation the Schrodinger Equation (4.249) rewrites under the so-called independent-electron problem: with the aid of effective electron Hamiltonian partitioning:

$$H_i^{eff}\psi_i = E_i\psi_i \tag{4.263}$$

$$H = \sum_i H_i^{eff} \tag{4.264}$$

and the correspondent molecular monoelectronic wave-functions (orbitals) fulfilling the conservation rule of probability:

$$\int \psi_i^2(\mathbf{r})d\mathbf{r} = 1 \tag{4.265}$$

However, when written as a linear combination over the atomic orbitals the resulted MO-LCAO wave-function:

$$\psi_i = \sum_v C_{vi}\phi_v \tag{4.266}$$

replaced in Eq. (4.263) followed by integration over the electronic space allows for matrix version of Eq. (4.263) of what are called as *Roothaan equations*:

$$\left(H^{eff}\right)\left(C\right) = \left(S\right)\left(C\right)\left(E\right) \tag{4.267}$$

Worth mentioning that while Eq. (4.267) unfolds under the eigen-function equations

$$\sum_v C_{vi}\int \phi_\mu^*(1)\hat{F}(1)\phi_v(1)d\tau_1 = \varepsilon_i \sum_v C_{vi}\int \phi_\mu^*(1)\phi_v(1)d\tau_1 \tag{4.268}$$

It corresponds with the effective Hamiltonian called as Fock operator driving the self-consistent equation (4.267), under the current form

$$\left(F\right)\left(C\right) = \left(S\right)\left(C\right)\left(E\right) \tag{4.269}$$

with the corresponding matrix elements

$$F_{\mu\nu} = \int \phi_{\mu}^{*}(1)\hat{F}(1)\phi_{\upsilon}(1)d\tau_1 \tag{4.270}$$

At the same time, Eq. (4.267) has the diagonal energy-matrix elements as the eigen-solution

$$(E)_{ij} = E_{ij} = E_i\delta_{ij} = \begin{cases} E_i ...i = j \\ 0 ...i \neq j \end{cases} \tag{4.271}$$

to be found in terms of the expansion coefficients matrix (*C*), the matrix of the Hamiltonian elements:

$$H_{\mu\nu} = \int \phi_{\mu} H^{eff} \phi_{\nu} d\tau \tag{4.272}$$

and the matrix of the (atomic) overlapping integrals:

$$S_{\mu\nu} = \int \phi_{\mu}\phi_{\nu} d\tau \tag{4.273}$$

where all indices in Eqs. (4.328)–(4.273) refers to matrix elements since the additional reference to the "*i*" electron was skipped for avoiding the risk of confusion.

The new one-electronic type equation (4.267) or (4.269) made the history of the computational chemistry in the second half of XX since they can be computed either from first principles (in which case one says that an *ab initio* approach was undertaken) or by resorting to experimental data (in which case the *semiempirical* approach was chosen).

It is worth noting that the particularization of Roothaan equations to spin up (alpha state) or spin down (beta state) through considering different spatial parts of the spin-orbitals generates the so-called *unrestricted Hartree-Fock* (UHF) frame of analysis. However, it describes the homolitic dissociation products or reactions in which the change in spin pairing is allowed. Otherwise, the so-called *restricted Hartree-Fock* (RHF) method can be employed whenever one would prefer to use orbital energy diagrams with two electrons rather than one electron per orbital.

Computationally, the procedure for solving the *HF* or Roothaan equations is *self-consistent* in the sense that the involved Fock operator depends implicitly upon the solutions. This feature is derived from the assumed one-electron picture in which a single electron would feel the potential influence coming from the fixed (or clamped) collection of nuclei and the average effects of all other $N-1$ electrons.

Therefore, the basic algorithm solves the one-electron problem (4.269) iteratively: guess the position for each electron, i.e., guess (C), then guess the average potential that an electron feels from the rest of electrons in the system, i.e., guess (F), solve the matrix equation, i.e., diagonalize to a new (C), form a new (F), repeat the procedure until the one-electronic wave function becomes consistent with the field produced by it and other electrons.

Regarding the *ab initio* methods, they are very effective since an arbitrary basis set of LCAO-MO produces accurate results without imposing additional approximations. Unfortunately, this method was criticized for this arbitrary degree of freedom, arguing that it produces a recipe in which "anything computes everything".

While this endeavor was made in the efforts to discredit the MO approach and the orbital concept in general, we believe that atomic orbitals and their linear combination provide the set of "elementary properties" of mater on which base the whole chemistry can be rationalized based on a single (i.e., the eigen-value problem) principle, either in Schrödinger, Hartree-Fock/ Roothaan or Kohn-Sham/Density Functional Theory (see below) approaches.

Yet, the solution of the matrix equation (4.267) may be unfolded through the Löwdin orthogonalization procedure (Löwdin, 1950; Löwdin, 1993), involving the diagonalization of the overlap matrix by means of a given unitary matrix (U), $(U)^{+}(U)$ =(4.249), by the resumed procedure:

$$(s)=(U)^{+}(S)(U) \tag{4.274}$$

$$\left(s^{-1/2}\right)_{ii}=\left[(s)_{ii}\right]^{-1/2} \tag{4.275}$$

$$\left(S^{-1/2}\right)=\left(U\right)\left(s^{-1/2}\right)\left(U\right)^{+} \tag{4.276}$$

$$\left(\left(S^{1/2}\right)\left(C\right)\right)^{+}\left(\left(S^{-1/2}\right)\left(H^{eff}\right)\left(S^{-1/2}\right)\right)\left(\left(S^{1/2}\right)\left(C\right)\right)=\left(E\right) \tag{4.277}$$

However, the solution given by Eq. (4.277) is based on the form of effective independent-electron Hamiltonians that can be quite empirically constructed – as in Extended Hückel Theory (Hoffmann, 1963); such "arbitrariness" can be nevertheless avoided by the so-called *self-consistent field* (SCF) in which the one-electron effective Hamiltonian is considered such that to depend by the solution of Eq. (4.266) itself, i.e., by the matrix of coefficients (*C*); this way we identify the resulted "Hamiltonian" as the Fock operator, while the associated eigen-problem rewrites the Hartree-Fock equation (4.267) under the mono-electronic wave-function representation:

$$F\psi_i = E_i\psi_i \tag{4.278}$$

The corresponding matrix representation actually gives further insight to the Eq. (4.269) now looking like:

$$\left(F\left(\left(C\right)\right)\right)\left(C\right)=\left(S\right)\left(C\right)\left(E\right) \tag{4.279}$$

Equation (4.279) may be iteratively solved through diagonalization procedure starting from an input (*C*) matrix or – more physically appealing – from a starting electronic distribution quantified by the density matrix:

$$P_{\mu\nu} = \sum_{i}^{occ} C_{\mu i} C_{i\nu} \tag{4.280}$$

with major influence on the Fock matrix elements:

$$F_{\mu\nu} = H_{\mu\nu} + \sum_{\lambda\sigma} P_{\lambda\sigma}\left[\left(\mu\nu\middle|\lambda\sigma\right)-\frac{1}{2}\left(\mu\lambda\middle|\nu\sigma\right)\right] \tag{4.281}$$

Note that now the one-electron Hamiltonian effective matrix components $H_{\mu V}$ differ from those of Eq. (4.272) in what they truly represent, this time the kinetic energy plus the interaction of a single electron with the core electrons around all the present nuclei. The other integrals appearing in Eq. (4.281) are generally called the two-electrons-multi-centers integrals and are written as:

$$(\mu\nu|\lambda\sigma)=\int\phi_\mu^A(\mathbf{r}_1)\phi_\nu^B(\mathbf{r}_1)\frac{1}{r_{12}}\phi_\lambda^C(\mathbf{r}_2)\phi_\sigma^D(\mathbf{r}_2)d\mathbf{r}_1d\mathbf{r}_2 \tag{4.282}$$

From definition (4.282), there is immediate to recognize the special integral $J=(\mu\mu|\nu\nu)$ as the Coulomb integral describing repulsion between two electrons with probabilities ϕ_μ^2 and ϕ_ν^2.

Moreover, the Hartree-Fock equation (4.279) with implementations given by Eqs. (4.280)–(4.281) are known as Roothaan equations (Roothaan, 1951) and constitute the basics for closed-shell (or restricted Hartree-Fock, RHF) molecular orbitals calculations. Their extension to the spin effects provides the equations for the open-shell (or unrestricted Hartree-Fock, UHF) known also as the Pople-Nesbet Unrestricted equations (Pople & Nesbet, 1954).

4.5.3 *INTRODUCING SEMI-EMPIRICAL APPROXIMATIONS*

The second level of approximation in molecular orbital computations regards the various ways the Fock matrix elements of Eq. (4.281) are considered, namely the approximations of the integrals (4.282) and of the effective one-electron Hamiltonian matrix elements $H_{\mu\nu}$.

The main route for such an endeavor is undertaken through neglecting at different degrees certain *differential* overlapping terms (integrals) – as an offset ansatz – although with limited physical justification – while the adjustment with experiment is done (post-factum) by fitting parameters – from where the semi-empirical name of such approximation. Practically, by emphasizing the (nuclear) centers in the electronic overlapping integral (4.273):

$$S_{\mu\nu}=\int\phi_\mu^A(\mathbf{r}_1)\phi_\nu^B(\mathbf{r}_1)d\mathbf{r}_1 \tag{4.283}$$

the differential overlap approximation may be considered by two situations.

- By *neglecting the differential overlap* (NDO) through the mono-atomic orbitalic constraint:

$$\phi_\mu \phi_\nu = \phi_\mu \phi_\mu \delta_{\mu\nu} \tag{4.284}$$

leaving with the simplified integrals:

$$S_{\mu\nu} = \delta_{\mu\nu} \int \phi_\mu^A(\mathbf{r}_1)\phi_\mu^A(\mathbf{r}_1)d\mathbf{r}_1 = \delta_{\mu\nu} \tag{4.285}$$

$$\begin{aligned}(\mu\nu|\lambda\sigma) &= \delta_{\mu\nu}\delta_{\lambda\sigma} \int \phi_\mu^A(\mathbf{r}_1)\phi_\mu^A(\mathbf{r}_1)\frac{1}{r_{12}}\phi_\lambda^B(\mathbf{r}_2)\phi_\lambda^B(\mathbf{r}_2)d\mathbf{r}_1 d\mathbf{r}_2 \\ &= \delta_{\mu\nu}\delta_{\lambda\sigma}(\mu_A\mu_A|\lambda_B\lambda_B) \equiv \delta_{\mu\nu}\delta_{\lambda\sigma}\gamma^{AB}\end{aligned} \tag{4.286}$$

thus reducing the number of bielectronic integrals, while the tri- and tetra-centric integrals are all neglected;
- By *neglecting the diatomic differential overlap* (NDDO) of the bi-atomic orbitals:

$$\phi_\mu^A \phi_\nu^B = \phi_\mu^A \phi_\nu^A \delta_{AB} \tag{4.287}$$

that implies the actual simplifications:

$$S_{\mu\nu} = \delta_{AB} \int \phi_\mu^A(\mathbf{r}_1)\phi_\nu^A(\mathbf{r}_1)d\mathbf{r}_1 = \delta_{AB}\delta_{\mu\nu} \tag{4.288}$$

$$(\mu\nu|\lambda\sigma) = \delta_{AB}\delta_{CD} \int \phi_\mu^A(\mathbf{r}_1)\phi_\nu^A(\mathbf{r}_1)\frac{1}{r_{12}}\phi_\lambda^C(\mathbf{r}_2)\phi_\sigma^C(\mathbf{r}_2)d\mathbf{r}_1 d\mathbf{r}_2 \tag{4.289}$$

when overlaps (or contractions) of atomic orbitals on different atoms are neglected.

For both groups of approximations specific methods are outlined below.

4.5.3.1 NDO Methods

The basic NDO approximation was developed by People and is known as the Complete Neglect of Differential Overlap CNDO semi-empirical method (Pople & Beveridge, 1970; Pople et al., 1965; Pople & Segal, 1965, 1966). It employs the approximations (4.284)–(4.286) such that the molecular rotational invariance is respected through the requirement the integral (4.286) depends only on the atoms A or B where the involved orbitals reside – and not by the orbitals themselves. That is the integral γ^{AB} in Eq. (4.285) is seen as the average electrostatic repulsion between an electron in any orbital of A and an electron in any orbital of B:

$$V_{AB} = Z_B \gamma^{AB} \tag{4.290}$$

In these conditions, the working Fock matrix elements of Eq. (4.281) become within RHF scheme:

$$F_{\mu\mu}^{CNDO} = H_{\mu\mu}^{CNDO} + \left(P_{AA} - \frac{1}{2} P_{\mu\mu} \right) \gamma^{AA} + \sum_{B \neq A} P_{BB} \gamma^{AB} \tag{4.291}$$

$$F_{\mu\nu}^{CNDO} = H_{\mu\nu}^{CNDO} - \frac{1}{2} P_{\mu\nu} \gamma^{AB} \tag{4.292}$$

From Eqs. (4.291) and (4.292) there follows that the core Hamiltonian has as well the diagonal and off-diagonal components; the diagonal one represents the energy of an electron in an atomic orbital of an atom (say A) written in terms of ionization potential and electron affinity of that atom (Oleari et al., 1966):

$$U_{\mu\mu}^{CNDO} = -\frac{1}{2}\left(I_\mu + A_\mu \right) - \left(Z_A - \frac{1}{2} \right) \gamma^{AA} \tag{4.293}$$

added to the attraction energy respecting the other (B) atoms to produce the one-center-one-electron integrals:

$$H_{\mu\mu}^{CNDO} = U_{\mu\mu}^{CNDO} - \sum_{B \neq A} V_{AB} \tag{4.294}$$

overall expressing the energy an electron in the atomic orbital φ_μ would have if all other *valence electrons* were removed to infinity. The non-diagonal terms (*the resonance integrals*) are parameterized in respecting the overlap integral and accounts (through β_{AB} parameter averaged over the atoms involved) on the diatomic bonding involved in overlapping:

$$H_{\mu\nu}^{CNDO} = \beta_{AB}^{CNDO} S_{\mu\nu} \tag{4.295}$$

The switch to the UHF may be eventually done through implementing the spin equivalence:

$$P^{T} \equiv P^{\uparrow+\downarrow} = \frac{1}{2}P^{\uparrow} = \frac{1}{2}P^{\downarrow} \tag{4.296}$$

although the spin effects are not at all considered since no exchange integral involved. This is in fact the weak point of the CNDO scheme and it is to be slightly improved by the next Semi-empirical methods.

The exchange effect due to the electronic spin accounted within the Intermediate Neglect of Differential Overlap (INDO) method (Slater, 1960) through considering in Eqs. (4.291) and (4.293) the exchange one-center integrals $\gamma^{AA} \equiv K = \left(\mu\nu\middle|\mu\nu\right)$ is evaluated as:

$$\left(sp_x\middle|sp_x\right)^{INDO} = \frac{1}{3}G^{1}, \left(p_x p_y\middle|p_x p_y\right)^{INDO} = \frac{3}{25}F^{2}, \ldots \tag{4.297}$$

in terms of the Slater-Condon parameters G^1, F^2, … usually used to describe atomic spectra.

The INDO method may be further modified in parameterization of the spin effects as developed by Dewar's group and led with the Modified Intermediate Neglect of Differential Overlap (MINDO) method (Baird & Dewar, 1969; Dewar & Hasselbach, 1970; Dewar & Lo, 1972; Bingham et al., 1975a-d; Dewar et al., 1975; Murrell & Harget, 1971) whose basic equations look like:

$$F_{\mu\nu}^{\uparrow(MINDO)} = \begin{cases} H_{\mu\nu}^{MINDO} - \left(\mu\mu\middle|\nu\nu\right)P_{\mu\nu}^{\uparrow} & \ldots\mu|_{A}, \nu|_{B\neq A} \\ \left(2P_{\mu\nu}^{\uparrow+\downarrow} - P_{\mu\nu}^{\uparrow}\right)\left(\mu\nu\middle|\mu\nu\right) - P_{\mu\nu}^{\uparrow}\left(\mu\mu\middle|\nu\nu\right) & \ldots\mu|_{A} \neq \nu|_{A} \end{cases} \tag{4.298}$$

$$F_{\mu\mu}^{\uparrow(MINDO)} = H_{\mu\mu}^{MINDO} + \sum_{\nu|A}\left[\left(\mu\mu\middle|\nu\nu\right)P_{\nu\nu}^{\uparrow+\downarrow} - \left(\mu\nu\middle|\mu\nu\right)P_{\nu\nu}^{\uparrow}\right]$$
$$+ \sum_{B}\gamma_{MINDO}^{AB}\sum_{A}^{B}P_{\mu\mu}^{\uparrow+\downarrow} \tag{4.299}$$

Apart from specific counting of spin effects, another particularity of MINDO respecting the CNDO/INDO is that all the non-zero two-center Coulomb integrals are set equal and parameterized by the appropriate one-center two electrons integrals A_A and A_B within the Ohno-Klopman expression (Ohno, 1964; Klopman, 1964):

$$\gamma_{MINDO}^{AB} = (s_A s_A | s_B s_B) = (s_A s_A | p_B p_B) = (p_A p_A | p_B p_B)$$
$$= \frac{1}{\sqrt{r_{AB}^2 + \frac{1}{4}\left(\frac{1}{A_A} + \frac{1}{A_B}\right)^2}} \quad (4.300)$$

The one-center-one-electron integral $H_{\mu\mu}$ is preserved from the CNDO/INDO scheme of computation, while the resonance integral (4.295) is modified as follows:

$$H_{\mu\nu}^{MINDO} = (I_\mu + I_\nu)\beta_{AB}^{MINDO} S_{\mu\nu} \quad (4.301)$$

with the parameter β_{AB}^{MINDO} being now dependent on the atoms-in-pair rather than the average of atomic pair involved. As in INDO, the exchange terms, i.e., the one-center-two-electron integrals, are computed employing the atomic spectra and the G^k, F^k, Slater-Condon parameters, see Eq. (4.297) (Pople et al., 1967). Finally, it is worth mentioning that the MINDO (also with its MINDO/3 version) improves upon the CNDO and INDO the molecular geometries, heats of formation, being particularly suited for dealing with molecules containing heteroatoms.

4.5.3.2 NDDO Methods

This second group of neglecting differential overlaps semi-empirical methods includes along the interaction quantified by the overlap of two orbitals centered on the same atom also the overlap of two orbitals belonging to different atoms. It is manly based on the Modified Neglect of Diatomic Overlap (MNDO) approximation of the Fock matrix, while introducing further types of integrals in the UHF framework (Dewar & Thiel, 1977;

Dewar & McKee, 1977; Dewar & Rzepa, 1978; Davis et al., 1981; Dewar & Storch, 1985; Thiel, 1988; Clark, 1985):

$$F_{\mu\nu}^{\uparrow(MNDO)} = \begin{cases} H_{\mu\nu}^{MNDO} - \sum_{\lambda|A}\sum_{\sigma|B}(\mu\lambda|\nu\sigma)P_{\lambda\sigma}^{\uparrow} & ...\mu|_A, \nu|_{B\neq A} \\ H_{\mu\nu}^{MNDO} + P_{\mu\nu}^{\uparrow}\left[3(\mu\nu|\mu\nu) - (\mu\mu|\nu\nu)\right] & \\ +\sum_{B}\sum_{\lambda|B}\sum_{\sigma|B}(\mu\nu|\lambda\sigma)P_{\lambda\sigma}^{\uparrow+\downarrow} & ...\mu|_A \neq \nu|_A \end{cases} \tag{4.302}$$

$$F_{\mu\mu}^{\uparrow(MNDO)} = H_{\mu\mu}^{MNDO} + \sum_{\nu|A}\left[(\mu\mu|\nu\nu)P_{\nu\nu}^{\uparrow+\downarrow} - (\mu\nu|\mu\nu)P_{\nu\nu}^{\uparrow}\right] + \sum_{B}\sum_{\lambda|B}\sum_{\sigma|B}(\mu\mu|\lambda\sigma)P_{\lambda\sigma}^{\uparrow+\downarrow} \tag{4.303}$$

Note that similar expressions can be immediately written within RHF once simply replacing:

$$P^{\uparrow(\downarrow)} = -\frac{1}{2}P^{\uparrow+\downarrow} \tag{4.304}$$

in above Fock (46a&b) expressions.

Now, regarding the (Coulombic) two-center-two-electron integrals of type (4.289) appearing in Eqs. (4.302) & (4.303) there were identified 22 unique forms for each pair of non-hydrogen atoms, *i.e.*, the rotational invariant 21 integrals $(ss|ss)$, $(ss|p_\sigma p_\sigma)$, $(ss|p_\pi p_\pi)$, ..., $(p_\sigma p_\sigma|p_\sigma p_\sigma)$, $(p_\pi p_\pi|p_\pi p_\pi)$, ..., $(sp_\sigma|sp_\sigma)$, $(sp_\pi|sp_\pi)$, ..., $(p_\pi p_\sigma|sp_\pi)$, $(p_\pi p_\sigma|p_\pi p_\sigma)$, and the 22nd one that is written as a combination of two of previously ones, namely $(p_\pi p_\pi'|p_\pi p_\pi') = 0.5\left[(p_\pi p_\pi|p_\pi p_\pi) - (p_\pi p_\pi|p_\pi' p_\pi')\right]$, with the typical integral approximation relaying on the Eq. (4.300) structure, however slightly modified as:

$$(ss|ss)^{MNDO} = \frac{1}{\sqrt{(r_{AB} + c_A + c_B)^2 + \frac{1}{4}\left(\frac{1}{A_A} + \frac{1}{A_B}\right)^2}} \tag{4.305}$$

where additional parameters c_A and c_B represent the distances of the multipole charges from their respective nuclei. The MNDO one-center one-electron integral has the same form as in NDO methods, *i.e.*, given by Eq. (4.294) with the average potential of Eq. (4.290) acting on concerned center; still, the resonance integral is modified as:

$$H_{\mu\nu}^{MNDO} = \frac{\beta_{\mu}^{MNDO} + \beta_{\nu}^{MNDO}}{2} S_{\mu\nu} \tag{4.306}$$

containing the atomic adjustable parameters β_{μ}^{MNDO} and β_{ν}^{MNDO} for the orbitals ϕ_{μ} and ϕ_{ν} of the atoms A and B, respectively. The exchange (one-center-two-electron) integrals are mostly obtained from data on isolated atoms. Basically, MNDO improves MINDO through the additional integrals considered the molecular properties such as the heats of formations, geometries, dipole moments, HOMO and LUMO energies, *etc.*, while problems still remaining with four-member rings (too stable), hypervalent compounds (too unstable) in general, and predicting out-of-plane nitro group in nitrobenzene and too short bond length (~ 0.17 Å) in peroxide – for specific molecules.

The MNDO approximation is further improved by aid of the Austin Model 1 (AM1) method (Dewar et al., 1985; Dewar & Dieter, 1986; Stewart, 1990) that refines the inter-electronic repulsion integrals:

$$\left(s_A s_A \middle| s_B s_B\right)^{AM1} = \frac{1}{\sqrt{r_{AB}^2 + \frac{1}{4}\left(\frac{1}{AM_A} + \frac{1}{AM_B}\right)^2}} \tag{4.307}$$

while correcting the one-center-two-electron atomic integrals of Eq. (4.300) by the specific (AM) monopole-monopole interaction parameters. In the same line, the nuclei-electronic charges interaction adds an energetic correction within the α_{AB} parameterized form:

$$\Delta E_{AB} = \sum_{A,B} \left\{ \begin{array}{l} Z_A Z_B \left(s_A s_A \middle| s_B s_B\right) \left[1 + \left(1 + \frac{1}{r_{AB}}\right) e^{-\alpha_{AB} r_{AB}}\right] \\ -Z_A Q_B \left(s_A s_A \middle| s_B s_B\right) \end{array} \right\} \tag{4.308}$$

The AM1 scheme, while furnishing better results than MNDO for some classes of molecules (e.g., for phosphorous compounds), still provides inaccurate modeling of phosphorous-oxygen bonds, too positive energy of nitro compounds, while the peroxide bond is still too short. In many case the reparameterization of AM1 under the Stewart's PM3 model (Stewart, 1989a,b) is helpful since it is based on a wider plethora of experimental data fitting with molecular properties. The best use of PM3 method lays in the organic chemistry applications.

To systematically implement the transition metal orbitals in semi-empirical methods the INDO method is augmented by Zerner's group either with non-spectroscopic and spectroscopic (*i.e.*, fitting with UV spectra) parameterization (Del Bene & Jaffé, 1968a-c), known as ZINDO/1 and ZINDO/S methods, respectively (Ridley & Zerner, 1976; Bacon & Zerner, 1979; Stavrev et al., 1995; Stavrev & Zerner, 1995; Cory et al., 1997; Anderson et al., 1986, 1991). The working equations are formally the same as those for INDO except for the energy of an atomic electron of Eq. (4.293) that now uses only the ionization potential instead of electronegativity of the concerned electron. Moreover, for ZINDO/S the core Hamiltonian elements $H_{\mu\mu}$ is corrected:

$$\Delta H_{\mu\mu}^{ZINDO} = \sum_{B}(Z_B - Q_B)\gamma_{(\mu\mu|ss)}^{AB(ZINDO)} \tag{4.309}$$

by the f_r parameterized integrals:

$$\gamma_{(\mu\mu|ss)}^{AB(ZINDO)} = \frac{f_r}{\dfrac{2f_r}{\gamma_{\mu\mu}^{A} + \gamma_{ss}^{B}} + r_{AB}},\ f_r = 1.2 \tag{4.310}$$

in terms of the one-center-two-electron Coulomb integrals $\gamma_{\mu\mu}^{A}, \gamma_{ss}^{B}$. Equation (4.310) conserves nevertheless the molecular rotational invariance through making the difference between the *s*- and *d*- Slater orbitals exponents. The same types of integrals correct also the nuclei-electronic interaction energy by quantity:

$$\Delta E_{AB} = \sum_{A,B}\left\{\frac{Z_A Z_B}{r_{AB}} - Z_A Q_B \gamma_{(\mu\mu|ss)}^{AB}\right\} \tag{4.311}$$

Since based on fitting with spectroscopic transitions the ZINDO methods are recommended in conjunction with single point calculation and not with geometry optimization, this should be consider by other off-set algorithms.

Beyond either NDO or NDDO methods, the self-consistent computation of molecular orbitals can be made by the so-called ab initio approach, directly relaying on the HF equation or on its density functional extension, as will be in next unfolded.

4.5.4 AB INITIO METHODS: THE HARTREE-FOCK APPROACH

The alternative to semi-empirical methods is the full self-consistent calculation or the so-called ab initio approach; it is based on computing of all integrals appearing on Eq. (4.281), yet with the atomic Slater type orbitals (STO), exp(−αr), being replaced by the Gaussian type orbitals (GTO) (Boys, 1950):

$$\phi_A^{GTO} = x_A^l y_A^m z_A^n \exp\left(-\alpha r_A^2\right) \tag{4.312}$$

in molecular orbitals expansion – a procedure allowing for much simplification in multi-center integrals computation. Nevertheless, at their turn, each GTO may be generalized to a contracted expression constructed upon the primitive expressions of Eq. (4.312):

$$\phi_\mu^{CGTO}\left(r_A\right) = \sum_p d_{p\mu}\phi_p^{GTO}\left(\alpha_p, r_A\right) \tag{4.313}$$

where $d_{p\mu}$ and α_A are called the exponents and the contraction coefficients of the primitives, respectively. Note that the primitive Gaussians involved may be chosen as approximate Slater functions (Szabo & Ostlund, 1996), Hartree-Fock atomic orbitals (Clementi & Roetti, 1974), or any other set of functions desired so that the computations become faster. In these conditions, a minimal basis set may be constructed with one function for H and He, five functions for Li to Ne, nine functions for Na to Ar, 13 functions for K and Ca, 18 functions for Sc to Kr, ..., etc., to describe the core and valence occupancies of atoms (Hehre et al., 1969; Collins et al., 1976; Stewart, 1970). Although such basis does not generally provide accurate results (because of its small cardinal), it contains the essential information regarding the chemical bond and may be useful for qualitative studies, as is the present case for aromaticity scales where the comparative trend is studied.

Actually, when simple ab initio method is referred it means that the Hartree-Fock equation (4.278) with full Fock matrix elements (Hartree, 1928a-b, 1957; Fock, 1930) of Eqs. (4.280) and (4.281) is solved for a Gaussian contracted basis (4.313). Actually, the method evaluates iteratively the kinetic energy and nuclear-electron attraction energy integrals – for the effective Hamiltonian, along the overlap and electron-electron repulsion energy integrals (for both the Coulomb and exchange terms), respectively written as:

$$T_{\mu\nu} = \langle\mu|\left(-\frac{1}{2}\nabla^2\right)|\nu\rangle \tag{4.314}$$

$$V_{\mu\nu} = \langle\mu|\frac{Z_A}{r_A}|\nu\rangle \tag{4.315}$$

$$S_{\mu\nu} = \langle\mu|\nu\rangle \tag{4.316}$$

$$(\mu\nu|\lambda\sigma) = (\mu\nu|\frac{1}{r_{12}}|\lambda\sigma) \tag{4.317}$$

until the consistency in electronic population of Eq. (4.280) between two consecutive steps is achieved.

Note that such calculation assumes the total wave function as a single Slater determinant, while the resultant molecular orbital is described as a linear combination of the atomic orbital basis functions (MO-LCAO). Multiple Slater determinants in MO description projects the configurationally and post-HF methods, and will not be discussed here.

4.5.4.1 Hartree-Fock Orbital Energy

Skipping the reference to the electronic (e) subscripts throughout Eqs. (4.250)–(4.262), the Hartree-Fock trial functional can firstly be arranged as by the optimization procedure (Putz & Chiriac, 2008)

$$E_0 \le E_{trial}^{HF}[\Psi^{HF}] = \left\langle \Psi^{HF} \middle| \hat{H} \middle| \Psi^{HF} \right\rangle = \left\langle \Psi^{HF} \middle| \hat{H}^{I} \middle| \Psi^{HF} \right\rangle + \left\langle \Psi^{HF} \middle| \hat{H}^{II} \middle| \Psi^{HF} \right\rangle \quad (4.318)$$

The one-electron (core) energetic component of Eq. (4.318) may be successively unfolded as:

$$\begin{aligned}
&\left\langle \Psi^{HF} \middle| \hat{H}^{I} \middle| \Psi^{HF} \right\rangle \\
&= N!\left\langle \Psi^{H} \middle| \hat{\wp}\hat{H}^{I}\hat{\wp} \middle| \Psi^{H} \right\rangle \\
&= N!\left\langle \Psi^{H} \middle| \hat{H}^{I}\hat{\wp}^{2} \middle| \Psi^{H} \right\rangle \\
&= N!\left\langle \Psi^{H} \middle| \hat{H}^{I}\hat{\wp} \middle| \Psi^{H} \right\rangle \\
&= \sum_{i=1}^{N}\sum_{P}(-1)^{P}\left\langle \Psi^{H} \middle| \left[-\frac{1}{2}\nabla_i^2 - \sum_{A}\frac{Z_A}{r_{iA}}\right]\hat{\wp} \middle| \Psi^{H} \right\rangle \\
&\equiv \sum_{i=1}^{N}\sum_{P}(-1)^{P}\left\langle \Psi^{H} \middle| \hat{h}_i(1)\hat{\wp} \middle| \Psi^{H} \right\rangle \\
&= \sum_{i=1}^{N}\left\langle \Psi^{H} \middle| \hat{h}_i(1) \middle| \Psi^{H} \right\rangle \\
&= \sum_{i=1}^{N}\left\langle \chi_i^{\sigma}(1) \middle| \hat{h}_i(1) \middle| \chi_i^{\sigma}(1) \right\rangle \\
&= \sum_{i=1}^{N}\int \chi_i^{\sigma *}(1)\hat{h}_i(1)\chi_i^{\sigma}(1)d\tau_1 \\
&\equiv \sum_{i=1}^{N} h_{ii} \\
&\equiv H_{ii}
\end{aligned} \quad (4.319)$$

where it was considered that the introduced one-electron effective operator $\hat{h}_i(1)$ selects from the Hartree wave function (4.259) the associate spin-orbital, for each electron, accordingly.

Similarly, the two-electron energetic component of Eq. (4.318) may be successively transformed as:

$$\begin{aligned}
&\left\langle \Psi^{HF}\left|\hat{H}^{II}\right|\Psi^{HF}\right\rangle \\
&= N!\left\langle \Psi^{H}\left|\hat{\wp}\hat{H}^{II}\hat{\wp}\right|\Psi^{H}\right\rangle \\
&= N!\left\langle \Psi^{H}\left|\hat{H}^{II}\hat{\wp}^{2}\right|\Psi^{H}\right\rangle \\
&= N!\left\langle \Psi^{H}\left|\hat{H}^{II}\hat{\wp}\right|\Psi^{H}\right\rangle \\
&= \sum_{\substack{i,j=1\\ i<j}}^{N}\sum_{P}(-1)^{P}\left\langle \Psi^{H}\left|\frac{1}{r_{ij}}\hat{\wp}\right|\Psi^{H}\right\rangle \\
&= \sum_{\substack{i,j=1\\ i<j}}^{N}\left\langle \Psi^{H}\left|\frac{1}{r_{ij}}(1-P_{ij})\right|\Psi^{H}\right\rangle \\
&= \sum_{\substack{i,j=1\\ i<j}}^{N}\left\langle \Psi^{H}\left|\frac{1}{r_{ij}}\right|\Psi^{H}\right\rangle - \sum_{\substack{i,j=1\\ i<j}}^{N}\left\langle \Psi^{H}\left|\frac{1}{r_{ij}}P_{ij}\right|\Psi^{H}\right\rangle \\
&= \sum_{\substack{i,j=1\\ i<j}}^{N}\left\langle \chi_i^{\sigma}(1)\chi_j^{\sigma}(2)\left|\frac{1}{r_{12}}\right|\chi_i^{\sigma}(1)\chi_j^{\sigma}(2)\right\rangle \\
&\quad - \sum_{\substack{i,j=1\\ i<j}}^{N}\left\langle \chi_i^{\sigma}(1)\chi_j^{\sigma}(2)\left|\frac{1}{r_{12}}\right|\chi_j^{\sigma}(1)\chi_i^{\sigma}(2)\right\rangle \\
&\equiv \sum_{\substack{i,j=1\\ i<j}}^{N}\left\langle \chi_i^{\sigma}(1)\left|\hat{J}_j(1)\right|\chi_i^{\sigma}(1)\right\rangle - \sum_{\substack{i,j=1\\ i<j}}^{N}\left\langle \chi_i^{\sigma}(1)\left|\hat{K}_j(1)\right|\chi_i^{\sigma}(1)\right\rangle \\
&= \sum_{\substack{i,j=1\\ i<j}}^{N}\iint \chi_i^{\sigma*}(1)\chi_j^{\sigma*}(2)\frac{1}{r_{ij}}\chi_i^{\sigma}(1)\chi_j^{\sigma}(2)d\tau_1 d\tau_2 \\
&\quad - \sum_{\substack{i,j=1\\ i<j}}^{N}\iint \chi_i^{\sigma*}(1)\chi_j^{\sigma*}(2)\frac{1}{r_{ij}}\chi_j^{\sigma}(1)\chi_i^{\sigma}(2)d\tau_1 d\tau_2
\end{aligned}$$

$$\begin{aligned}
&\equiv \sum_{\substack{i,j=1\\ i<j}}^{N} J_{ij} - \sum_{\substack{i,j=1\\ i<j}}^{N} K_{ij} \\
&= \frac{1}{2}\sum_{i,j=1}^{N}\left[J_{ij} - K_{ij}\right] \\
&\equiv U_{ee}
\end{aligned} \tag{4.320}$$

resulting in the effective electron-electron repulsion energy once the quantum exchange terms K_{ij} are subtracted from the classical Coulombic ones J_{ij}. Here we recognize the combined classical (Coulombic) – quantum (exchange) effects that appear in the inter-electronic repulsion Hamiltonian term (4.255).

All together, with the results of Eqs. (4.319) and (4.320) back in Eq. (4.318), we get for the trial Hartree-Fock functional the expression:

$$E_{trial}^{HF}[\Psi^{HF}] = \sum_{i=1}^{N} h_{ii} + \frac{1}{2}\sum_{i,j=1}^{N}\left[J_{ij} - K_{ij}\right] \tag{4.321}$$

In next, we are going to apply the variational principle respecting the variations of the spin-orbitals in terms of Lagrange multipliers ε_{ij} that widely demands that:

$$\delta\left\{E_{trial}^{HF}[\Psi^{HF}] - \sum_{i,j=1}^{N} \varepsilon_{ij}^{HF}\left(\left\langle \chi_i^{\sigma}(1)\middle|\chi_j^{\sigma}(1)\right\rangle - 1\right)\right\} = 0 \tag{4.322a}$$

However, by employing the canonical transformation, i.e., the N^2 parameters may be considered as the elements of a Hermitian matrix which through a unitary transformation become a diagonal matrix, the outset form of the variational principle (4.322) now reads:

$$\delta\left\{E_{trial}^{HF}[\Psi^{HF}] - \sum_{i=1}^{N} \varepsilon_{i}^{HF}\left(\left\langle \chi_i^{\sigma}(1)\middle|\chi_i^{\sigma}(1)\right\rangle - 1\right)\right\} = 0 \tag{4.322b}$$

Note that performing a unitary transformation will not affect the average of the electronic Hamiltonian but only the *HF* wave function by a phase factor of unity modulus. Under these circumstances, the famous Hartree-Fock equation results from the successive equivalent forms:

$$\sum_{i=1}^{N}\frac{\delta}{\delta\chi_i^{\sigma*}(1)}\left\{\begin{array}{l}-\frac{1}{2}\int\chi_i^{\sigma*}(1)\nabla_i^2\chi_i^{\sigma}(1)d\tau_1-\sum_A Z_A\int\chi_i^{\sigma*}(1)\frac{1}{r_{iA}}\chi_i^{\sigma}(1)d\tau_1\\+\frac{1}{2}\sum_{j=1}^{N}\iint\chi_i^{\sigma*}(1)\chi_j^{\sigma*}(2)\frac{1}{r_{ij}}\chi_i^{\sigma}(1)\chi_j^{\sigma}(2)d\tau_1 d\tau_2\\-\frac{1}{2}\sum_{j=1}^{N}\iint\chi_i^{\sigma*}(1)\chi_j^{\sigma*}(2)\frac{1}{r_{ij}}\chi_j^{\sigma}(1)\chi_i^{\sigma}(2)d\tau_1 d\tau_2\\-\varepsilon_i^{HF}\int\chi_i^{\sigma*}(1)\chi_i^{\sigma}(1)d\tau_1\end{array}\right\}=0$$

$$\Leftrightarrow\left[-\frac{1}{2}\nabla_i^2-\sum_A\frac{Z_A}{r_{iA}}\right]\chi_i^{\sigma}(1)+\left[\sum_{j=1}^{N}\int\chi_j^{\sigma*}(2)\frac{1}{r_{ij}}\chi_j^{\sigma}(2)d\tau_2\right]\chi_i^{\sigma}(1)$$
$$-\left[\sum_{j=1}^{N}\int\chi_j^{\sigma*}(2)\frac{1}{r_{ij}}\chi_i^{\sigma}(2)d\tau_2\right]\chi_j^{\sigma}(1)=\varepsilon_i^{HF}\chi_i^{\sigma}(1) \quad (4.322c)$$

Still, a more compact form of *HF* equation (4.322c) may be achieved since specific potential notations are introduced. For instance, the electrostatic repulsion potential (i.e., the Coulombic interaction) can be shortened as:

$$V_j^{ee}(1)=\int\chi_j^{\sigma*}(2)\frac{1}{r_{ij}}\chi_j^{\sigma}(2)d\tau_2 \quad (4.323)$$

while for the exchange potential (i.e., non-local interaction) we can define it as satisfying the relation:

$$V_j^{ex}(1)f(1)=\left[\int\chi_j^{\sigma*}(2)\frac{1}{r_{ij}}f(2)d\tau_2\right]\chi_j^{\sigma}(1) \quad (4.324)$$

With these the above *HF* equation (4.322c) reduces to its most simple form:

$$\hat{F}(1)\chi_i^{\sigma}(1)=\varepsilon_i^{HF}\chi_i^{\sigma}(1) \quad (4.325)$$

where the one-electronic Fock operator

$$\hat{F}(1)=-\frac{1}{2}\nabla_i^2+V_i^{HFeff}(1) \quad (4.326)$$

was introduced in terms of the effective-one potential

$$V_i^{HFeff}(1) = -\sum_A \frac{Z_A}{r_{iA}} + \sum_{j=1}^{N} V_j^{ee}(1) - \sum_{j=1}^{N} V_j^{ex}(1) \quad (4.327)$$

Now, since the spin-orbitals satisfies the normalization condition

$$\int \left|\chi_i^\sigma(1)\right|^2 d\tau_1 = 1 \quad (4.328)$$

the orbital energies look like:

$$\varepsilon_i^{HF} = \int \chi_i^{\sigma *}(1)\hat{F}(1)\chi_i^\sigma(1)d\tau_1 = h_{ii} + \sum_{j=1}^{N}\left[J_{ij} - K_{ij}\right] \quad (4.329)$$

while the total HF energy will take the form:

$$E^{HF}[\Psi^{HF}] = \sum_{i=1}^{N} \varepsilon_i^{HF} - U_{ee} \quad (4.330)$$

where

$$\sum_{i=1}^{N} \varepsilon_i^{HF} = \sum_{i=1}^{N} h_{ii} + \sum_{i,j=1}^{N}\left[J_{ij} - K_{ij}\right] \quad (4.331)$$

Remarkably, one can clearly see that the predicted *HF* energy (4.330) differs from the simple sum over the *HF* orbital energies (4.331) by the effective electron-electron interaction energy U_{ee}. We will return on this matter with more subtle consequences on Section 4.4.4.3.

4.5.4.2 About Correlation Energy

The post self-consistent era was mainly dedicated to the implementation of the so nominated *correlation energy* in the computation (Putz & Chiriac, 2008).

Firstly, it was noticed that a single Slater determinant (on which base the current *HF* analysis was exposed) can never account for a complete description of the many-electronic interaction. That is, the correlation energy can be introduced as the difference between the exact eigen-value and the Hartree-Fock energy of the same Hamiltonian for the concerning state:

$$E_{corr} = E - E^{HF} \tag{4.332}$$

The next step was sustained by the assumption that the correlation energy can be seen as the perturbation of the self-consistent-field energy, which is associated with a wave function derived for a single electronic configuration. At this point the basic methods of approximation used in quantum chemistry, namely the *perturbation* and *variational*, can be considered.

In the case that perturbation method is employed, assuming the unperturbed wave function and energy as the HF solutions the exact eigen-functions and eigen-values can be written as expanded series

$$\Psi_e = \Psi_e^{HF} + \ell\Psi_e^{(1)} + \ell^2\Psi_e^{(2)} + ... \tag{4.333}$$

$$E_e = E^{HF} + \ell E_e^{(1)} + \ell^2 E_e^{(2)} + ... \tag{4.334}$$

by introducing the ordering parameter ℓ. Through truncating the series in the second, third or fourth order generates the so-called Møller-Plesset MP2, MP3, and MP4 perturbative approximations, respectively.

On the other side, the linear variational method can be practiced within the *configuration interaction (CI)* approach of the many-electronic wave-function:

$$\Psi_e^{CI} = c_0\Psi_e^{HF} + \sum_s c_a^s\Psi_a^s + c_{ab}^{sd}\Psi_{ab}^{sd} + c_{abc}^{sdt}\Psi_{abc}^{sdt} + ... \tag{4.335}$$

where the Ψ_0, Ψ_a^s, Ψ_{ab}^{sd}, Ψ_{abc}^{sdt} stands for the ground, single excited, double excited, and triple excited *N*-electron trial wave functions, respectively, for a given spin state. While the *CI* wave function is the subject of the eigen-problem:

$$\hat{H}_e\Psi_e^{CI} = \tilde{E}_0\Psi_e^{CI} \tag{4.336}$$

the correlation correction to *HF* energy can be achieved through subtracting the *HF* energy from last equation

$$\left(\hat{H}_e - E^{HF}\right)\Psi_e^{CI} = \left(\tilde{E}_0 - E^{HF}\right)\Psi_e^{CI} = E_{corr}\Psi_e^{CI} \tag{4.337}$$

However, although, starting from this point, many sophisticated methods for wave function expansion, for example, the coupled cluster approach, multi-configuration self-consistent-field method or multi-reference *CI* methods, have been developed, the correlation problem faced many computational limitation, some of them almost insurmountable, due to the immense number of integrals to be evaluated.

4.5.4.3 Koopman's Orbital Theorem with Hartree-Fock Picture

Now one will make the essential difference between (Putz, 2013):

- the entire orbital spectrum available to a many –body systems, which include *occupied orbitals + unoccupied orbitals (up to infinity)*, denoted by $\left|\psi_i\right\rangle_{i=1,\dots,N\to\infty}$ which generates the Hartree-Fock energy
- the occupied orbitals on the many-body system, which will determine the total energy of the system, denoted by $\left|\psi_a\right\rangle_{i=1,\dots,N}$

All together, we can deal with the first lowest N spin-orbitals occupied in the overall wave-function $\left|\Psi_0^{(N)}\right\rangle = \left|\psi_1 ... \psi_a ... \psi_N\right\rangle$, while the rest (from N up to infinity) virtual of unoccupied orbitals, formally denoted as $\psi_r, \psi_s, ...$ (see Figure 4.3)
The conceptual difference consist in the fact that when dealing with infinite number of orbitals one does not avoid the double counting since there will be always available virtual orbitals to be occupied since the infinite cardinal of this set of orbitals; so we have with the actual subtle (mixed) notations:

$$\begin{aligned}\varepsilon_{i=1,\dots,\infty}^{HF} &= \left\langle\psi_i\right|\hat{f}\left|\psi_i\right\rangle \\ &= \left\langle\psi_i\right|\left(\hat{h} + \sum_{b=1}^{N}\left[\hat{J}_b - \hat{K}_b\right]\right)\left|\psi_i\right\rangle\end{aligned}$$

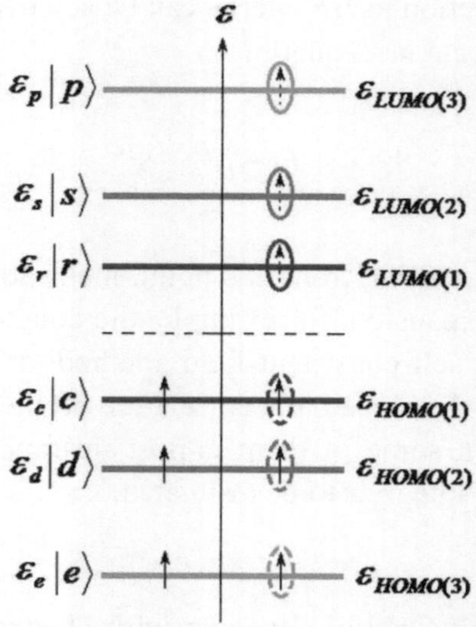

FIGURE 4.3 The paradigmatic *in silico* spectra of the first three highest occupied and lowest unoccupied molecular orbitals HOMOs and LUMOs illustrating the respective, successive, ionization and affinities energies as provided by Koopmans' theorem. Note KT implies ionization and affinity of one electron on successive levels and not of successive electrons on levels- see the marked occupied and virtual spin-orbitals (Putz, 2013).

$$=\langle\psi_i|\hat{h}|\psi_i\rangle+\sum_{b=1}^{N}\left[\langle\psi_i|\hat{J}_b|\psi_i\rangle-\langle\psi_i|\hat{K}_b|\psi_i\rangle\right]$$

$$\equiv\langle i|\hat{h}|i\rangle+\sum_{b=1}^{N}\left[\langle ii|bb\rangle-\langle ib|bi\rangle\right]$$

$$\equiv\langle i|\hat{h}|i\rangle+\sum_{b=1}^{N}\langle ib\|ib\rangle \tag{4.338}$$

Here one remarks that the Coulombic inter-electronic

$$J_b(1)=\int d2\psi_b^*(2)r_{12}^{-1}\psi_b(2) \tag{4.339}$$

and exchange terms

$$K_b(1)=\int d2\psi_b^*(2)r_{12}^{-1}\psi_b(1)=\int d2\psi_b^*(2)r_{12}^{-1}\wp_{12}\psi_b(2) \tag{4.340}$$

were remained with occupied orbitals' notation since they are readily computed among existing electrons.

Interesting, when the orbital energy (4.338) is summed just over occupied Hartree-Fock orbitals, as done in Eq. (4.331), now we equivalently obtain, yet under new notation revealing the restrain to the occupied orbitals

$$\sum_{a=1}^{N}\varepsilon_a = \sum_{a=1}^{N}\langle a|\hat{h}|a\rangle + \sum_{a=1,b=1}^{N}\langle ab|ab\rangle \tag{4.341}$$

Instead, when searched for total energy of the system one should avoid double counting and deal with occupied only orbitals to successively get within the actual notations

$$\begin{aligned}
E_N &= \left\langle \Psi_0^{(N)} \middle| \widehat{H} \middle| \Psi_0^{(N)} \right\rangle \\
&= \sum_{a=1}^{N} \langle \psi_a | \left(\hat{h} + \sum_{b=1}^{N} \left[\widehat{J}_b - \widehat{K}_b \right] \right) | \psi_a \rangle \\
&= \sum_{a=1}^{N} \langle \psi_a | \hat{h} | \psi_a \rangle + \frac{1}{2} \sum_{a=1,b=1}^{N} \begin{bmatrix} \langle \psi_a | \int d2 \psi_b^*(2) r_{12}^{-1} \psi_b(2) | \psi_a \rangle \\ -\langle \psi_a | \int d2 \psi_b^*(2) r_{12}^{-1} \wp_{12} \psi_b(2) | \psi_a \rangle \end{bmatrix} \\
&= \sum_{a=1}^{N} \langle \psi_a | \hat{h} | \psi_a \rangle + \frac{1}{2} \sum_{a=1,b=1}^{N} \begin{bmatrix} \int d1 d2 \psi_a^*(1) \psi_b^*(2) r_{12}^{-1} \psi_b(2) \psi_a(1) \\ -\int d1 d2 \psi_a^*(1) \psi_b^*(2) r_{12}^{-1} \wp_{12} \psi_b(2) \psi_a(1) \end{bmatrix} \\
&= \sum_{a=1}^{N} \langle \psi_a | \hat{h} | \psi_a \rangle + \frac{1}{2} \sum_{a=1,b=1}^{N} \begin{bmatrix} \int d1 d2 \psi_a^*(1) \psi_b^*(2) r_{12}^{-1} \psi_b(2) \psi_a(1) \\ -\int d1 d2 \psi_a^*(1) \psi_b^*(2) r_{12}^{-1} \psi_b(1) \psi_a(2) \end{bmatrix} \\
&= \sum_{a=1}^{N} \langle \psi_a | \hat{h} | \psi_a \rangle + \frac{1}{2} \sum_{a=1,b=1}^{N} \begin{bmatrix} \int d1 d2 \psi_a^*(1) \psi_a(1) r_{12}^{-1} \psi_b^*(2) \psi_b(2) \\ -\int d1 d2 \psi_a^*(1) \psi_b(1) r_{12}^{-1} \psi_b^*(2) \psi_a(2) \end{bmatrix} \\
&\equiv \sum_{a=1}^{N} \langle \psi_a | \hat{h} | \psi_a \rangle + \frac{1}{2} \sum_{a=1,b=1}^{N} \left[\langle aa|bb \rangle - \langle ab|ba \rangle \right] \\
&\equiv \sum_{a=1}^{N} \langle a | \hat{h} | a \rangle + \frac{1}{2} \sum_{a=1,b=1}^{N} \langle ab|ab \rangle
\end{aligned} \tag{4.342}$$

In obvious difference respecting Eq. (4.341), as already anticipated from the Eqs. (4.330) & (4.331).

Eq. (4.341) does not exactly recovering the above total energy of the N-occupied spin-orbitals Eq. (4.342), when they where considered "free (not depending)" of computation (basis set); however, this may be considered as *in silico* manifestation of quantum "observability" (once a basis set representation applies) which destroys the quantum system in itself's (or eigen) manifestation. Here the mathematical properties of eigen-function computed upon a given basis on Hilbert-Banach spaces determine the "shift" or the "unrealistic" energies of orbitals since spanning those occupied and unoccupied alike; from the present dichotomy basically follows all critics on the Hartree-Fock formalism and of allied molecular orbital theory, Koopmans' "theorem" included (see below); instead, there seems that such departure of the computed from the exact energy orbitals is inherent to quantum formalism and not necessary a weakness of the Hartree-formalism itself, since it will appear to any quantum many-particle problem involving eigen-problems.

Now, returning to the above occupied and unoccupied orbital energy, one may assume (*Koopmans' ansatz*) that, on the frontier levels of a many-electronic system, extracting or adding of an electron (or even few of them, but lesser than the total number of valence electrons) will not affect the remaining (or $N \pm 1, N \pm 1 \pm 1', \ldots$ electronic orbitals) states, on successive levels and not successive electrons on levels (see Figure 4.3).

This approach allows simplifying of the common terms and emphasizing only on the involving frontier orbitals participating in chemical reactivity. Accordingly, for the first ionization potential one successively obtains the first highest occupied molecular orbital (HOMO), see Figure 4.3:

$$\begin{aligned}
IP_1 &= E_{N-1} - E_N \\
&= \left\langle \Psi_c^{(N-1)} \middle| \widehat{H} \middle| \Psi_c^{(N-1)} \right\rangle - \left\langle \Psi_0^{(N)} \middle| \widehat{H} \middle| \Psi_0^{(N)} \right\rangle \\
&= \left\{ \sum_{\substack{a=1 \\ a \neq c}}^{N} \langle a | \hat{h} | a \rangle + \frac{1}{2} \sum_{\substack{a=1, b=1 \\ a \neq c, b \neq c}}^{N} \langle ab | ab \rangle \right\}
\end{aligned}$$

$$
\begin{aligned}
&-\left\{\left[\sum_{\substack{a=1\\a\neq c}}^{N}\langle a|\hat{h}|a\rangle+\langle c|\hat{h}|c\rangle\right]+\frac{1}{2}\left[\begin{array}{l}\sum\limits_{\substack{a=1,b=1\\a\neq c,b\neq c}}^{N}\langle ab|ab\rangle+\sum\limits_{\substack{a=1,b=1\\a\neq c,b=c}}^{N}\langle ac|ac\rangle\\+\sum\limits_{\substack{a=1,b=1\\a=c,b\neq c}}^{N}\langle cb|cb\rangle\end{array}\right]\right\}\\
&=-\langle c|\hat{h}|c\rangle-\sum_{b=1}^{N}\langle cb|cb\rangle\\
&=-\varepsilon_{c}\\
&=-\varepsilon_{HOMO(1)}
\end{aligned}
\tag{4.343}
$$

Remarkable, in this analytics, one starts with *in se* quantum expression of total energies of the N and $(N–1)$ systems and ends up with a result characteristic to the computational (shifted) realm since recovering the orbital energy of the *in silico* state from which the electron was removed. Yet, one may ask how such *in se–to–in silico* quantum chemical passage is possible; the answer is naturally positive since the above derivation associates with the ionization process which is basically an observer intervention to the genuine quantum system, from where the final result will reflect the energetic deviation from *in se–to–in silico* as an irrefutable quantum manifestation of electronic system.

Similarly, for electronic affinity, one will act on the *in se* quantum system to add an electron at the frontier level and, under the "frozen spin-orbitals" physical-chemical assumption, one gets the energetic turn from the genuine HF expression to the *in silico* orbital energy on which the "action" was undertaken towards the first lowest occupied molecular orbital (LUMO), see Figure 4.3:

$$
\begin{aligned}
EA_1&=E_N-E_{N+1}\\
&=-\langle r|\hat{h}|r\rangle-\sum_{b=1}^{N}\langle rb|rb\rangle\\
&=-\varepsilon_{r}\\
&=-\varepsilon_{LUMO(1)}
\end{aligned}
\tag{4.344}
$$

These results are usually considered as defining the popular Koopmans theorem, used for estimating the observable quantities as ionization potential and electronic affinity in terms of "artefactual" computed orbital energies (first approximation) and in the spin-orbitalic frozen framework during the electronic extraction or addition (the second approximation).

However one may ask whether this approximation is valid and in which conditions. This can be achieved by reconsidering the above Koopmans first order IP and EA to the superior differences within Hartree-Fock framework; as such, for the second order of ionization potential one gets the second highest occupied molecular orbital ($HOMO_2$), see Figure 4.3:

$$
\begin{aligned}
IP_2 &= E_{N-2} - E_{N-1} \\
&= \left\{ \begin{array}{l} \sum\limits_{\substack{a=1\\a\neq c\\a\neq d}}^{N} \langle a|\hat{h}|a\rangle \\ +\dfrac{1}{2}\sum\limits_{\substack{a=1,b=1\\a\neq c,b\neq c\\a\neq d,b\neq d}}^{N} \langle ab|ab\rangle \end{array} \right\} - \left\{ \begin{array}{l} \sum\limits_{\substack{a=1\\a\neq c}}^{N} \langle a|\hat{h}|a\rangle \\ +\dfrac{1}{2}\sum\limits_{\substack{a=1,b=1\\a\neq c,b\neq c}}^{N} \langle ab|ab\rangle \end{array} \right\} \\
&= \left\{ \sum\limits_{\substack{a=1\\a\neq c\\a\neq d}}^{N} \langle a|\hat{h}|a\rangle + \frac{1}{2}\sum\limits_{\substack{a=1,b=1\\a\neq c,b\neq c\\a\neq d,b\neq d}}^{N} \langle ab|ab\rangle \right\} \\
&\quad - \left\{ \begin{array}{l} \sum\limits_{\substack{a=1\\a\neq c\\a\neq d}}^{N} \langle a|\hat{h}|a\rangle + \langle d|\hat{h}|d\rangle + \dfrac{1}{2}\sum\limits_{\substack{a=1,b=1\\a\neq c,b\neq c\\a\neq d,b\neq d}}^{N} \langle ab|ab\rangle \\ +\dfrac{1}{2}\sum\limits_{\substack{a=1,b=1\\a\neq c,b\neq c\\a=d,b\neq d}}^{N} \langle db|db\rangle + \dfrac{1}{2}\sum\limits_{\substack{a=1,b=1\\a\neq c,b\neq c\\a\neq d,b=d}}^{N} \langle ad|ad\rangle \end{array} \right\} \\
&= -\langle d|\hat{h}|d\rangle - \sum\limits_{\substack{a=1,b=1\\b\neq d}}^{N} \langle db|db\rangle \\
&= -\varepsilon_d \\
&= -\varepsilon_{HOMO(2)}
\end{aligned}
\tag{4.345}
$$

Note that this derivation eventually employs the equivalency for the Coulombic and exchange terms for orbitals of the same nature (with missing the same number of spin-orbitals, see Figure 4.3). However, in the case this will be further refined to isolate the first two orders of highest occupied molecular orbitals, the last expression will be corrected with $HOMO_1$/ $HOMO_2$ (Coulombic and exchange) interaction to successively become

$$\begin{aligned} IP_2 &= E_{N-2} - E_{N-1} \\ &= -\langle d|\hat{h}|d\rangle - \left\{ \begin{aligned} &\frac{1}{2}\sum_{\substack{a=1,b=1\\ a\neq c,\\ a=d,b\neq d}}^{N} \langle db|db\rangle + \frac{1}{2}\langle dc|dc\rangle \\ &+\frac{1}{2}\sum_{\substack{a=1,b=1\\ b\neq c,\\ a\neq d,b=d}}^{N} \langle ad|ad\rangle + \frac{1}{2}\langle cd|cd\rangle \end{aligned} \right\} \\ &= -\langle d|\hat{h}|d\rangle - \sum_{\substack{a=1,b=1\\ b\neq d}}^{N} \langle db|db\rangle + \langle cd|cd\rangle \\ &= -\varepsilon_d + \langle cd|cd\rangle \\ &= -\varepsilon_{HOMO(2)} + \langle HOMO_1\, HOMO_2|HOMO_1\, HOMO_2\rangle \end{aligned} \tag{4.346}$$

However, reloading this procedure for electronic affinity process too, one gets

$$\begin{aligned} EA_2 &= E_{N+1} - E_{N+2} \\ &= -\varepsilon_{LUMO(2)} + \langle LUMO_1\, LUMO_2|LUMO_1\, LUMO_2\rangle \end{aligned} \tag{4.347}$$

When combining Eq. (4.347) with its IP counterpart (4.346) there appears that the simple Koopmans' orbitals energy difference is corrected by the $HOMO_1/HOMO_2$ vs. $LUMO_1/LUMO_2$

$$\begin{aligned} IP_2 - EA_2 &= \varepsilon_{LUMO(2)} - \varepsilon_{HOMO(2)} \\ &+ \begin{pmatrix} \langle HOMO_1\, HOMO_2|HOMO_1\, HOMO_2\rangle \\ -\langle LUMO_1\, LUMO_2|LUMO_1\, LUMO_2\rangle \end{pmatrix} \end{aligned} \tag{4.348}$$

This expression is usually reduced to the superior order LUMO-HOMO difference

$$IP_2 - EA_2 \cong \varepsilon_{LUMO(2)} - \varepsilon_{HOMO(2)} \tag{4.349}$$

due to the energetic spectra symmetry of Figure 4.3 relaying on the bonding vs. anti-bonding displacements of orbitals, specific to molecular orbital theory. Therefore, with the premise that molecular orbital theory itself is correct, or at least a reliable quantum undulatory modeling of multi-electronic systems moving in a nuclei potential, the above IP-EA differences in terms of Koopmans' *in silico* LUMO-HOMO energetic gaps holds also for superior orders.

4.5.4.4 Chemical Reactivity Indices in Orbital Energy Representation

Koopmans' theorem entered on the quantum chemistry as a versatile tool for estimating the ionization potentials for closed-shells systems, and it was widely confirmed for organic molecular systems, due to the inner usually separation between sigma (core) and pi (valence) sub-electronic systems, allowing to treat the "frozen spin-orbitals" as orbitals not essentially depending on the number of electrons in the valence shells, when some of them are extracted (via ionization) or added (via negative attachments); this approximation ultimately works for Hartree-Fock systems when electronic correlation may be negligible or cancels with the orbital relaxations during ionization or affinity processes, respectively; naturally, it works less when correlation is explicitly counted, as in Density Functional Theory, where instead the exchange energies are approximated or merely parameterized so that "loosing" somehow on the genuine spin-orbital nature of the mono-determinantal approach of the Hartree-Fock, with a natural energetic hierarchy included.

Beside the many concepts in modeling the chemical reactivity and interaction electronegativity and chemical hardness are by far the most versatile measures, to be detailed in the next volumes of this set, since their direct connection with total, valence or orbital energies of atoms and molecules via the first and the second derivative of such energies with respecting the available or concerned electrons therein. Actually, such derivatives, may use the molecular frontier orbitals when based on differential expansion of the energy around its isolated value to account both for the electrophilic (electrons accepting) and nucleophilic (electrons donating) states.

Starting from the general mathematical framework, given the values of a function *f*(*n*) on a set of nodes $\{...,n-3,n-2,n-1,n,n+1,n+2,n+3,...\}$ the finite difference approximations of the first $f_n^{'}$ and second $f_n^{''}$ derivatives in the node *n*, will spectrally depend on the all the nodal values. However, the compact finite differences, or Padé, schemes that mimic this global dependence write as (Lele, 1992):

$$\beta_1 f_{n-2}^{'} + \alpha_1 f_{n-1}^{'} + f_n^{'} + \alpha_1 f_{i+1}^{'} + \beta_1 f_{n+2}^{'}$$
$$= c_1 \frac{f_{n+3} - f_{n-3}}{6} + b_1 \frac{f_{n+2} - f_{n-2}}{4} + a_1 \frac{f_{n+1} - f_{n-1}}{2} \tag{4.350}$$

$$\beta_2 f_{n-2}^{''} + \alpha_2 f_{n-1}^{''} + f_n^{''} + \alpha_2 f_{i+1}^{''} + \beta_2 f_{n+2}^{''}$$
$$= c_2 \frac{f_{n+3} - 2f_n + f_{n-3}}{9} + b_2 \frac{f_{n+2} - 2f_n + f_{n-2}}{4}$$
$$+ a_2 \left(f_{n+1} - 2f_n + f_{n-1}\right) \tag{4.351}$$

The involved sets of coefficients, $\{a_1, b_1, c_1, \alpha_1, \beta_1\}$ and $\{a_2, b_2, c_2, \alpha_2, \beta_2\}$ are derived by matching Taylor series coefficients of various orders. This way, their particularizations can be reached as the second (2C)-, fourth (4C)- and sixth (6C)-order central differences; standard Pade (SP) schemes; sixth (6T)- and eight (8T)-order tridiagonal schemes; eighth (8P)- and tenth (10P)- order pentadiagonal schemes up to spectral-like resolution (SLR) ones, see Table 4.1.

Assuming that the function *f*(*n*) is the total energy *E*(*N*) in the actual node that corresponds to the number of electrons, the compact finite difference, the derivatives of Eqs. (4.350) and (4.351) may be accurately evaluated through considering the states with *N*-3, *N*-2, *N*-1, *N*+1, *N*+2, *N*+3 electrons, whereas the derivatives in the neighbor states will be taken only as their most neighboring dependency. This way, the working formulas for electronegativity will be (Putz, 2010a):

$$-\chi = \left.\frac{\partial E}{\partial N}\right|_{|N\rangle}$$
$$\cong a_1 \frac{E_{N+1} - E_{N-1}}{2} + b_1 \frac{E_{N+2} - E_{N-2}}{4} + c_1 \frac{E_{N+3} - E_{N-3}}{6}$$

TABLE 4.1 Numerical Parameters for the Compact Finite Second (2C)-, Fourth (4C)- and Sixth (6C)-Order Central Differences; Standard Padé (SP) Schemes; Sixth (6T)- and Eight (8T)-Order Tridiagonal Schemes; Eighth (8P)- and Tenth (10P)-Order Pentadiagonal Schemes up to Spectral-Like Resolution (SLR) Schemes Unfolding the Numerical Derivatives (4.350) and (4.350) Then Used for the Electronegativity and Chemical Hardness of Eqs. (4.352) and (4.353) and the Subsequent of Their Respective Formulations: Eqs. (4.362) and (4.363); (4.368) and (4.369)

Scheme	Electronegativity					Chemical Hardness				
	a_1	b_1	c_1	α_1	β_1	a_2	b_2	c_2	α_2	β_2
2C	1	0	0	0	0	1	0	0	0	0
4C	$\frac{4}{3}$	$-\frac{1}{3}$	0	0	0	$\frac{4}{3}$	$-\frac{1}{3}$	0	0	0
6C	$\frac{3}{2}$	$-\frac{3}{5}$	$\frac{1}{10}$	0	0	$\frac{12}{11}$	$\frac{3}{11}$	0	$\frac{2}{11}$	0
SP	$\frac{5}{3}$	$\frac{1}{3}$	0	$\frac{1}{2}$	0	$\frac{6}{5}$	0	0	$\frac{1}{10}$	0
6T	$\frac{14}{9}$	$\frac{1}{9}$	0	$\frac{1}{3}$	0	$\frac{3}{2}$	$-\frac{3}{5}$	$\frac{1}{5}$	0	0
8T	$\frac{19}{12}$	$\frac{1}{6}$	0	$\frac{3}{8}$	0	$\frac{147}{152}$	$\frac{51}{95}$	$-\frac{23}{760}$	$\frac{9}{38}$	0
8P	$\frac{40}{27}$	$\frac{25}{54}$	0	$\frac{4}{9}$	$\frac{1}{36}$	$\frac{320}{393}$	$\frac{310}{393}$	0	$\frac{344}{1179}$	$\frac{23}{2358}$
10P	$\frac{17}{12}$	$\frac{101}{150}$	$\frac{1}{100}$	$\frac{1}{2}$	$\frac{1}{20}$	$\frac{1065}{1798}$	$\frac{1038}{899}$	$\frac{79}{1798}$	$\frac{334}{899}$	$\frac{43}{1798}$
SLR	1.303	0.994	0.038	0.577	0.09	0.216	1.723	0.177	0.502	0.056

Adapted from Rubin & Khosla (1977), Putz (2010a, 2011), and Putz et al. (2004).

$$-\alpha_1\left(\frac{\partial E}{\partial N}\bigg|_{|N-1\rangle}+\frac{\partial E}{\partial N}\bigg|_{|N+1\rangle}\right)-\beta_1\left(\frac{\partial E}{\partial N}\bigg|_{|N-2\rangle}+\frac{\partial E}{\partial N}\bigg|_{|N+2\rangle}\right)$$

$$=a_1\frac{E_{N+1}-E_{N-1}}{2}+b_1\frac{E_{N+2}-E_{N-2}}{4}+c_1\frac{E_{N+3}-E_{N-3}}{6}$$

$$-\alpha_1\left(a_1\frac{E_N-E_{N-2}}{2}+a_1\frac{E_{N+2}-E_N}{2}\right)-\beta_1\left(a_1\frac{E_{N-1}-E_{N-3}}{2}+a_1\frac{E_{N+3}-E_{N+1}}{2}\right)$$

$$
\begin{aligned}
&= a_1\left(1+\beta_1\right)\frac{E_{N+1}-E_{N-1}}{2}+\left(b_1-2a_1\alpha_1\right)\frac{E_{N+2}-E_{N-2}}{4}\\
&+\left(c_1-3a_1\beta_1\right)\frac{E_{N+3}-E_{N-3}}{6}
\end{aligned}
\tag{4.352}
$$

and respectively for the chemical hardness as (Putz, 2011; Putz, 2010a; Putz et al., 2004):

$$
\begin{aligned}
2\eta &= \left.\frac{\partial^2 E}{\partial N^2}\right|_{|N\rangle}\\
&\cong 2a_2\frac{E_{N+1}-2E_N+E_{N-1}}{2}+b_2\frac{E_{N+2}-2E_N+E_{N-2}}{4}\\
&+c_2\frac{E_{N+3}-2E_N+E_{N-3}}{9}\\
&-\alpha_2\left(\left.\frac{\partial^2 E}{\partial N^2}\right|_{|N-1\rangle}+\left.\frac{\partial^2 E}{\partial N^2}\right|_{|N+1\rangle}\right)-\beta_2\left(\left.\frac{\partial^2 E}{\partial N^2}\right|_{|N-2\rangle}+\left.\frac{\partial^2 E}{\partial N^2}\right|_{|N+2\rangle}\right)\\
&= 2a_2\frac{E_{N+1}-2E_N+E_{N-1}}{2}+b_2\frac{E_{N+2}-2E_N+E_{N-2}}{4}\\
&+c_2\frac{E_{N+3}-2E_N+E_{N-3}}{9}\\
&-\alpha_2\left(2a_2\frac{E_N-2E_{N-1}+E_{N-2}}{2}+2a_2\frac{E_{N+2}-2E_{N+1}+E_N}{2}\right)\\
&-\beta_2\left(2a_2\frac{E_{N-1}-2E_{N-2}+E_{N-3}}{2}+2a_2\frac{E_{N+3}-2E_{N+2}+E_{N+1}}{2}\right)\\
&= 2a_2\left(1+2\alpha_2-\beta_2\right)\frac{E_{N+1}+E_{N-1}}{2}+\left(8a_2\beta_2+b_2-4a_2\alpha_2\right)\frac{E_{N+2}+E_{N-2}}{4}\\
&+\left(c_2-9a_2\beta_2\right)\frac{E_{N+3}+E_{N-3}}{9}-\left(2a_2+\frac{1}{2}b_2+\frac{2}{9}c_2+2a_2\pm_2\right)E_N
\end{aligned}
\tag{4.353}
$$

where the involved parameters discriminate between various schemes of computations and the spectral-like resolution (SLR), see Table 4.1 (Rubin & Khosla, 1977; Putz, 2011; Putz, 2010a; Putz et al., 2004).

Next, the Eqs. (4.352) and (4.353) may be rewritten in terms of the observational quantities, as the ionization energy and electronic affinity are with the aid of their basic definitions from the involved eigen-energies of *i*-th (*i*=1,2,3) order

$$I_i = E_{N-i} - E_{N-i+1} \tag{4.354}$$

$$A_i = E_{N+i-1} - E_{N+i} \tag{4.355}$$

As such they allow the energetic equivalents for the differences

$$E_{N+1} - E_{N-1} = -(I_1 + A_1) \tag{4.356}$$

$$E_{N+2} - E_{N-2} = -(I_1 + A_1) - (I_2 + A_2) \tag{4.357}$$

$$E_{N+3} - E_{N-3} = -(I_1 + A_1) - (I_2 + A_2) - (I_3 + A_3) \tag{4.358}$$

and for the respective sums (Putz, 2011; Putz, 2010a; Putz et al., 2004)

$$E_{N+1} + E_{N-1} = (I_1 - A_1) + 2E_N \tag{4.359}$$

$$E_{N+2} + E_{N-2} = (I_1 - A_1) + (I_2 - A_2) + 2E_N \tag{4.360}$$

$$E_{N+3} + E_{N-3} = (I_1 - A_1) + (I_2 - A_2) + (I_3 - A_3) + 2E_N \tag{4.361}$$

being then implemented to provide the associate "spectral" molecular analytical forms of electronegativity

$$\begin{aligned} \chi_{CFD} = & \left[a_1 (1 - \alpha_1) + \frac{1}{2} b_1 + \frac{1}{3} c_1 \right] \frac{I_1 + A_1}{2} \\ & + \left[b_1 + \frac{2}{3} c_1 - 2a_1 (\alpha_1 + \beta_1) \right] \frac{I_2 + A_2}{4} \\ & + (c_1 - 3a_1 \beta_1) \frac{I_3 + A_3}{6} \end{aligned} \tag{4.362}$$

and for chemical hardness (Putz, 2011; Putz, 2010a; Putz et al., 2004):

$$\eta_{CFD} = \left[a_2(1-\alpha_2+2\beta_2)+\frac{1}{4}b_2+\frac{1}{9}c_2 \right]\frac{I_1 - A_1}{2} + \left[\frac{1}{2}b_2+\frac{2}{9}c_2+2a_2(\beta_2-\alpha_2) \right]\frac{I_2 - A_2}{4} + \left[\frac{1}{3}c_2 - 3a_2\beta_2 \right]\frac{I_3 - A_3}{6} \quad (4.363)$$

It is worth remarking that when particularizing these formulas for the fashioned two-point central finite difference, i.e., when having $a_1 = 1, b_1 = c_1 = \alpha_1 = \beta_1 = 0$ and $a_2 = 1, b_2 = c_2 = \alpha_2 = \beta_2 = 0$ of Table 4.1, one recovers the consecrated Mulliken (spectral) electronegativity (Mulliken, 1934)

$$\chi_{2C} = \frac{I_1 + A_1}{2} \quad (4.364)$$

and the chemical hardness basic form relating with the celebrated Pearson nucleophilic-electrophilic reactivity gap (Parr & Yang, 1989; Pearson, 1997)

$$\eta_{2C} = \frac{I_1 - A_1}{2} \quad (4.365)$$

already used as measuring the aromaticity through the molecular stability against the reaction propensity (Ciesielski et al., 2009; Chattaraj et al., 2007).

Finally, for computational purposes, Eqs. (4.362) and (4.363) may be once more reconsidered within the Koopmans' frozen core approximation (Koopmans, 1934), in which various orders of ionization potentials and electronic affinities are replaced by the corresponding frontier energies

$$I_i = -\varepsilon_{HOMO(i)} \quad (4.366)$$

$$A_i = -\varepsilon_{LUMO(i)} \tag{4.367}$$

so that the actual working compact finite difference (CFD) orbital molecular electronegativity unfolds as (Putz, 2011; Putz, 2010a; Putz et al., 2004):

$$\begin{aligned}\chi_{CFD} = &-\left[a_1(1-\alpha_1) + \frac{1}{2}b_1 + \frac{1}{3}c_1\right]\frac{\varepsilon_{HOMO(1)} + \varepsilon_{LUMO(1)}}{2} \\ &-\left[b_1 + \frac{2}{3}c_1 - 2a_1(\alpha_1 + \beta_1)\right]\frac{\varepsilon_{HOMO(2)} + \varepsilon_{LUMO(2)}}{4} \\ &-(c_1 - 3a_1\beta_1)\frac{\varepsilon_{HOMO(3)} + \varepsilon_{LUMO(3)}}{6}\end{aligned} \tag{4.368}$$

along with the respective chemical hardness formulation

$$\begin{aligned}\eta_{CFD} = &\left[a_2(1-\alpha_2+2\beta_2) + \frac{1}{4}b_2 + \frac{1}{9}c_2\right]\frac{\varepsilon_{LUMO(1)} - \varepsilon_{HOMO(1)}}{2} \\ &+\left[\frac{1}{2}b_2 + \frac{2}{9}c_2 + 2a_2(\beta_2 - \alpha_2)\right]\frac{\varepsilon_{LUMO(2)} - \varepsilon_{HOMO(2)}}{4} \\ &+\left[\frac{1}{3}c_2 - 3a_2\beta_2\right]\frac{\varepsilon_{LUMO(3)} - \varepsilon_{HOMO(3)}}{6}\end{aligned} \tag{4.369}$$

Note that the actual CFD electronegativity and chemical hardness expressions do not distinguish for the atoms-in-molecule contributions, while providing post-bonding information and values, i.e., for characterizing the already stabilized/optimized molecular structure towards its further reactive engagements. The difference between the atoms-in-molecule pre-bonding stage and the molecular post-bonding one constitutes the basis of the actual absolute aromaticity as will be elsewhere introduced (see Volume III/Chapter 4 of the present five-volume set).

An illustrative analysis for homologues organic aromatic hydrocarbons regarding how much the second, respectively the third order of the IP-EA or LUMO-HOMO gaps affect the chemical hardness hierarchies, and therefore their ordering aromaticity, will be in the next section exposed and discussed.

4.5.4.5 Testing Koopmans Theorem by Chemical Harness Reactivity Index

It is true Koopmans theorem seems having some limitation for small molecules and for some inorganic complexes (Duke & O'Leary, 1995; Angeli, 1998).

However, one is interested here for testing the Koopmans' superior orders' HOMO-LUMO behavior on the systems that work, such as the aromatic hydrocarbons. Accordingly, in Table 4.2 a short series of paradigmatic organics are considered, with one and two rings and various basic ring substitutions or additions, respectively (Putz, 2010b). For them, the HOMO and LUMO are computed, within semi-empirical AM1 framework (Hypercube, 2002), till the third order of Koopmans frozen spin-orbitals' approximation; they are then combined into the various finite difference forms (from 2C to SLR) of chemical hardness as above, see Table 4.1, grouped also in sequential order respecting chemical hardness gap contributions (i.e., separately for {LUMO1-HOMO1}, {LUMO1-HOMO1, LUMO2-HOMO2}, {LUMO1-HOMO1, LUMO2-HOMO2, LUMO3-HOMO3}): the results are systematically presented in Tables 4.3–4.5.

The results of Tables 4.3-4.5 reveals very interesting features, in the light of considering the aromaticity as being reliably measured by chemical hardness alone, sine both associate with chemical resistance to reactivity or the terminus of a chemical reaction according with the maximum chemical hardness principle (Chattaraj et al., 1991,1995).

Moreover, the benchmark ordering hierarchy was chosen as produced by Hückel theory (since being an approximate approach for quantum chemical modeling of chemical bonding is let to be exposed in the Volume III of this work (Putz, 2016a), dedicated to quantum molecule and chemical reactivity) and approximation since closely related with pi-electrons delocalized at the ring level as the main source of the experimentally recorded aromaticity of organic compounds under study (Putz et al., 2010).

Note that although computational method used here is of low level it nevertheless responds to present desiderate having an non (orbitalic) basis dependent computational output and discussion, whereas further (Hartree-Fock) ab initio, (Møller–Plesset) perturbation methods and basis set dependency considerations, as HF, MP2, and DFT, respectively, for instance, can be further considered for comparative analysis.

TABLE 4.2 Molecular Structures of Paradigmatic Aromatic Hydrocarbons (Putz, 2010b), Ordered Downwards According with Their Hückel First Order HOMO-LUMO Gap (Putz et al., 2010), along Their First Three Highest Occupied (HOMOs) and Lowest Unoccupied (LUMOs) (in electron-volts, eV) Computationally Recorded Levels Within Semi-Empirical AM1 Method (Hypercube, 2002)

Formula **Name** **CAS** **Index (mw[g/mol])**	**Molecular Structure**	**HOMO (1)**	**HOMO (2)**	**HOMO (3)**	**LUMO (1)**	**LUMO (2)**	**LUMO (3)**
C_6H_6 Benzene 71-43-2 I (78.11)	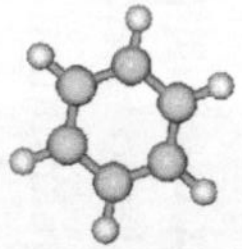	–9.652904	–9.653568	–11.887457	0.554835	0.555246	2.978299
$C_4H_4N_2$ Pyrimidine 289-95-2 II (80.088)	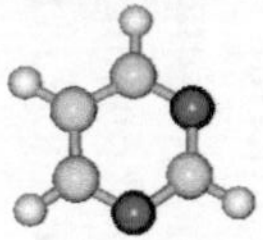	–10.578436	–10.614932	–11.602985	–0.234993	–0.081421	2.543489
C_5H_5N Pyridine 110-86-1 III (79.10)	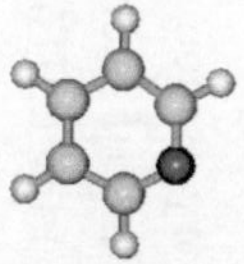	–9.932324	–10.642881	–10.716373	0.138705	0.278273	2.791518
C_6H_6O Phenol 108-95-2 IV (94.11)	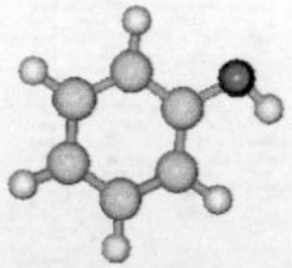	–9.114937	–9.851116	–11.940266	0.397517	0.507986	2.839472

TABLE 4.2 Continued

Formula Name CAS Index (mw[g/mol])	Molecular Structure	HOMO (1)	HOMO (2)	HOMO (3)	LUMO (1)	LUMO (2)	LUMO (3)
C_6H_7N Aniline 62-53-3 V (93.13)	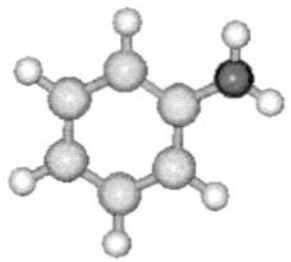	–8.213677	–9.550989	–11.501620	0.758436	0.888921	2.828224
$C_{10}H_8$ Naphthalene 91-20-3 VI (128.17)		–8.710653	–9.340973	–10.658237	–0.265649	0.180618	1.210350
$C_{10}H_8O$ 2-Naphthol 135-19-3 VII (144.17)	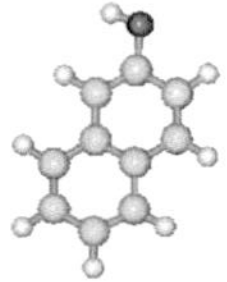	–8.641139	–9.194596	–10.673578	–0.348490	0.141728	1.117961

TABLE 4.2 Continued

Formula Name CAS Index (mw[g/mol])	Molecular Structure	HOMO (1)	HOMO (2)	HOMO (3)	LUMO (1)	LUMO (2)	LUMO (3)
$C_{10}H_8O$ 1-Naphthol 90-15-3 VIII (144.17)	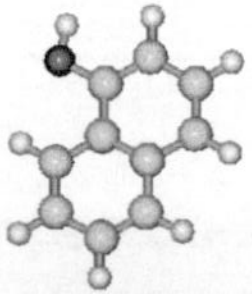	–8.455599	–9.454717	–10.294406	–0.247171	0.100644	1.184179
$C_{10}H_9N$ 2-Naphthalenamine 91-59-8 IX (143.19)	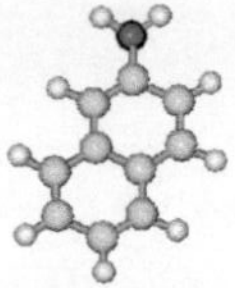	–8.230714	–8.984826	–10.346699	–0.177722	0.278785	1.298534
$C_{10}H_9N$ 1-Naphthalenamine 134-32-7 X (143.19)	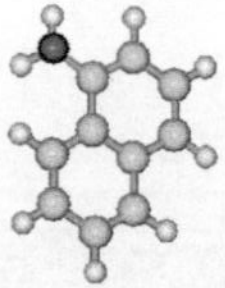	–8.109827	–9.343444	–9.940875	–0.176331	0.230424	1.235745

TABLE 4.3 Chemical Hardness Values (in eV) as Computed for Molecules of Table 4.2 with First LUMO(1)–HOMO(1) Gap Order of Eq. (4.326) with Parameters of Table 4.1 (Putz, 2013)

Molecule	η_{2C}	η_{4C}	η_{6C}	η_{SP}	η_{6T}	η_{8T}	η_{8P}	η_{10P}	η_{SLR}
I	5.10387	6.379837	4.903511	5.512179	7.003643	4.434762	4.030827	3.542746	2.971354
II	5.171722	6.464652	4.968699	5.585459	7.096751	4.493719	4.084414	3.589844	3.010856
III	5.035515	6.294393	4.837839	5.438356	6.909845	4.375368	3.976843	3.495299	2.931559
IV	4.756227	5.945284	4.569516	5.136725	6.5266	4.132695	3.756273	3.301437	2.768964
V	4.486057	5.607571	4.309951	4.844941	6.155866	3.897943	3.542904	3.113904	2.611677
VI	4.222502	5.278128	4.056743	4.560302	5.794211	3.66894	3.334759	2.930963	2.458242
VII	4.146325	5.182906	3.983556	4.47803	5.689679	3.60275	3.274597	2.878086	2.413893
VIII	4.104214	5.130268	3.943098	4.432551	5.631894	3.56616	3.24134	2.848856	2.389378
IX	4.026496	5.03312	3.868431	4.348616	5.525247	3.49863	3.179962	2.794909	2.344132
X	3.966748	4.958435	3.811029	4.284088	5.44326	3.446715	3.132775	2.753437	2.309348

TABLE 4.4 Chemical Hardness Values (in eV) as Computed for Molecules of Table 4.2 with First LUMO(1)–HOMO(1) and Second Order LUMO(2)–HOMO(2) Gaps of Eq. (4.326) with Parameters of Table 4.1 (Putz, 2013)

Molecule	η_{2C}	η_{4C}	η_{6C}	η_{SP}	η_{6T}	η_{8T}	η_{8P}	η_{10P}	η_{SLR}
I	5.10387	5.95447	4.239094	4.89965	6.351413	3.933493	3.865279	3.990091	4.778726
II	5.171722	6.025756	4.283151	4.953449	6.423777	3.976506	3.9136	4.051417	4.875712
III	5.035515	5.839345	4.127062	4.783086	6.212105	3.839122	3.799743	3.973858	4.865044
IV	4.756227	5.513655	3.895318	4.515179	5.864769	3.624046	3.588288	3.755367	4.602943
V	4.486057	5.172574	3.630494	4.218546	5.488872	3.385327	3.373608	3.571375	4.459963
VI	4.222502	4.881395	3.437052	3.989007	5.185887	3.201415	3.180355	3.348194	4.143948
VII	4.146325	4.793892	3.375923	3.917851	5.093191	3.144321	3.123197	3.287199	4.066799
VIII	4.104214	4.732127	3.32121	3.859229	5.021412	3.096976	3.086388	3.267567	4.081062
IX	4.026496	4.647136	3.265531	3.792799	4.933405	3.043772	3.029741	3.200836	3.984165
X	3.966748	4.559524	3.187936	3.709656	4.831596	2.976622	2.977523	3.172958	4.004309

TABLE 4.5 Chemical Hardness Values (in eV) as Computed for Molecules of Table 4.2 with First LUMO(1)–HOMO(1), Second LUMO(2)–HOMO(2) and Third Order LUMO(3)–HOMO(3) Gaps of Eq. (4.326) with Parameters of Table 4.1 (Putz, 2013)

Molecule	η_{2C}	η_{4C}	η_{6C}	η_{SP}	η_{6T}	η_{8T}	η_{8P}	η_{10P}	η_{SLR}
I	5.10387	5.95447	4.239094	4.89965	6.516588	3.908499	3.806245	3.921086	4.834997
II	5.171722	6.025756	4.283151	4.953449	6.58096	3.952722	3.857423	3.985751	4.929261
III	5.035515	5.839345	4.127062	4.783086	6.362192	3.816411	3.746102	3.911156	4.916176
IV	4.756227	5.513655	3.895318	4.515179	6.028988	3.599197	3.529596	3.686762	4.658889
V	4.486057	5.172574	3.630494	4.218546	5.648093	3.361234	3.316702	3.504858	4.514206
VI	4.222502	4.881395	3.437052	3.989007	5.31776	3.18146	3.133223	3.293101	4.188874
VII	4.146325	4.793892	3.375923	3.917851	5.224208	3.124496	3.076372	3.232464	4.111434
VIII	4.104214	4.732127	3.32121	3.859229	5.148952	3.077677	3.040805	3.214284	4.124512
IX	4.026496	4.647136	3.265531	3.792799	5.062797	3.024193	2.983496	3.14678	4.028246
X	3.966748	4.559524	3.187936	3.709656	4.955781	2.957831	2.933139	3.121078	4.046616

In these conditions, the main Koopmans' analysis of chemical hardness or aromaticity behavior for the envisaged molecules leaves with relevant observations:

- In absolutely all cases, analytical or computational, the first two molecules, Benzene (I) and Pyrimidine (II) are inversed for their chemical hardness/aromaticity hierarchies respecting the benchmarking Hückel one, meaning that even in the most simple case, say 2C/{LUMO1-HOMO1}, double substitution of carbon with nitrogen increases the ring stability, most probably due to the additional pairing of electrons entering the pi-system as coming from the free valence of *N* atoms (equivalently with *N* pi-valence electrons) in molecular ring. This additional pair of electrons eventually affects by shielding also the core of the hydrocarbon rings, i.e., the sigma-system of Pyrimidine (II), in a specific quantum way, not clearly accounted by the Hückel theory.
- The same behavior is recorded also for the couple of molecules I and III (Pyridine), however, only for the SLR of chemical hardness computed with second and the third orders of Koopmans frozen spin-orbitals; this suggest the necessary insight the spectral like resolution analysis may provide respecting the other forms of finite compact differences in chemical hardness computation – yet only when it is combined with higher Koopmans HOMO and LUMO orbitals.
- In the same line of discussion, only for the second and the third Koopmans order and only for the SLR chemical hardness development, i.e., the last columns of Tables 4.4 and 4.5, one record similar reserve order of the molecules 2-Napthol (VII) and 1-Naphtol (VIII), with the more aromatic character for the last case when having the OH group more closely to the middy of the naphthalene structure; it is explained as previously, due to the electronic pair of chemical bonding contribution more close to the "core" of the system with direct influence to increase the shielding electrons of the sigma systems, while leading with smoothly increased stabilization contribution (enlarging also the sigma-pi chemical gap); yet this is manifested when all the spectral like resolution complexity is considered in chemical hardness expression and only in superior Koopmans orders (second and third), otherwise not being recorded.

However, this result advocates the meaningful of considering of the SLR coupled with superior Koopmans analysis in revealing subtle effects in sigma-pi aromatic systems.

- In the rest of cases the Hückel downward hierarchy of Table 4.2 is recovered in Tables 4.3–4.5 in a systematic way.
- When going from 2C to SLR chemical hardness analytical forms of any of Koopmans orders, on the horizontal axis through the Tables 4.3–4.5, one systematically record an increasing of the average chemical hardness/aromaticity values from 2C to 6T schemes of computations while going again down towards SLR scheme of Table 4.1.

All in all, one may compare the extreme 2C and SLR outputs of Tables 4.3–4.5 for a global view for the Koopmans' behavior respecting various orders and chemical hardness schemes of (compact finite forms) computations: the result is graphically presented in Figure 4.4. The analysis of Figure 4.4 yields a fundamental result for the present study, i.e., the practical identity among:

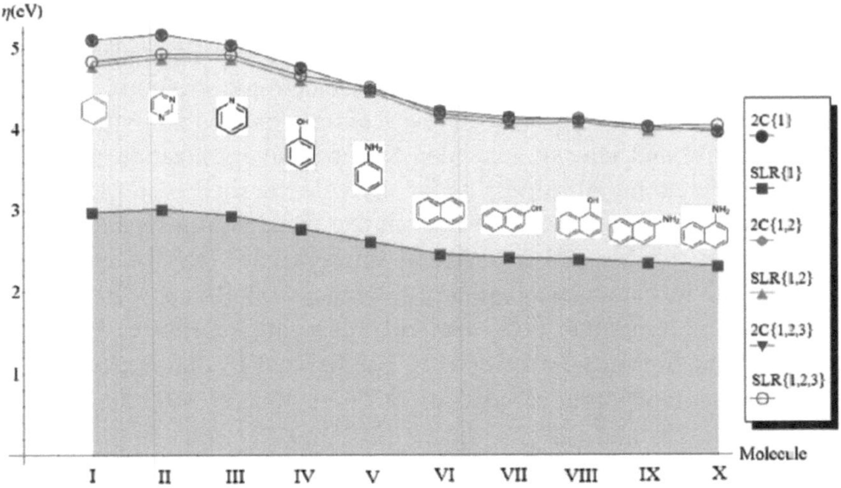

FIGURE 4.4 Representation of the 2C and SLR chemical hardness hierarchies for the set of molecules of Table 4.2 upon the first, second and third order of the Koopmans' theorem applications as presented din Tables 4.3–4.5, respectively (Putz, 2013).

- All Koopmans superior orbitals based chemical hardness computations;
- The simplest 2C and the complex SLR analytical forms for compact finite difference schemes of chemical hardness for the superior HOMO-LUMO gap extensions;

By contrary to someone expecting the first order of Koopmans theorem being more systematic, only in this order 2C result is practically doubled respecting SLR counterpart; such double behavior becomes convergent when superior Koopmans orders of valence orbitals are considered either in simpler or complex forms of 2C and SLR, respectively.

Despite the debating context in which Koopmans theorem is valid, or associates with a physical-chemical sense, the present work give some insight in this matter by clarifying upon some key features of Koopmans analysis, namely:

- The Hartree-Fock spin-orbitals involved in Koopmans' theorem are of computational nature, emerged through solving an eigen-problem in a given basis set so that being characterized by a sort of "quantum shift" related with quantum uncertainty when the free system is affected by observation – here by computation; so this behavior is at its turn computationally naturally and not viewed as a conceptual error in structurally assessing a many-electronic structure;
- The Koopmans' theorem not restrictedly refers to the first ionization potential and may be extended to successive ionization potentials (and electronic affinities) as far the valence shell is not exhausted by the pi-collective electrons, such that the sigma-pi separation may be kept reliable and the "frozen spin-orbitals" may be considered as such through cancellation of the relaxation effects with the electronic correlations, both explicitly escaping to Hartree-Fock formalism; this was however here emphasized by the appearance of the quantum terms of type $\langle HOMO_1 HOMO_2 | HOMO_1 HOMO_2 \rangle$ in Eq. (4.328) and $\langle LUMO_1 LUMO_2 | LUMO_1 LUMO_2 \rangle$ in Eq. (4.329) which were considered as reciprocal annihilating in chemical hardness' IP-EA differences in Eq. (4.330) due to symmetrical bonding vs. anti-bonding spectra displacements in molecular orbital theory – as a simplified version of Hartree-Fock theory;

- The Koopmans theorem goes at best with chemical harness or aromaticity evaluation by means of LUMO-HOMO gaps when they manifested surprisingly the same for superior orders of IPs-EAs, this way confirming the previous point.

Application on a paradigmatic set of mono and double benzoic rings molecules supported these conclusions, yet leaving enough space for further molecular set extensions and computational various frameworks comparison.

This may lead with the fruitful result according which the Koopmans theorem works better when superior HOMO-LUMO frozen spin-orbitals are considered, probably due to compensating correlating effects such extension implies, see the last section' analytical discussion. In any case, the present molecular illustration of Koopmans' approximations to chemical harness computation clearly shows that, at least for organic aromatic molecules, it works better for superior orders of "freezing" spin-orbitals and is not limitative to the first valence orbitals, as would be the common belief. Moreover, it was also clear the Koopmans theorem finely accords also with more complex ponder of its superior order orbitals in chemical hardness expansions Eq. (4.326), when subtle effects in lone pairing electrons (since remained orbital is frozen upon successive electronic attachment/removals on/from it) or chemical bonding pair of electrons influence the aromatic ring core towards increasing its shielding and the overall molecular reactivity resistance. All these conceptual and computational results should be further extended and tested on increased number of molecules, enlarging their variety too, as well as by considering more refined quantum computational frameworks as the Density Functional Theory and (Hartree-Fock) ab initio schemes are currently compared and discussed for various exchange-correlation and parameterization limits and refutations.

4.6 DENSITY FUNCTIONAL THEORY: OBSERVABLE QUANTUM CHEMISTRY

The main weakness of the Hartree-Fock method, namely the lack in correlation energy, is ingeniously restored by the Density Functional method through introducing of the so-called effective one-electron exchange-correlation potential, yet with the price of not knowing its analytical form. However, the working equations have the simplicity of the HF ones, while replacing the exchange term in Eq. (4.281) by the exchange-correlation ("XC") contribution; there resulted the (general) unrestricted matrix form of the Kohn-Sham equations (Kohn & Sham, 1965):

$$F_{\mu\nu}^{\uparrow} = H_{\mu\nu}^{\uparrow} + \sum_{\lambda\sigma} P_{\lambda\sigma}^{T} (\mu\nu|\lambda\sigma) + F_{\mu\nu}^{XC\uparrow} \tag{4.370}$$

$$F_{\mu\nu}^{\downarrow} = H_{\mu\nu}^{\downarrow} + \sum_{\lambda\sigma} P_{\lambda\sigma}^{T} (\mu\nu|\lambda\sigma) + F_{\mu\nu}^{XC\downarrow} \tag{4.371}$$

$$P^{T} \equiv P^{\uparrow+\downarrow} = P^{\uparrow} + P^{\downarrow} \tag{4.372}$$

in a similar fashion with the Pople-Nesbet equations of Hartree-Fock theory. The restricted (closed-shell) variant is resembled by the density constraint:

$$\rho^{\uparrow} = \rho^{\downarrow} \tag{4.373}$$

in which case the Roothaan analogous equations (for exchange-correlation potential) are obtained.

Either Eq. (4.370) or (4.371) fulfills the general matrix equation of type (4.279) for the energy solution:

$$E = \sum_{\mu\nu} P_{\mu\nu} H_{\mu\nu} + \frac{1}{2} \sum_{\mu\nu\lambda\sigma} P_{\mu\nu} P_{\lambda\sigma} (\mu\nu|\lambda\sigma) + E_{XC} \tag{4.374}$$

that can be actually regarded as the solution of the Kohn-Sham equations themselves. The appeared exchange-correlation energy E_{XC} may be at its turn conveniently expressed through the energy density (per unit volume) by the integral formulation:

$$E_{XC} = E_{XC}\left[\rho^{\uparrow}, \rho^{\downarrow}\right] = \int f\left(\rho^{\uparrow}, \rho^{\downarrow}\right) d\tau \tag{4.375}$$

once the Fock elements of exchange-correlation are recognized to be of density gradient form (Johnson et al., 1994):

$$F_{\mu\nu}^{XC\uparrow(\downarrow)} = \int \frac{\partial f}{\partial \rho^{\uparrow(\downarrow)}} \phi_\mu \phi_\nu d\tau \tag{4.376}$$

The quest for various approximations for the exchange-correlation energy density $f(\rho)$ had spanned the last decades in quantum chemistry, and will be in the next reviewed (Putz, 2008). Here we will thus present the "red line" of its implementation as will be further used for the current aromaticity applications.

4.6.1 HOHENBERG-KOHN THEOREMS

Unlike the Hartree-Fock method, a completely different approach was invented to overcome from a single shoot both the *exchange and correlation* terms to the total electronic energy. That was possible, however with the price of revisiting the wave function concept, through contracting it into the electronic density:

$$\rho(r) = \sum_i n_i \sum_{\sigma=\alpha,\beta} \left|\chi_i^\sigma(r)\right|^2 \tag{4.377}$$

written in general terms of the fractional occupancy numbers $n_i \in [0,1]$ so that (Nagy, 1998)

$$N[\rho] = \int \rho(r) dr = \sum_i n_i \tag{4.378}$$

Worth noting that by introducing of the fractional occupation numbers both the concepts of one-orbitals as well as exact N-one-orbitals become generalized to fractionally occupied orbitals and to an arbitrary number of orbitals, hereafter called as *Kohn-Sham orbitals*. This way the distinction respecting the Hartree-Fock approach is made in clear.

The first Hohenberg-Kohn (HK1) theorem gives space to the concept of *electronic density of the system* $\rho(\mathbf{r})$ in terms of the extensive relation with the N electrons from the system that it characterizes (Bamzai & Deb, 1981):

$$\int \rho(\mathbf{r})dr = N \tag{4.379}$$

The relation (4.379) as much simple it could appears stands as the decisive passage from the eigen-wave function level to the level of total electronic density (Parr & Young, 1989; Putz, 2003):

$$\rho(\mathbf{r}) = N\int \Psi^*(\mathbf{r},\mathbf{r}_2,...,\mathbf{r}_N)\Psi(\mathbf{r},\mathbf{r}_2,...,\mathbf{r}_N)d\mathbf{r}_2...d\mathbf{r}_N \tag{4.380}$$

Firstly, Eq. (4.380) satisfies Eq. (4.379); this can be used also as simple immediate proof of the relation (4.379) itself. Then, the dependency from the 3N-dimensions of configuration space was reduced at 3 coordinates in the real space, physically measurable.

However, still remains the question: what represents the electronic density of Eq. (4.380)? Definitely, it neither represents the electronic density in the configuration space nor the density of a single electron, since the N-electronic dependency as multiplication factor of the multiple integral in Eq. (4.380). What remains is that $\rho(r)$ *is simple the electronic density (of the whole concerned system) in* "**r**" space point. Such simplified interpretation, apparently classics, preserves its quantum roots through the averaging (integral) over the many-electronic eigenfunction $\Psi(\mathbf{r}_1,...,\mathbf{r}_N)$ in Eq. (4.380). Alternatively, the explicit non-dependency of density on the wave function is also possible within the quantum statistical approach where the relation with partition function of the system (the global measure of the distribution of energetic states of a system) is mainly considered.

The major consequence of this theorem consists in defining of the total energy of a system as a function of the electronic density function in what is known as the density functional (Parr & Young, 1989; Putz, 2003):

$$E[\rho] = F_{HK}[\rho] + C_A[\rho] \tag{4.381}$$

from where the name of the theory. The terms of energy decomposition in (4.381) are identified as: the Hohenberg-Kohn density functional (Hohenberg & Kohn, 1964)

$$F_{HK}[\rho]=T[\rho]+V_{ee}[\rho] \tag{4.382}$$

viewed as the summed electronic kinetic $T[\rho]$ and electronic repulsion $V_{ee}[\rho]$, and the so-called *chemical action* term (Putz, 2007a):

$$C_A[\rho]=\int \rho(\mathbf{r})V(\mathbf{r})d\mathbf{r} \tag{4.383}$$

being the only explicit functional of total energy.

Although not entirely known the HK functional has a remarkably property: it is universally, in a sense that both the kinetic and inter-electronic repulsion are independent of the concerned system. The consequence of such universal nature offers the possibility that once it is exactly or approximately knew the HK functional *for a given external potential* $V(\mathbf{r})$ remain valuable for any other type of potential $V'(\mathbf{r})$ applied on the concerned many-electronic system. Let's note the fact that $V(\mathbf{r})$ *should be not reduced only to the Coulombic type* of potentials but is carrying the role of the generic potential applied, that could beg of either an electric, magnetic, nuclear, or even electronic nature as far it is external to the system fixed by the N electrons in the investigated system.

Once "in game" the external applied potential provides the second Hohenberg-Kohn (HK2) theorem. In short, HK2 theorem says that "the external applied potential is determined up to an additive constant by the electronic density of the N-electronic system ground state". In mathematical terms, the theorem assures the validity of the variational principle applied to the density functional (4.381) relation, i.e., (Ernzerhof, 1994)

$$E[\overline{\rho}]\geq E[\rho]\Leftrightarrow \delta E[\rho]=0 \tag{4.384}$$

for every electronic test density $\overline{\rho}$ around the real density ρ of the ground state.

The proof of variational principle in Eq. (4.384), or, in other words, the one-to-one correspondence between the applied potential and the ground state electronic density, employs the *reduction ad absurdum* procedure.

That is to assume that the ground state electronic density $\rho(\mathbf{r})$ corresponds to two external potentials (V_1, V_2) fixing two associate Hamiltonians (H_1, H_2) to which two eigen-total energy (E_1, E_2) and two eigen-wave functions (Ψ_1, Ψ_2) are allowed. Now, if eigen-function Ψ_1 is considered as the true one for the ground state the variational principle (4.384) will cast as the inequality:

$$E_1[\rho] = \int \Psi_1^* \hat{H}_1 \Psi_1 d\tau < \int \Psi_2^* \hat{H}_1 \Psi_2 d\tau = \int \Psi_2^* \left[\hat{H}_2 + \left(\hat{H}_1 - \hat{H}_2 \right) \right] \Psi_2 d\tau \tag{4.385}$$

which is further reduced, on universality reasons of the HK functional in (4.381), to the form:

$$E_1[\rho] < E_2[\rho] + \int \rho(\mathbf{r}) [V_1(\mathbf{r}) - V_2(\mathbf{r})] d\mathbf{r} \tag{4.386}$$

On another way, if the eigen-function Ψ_2 is assumed as being the one true ground state wave-function, the analogue inequality springs out as:

$$E_2[\rho] < E_1[\rho] + \int \rho(\mathbf{r}) [V_2(\mathbf{r}) - V_1(\mathbf{r})] d\mathbf{r} \tag{4.387}$$

Taken together relations (4.386) and (4.387) generate, by direct summation, the evidence of the contradiction:

$$E_1[\rho] + E_2[\rho] < E_1[\rho] + E_2[\rho] \tag{4.388}$$

The removal of such contradiction could be done in a single way, namely, by abolishing, in a reverse phenomenologically order, the fact that two eigen-functions, two Hamiltonians and respectively, two external potential exist for characterizing the same ground state of a given electronic system. With this statement the HK2 theorem is formally proofed.

Yet, there appears the so-called *V-representability* problem signaling the impossibility of an *a priori* selection of the external potentials types that are in bi-univocal relation with ground state of an electronic system (Chen & Stott, 1991a,b; Kryachko & Ludena, 1991a,b). The problem was revealed as very difficult at mathematical level due to the equivocal potential intrinsic behavior that is neither of universal nor of referential independent value. Fortunately, such principal limitation does not affect the general validity of the variational principle (4.384) regarding the selection of the energy of ground state level from a collection of states with different associated external potentials.

That because, the problem of *V*-representability can be circumvented by the so-called *N-contingency features of ground state electronic density* assuring that, aside of the N – *integrability condition* (4.379), the candidate ground state densities should fulfill the *positivity condition* (an electronic density could not be negative) (Kryachko & Ludena, 1991a,b):

$$\rho(\mathbf{r}) \geq 0 \ , \ \forall |\mathbf{r}| \in \Re \tag{4.389}$$

as well as the *non-divergent integrability condition* on the real domain (in relation with the fact that the kinetic energy of an electronic system could not be infinite – since the light velocity restriction):

$$\int_{\Re} \left| \nabla \rho(\mathbf{r})^{1/2} \right|^2 d\mathbf{r} < \infty \tag{4.390}$$

Both Eqs. (4.389) and (4.390) conditions are easy accomplished by every reasonable density, allowing the employment of the variational principle (4.384) in two steps, according to the so-called *Levy-Lieb double minimization algorithm* (Levy & Perdew, 1985): one regarding the intrinsic minimization procedure of the energetic terms respecting all possible eigen-functions folding a trial electronic density followed by the external minimization over all possible trial electronic densities yielding the correct ground state (GS) energy density functional

$$\begin{aligned} E_{GS} &= \min_{\rho} \left[\min_{\Psi \to \rho} \left(\int \Psi^* (T + V_{ee} + V) \Psi d\tau \right) \right] \\ &= \min_{\rho} \left[\min_{\Psi \to \rho} \left(\int \Psi^* (T + V_{ee}) \Psi d\tau \right) + \int \rho(\mathbf{r}) V(\mathbf{r}) d\mathbf{r} \right] \\ &= \min_{\rho} \left(F_{HK}[\rho] + C_A[\rho] \right) = \min_{\rho} \left(E[\rho] \right) \end{aligned} \tag{4.391}$$

One of the most important consequences of the HK2 conveys the rewriting of the variational principle (4.384) in the light of above *N*-contingency conditions of the trial densities as the working Euler type equation:

$$\delta\{E[\rho]-\mu N[\rho]\}=0 \tag{4.392}$$

from where, there follows the Lagrange multiplication factor with the functional definition:

$$\mu=\left(\frac{\delta E[\rho]}{\delta\rho}\right)_{\rho=\rho(V)} \tag{4.393}$$

this way introducing the *chemical potential* as the fundamental quantity of the theory. At this point, the whole chemistry can spring out since identifying the electronic systems electronegativity with the negative of the density functional chemical potential (Parr & Young, 1989):

$$\chi=-\mu \tag{4.394}$$

making thus the DFT approach compatible with Hartree-Fock-Koopmans previous formulation of electronagivity for frontier orbital energies, see Eq. (4.352).

However, the Hohenberg-Kohn theorems give new conceptual quantum tools for physico-chemical characterization of an electronic sample by means of electronic density and its functionals, the total energy and chemical potential (electronegativity). Such density functional premises are in next analyzed towards elucidating of the quantum nature of the chemical bond as driven by chemical reactivity (Putz, 2007b).

4.6.2 OPTIMIZED ENERGY-ELECTRONEGATIVITY CONNECTION

Back from Paris, in the winter of 1964, Kohn met at the San Diego University of California his new post-doc Lu J. Sham with who propose to extract from HK1 & 2 theorems the equation of total energy of the ground state. In fact, they propose themselves to find the correspondent of the stationary eigen-equation of Schrödinger type, employing the relationship between the electronic density and the wave function.

Their basic idea consists in assuming a so-called orbital basic set for the *N*-electronic system by replacing the integration in the relation (4.380) with summation over the virtual uni-electronic orbitals φ_i, $i=\overline{1,N}$, in accordance with Pauli principle, assuring therefore the HK1 frame with maximal spin/orbital occupancy (Janak, 1978):

$$\rho(\mathbf{r})=\sum_i^N n_i\left|\varphi_i(\mathbf{r})\right|^2 \quad ,0\le n_i\le 1\ ,\ \sum_i n_i=N \tag{4.395}$$

Then, the *trial* total eigen-energy may be rewritten as density functional of Eq. (4.381) nature expanded in the original form (Moscardo & San-Fabian, 1991; Neal, 1998):

$$\begin{aligned}E[\rho]&=F_{HK}[\rho]+C_A[\rho]\\&=T[\rho]+V_{ee}[\rho]+C_A[\rho]\\&=T_s[\rho]+J[\rho]+\{(T[\rho]-T_s[\rho])+(V_{ee}[\rho]-J[\rho])\}+C_A[\rho]\\&=\sum_i^N\int n_i\varphi_i^*(\mathbf{r})\left[-\frac{1}{2}\nabla^2\right]\varphi_i(\mathbf{r})d\mathbf{r}+\frac{1}{2}\iint\frac{\rho(\mathbf{r}_1)\rho(\mathbf{r}_2)}{r_{12}}d\mathbf{r}_1d\mathbf{r}_2+E_{xc}[\rho]\\&\quad+\int V(\mathbf{r})\rho(\mathbf{r})d\mathbf{r}\end{aligned} \tag{4.396}$$

where, the contribution of the referential uniform kinetic energy contribution

$$T_s[\rho]=\sum_i^N\int n_i\varphi_i^*(\mathbf{r})\left[-\frac{1}{2}\nabla^2\right]\varphi_i(\mathbf{r})d\mathbf{r} \tag{4.397}$$

with the inferior index "s" referring to the "spherical" or homogeneous attribute together with the classical energy of Coulombic inter-electronic repulsion

$$J[\rho]=\frac{1}{2}\iint\frac{\rho(\mathbf{r}_1)\rho(\mathbf{r}_2)}{r_{12}}d\mathbf{r}_1d\mathbf{r}_2 \tag{4.398}$$

were used as the analytical vehicles to elegantly produce the exchange-correlation energy E_{xc} containing exchange $(V_{ee}[\rho]-J[\rho])$ and correlation $(T[\rho]-T_s[\rho])$ heuristically introduced terms as the quantum effects of spin anti-symmetry over the classical interelectronic potential and of corrected homogeneous electronic movement, respectively.

Next, the trial density functional energy (4.396) will be optimized in the light of variational principle (4.392) as prescribed by the HK2 theorem. The combined result of the HK theorems will eventually furnish the new quantum energy expression of multi-electronic systems beyond the exponential wall of the wave function.

An instructive method for deriving such equation assume the same types of orbitals for the density expansion (4.395),

$$\rho(\mathbf{r}) = N\varphi^*(\mathbf{r})\varphi(\mathbf{r}) \tag{4.399}$$

that, without diminishing the general validity of the results, since preserving the N-electronic character of the system, highly simplifies the analytical discourse.

Actually, with the trial density (4.399) replaced throughout the energy expression in Eq. (4.396) has to undergo the minimization procedure (4.392) with the practical equivalent integral variant:

$$\int \frac{\delta\left(E[\rho]-\mu N[\rho]\right)}{\delta\varphi^*}\delta\varphi^* d\mathbf{r} = 0 \tag{4.400}$$

Note that, in fact, we chose the variation in the conjugated uni-orbital $\varphi^*(\mathbf{r})$ in (4.400) providing from (4.399) the useful differential link:

$$\delta\rho(\mathbf{r}) = N\varphi(\mathbf{r})\delta\varphi^*(\mathbf{r}) \Rightarrow \delta\varphi^*(\mathbf{r}) = N\varphi(\mathbf{r})\frac{1}{\delta\rho(\mathbf{r})} \tag{4.401}$$

Now, unfolding the Eq. (4.400) with the help of relations (4.396) and (4.399), together with fundamental density functional prescription (4.379), one firstly gets (Putz & Chiriac, 2008):

$$\frac{\delta}{\delta\phi^*(\mathbf{r})}\left\{\begin{array}{l} -\frac{N}{2}\int \varphi^*(\mathbf{r})\nabla^2\varphi(\mathbf{r})d\mathbf{r} + J[\rho] + E_{xc}[\rho] \\ +N\int V(\mathbf{r})\varphi^*(\mathbf{r})\varphi(\mathbf{r})d\mathbf{r} - \mu N\int \varphi^*(\mathbf{r})\varphi(\mathbf{r})d\mathbf{r} \end{array}\right\} = 0 \tag{4.402}$$

By performing the required partial functional derivations respecting the uni-orbital $\varphi^*(\mathbf{r})$ and by taking account of the equivalence (4.401) in derivatives relating $J[\rho]$ and $E_{xc}[\rho]$ terms, Eq. (4.402) takes the further form:

$$-\frac{N}{2}\nabla^2\varphi(\mathbf{r})+N\varphi(\mathbf{r})\frac{\delta J[\rho]}{\delta\rho}+N\varphi(\mathbf{r})\frac{\delta E_{xc}}{\delta\rho}+NV(\mathbf{r})\varphi(\mathbf{r})-\mu N\varphi(\mathbf{r})=0 \tag{4.403}$$

After immediate suppressing of the N factor in all the terms and by considering the exchange-correlation potential with the formal definition:

$$V_{xc}(\mathbf{r})=\left(\frac{\delta E_{xc}[\rho]}{\delta\rho(r)}\right)_{V(\mathbf{r})} \tag{4.404}$$

Equation (4.403) simplifies as (Flores & Keller, 1992; Keller, 1986):

$$\left[-\frac{1}{2}\nabla^2+\left(V(\mathbf{r})+\int\frac{\rho(\mathbf{r}_2)}{|\mathbf{r}-\mathbf{r}_2|}d\mathbf{r}_2+V_{xc}(\mathbf{r})\right)\right]\varphi(\mathbf{r})=\mu\varphi(\mathbf{r}) \tag{4.405}$$

Moreover, once introducing the so-called *effective potential*:

$$V_{eff}(\mathbf{r})=V(\mathbf{r})+\int\frac{\rho(\mathbf{r}_2)}{|\mathbf{r}-\mathbf{r}_2|}d\mathbf{r}_2+V_{xc}(\mathbf{r}) \tag{4.406}$$

the resulted equation recovers the traditional Schrödinger shape:

$$\left[-\frac{1}{2}\nabla^2+V_{eff}\right]\varphi(\mathbf{r})=\mu\varphi(\mathbf{r}) \tag{4.407}$$

The result (4.407) is fundamental and equally subtle. Firstly, it was proved that the joined Hohenberg-Kohn theorems are compatible with consecrated quantum mechanical postulates, however, still offering a generalized view of the quantum nature of electronic structures, albeit the electronic density was assumed as the foreground reality. In these conditions, the meaning of functions $\varphi(\mathbf{r})$ is now unambiguously producing the analytical passage from configuration (3N-D) to real (3D) space for the whole system under consideration. Nevertheless, the debate may still remain because once equation (4.407) is solved the basic functions $\varphi(\mathbf{r})$ generating the electronic density (4.399) and not necessarily the eigen-functions of the original system due to the practical approximations of the exchange and correlation terms appearing in the effective potential (4.406). This is why the functions $\varphi(\mathbf{r})$ are used to be called as *Kohn-Sham (KS) orbitals*; they provide the orbital set solutions of the associate KS equations (Kohn & Sham, 1965):

$$\left[-\frac{1}{2}\nabla^2 + V_{eff}\right]\varphi_i(\mathbf{r}) = \mu_i\varphi_i(\mathbf{r})\ ,\ i = \overline{1,N} \tag{4.408}$$

once one reconsiders electronic density (4.399) back with general case (4.395). Yet, Eq. (4.408), apart of delivering the KS wave-functions $\varphi_i(\mathbf{r})$, associate with another famous physico-chemical figure, the orbital chemical potential μ_i, which in any moment can be seen as the negative of the orbital electronegativities on the base of the relation (4.394). Going now to a summative characterization of the above optimization procedure worth observing that the N-electronic in an arbitrary external V-potential problem is conceptual-computationally solved by means of the following self-consistent algorithm:

1. It starts with a trial electronic density (4.395) satisfying the N-contingency conditions (4.389) and (4.390);
2. With trial density the effective potential (4.406) containing exchange and correlation is calculated;
3. With computed V_{eff} the Eq. (4.408) are solved for $\varphi_i(\mathbf{r})$, $i = \overline{1,N}$;
4. With the set of functions $\{\varphi_i(\mathbf{r})\}_{i=\overline{1,N}}$ the new density (4.395) is recalculated;
5. The procedure is repeated until the difference between two consecutive densities approaches zero;
6. Once the last condition is achieved one retains the last set $\{\varphi_i(\mathbf{r}), \mu_i = -\chi_i\}_{i=\overline{1,N}}$;
7. The electronegativity orbital observed contributions are summed up from Eq. (4.408) with the expression:

$$-\sum_i^N \langle \chi_i \rangle = \sum_i^N \int n_i\varphi_i^*(\mathbf{r})\left[-\frac{1}{2}\nabla^2 + V_{eff}(\mathbf{r})\right]\varphi_i(\mathbf{r})d\mathbf{r} = T_s[\rho] + \int V_{eff}\,\rho(\mathbf{r})d\mathbf{r} \tag{4.409}$$

8. Replacing in Eq. (4.409) the uniform kinetic energy, $T_s[\rho]$ from the general relation (4.396) the density functional of the total energy for the N-electronic system will take the final figure (Putz, 2008):

$$E[\rho] = -\sum_{i}^{N}\langle\chi_i\rangle - \frac{1}{2}\iint\frac{\rho(\mathbf{r}_1)\rho(\mathbf{r}_2)}{|\mathbf{r}_{12}|}d\mathbf{r}_1 d\mathbf{r}_2 + \left\{E_{xc}[\rho] - \int V_{xc}(\mathbf{r})\rho(\mathbf{r})d\mathbf{r}\right\} \tag{4.410}$$

showing that the optimized many-electronic ground state energy is directly related with global or summed over observed or averaged or expected orbital electronegativities. One can observe from Eq. (4.410) that even in the most optimistic case when the last two terms are hopefully canceling each other there still remains a (classical) correction to be added on global electronegativity in total energy. Or, in other terms, electronegativity alone is not enough to better describe the total energy of a many-electronic system, while its correction can be modeled in a global (almost classical) way. Such considerations stressed upon the accepted semiclassical behavior of the chemical systems, at the edge between the full quantum and classical treatments.

However, analytical expressing the total energy requires the use of suitable approximations, whereas for chemical interpretation of bonding the electronic localization information extracted from energy is compulsory. This subject is in next focused followed by a review of the popular energetic density functionals and approximations.

4.6.3 POPULAR ENERGETIC DENSITY FUNCTIONALS

Since the terms of total energy are involved in bonding and reactivity states of many-electronic systems, i.e., the kinetic energetic terms in ELF topological analysis or the exchange and correlation density functionals in chemical reactivity in relation with either localization and chemical potential or electronegativity, worth presenting various schemes of quantification and approximation of these functionals for better understanding their role in chemical structure and dynamics.

4.6.3.1 Density Functionals of Kinetic Energy

When the electronic density is seen as the diagonal element $\rho(\mathbf{r}_1) = \rho(\mathbf{r}_1, \mathbf{r}_1)$ the kinetic energy may be generally expressed from the Hartree-Fock model, through employing the single determinant $\rho(\mathbf{r}_1, \mathbf{r}'_1)$, as the quantity (Lee & Parr, 1987):

$$T[\rho] = -\frac{1}{2}\int \left[\nabla^2_{\mathbf{r}'_1} \rho(\mathbf{r}_1, \mathbf{r}'_1)\right]_{\mathbf{r}_1=\mathbf{r}'_1} d\mathbf{r}_1 \tag{4.411}$$

it may eventually be further written by means of the thermodynamical (or statistical) density functional:

$$T_\beta = \frac{3}{2}\int \rho(\mathbf{r}) k_B T(\mathbf{r}) d\mathbf{r} = \frac{3}{2}\int \rho(\mathbf{r}) \frac{1}{\beta(\mathbf{r})} d\mathbf{r} \tag{4.412}$$

that supports various specializations depending on the statistical factor particularization β.

For instance, in LDA approximation, the temperature at a point is assumed as a function of the density in that point, $\beta(\mathbf{r}) = \beta(\rho(\mathbf{r}))$; this may be easily reached out by employing the scaling transformation to be (Ou-Yang & Levy, 1990)

$$\rho_\lambda(\mathbf{r}) = \lambda^3 \rho(\lambda \mathbf{r}) \Rightarrow T[\rho_\lambda] = \lambda^2 T[\rho], \lambda = ct \tag{4.413}$$

providing that

$$\beta(\mathbf{r}) = \frac{3}{2} C \rho^{-2/3}(\mathbf{r}) \tag{4.414}$$

a result that helps in recovering the traditional (Thomas-Fermi) energetic kinetic density functional form

$$T[\rho] = C\int \rho^{5/3}(\mathbf{r}) d\mathbf{r} \tag{4.415}$$

while the indeterminacy remained is smeared out in different approximation frames in which also the exchange energy is evaluated. Note that the kinetic energy is generally foreseen as having an intimate relation with the exchange energy since both are expressed in Hartree-Fock model as determinantal values of $\rho(\mathbf{r}_1,\mathbf{r}'_1)$, see below.

Actually, the different LDA particular cases are derived by equating the total number of particle N with various realization of the integral

$$N = \frac{1}{2}\iint \left|\rho(\mathbf{r}_1,\mathbf{r}'_1)\right|^2 d\mathbf{r}_1 d\mathbf{r}_1{}' \tag{4.416}$$

by rewriting it within the inter-particle coordinates frame:

$$\mathbf{r} = 0.5(\mathbf{r}_1 + \mathbf{r}_1{}'), \mathrm{s} = \mathbf{r}_1 - \mathbf{r}_1{}' \tag{4.417}$$

as:

$$N = \frac{1}{2}\iint \left|\rho(\mathbf{r} + \mathrm{s}/2, \mathbf{r} - \mathrm{s}/2)\right|^2 d\mathbf{r}d\mathrm{s} \tag{4.418}$$

followed by spherical averaged expression:

$$N = 2\pi \iint \rho^2(\mathbf{r})\Gamma(\mathbf{r},s)d\mathbf{r}s^2 ds \tag{4.419}$$

with

$$\Gamma(\mathbf{r},s) = 1 - \frac{s}{\beta(\mathbf{r})} + \ldots \tag{4.420}$$

The option in choosing the $\Gamma(\mathbf{r},s)$ series (4.420) so that to converge in the sense of charge particle integral (4.419) fixes the possible cases to be considered (Lee & Parr, 1987):

1. the Gaussian resummation uses:

$$\Gamma(\mathbf{r},s) \cong \Gamma_G(\mathbf{r},s) = \exp\left(-\frac{s^2}{\beta(\mathbf{r})}\right) \tag{4.421}$$

2. the trigonometric (uniform gas) approximation looks like:

$$\Gamma(\mathbf{r},s) \cong \Gamma_T(\mathbf{r},s) = 9\frac{(\sin t - t\cos t)^2}{t^6}, t = s\sqrt{\frac{5}{\beta(\mathbf{r})}} \tag{4.422}$$

In each of (4.421) and (4.422) cases the LDA-β function (4.414) is firstly replaced; then, the particle integral (4.419) is solved to give the constant C and then the respective kinetic energy density functional of Eq. (4.415) type is delivered; the results are (Lee & Parr, 1987):

1. in Gaussian resummation:

$$T_G^{LDA} = \frac{3\pi}{2^{5/3}} \int \rho^{5/3}(\mathbf{r}) d\mathbf{r} \tag{4.423}$$

2. whereas in trigonometric approximation

$$T_{TF}^{LDA} = \frac{3}{10}\left(3\pi^2\right)^{2/3} \int \rho^{5/3}(\mathbf{r}) d\mathbf{r} \tag{4.424}$$

one arrives to the Thomas-Fermi original density functional formulation.

Next on, one will consider the non-local functionals; this can be achieved through the gradient expansion in the case of slowly varying densities – that is assuming the expansion (Murphy, 1981):

$$\begin{aligned} T &= \int d\mathbf{r}\left[\tau(\rho_\uparrow) + \tau(\rho_\downarrow)\right] \\ &= \int d\mathbf{r} \sum_{m=0}^{\infty} \left[\tau_{2m}(\rho_\uparrow) + \tau_{2m}(\rho_\downarrow)\right] \\ &= \int d\mathbf{r} \sum_{m=0}^{\infty} \tau_{2m}(\rho) \\ &= \int d\mathbf{r}\, \tau(\rho) \end{aligned} \tag{4.425}$$

The first two terms of the series respectively covers: the Thomas Fermi typical functional for the homogeneous gas

$$\tau_0(\rho) = \frac{3}{10}\left(6\pi^2\right)^{2/3} \rho^{5/3} \tag{4.426}$$

and the Weizsäcker related first gradient correction:

$$\tau_2(\rho) = \frac{1}{9}\tau_W(\rho) = \frac{1}{72}\frac{\left|\nabla\rho\right|^2}{\rho} \tag{4.427}$$

They both correctly behave in asymptotic limits:

$$\tau(\rho)=\begin{cases}\tau_0(\rho)=\tau_2(\rho) & \ldots\ \nabla\rho << \ (\textit{far from nucleus})\\ 9\tau_2(\rho)=\tau_W(\rho)=\dfrac{1}{8}\dfrac{|\nabla\rho|^2}{\rho} & \ldots\ \nabla\rho >> \ (\textit{close to nucleus})\end{cases} \tag{4.428}$$

However, an interesting resummation of the kinetic density functional gradient expansion series (4.425) may be formulated in terms of the Padé-approximant model (DePristo & Kress, 1987):

$$\tau(\rho)=\tau_0(\rho)P_{4,3}(x) \tag{4.429}$$

with

$$P_{4,3}(x)=\frac{1+0.95x+a_2x^2+a_3x^3+9b_3x^4}{1-0.05x+b_2x^2+b_3x^3} \tag{4.430}$$

and where the x-variable is given by

$$x=\frac{\tau_2(\rho)}{\tau_0(\rho)}=\frac{5}{108}\frac{1}{\left(6\pi^2\right)^{2/3}}\frac{|\nabla\rho|^2}{\rho^{8/3}} \tag{4.431}$$

while the parameters a_2, a_3, b_2, and b_3 are determined by fitting them to reproduce Hartree-Fock kinetic energies of He, Ne, Ar, and Kr atoms, respectively (Liberman et al., 1994). Note that Padé function (4.430) may be regarded as a sort of generalized electronic localization function (ELF) susceptible to be further used in bonding characterizations.

4.6.3.2 Density Functionals of Exchange Energy

Starting from the Hartree-Fock framework of exchange energy definition in terms of density matrix (Levy et al., 1996),

$$K[\rho] = -\frac{1}{4}\iint \frac{|\rho(\mathbf{r}_1, \mathbf{r}'_1)|^2}{|\mathbf{r}_1 - \mathbf{r}'_1|} d\mathbf{r}_1 d\mathbf{r}_1' \tag{4.432}$$

within the same consideration as before, we get that the spherical averaged exchange density functional

$$K = \pi \iint \rho^2(\mathbf{r})\Gamma(\mathbf{r}, s) d\mathbf{r} s ds \tag{4.433}$$

takes the particular forms (Lee & Parr, 1987):

1. in Gaussian resummation:

$$K_G^{LDA} = -\frac{1}{2^{1/3}} \int \rho^{4/3}(\mathbf{r}) d\mathbf{r} \tag{4.434}$$

2. and in trigonometric approximation (recovering the Dirac formula):

$$K_D^{LDA} = -\frac{3}{4}\left(\frac{3}{\pi}\right)^{1/3} \int \rho^{4/3}(\mathbf{r}) d\mathbf{r} \tag{4.435}$$

Alternatively, by paralleling the kinetic density functional previous developments the gradient expansion for the exchange energy may be regarded as the density dependent series (Cedillo et al., 1988):

$$\begin{aligned} K &= \sum_{n=0}^{\infty} K_{2n}(\rho) \\ &= \int d\mathbf{r} \sum_{n=0}^{\infty} k_{2n}(\rho) \\ &= \int d\mathbf{r} k(\rho) \end{aligned} \tag{4.436}$$

while the first term reproduces the Dirac LDA term (Perdew & Yue, 1986; Manoli & Whitehead, 1988):

$$k_0(\rho) = -\frac{3}{2}\left(\frac{3}{4\pi}\right)^{1/3} \rho^{4/3} \tag{4.437}$$

and the second term contains the density gradient correction, with the Becke proposed approximation (Becke, 1986):

$$k_2(\rho) = -b\frac{\dfrac{|\nabla\rho|^2}{\rho^{4/3}}}{\left(1+d\dfrac{|\nabla\rho|^2}{\rho^{8/3}}\right)^a} \tag{4.438}$$

where the parameters b and d are determined by fitting the k_0+k_2 exchange energy to reproduce Hartree-Fock counterpart energy of He, Ne, Ar, and Kr atoms, and where for the a exponent either 1.0 or 4/5 value furnishes excellent results. However, worth noting that when analyzing the asymptotic exchange energy behavior, we get in small gradient limit (Becke, 1986):

$$k(\rho)\xrightarrow{\nabla\rho<<}k_0(\rho) - \frac{7}{432\pi\left(6\pi^2\right)^{1/3}}\frac{|\nabla\rho|^2}{\rho^{4/3}} \tag{4.439}$$

whereas the adequate large-gradient limit is obtained by considering an arbitrary damping function as multiplying the short-range behavior of the exchange-hole density, with the result:

$$k(\rho)\xrightarrow{\nabla\rho>>}c\rho^{4/5}|\nabla\rho|^{2/5} \tag{4.440}$$

where the constant c depends of the damping function choice.

Next, the Padé-resummation model of the exchange energy prescribes the compact form (Cedillo et al., 1988):

$$k(\rho) = \frac{10}{9}\frac{k_0(\rho)}{P_{4,3}(x)} \tag{4.441}$$

with the same Padé-function (4.430) as previously involved when dealing with the kinetic functional resummation. Note that when x=0, one directly obtains the Ghosh-Parr functional (Ghosh & Parr, 1986):

$$k(\rho) = \frac{10}{9}k_0(\rho) \tag{4.442}$$

Moreover, the asymptotic behavior of Padé exchange functional (4.441) leaves with the convergent limits:

$$k(\rho)=\begin{cases}\dfrac{10}{9}\left(\dfrac{k_0+\dfrac{15}{17}\dfrac{7}{432\pi\left(6\pi^2\right)^{1/3}}}{\dfrac{|\nabla\rho|^2}{\rho^{4/3}}}\right) & \ldots\ x\to 0\ (SMALL\quad GRADIENTS)\\ -12\pi\dfrac{\rho^2}{|\nabla\rho|^2} & \ldots\ x\to\infty\,(LARGE\quad GRADIENTS)\end{cases} \tag{4.443}$$

Once again, note that when particularizing small or large gradients and fixing asymptotic long or short range behavior, we are discovering the various cases of bonding modeled by the electronic localization recipe as provided by electronic localization function limits, see Volume II of the present five-volumes set (Putz, 2016b).

Another interesting approach of exchange energy in the gradient expansion framework was given by Bartolotti through the two-component density functional (Bartolotti, 1982):

$$K[\rho]=C(N)\int\rho(\mathbf{r})^{4/3}d\mathbf{r}+D(N)\int\mathbf{r}^2\frac{|\nabla\rho|^2}{\rho^{2/3}}d\mathbf{r} \tag{4.444}$$

where the N-dependency is assumed to behave like:

$$C(N)=C_1+\frac{C_2}{N^{2/3}},\ D(N)=\frac{D_2}{N^{2/3}} \tag{4.445}$$

while the introduced parameters C_1, C_2, and D_2 were fond with the exact values (Perdew et al., 1992; Wang et al., 1990; Alonso & Girifalco, 1978):

$$C_1=-\frac{3}{4}\pi^{1/3},\ C_2=-\frac{3}{4}\pi^{1/3}\left[1-\left(\frac{3}{\pi^2}\right)^{1/3}\right],\ D_2=\frac{\pi^{1/3}}{729} \tag{4.446}$$

Worth observing that the exchange Bartolotti functional (4.444) has some important phenomenological features: it scales like potential energy, fulfills the non-locality behavior through the powers of the electron and powers of the gradient of the density, while the atomic cusp condition is preserved (Levy & Gorling, 1996).

However, density functional exchange-energy approximation with correct asymptotic (long range) behavior, i.e., satisfying the limits for the density

$$\lim_{r\to\infty} \rho_\sigma = \exp(-a_\sigma r) \tag{4.447}$$

and for the Coulomb potential of the exchange charge, or Fermi hole density at the reference point **r**

$$\lim_{r\to\infty} U_X^\sigma = -\frac{1}{r}, \sigma = \alpha\,(or\uparrow), \beta\,(or\downarrow)...spin\;\;states \tag{4.448}$$

in the total exchange energy

$$K[\rho] = \frac{1}{2}\sum_\sigma \int \rho_\sigma U_X^\sigma d\mathbf{r} \tag{4.449}$$

was given by Becke *via* employing the so-called semiempirical (SE) modified gradient-corrected functional (Becke, 1986):

$$K^{SE} = K_0 - \beta\sum_\sigma \int \rho_\sigma^{4/3} \frac{x_\sigma^2(\mathbf{r})}{1+\gamma x_\sigma^2(\mathbf{r})} d\mathbf{r},\; K_0 = \int d\mathbf{r} k_0[\rho(\mathbf{r})],\; x_\sigma(\mathbf{r}) = \frac{|\nabla\rho_\sigma(\mathbf{r})|}{\rho_\sigma^{4/3}(\mathbf{r})} \tag{4.450}$$

to the working single-parameter dependent one (Becke, 1988):

$$K^{B88} = K_0 - \beta\sum_\sigma \int \rho_\sigma^{4/3}(\mathbf{r}) \frac{x_\sigma^2(\mathbf{r})}{1+6\beta x_\sigma(\mathbf{r})\sinh^{-1} x_\sigma(\mathbf{r})} d\mathbf{r} \tag{4.451}$$

where the value $\beta = 0.0042[a.u.]$ was found as the best fit among the noble gases (He to Rn atoms) exchange energies; the constant a_σ is related to the ionization potential of the system.

Still, having different exchange approximation energetic functionals as possible worth explaining from where such ambiguity eventually comes.

To clarify this, it helps in rewriting the starting exchange energy (4.432) under the formally exact form (Taut, 1996):

$$K[\rho]=\sum_{\sigma}\int \rho_{\sigma}(\mathbf{r})k[\rho_{\sigma}(\mathbf{r})]g[x_{\sigma}(\mathbf{r})]d\mathbf{r} \tag{4.452}$$

where the typical components are identified as:

$$k[\rho]=-A_X\rho^{1/3},\ A_X=\frac{3}{2}\left(\frac{3}{4\pi}\right)^{1/3} \tag{4.453}$$

while the gradient containing correction $g(x)$ is to be determined.

Firstly, one can notice that a sufficiency condition for the two exchange integrals (4.449) and (4.452) to be equal is that their integrands, or the exchange potentials, to be equal; this provides the leading gradient correction:

$$g_0(x)=\frac{1}{2}\frac{U_X(\mathbf{r}(x))}{k[\rho(\mathbf{r}(x))]} \tag{4.454}$$

with $\mathbf{r}(x)$ following from $x(\mathbf{r})$ by (not unique) inversion.

Unfortunately, the above "integrity" condition for exchange integrals to be equal is not also necessary, since any additional gradient correction

$$g(x)=g_0(x)+\Delta g(x) \tag{4.455}$$

fulfills the same constraint if it is chosen so that

$$\int \rho^{4/3}(\mathbf{r})\Delta g(x(\mathbf{r}))d\mathbf{r}=0 \tag{4.456}$$

or, with the general form:

$$\Delta g(x)=f(x)-\frac{\int \rho^{4/3}(\mathbf{r})f(x(\mathbf{r}))d\mathbf{r}}{\int \rho^{4/3}(\mathbf{r})d\mathbf{r}} \tag{4.457}$$

being $f(x)$ an arbitrary function.

Nonetheless, if, for instance, the function $f(x)$ is specialized so that

$$f(x) = -g_0(x) \tag{4.458}$$

the gradient correcting function (4.455) becomes:

$$g(x) = -\frac{1}{2A_X}\frac{\int \rho(\mathbf{r})U_X(\mathbf{r})d\mathbf{r}}{\int \rho^{4/3}(\mathbf{r})d\mathbf{r}} \equiv \alpha_X \tag{4.459}$$

recovering the Slater's famous X_α method for exchange energy evaluations (Slater, 1951; Slater & Johnson, 1972):

$$K[\rho] = -\alpha_X A_X \int \rho^{4/3}(\mathbf{r})d\mathbf{r} \tag{4.460}$$

Nevertheless, the different values of the multiplication factor α_X in Eq. (4.460) can explain the various forms of exchange energy coefficients and forms above. Moreover, following this conceptual line the above Becke'88 functional (4.451) can be further rearranged in a so-called Xα-Becke88 form (Lee & Zhou, 1991):

$$K^{XB88} = \alpha_{XB}\sum_{\sigma}\int \rho_\sigma^{4/3}(\mathbf{r})\left[2^{1/3} + \frac{x_\sigma^2(\mathbf{r})}{1+6\beta_{XB}x_\sigma(\mathbf{r})\sinh^{-1}x_\sigma(\mathbf{r})}\right]d\mathbf{r} \tag{4.461}$$

where the parameters α_{XB} and β_{XB} are to be determined, as usually, throughout atomic fitting; it may lead with a new workable valuable density functional in exchange family.

4.6.3.3 Density Functionals of Correlation Energy

The first and immediate definition of energy correlation may be given by the difference between the exact and Hartree-Fock (HF) total energy of a poly-electronic system (Senatore & March, 1994):

$$E_c[\rho] = E[\rho] - E_{HF}[\rho] \tag{4.462}$$

Instead, in density functional theory the correlation energy can be seen as the gain of the kinetic and electron repulsion energy between the full interacting ($\lambda=1$) and non-interacting ($\lambda=0$) states of the electronic systems (Liu et al., 1999):

$$E_c^\lambda[\rho]=\left\langle\psi^\lambda\left|\left(\hat{T}+\lambda\hat{V}_{ee}\right)\right|\psi^\lambda\right\rangle-\left\langle\psi^{\lambda=0}\left|\left(\hat{T}+\lambda\hat{V}_{ee}\right)\right|\psi^{\lambda=0}\right\rangle \quad (4.463)$$

In this context, taking the variation of the correlation energy (4.463) respecting the coupling parameter λ (Ou-Yang & Levy, 1991; Nagy et al., 1999),

$$\lambda\frac{\partial E_c^\lambda[\rho]}{\partial\lambda}=E_c^\lambda[\rho]+\int\rho(\mathbf{r})\mathbf{r}\cdot\nabla\frac{\delta E_c^\lambda[\rho]}{\delta\rho(\mathbf{r})}d\mathbf{r} \quad (4.464)$$

by employing it through the functional differentiation with respecting the electronic density,

$$\lambda\frac{\partial V_c^\lambda[\rho]}{\partial\lambda}-V_c^\lambda[\rho]=\mathbf{r}\cdot\nabla V_c^\lambda+\int\rho(\mathbf{r}_1)\mathbf{r}_1\cdot\nabla_1\frac{\delta^2E_c^\lambda[\rho]}{\delta\rho(\mathbf{r})\delta\rho(\mathbf{r}_1)}d\mathbf{r}_1 \quad (4.465)$$

one obtains the equation to be solved for correlation potential $V_c^\lambda=\delta E_c^\lambda[\rho]/\delta\rho$; then the correlation energy is yielded by back integration:

$$E_c^\lambda[\rho]=\int V_c^\lambda(\mathbf{r},[\rho])\rho(\mathbf{r})d\mathbf{r} \quad (4.466)$$

from where the full correlation energy is reached out by finally setting $\lambda=1$.

When restricting to atomic systems, i.e., assuming spherical symmetry, and neglecting the last term of the correlation potential equation above, believed to be small (Liu et al., 1999), the equation to be solved simply becomes:

$$\lambda\frac{\partial V_c^\lambda[\rho]}{\partial\lambda}-V_c^\lambda[\rho]=r\nabla V_c^\lambda \quad (4.467)$$

that can really be solved out with the solution:

$$V_c^\lambda=A_p\lambda^{p+1}r^p \quad (4.468)$$

with the integration constants A_p and p.

However, since the Eq. (4.467) is a homogeneous differential one, the linear combination of solutions gives a solution as well. This way, the general form of correlation potential looks like:

$$V_c^\lambda = \sum_p A_p \lambda^{p+1} r^p \tag{4.469}$$

This procedure can be then iterated by taking further derivative of Eq. (4.465) with respect to the density, solving the obtained equation until the second order correction over above first order solution (4.469),

$$V_c^\lambda = \sum_{p1} A_{p1} \lambda^{p1+1} r^{p1} + \sum_{p2} A_{p2} \lambda^{2p1+1} r^{p2} \left\langle r^{p2} \rho \right\rangle \tag{4.470}$$

By mathematical induction, when going to higher orders the K-truncated solution is iteratively founded as:

$$V_c^\lambda = \sum_p \sum_{k=1}^{K} A_{pk} \lambda^{pk+1} r^p \left\langle r^p \rho \right\rangle^{k-1} \tag{4.471}$$

producing the λ-related correlation functional:

$$E_c^\lambda[\rho] = \sum_p \sum_{k=1}^{K} \frac{1}{k} A_{pk} \lambda^{pk+1} \left\langle r^p \rho \right\rangle^k \tag{4.472}$$

and the associate full correlation energy functional (λ=1) expression:

$$E_c[\rho] = \sum_p \sum_{k=1}^{K} \frac{1}{k} A_{pk} \left\langle r^p \rho \right\rangle^k \tag{4.473}$$

As an observation, the correlation energy (4.473) supports also the immediate not spherically (molecular) generalization:

$$E_c[\rho] = \sum_{lmn} \sum_{k=1}^{K} \frac{1}{k} A_{lmnk} \left\langle x^l x^m x^n \rho \right\rangle^k \tag{4.474}$$

Nevertheless, for atomic systems, the simplest specialization of the relation (4.473) involves the simplest density moments $\langle \rho \rangle = N$ and $\langle r\rho \rangle$ that gives:

$$E_c[\rho] = A_{c0} N + A_{c1} \left\langle r\rho \right\rangle \tag{4.475}$$

Unfortunately, universal atomic values for the correlation constants A_{c0} and A_{c1} in Eq. (4.475) are not possible; they have to be related with the atomic number Z that on its turn can be seen as functional of density as well. Therefore, with the settings

$$A_{c0} = C_{c0} \ln Z, A_{c1} = C_{c1} Z \tag{4.476}$$

the fitting of Eq. (4.475) with the HF related correlation energy (4.462) reveals the atomic-working correlation energy with the form (Liu et al., 1999):

$$E_c = -0.16569 N \ln Z + 0.000401 Z \langle r\rho \rangle \tag{4.477}$$

The last formula is circumvented to the high-density total correlation density approaches rooting at their turn on the Thomas-Fermi atomic theory. Very interesting, the relation (4.477) may be seen as an atomic reflection of the (solid state) high-density regime ($r_s < 1$) given by Perdew et al. (Perdew, 1986; Wang & Perdew, 1989; Seidl et al., 1999; Perdew et al., 1996):

$$E_c^{PZ\infty}[\rho] = \int d\mathbf{r}\rho(\mathbf{r})\left(-0.048 - 0.0116 r_s + 0.0311 \ln r_s + 0.0020 r_s \ln r_s\right) \tag{4.478}$$

in terms of the dimensionless ratio

$$r_s = \frac{r_0}{a_0} \tag{4.479}$$

between the Wigner-Seitz radius $r_0 = (3/4\pi\rho)^{1/3}$ and the first Bohr radius $a_0 = \hbar^2 / me^2$.

Instead, within the low density regime ($r_s \geq 1$) the first approximation for correlation energy goes back to the Wigner jellium model of electronic fluid in solids thus providing the LDA form (Perdew et al., 1998; Wilson & Levy, 1990):

$$E_c^{W-LDA}[\rho] = \int \varepsilon_c[\rho(\mathbf{r})]\rho(\mathbf{r}) d\mathbf{r} \tag{4.480}$$

where

$$\varepsilon_c[\rho(\mathbf{r})] = -\frac{0.44}{7.8 + r_s} \tag{4.481}$$

is the correlation energy per particle of the homogeneous electron gas with density ρ (Zhao et al., 1994; Gritsenko et al., 2000; Zhao & Parr, 1992; Lam et al., 1998; Gaspar & Nagy, 1987; Levy, 1991).

However, extended parameterization of the local correlation energy may be unfolded since considering the fit with an LSDA ($\rho_\uparrow$ and $\rho_\downarrow$) analytical expression by *Vosko, Wilk and Nusair* (VWN) (Vosko et al., 1980),

$$E_c^{VWN}[\rho_\uparrow, \rho_\downarrow] = \int \varepsilon_c[\rho_\uparrow(\mathbf{r}), \rho_\downarrow(\mathbf{r})]\rho(\mathbf{r})d\mathbf{r} \tag{4.482}$$

while further density functional *gradient corrected Perdew* (GCP) expansion will look like:

$$E_c^{GCP}[\rho_\uparrow, \rho_\downarrow] = \int d\mathbf{r}\varepsilon_c[\rho_\uparrow(\mathbf{r}), \rho_\downarrow(\mathbf{r})]\rho(\mathbf{r}) + \int d\mathbf{r}B[\rho_\uparrow(\mathbf{r}), \rho_\downarrow(\mathbf{r})]\left|\nabla\rho(\mathbf{r})\right|^2 + \ldots \tag{4.483}$$

where the Perdew recommendation for the gradient integrant has the form (Perdew, 1986):

$$B_c^P[\rho_\uparrow(\mathbf{r}), \rho_\downarrow(\mathbf{r})] = B_c[\rho]\frac{\exp\left(-b[\rho]f\left|\nabla\rho\right|\rho^{-7/6}\right)}{d(m)} \tag{4.484}$$

with

$$B_c[\rho] = \rho^{-4/3}C[\rho] \tag{4.485}$$

being the electron gas expression for the coefficient of the gradient expansion. The normalization in Eq. (4.484) is to the spin degeneracy:

$$d(m) = 2^{1/3}\left[\left(\frac{1+m}{2}\right)^{5/3} + \left(\frac{1-m}{2}\right)^{5/3}\right]^{1/2}$$

$$m = \frac{\rho_\uparrow - \rho_\downarrow}{\rho}, \; \rho = \rho_\uparrow + \rho_\downarrow \tag{4.486}$$

while the exponent containing functional

$$b[\rho] = (9\pi)^{1/6} \frac{C[\rho \to \infty]}{C[\rho]} \tag{4.487}$$

is written as the ratio of the asymptotic long-range density behavior to the current one, and is controlled by the cut-off f exponential parameter taking various values depending of the fitting procedures it subscribes (0.17 for closed shells atoms and 0.11 for Ne particular system (Savin et al., 1986, 1987)).

More specifically, we list bellow some nonlocal correlation density functionals in the low density (gradient corrections over LDA) regime:

- the Rasolt and Geldar paramagnetic case ($\rho_\uparrow = \rho_\downarrow = \rho / 2$) is covered by correlation energy (Rasolt & Geldart, 1986):

$$E_c^{RG}[\rho] = c_1 + \frac{c_2 + c_3 r_s + c_4 r_s^2}{1 + c_5 r_s + c_6 r_s^2 + c_7 r_s^3} \tag{4.488}$$

  with c_1=1.667×10^{-3}, c_1=2.568×10^{-3}, c_3=2.3266×10^{-2}, c_4=7.389×10^{-6}, c_5=8.723, c_6=0.472, c_7=7.389×10^{-2} (in atomic units).
- The gradient corrected correlation functional reads as (Savin et al., 1984):

$$E_c^{GC} = \int d\mathbf{r}\varepsilon_c[\rho_\uparrow, \rho_\downarrow]\rho(\mathbf{r}) + \int d\mathbf{r} B_c^P[\rho_\uparrow, \rho_\downarrow]_{C[\rho]=\sqrt{2}\pi/4\left(6\pi^2\right)^{4/3}, f=0.17} \left|\nabla\rho(\mathbf{r})\right|^2$$
$$+9\frac{\pi}{4\left(6\pi^2\right)^{4/3}}(0.17)^2 \int d\mathbf{r}\left(\left|\nabla\rho_\uparrow\right|^2 \rho_\uparrow^{-4/3} + \left|\nabla\rho_\downarrow\right|^2 \rho_\downarrow^{-4/3}\right) \tag{4.489}$$

- The *Lee, Yang, and Parr* (LYP) functional within Colle-Salvetti approximation unfolds like (Lee et al., 1988):

$$E_c^{LYP} = -a_c b_c \int d\mathbf{r}\gamma(\mathbf{r})\xi(\mathbf{r}) \begin{pmatrix} \sum_\sigma \rho_\sigma(\mathbf{r}) \sum_i \left|\nabla\varphi_{i\sigma}(\mathbf{r})\right|^2 - \frac{1}{4}\sum_\sigma \rho_\sigma(\mathbf{r})\Delta\rho_\sigma(\mathbf{r}) \\ -\frac{1}{4}\left|\nabla\rho(\mathbf{r})\right|^2 + \frac{1}{4}\rho(\mathbf{r})\Delta\rho(\mathbf{r}) \end{pmatrix}$$
$$-a_c \int d\mathbf{r}\frac{\gamma(\mathbf{r})}{\eta(\mathbf{r})}\rho(\mathbf{r}) \tag{4.490}$$

where

$$\gamma(\mathbf{r}) = 4\frac{\rho_\uparrow(\mathbf{r})\rho_\downarrow(\mathbf{r})}{\rho(\mathbf{r})^2}, \eta(\mathbf{r}) = 1 + d_c\rho(\mathbf{r})^{-1/3}, \xi(\mathbf{r}) = \frac{\rho(\mathbf{r})^{-5/3}}{\eta(\mathbf{r})}\exp\left[-c_c\rho(\mathbf{r})^{-1/3}\right] \tag{4.491}$$

and the constants: a_c=0.04918, b_c=0.132, c_c=0.2533, d_c=0.349.

- The open-shell (OS) case provides the functional (Wilson & Levy, 1990):

$$E_c^{OS} = \int d\mathbf{r}\frac{a_s\rho(\mathbf{r}) + b_s|\nabla\rho(\mathbf{r})|\rho(\mathbf{r})^{-1/3}}{c_s + d_s\left(|\nabla\rho_\uparrow|\rho_\uparrow^{-4/3} + |\nabla\rho_\downarrow|\rho_\downarrow^{-4/3}\right) + r_s}\sqrt{1-\zeta^2} \tag{4.492}$$

with the spin-dependency regulated by the factor $\zeta = (\rho_\uparrow - \rho_\downarrow)/(\rho_\uparrow + \rho_\downarrow)$, approaching zero for closed-shell case, while the specific coefficients are determined through a scaled-minimization procedure yielding the values: a_s=–0.74860, b_s=–0.06001, c_s=3.60073, d_s=0.900000.

- Finally, Perdew and Zunger (PZ) recommend the working functional (Perdew & Zunger, 1981):

$$E_c^{PZ0}[\rho] = \int d\mathbf{r}\rho(\mathbf{r})\frac{\alpha_p}{1 + \beta_{1p}\sqrt{r_s} + \beta_{2p}r_s} \tag{4.493}$$

with the numerical values for the fitting parameters founded as: α_p=-0.1423, β_{1p}=1.0529, β_{2p}=0.3334.

4.6.3.4 Density Functionals of Exchange-Correlation Energy

Another approach in questing exchange and correlation density functionals consists in finding them both at once in what was defined as exchange-correlation density functional (4.404). In this regard, following the Lee and Parr approach (Lee & Parr, 1990), the simplest starting point is to rewrite the inter-electronic interaction potential

$$V_{ee} = \iint\frac{\rho_2(\mathbf{r}_1, \mathbf{r}_2)}{r_{12}}d\mathbf{r}_1 d\mathbf{r}_2 \tag{4.494}$$

and the classical (Coulombic) repulsion

$$J = \frac{1}{2}\iint \frac{\rho(\mathbf{r}_1)\rho(\mathbf{r}_2)}{r_{12}} d\mathbf{r}_1 d\mathbf{r}_2 \tag{4.495}$$

appeared in the formal exchange energy $(V_{ee} - J)$ in Eq. (4.396), by performing the previously introduced coordinate transformation (4.417), followed by integration of the averaged pair and coupled densities (denoted with over-bars) over the angular components of **s**:

$$V_{ee} = 4\pi \int d\mathbf{r} \int s ds \overline{\rho}_2(\mathbf{r}, s) \tag{4.496}$$

$$J = 2\pi \int d\mathbf{r} \int s ds \overline{\rho(\mathbf{r} + \mathbf{s}/2)\rho(\mathbf{r} - \mathbf{s}/2)} \tag{4.497}$$

Now, the second order density matrix in Eq. (4.496) can be expressed as

$$\overline{\rho}_2(\mathbf{r}, s) = \frac{1}{2}\overline{\rho(\mathbf{r} + \mathbf{s}/2)\rho(\mathbf{r} - \mathbf{s}/2)}\left[1 + F_1(\mathbf{r}, s)\right] \tag{4.498}$$

with the help of the introduced function $F_1(\mathbf{r}, s)$ carrying the form

$$F_1(\mathbf{r}, s) = -\frac{\exp[-\alpha(\mathbf{r})s]}{1 + \alpha(\mathbf{r})}\left\{1 + [\alpha(\mathbf{r})s]^2 F_2(\mathbf{r}, s)\right\} \tag{4.499}$$

so that the cusp condition for $\overline{\rho}_2(\mathbf{r}, s)$

$$\left.\frac{\partial \ln \overline{\rho}_2(\mathbf{r}, s)}{\partial s}\right|_{s=0} = 1 \tag{4.500}$$

to be satisfied for a well behaved function of a Taylor series expansion type

$$F_2(\mathbf{r}, s) = \sum_{k=0}^{\infty} a_k(\mathbf{r})[\alpha(\mathbf{r})s]^k \tag{4.501}$$

when $\alpha(\mathbf{r})$ stands for a suitable function of **r** as well, see bellow.

On the other side, the average $\overline{\rho(\mathbf{r}+\mathbf{s}/2)\rho(\mathbf{r}-\mathbf{s}/2)}$ in (4.497) and (4.498) supports a Taylor expansion (Berkowitz, 1986):

$$\overline{\rho(\mathbf{r}+\mathbf{s}/2)\rho(\mathbf{r}-\mathbf{s}/2)}=\rho^2(\mathbf{r})\left[1-\frac{2\tau_w(\mathbf{r})}{3\rho(\mathbf{r})}s^2+...\right] \tag{4.502}$$

with

$$\tau_w(\mathbf{r})=\frac{1}{8}\frac{|\nabla\rho(\mathbf{r})|^2}{\rho(\mathbf{r})}-\frac{1}{8}\nabla^2\rho(\mathbf{r}) \tag{4.503}$$

being the Parr modified kinetic energy of Weizsäcker type (Parr & Young, 1989).

Inserting relations (4.496)–(4.503) in $(V_{ee}-J)$ difference it is eventually converted from the "genuine" exchange meaning into practical exchange-correlation energy characterized by the density functional form:

$$\begin{aligned}E_{xc}&=2\pi\int d\mathbf{r}\int sds\,\overline{\rho(\mathbf{r}+\mathbf{s}/2)\rho(\mathbf{r}-\mathbf{s}/2)}F_1(\mathbf{r},s)\\&=-2\pi\int d\mathbf{r}\frac{\rho^2(\mathbf{r})}{1+\alpha(\mathbf{r})}\int ds\,s\exp[-\alpha(\mathbf{r})s]\left\{1-\frac{2\tau_w(\mathbf{r})}{3\rho(\mathbf{r})}s^2+...\right\}\\&\left\{1+[\alpha(\mathbf{r})s]^2\sum_{k=0}^{\infty}a_k(\mathbf{r})[\alpha(\mathbf{r})s]^k\right\}\end{aligned} \tag{4.504}$$

Making use of the two possible multiplication of the series in Eq. (4.504), i.e., either by retaining the $\alpha(\mathbf{r})$ containing function only or by including also the density gradient terms in the first curled brackets, thus retaining also the term containing $\tau_w(\mathbf{r})$ function, the so-called *I-xc or II-xc type functionals* are respectively obtained.

Now, laying aside other variants and choosing the simple (however meaningfully) density dependency

$$\alpha(\mathbf{r})=\kappa\rho^{1/3}(\mathbf{r}),\ \kappa=\text{constant} \tag{4.505}$$

the provided exchange-correlation functionals are generally shaped as (Lee & Parr, 1990):

$$E_{xc}^{I} = -\frac{1}{\kappa^2}\int d\mathbf{r}\rho^{4/3}(\mathbf{r})\frac{A_{xc}(\mathbf{r})}{1+\kappa\rho^{1/3}(\mathbf{r})}$$
$$E_{xc}^{II} = -\frac{1}{\kappa^2}\int d\mathbf{r}\frac{\rho^{4/3}(\mathbf{r})}{1+\kappa\rho^{1/3}(\mathbf{r})}\left[B_{xc}(\mathbf{r})+\frac{2}{3}\frac{\tau_w(\mathbf{r})}{\rho^{5/3}(\mathbf{r})}C_{xc}(\mathbf{r})\right] \quad (4.506)$$

These functionals are formally exact for any κ albeit the resumed functions $A_{xc}(\mathbf{r})$, $B_{xc}(\mathbf{r})$, and $C_{xc}(\mathbf{r})$ are determined for each particular specialization.

Going now to the specific models, let's explore the type I of exchange-correlation functionals (4.506). Firstly, they can further undergo simplification since the reasonable (atomic) assumption according which

$$\kappa\rho^{1/3}(\mathbf{r}) << 1, \forall\mathbf{r} \quad (4.507)$$

Within this frame the best provided model is of $X\alpha$-Padé approximation type, containing N-dependency (Lee & Parr, 1990):

$$E_{xc}^{I(X\alpha)} = -a_0^{X\alpha}\frac{1+a_1^{X\alpha}/N}{1+a_2^{X\alpha}/N}\int\rho^{4/3}(\mathbf{r})d\mathbf{r} \quad (4.508)$$

with $a_0^{X\alpha}$=0.7475, $a_1^{X\alpha}$ =17.1903, and $a_2^{X\alpha}$ =14.1936 (atomic units).

When the condition (4.507) for κ is abolished the Wigner-like model results, again, having the best approximant exchange-correlation model as the Padé form (Lee & Parr, 1990):

$$E_{xc}^{I(Wig)} = -a_0^{Wig}\frac{1+a_1^{Wig}/N}{1+a_2^{Wig}/N}\int\frac{\rho^{4/3}(\mathbf{r})}{1+\kappa^{I(Wig)}\rho^{1/3}(\mathbf{r})}d\mathbf{r} \quad (4.509)$$

with a_0^{Wig}=0.76799, a_1^{Wig} =17.5943, a_2^{Wig} =14.8893, and $\kappa^{I\,(Wig)}$=4.115·10^{-3} (atomic units).

Turning to the II-type of exchange-correlation functionals, the small density condition (4.507) delivers the gradient corrected $X\alpha$ model, taking its best fitting form as the N-dependent Padé approximant (Lee & Parr, 1990):

$$E_{xc}^{II(X\alpha)} = -b_0^{X\alpha} \frac{1 + b_1^{X\alpha} / N}{1 + b_2^{X\alpha} / N} \int \rho^{4/3}(\mathbf{r}) d\mathbf{r} - c_0^{X\alpha} \int \rho^{-1/3}(\mathbf{r}) \tau_w(\mathbf{r}) d\mathbf{r} \quad (4.510)$$

with $b_0^{X\alpha}$=0.7615, $b_1^{X\alpha}$=1.6034, $b_2^{X\alpha}$=2.1437, and $c_2^{X\alpha}$=6.151×10^{-2} (atomic units), while when laying outside the Eq. (4.507) condition the gradient corrected Wigner-like best model is proved to be without involving the N-dependency (Lee & Parr, 1990):

$$E_{xc}^{II(Wig)} = -b_0^{Wig} \int \frac{\rho^{4/3}(\mathbf{r})}{1 + \kappa^{II(Wig)} \rho^{1/3}(\mathbf{r})} d\mathbf{r} - c_0^{Wig} \int \frac{\rho^{-1/3}(\mathbf{r}) \tau_w(\mathbf{r})}{1 + \kappa^{II(Wig)} \rho^{1/3}(\mathbf{r})} d\mathbf{r} \quad (4.511)$$

with b_0^{Wig} =0.80569, c_0^{Wig} =3.0124×10^{-3}, and $\kappa^{II(Wig)}$=4.0743×10^{-3} (atomic units).

Still, a Padé approximant for the gradient-corrected Wigner-type exchange-correlation functional exists and it was firstly formulated by (Rasolt & Geldar, 1986) with the working form (Lee & Bartolotti, 1991):

$$E_{xc}^{RG} = E_{xc}^{LDA(or\,X\alpha)} + \int B_{xc}^{RG}[\rho(\mathbf{r})] \frac{|\nabla \rho(\mathbf{r})|^2}{\rho^{1/3}(\mathbf{r})} d\mathbf{r} \quad (4.512)$$

with B_{xc}^{RG} given with the Padé form:

$$B_{xc}^{RG}[\rho(\mathbf{r})] = -1 \times 10^{-3} c_1^{RG} \frac{1 + c_2^{RG} r_s + c_3^{RG} r_s^2}{1 + c_4^{RG} r_s + c_5^{RG} r_s^2 + c_6^{RG} r_s^3} \quad (4.513)$$

having the fitted coefficients c_1^{RG}=2.568, c_2^{RG}=9.0599, c_3^{RG}=2.877×10^{-3}, c_4^{RG}=8.723, c_5^{RG}=0.472, and c_3^{RG}=7.389×10^{-2} (atomic units). Some studies also consider the nonlocal correction in Eq. (4.512) premultiplied by the 10/7 factor, which was found as appropriate procedure for atomic systems.

Finally, worth noting the Tozer and Handy general form for exchange-correlation functionals viewed as a sum of products of powers of density and gradients (Tozer & Handy, 1998):

$$E_{xc}^{TH} = \int F_{xc}\left(\rho_{\uparrow},\rho_{\downarrow},\zeta_{\uparrow},\zeta_{\downarrow},\zeta_{\uparrow\downarrow}\right)d\mathbf{r} \tag{4.514}$$

with

$$F_{xc} = \sum_{abcd}\omega_{abcd}R^{a}S^{b}X^{c}Y^{d} = \sum_{abcd}\omega_{abcd}f_{abcd}(\mathbf{r}) \tag{4.515}$$

Where $R^{a} = \rho_{\uparrow}^{a} + \rho_{\downarrow}^{a}$, $S^{b} = m^{2b}$, see Eq. (4.486) for m definition, along the notations

$$X^{c} = \frac{\zeta_{\uparrow}^{c} + \zeta_{\downarrow}^{c}}{2\rho^{c4/3}},\ Y^{d} = \left(\frac{\zeta_{\uparrow}^{2} + \zeta_{\downarrow}^{2} - 2\zeta_{\uparrow\downarrow}}{\rho^{8/3}}\right)^{d} \tag{4.516}$$

and

$$\zeta_{\uparrow} = \left|\nabla\rho_{\uparrow}\right|,\ \zeta_{\downarrow} = \left|\nabla\rho_{\downarrow}\right|,\ \zeta_{\uparrow\downarrow} = \nabla\rho_{\uparrow}\cdot\nabla\rho_{\downarrow},\ \rho = \rho_{\uparrow} + \rho_{\downarrow} \tag{4.517}$$

The coefficients ω_{abcd} of Eq. (4.515) are determined through minimization procedure involving the associated exchange-correlation potentials $V_{xc\uparrow(\downarrow)}^{abcd}(\mathbf{r}) = \delta f_{abcd}(\mathbf{r}) / \delta\rho_{\uparrow(\downarrow)}(\mathbf{r})$ in Eq. (4.514) functional. The results would depend upon the training set of atoms and molecules but presents the advantage of incorporating the potential information in a non-vanishing asymptotical manner, with a semi-empirical value. Moreover, its exact asymptotic exchange-correlation potential equals chemical hardness (Putz, 2003, 2007a,b) for open-shell being less than that for closed shell systems, thus having the merit of including chemical hardness as an intrinsic aspect of energetic approach, a somewhat absent aspect from conventional functionals so far.

However, since electronegativity and chemical hardness closely relate with chemical bonding, their relation with the total energy and component functionals is in next at both conceptual and applied levels explored.

4.7 OBSERVABLE QUANTUM CHEMISTRY: EXTENDING HEISENBERG'S UNCERTAINTY

4.7.1 *PERIODIC PATH INTEGRALS*

4.7.1.1 Effective Partition Function

As previously shown, see Section 4.3.2, for instance, considering the path integral propagator that underlies the canonical density in the quantum statistical algorithm, see Eqs. (4.243)–(4.248), accounts for the quantum effects (fluctuations) induced on single particle paths by the presence of an external potential, while being analytically computed by averaging these over all possible configurations. Yet, one could observe that for periodic paths, *i.e.*, when the final and initial space-points coincide (Feynman, 1948; Feynman & Hibbs, 1965; Feynman & Kleinert, 1986)

$$x_a = x_b \tag{4.518}$$

the particle travels in very short time not far away from the initial position and then is back on the initial point; such picture has the physical measurable consequence a particle is observed on the initial point, *i.e.*, it is found on a stationary state/orbit, while the quantum fluctuations are oscillating around the equilibrium (initial=final) space-point. Even clearer, the situation corresponds to the classical picture in which a particle behaves, being accommodated in an equilibrium state/stationary orbit under external potential influence. This means that the external influence itself is observable in (initial=final) concerned/measured state, thus being no longer a path parameterized function, but a constant:

$$V(x(\tau),\tau) \to V(x_a) \tag{4.519}$$

Therefore, the associated periodic propagator (4.34) becomes (Kleinert, 2004)

$$(x_a \hbar\beta; x_a 0) = e^{-\beta V(x_a)} \underbrace{\int_{x(0)=x_a}^{x(\hbar\beta)=x_a} \mathrm{D}'x(\tau) \exp\left\{-\frac{1}{\hbar}\int_0^{\hbar\beta}\left[\frac{m}{2}\dot{x}^2(\tau)\right]d\tau\right\}}_{FREE\ MOTION\ PROPAGATOR}$$

$$\left| \begin{array}{l} t := -i\tau \ , \ \tau = \hbar\beta \\ dt = -id\tau \\ \dfrac{d}{dt} = i\dfrac{d}{d\tau} \end{array} \right. \downarrow$$

$$= e^{-\beta V(x_a)} \int_{x(t_a)=x_a}^{x(t_b)=x_a} \mathrm{D}' x(t) \exp\left\{ \frac{i}{\hbar} \int_{t_a}^{t_b} \left[\frac{m\dot{x}^2}{2} \right] dt \right\}_{x_{cl}(t)=x_a+\frac{x_b-x_a}{t_b-t_a}(t-t_a)}$$

$$= e^{-\beta V(x_a)} \left[\sqrt{\frac{m}{2\pi i\hbar (t_b - t_a)}} \exp\left\{ \frac{i}{\hbar} \frac{m}{2} \frac{(x_b - x_a)^2}{t_b - t_a} \right\} \right]_{\substack{x_b = x_a \\ t_b - t_a \to -i\hbar\beta}}$$

$$= e^{-\beta V(x_a)} \sqrt{\frac{m}{2\pi\beta\hbar^2}} \tag{4.520}$$

where the recognized path integral of free motion was solved by plugging into its quantum mechanical solution (4.102) the present conditions (4.518) and the Wick transformation (4.33) for accounting of the path periodicity and quantum statistics, respectively.

At the same time there is clear that the periodic path condition (4.518) is not arbitrarily but a compulsory step since characteristic in passing from density matrix to partition function and then to the real (measurable or workable) canonical and *N*-particle density, according with the density matrix algorithm (4.243)–(4.248). Therefore, the resulting partition function built from the un-normalized canonical density (4.520) assumes the simple form

$$Z_{cl} = \int (x_a \hbar\beta; x_a 0) dx_a = \sqrt{\frac{m}{2\pi\beta\hbar^2}} \int e^{-\beta V(x_a)} dx_a \tag{4.521}$$

while being susceptible of universal reliability if not limited by the degree the periodicity between the final and initial space-point is achieved through condition (4.518). However, looking to free motion path integral solution (4.102) we see that the classical observation is readily valid for the coordinate departure not exceeding the critical value

$$x_b - x_a = \Delta x_{cl} = \hbar\sqrt{\frac{\beta}{m}} \tag{4.522}$$

in which case the exponential limit

$$\exp\left\{-\frac{m}{2\hbar\beta}(x_b - x_a)^2\right\} \xrightarrow{x_a - x_b = \Delta\xi_{cl}} e^{-1/2} = 0.607 \tag{4.523}$$

is approximated with unity in expression (4.520), thus with an error of 40% at the maximum displacement of Eq. (4.522) value; as the classical displacement (4.522) tends to zero as the expressions (4.520) and (4.521) become more accurate. Following the Feynman standard example, for a crystal with atoms of typical atomic mass (A) about 20, at room temperature, the classical limit of displacement (4.522) gives about 0.1 Å; this is the maximum displacement of those atoms around their equilibrium position in the lattice when the thermodynamic properties of the solid can be evaluated through considering the classical form of partition function (4.521). Just in passing worth noting that the partition function (4.521) is called "classical" despite carrying the exponential pre-factor with the quantum Planck constant since the configuration integral $\int \exp(-\beta V)$ was historically anticipated and worked out by Boltzmann, in the pre-quantum era with a non-specified multiplying constant, known today as the inverse of the so-called *thermal length*

$$\lambda_{th} = \sqrt{\frac{2\pi\beta\hbar^2}{m}} \tag{4.524}$$

With these considerations there appears as natural the generalization of the classical partition form Eq. (4.521) into the more comprehensive one known as the effective classical partition function (Kleinert, 1986; Giachetti et al., 1986; Janke & Cheng, 1988; Voth, 1991; Cuccoli et al., 1992)

$$Z_{eff-cl} = \int_{-\infty}^{+\infty} \frac{dx_0}{\sqrt{2\pi\hbar^2\beta / m}} \exp\left[-\beta V_{eff-cl}(x_0)\right] \tag{4.525}$$

with the integration variable defined as the thermal average of the periodic quantum paths

$$x_0 \equiv \overline{x} = \frac{1}{\hbar\beta}\int_0^{\hbar\beta} x(\tau)d\tau \tag{4.526}$$

sometimes called as the *Feynman centroid*, while the notation is to be right bellow justified.

Moreover, the search for the best approximation of effective-classical partition function (4.524) will be conducted as such the quantum fluctuations be not dependent on the classical displacement (4.522), abstracted from the free motion, but being driven by the quantum harmonic oscillations – through they constitute a generalization of the free motion itself, see for instance the equivalence of classical paths or propagators of free with harmonic motion in the zero-frequency limit, see (Putz, 2009).

However, the periodicity condition (4.518) for paths is to be maintained and properly implemented in approximating the effective-classical partition function (4.525) being, nevertheless, closely and powerfully related with the quantum beloved concept of stationary orbits defined/described by periodic quantum waves/paths. This way, the effective-classical path integral approach appears as the true quantum justification of the quantum atom and of the quantum stabilization of matter in general, providing reliable results without involving observables or operators relaying on special quantum postulates other than the variational principles – with universal (classical or quantum) value.

4.7.1.2 Periodic Quantum Paths

As always done when a new type of path integral is under consideration the reconsideration of the quantum paths, and in fact the quantum fluctuations, is undertaken so that facilitating the best way for solving it. Yet, this time due to the periodicity condition of paths the propagator is hidden by the associated partition function. Therefore, the optimum approximation for the effective classical potential in Eq. (4.525) will provide the periodic evolution amplitude as well, *i.e.*, the un-normalized density, which by normalization with partition function will lead with the searched canonical density counterpart.

Going to characterize the periodic paths, they will be seen as the Fourier series (Feynman & Hibbs, 1965; Feynman, 1972; Schulman, 1981; Wiegel, 1986; Kleinert, 2004)

$$x(\tau)=\sum_{m=-\infty}^{+\infty} x_m \exp(i\omega_m \tau) \tag{4.527}$$

in terms of the so-called Matsubara frequencies ω_m; they are explicitly found through specializing the condition (4.518) into the actual statistical one, see Figure 4.5 (Putz, 2009)

$$x_a = x(0) = x(\hbar\beta) = x_b \tag{4.528}$$

resulting in the equality

$$1 = \exp(i\omega_m \hbar\beta) \tag{4.529}$$

with the solution

$$\omega_m = \frac{2\pi}{\hbar\beta} m \text{ , } m \in \mathbf{Z} \tag{4.530}$$

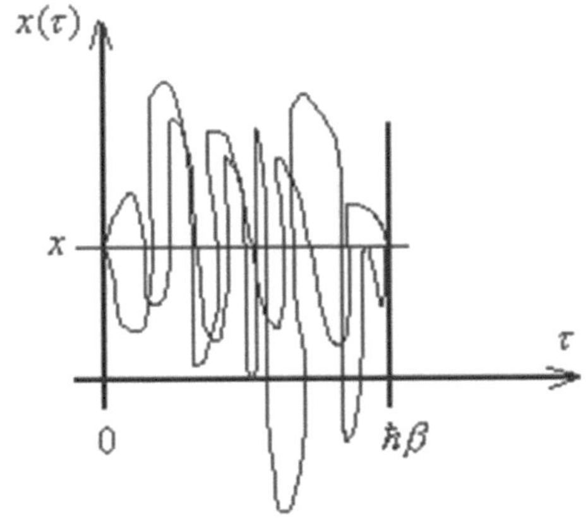

FIGURE 4.5 The representation of the periodic paths (Putz, 2009).

which certifies the quantization of the paths (4.527). Moreover, under the condition the quantum paths (4.527) are real

$$x^*(\tau) = x(\tau) \tag{4.531}$$

the equivalent expanded form with the conjugated path

$$x^*(\tau) = \sum_{m=-\infty}^{+\infty} x_m^* \exp(-i\omega_m \tau) = \sum_{m=-\infty}^{+\infty} x_m \exp(i\omega_m \tau) \tag{4.532}$$

yields for the coefficients of the periodical paths the relationship

$$x_m^* = x_{-m} = x_m \tag{4.533}$$

With this, the quantified form of periodic path frequencies, Eq. (4.530), allows separating the paths (4.527) into the constant and complex conjugated oscillating contributions

$$\begin{aligned} x(\tau) &= x_0 + \sum_{m=1}^{+\infty} x_m \exp(i\omega_m \tau) + \sum_{m=-1}^{-\infty} x_m \exp(i\omega_m \tau) \\ &= x_0 + \sum_{m=1}^{+\infty} x_m \exp(i\omega_m \tau) + \sum_{m=1}^{+\infty} x_{-m} \exp(-i\omega_m \tau) \\ &= x_0 + \sum_{m=1}^{+\infty} x_m \exp(i\omega_m \tau) + c.c. \end{aligned} \tag{4.534}$$

with the 0th terms viewed more than the "zero-oscillating" or free motion path but the thermal averaged path over entire quantum paths (4.527)

$$\begin{aligned} \frac{1}{\hbar\beta}\int_0^{\hbar\beta} x(\tau)d\tau &= \frac{1}{\hbar\beta}\int_0^{\hbar\beta}\left[x_0 + \sum_{m=1}^{+\infty} x_m \exp(i\omega_m \tau) + c.c.\right]d\tau \\ &= x_0 + \sum_{m=1}^{+\infty} x_m \frac{1}{\hbar\beta}\int_0^{\hbar\beta} \exp(i\omega_m \tau)d\tau \\ &\quad + \sum_{m=1}^{+\infty} x_{-m} \frac{1}{\hbar\beta}\int_0^{\hbar\beta} \exp(-i\omega_m \tau)d\tau \end{aligned}$$

$$
\begin{aligned}
&= x_0 + \frac{1}{\hbar\beta}\sum_{m=1}^{+\infty} x_m \int_0^{\hbar\beta} \left[\exp(i\omega_m\tau) + \exp(-i\omega_m\tau)\right] d\tau \\
&= x_0 + \frac{2}{\hbar\beta}\sum_{m=1}^{+\infty} x_m \int_0^{\hbar\beta} \cos(\omega_m\tau) d\tau \\
&= x_0 + \frac{2}{\hbar\beta}\sum_{m=1}^{+\infty} \frac{x_m}{\omega_m} \sin(\omega_m\hbar\beta) \\
&= x_0 + \underbrace{\frac{2}{\hbar\beta}\sum_{m=1}^{+\infty} \frac{x_m}{\omega_m} \underbrace{\sin(2\pi m)_{m\in Z}}_{0}}_{0}
\end{aligned}
\tag{4.535}
$$

thus resulting in the Feynman centroid formula (4.526).

4.7.2 HEISENBERG UNCERTAINTY RELOADED

The actual philosophy is to introduce appropriately the quantum fluctuation information $a = a(x_0)$ respecting the average of the observed coordinate (x_0), by the Feynman integration rule founded in the ordinary quantum average

$$
\langle f \rangle_{a^2(x_0)} = \int_{-\infty}^{+\infty} dx \psi^*\left(x, a^2(x_0)\right) f \psi\left(x, a^2(x_0)\right) \tag{4.536}
$$

for the normalized Gaussian wave-function

$$
\psi\left(x, a^2(x_0)\right) = \frac{1}{\left[2\pi a^2(x_0)\right]^{1/4}} \exp\left[-\frac{(x - x_0)^2}{4a^2(x_0)}\right] \tag{4.537}
$$

recovering the de Broglie wave-packet, the 1D version of Eq. (2.20), upon which a quantum property may be estimated (Feynman & Kleinert, 1986; de Broglie, 1987).

It is obvious that the Eqs. (4.536) and (4.537) fulfill the necessary (natural) condition according which the average of the coordinate over the quantum fluctuations recovers the observed quantity of Eq. (4.535), the Feynman centroid, based on simple Poisson integration rules (see Appendix A.2)

$$\begin{aligned}
\langle x \rangle_{a^2(x_0)} &= \frac{1}{\sqrt{2\pi a^2(x_0)}} \int_{-\infty}^{+\infty} dx [x - x_0 + x_0] \exp\left[-\frac{(x-x_0)^2}{2a^2(x_0)}\right] \\
&= \underbrace{\frac{1}{\sqrt{2\pi a^2(x_0)}} \int_{-\infty}^{+\infty} dx [x - x_0] \exp\left[-\frac{(x-x_0)^2}{2a^2(x_0)}\right]}_{0} \\
&\quad + \underbrace{\frac{1}{\sqrt{2\pi a^2(x_0)}} \int_{-\infty}^{+\infty} dx [x_0] \exp\left[-\frac{(x-x_0)^2}{2a^2(x_0)}\right]}_{\langle x_0 \rangle_{a^2(x_0)}} \\
&= x_0 \underbrace{\frac{1}{\sqrt{2\pi a^2(x_0)}} \int_{-\infty}^{+\infty} dx \exp\left[-\frac{(x-x_0)^2}{2a^2(x_0)}\right]}_{1} = x_0; \\
\langle x \rangle_{a^2(x_0)} &= \langle x_0 \rangle_{a^2(x_0)} = x_0
\end{aligned} \tag{4.538}$$

The next test is about the validity of the Heisenberg uncertainty relationship (HUR) itself. To this end the standard deviation of coordinate (x) and momentum (p)

$$\Delta x = \sqrt{\langle x^2 \rangle - \langle x \rangle^2}, \ \Delta p = \sqrt{\langle p^2 \rangle - \langle p \rangle^2} \tag{4.539}$$

are computed with the aid of Feynman-de Broglie rule (4.537); firstly, one gets

$$\left\langle (x - x_0)^2 \right\rangle_{a^2(x_0)} = \frac{1}{\sqrt{2\pi a^2(x_0)}} \int_{-\infty}^{+\infty} dx (x - x_0)^2 \exp\left[-\frac{(x-x_0)^2}{2a^2(x_0)}\right] = a^2 \tag{4.540}$$

Then, through combining the expression

$$a^2 = \left\langle (x - x_0)^2 \right\rangle_{a^2(x_0)} = \langle x^2 \rangle_{a^2(x_0)} - 2\langle x \rangle_{a^2(x_0)} \langle x_0 \rangle_{a^2(x_0)} + \langle x_0^2 \rangle_{a^2(x_0)} \tag{4.541}$$

with the prescription (4.538) we are left with the actual result

$$\left\langle x^2 \right\rangle_{a^2(x_0)} = a^2 + x_0^2 \tag{4.542}$$

that, when plugged in the basic Eq. (4.539) alongside the information of Eq. (4.538), yields the coordinate dispersion

$$\Delta x = a \tag{4.543}$$

featuring it in a direct relationship with the quantum fluctuation width.

In the same manner, the evaluations for the integrals of the first and second orders of kinetic moment unfold as

$$\begin{aligned}\left\langle p \right\rangle_{a^2(x_0)} &= \frac{1}{\sqrt{2\pi a^2(x_0)}} \int_{-\infty}^{+\infty} dx \exp\left[-\frac{(x-x_0)^2}{4a^2(x_0)}\right] \left(-i\hbar\partial_x\right) \exp\left[-\frac{(x-x_0)^2}{4a^2(x_0)}\right] \\ &= \frac{i\hbar}{2a^2(x_0)} \frac{1}{\sqrt{2\pi a^2(x_0)}} \int_{-\infty}^{+\infty} dx (x-x_0) \exp\left[-\frac{(x-x_0)^2}{2a^2(x_0)}\right] = 0\end{aligned} \tag{4.544}$$

$$\begin{aligned}\left\langle p^2 \right\rangle_{a^2(x_0)} &= \frac{1}{\sqrt{2\pi a^2(x_0)}} \int_{-\infty}^{+\infty} dx \exp\left[-\frac{(x-x_0)^2}{4a^2(x_0)}\right] \left(-\hbar^2\partial_x^2\right) \exp\left[-\frac{(x-x_0)^2}{4a^2(x_0)}\right] \\ &= -\frac{\hbar^2}{a^2(x_0)\sqrt{2\pi a^2(x_0)}} \int_{-\infty}^{+\infty} dx \left[\frac{(x-x_0)^2}{4a^2(x_0)} - \frac{1}{2}\right] \exp\left[-\frac{(x-x_0)^2}{2a^2(x_0)}\right] = \frac{\hbar^2}{4a^2}\end{aligned} \tag{4.545}$$

while when plugging them in Eq. (4.539) produce the momentum dispersion expression

$$\Delta p = \frac{\hbar}{2a} \tag{4.546}$$

Worth noting is that from the coordinate and momentum dispersions, Eqs. (4.543) and (4.546), it appears that the dependency of Planck constant is restricted only to the latter, whereas the quantum fluctuations are in both present, in a direct and inverse manner, respectively.

However, when multiplying the expressions (4.543) and (4.546) the Heisenberg uncertainty is naturally approached by exact specialization of Heisenberg Equation, see Eq. (2.99)

$$\Delta x \Delta p = \frac{\hbar}{2} \tag{4.547}$$

this way resembling in an elegant manner the previous result of statistical complementary observables of position and momentum (Hall, 2001).

For the sake of experimental precision it is worth noting that the error in coordinate localization is given *at least* by one quantum fluctuation "leap" in Eq. (4.543), *i.e.*, by the width in the de Broglie wave packet of Eq. (4.537) that may be naturally exceeded in certain (large) coordinate observations – from where the general HUR emerges. Remarkably, the HUR validity was here proved using only the wave-packet properties, including the quantum fluctuation $a = a(x_0)$ that appears in the final coordinate-momentum multiplied dispersions—being therefore incorporated in the HUR result—a feature not obviously revealed by earlier demonstrations.

Yet, another important idea was raised, namely that the coordinate and momentum dispersions, although in reciprocal relationship with quantum fluctuation, *i.e.*, when during an experiment the quantum fluctuation may be set out in coordinate or momentum it acts larger in the other – and *vice versa,* may be treated somehow *separated,* from where the possibility of different realizations for coordinate dispersion through relations (4.538) and (4.543), with consequences for HUR reformulations. Such possibilities and the inter-connection with the wave-particle quantum issue are next explored.

4.7.3 EXTENDING HEISENBERG UNCERTAINTY

4.7.3.1 Averaging Quantum Fluctuations

We like to identify the general quantum fluctuation conditions in which the HUR is valid and when it is eventually extended. We already note that, whereas the momentum dispersion computation is fixed by relations (4.544)–(4.546), the evaluation of the coordinate dispersion has more freedom in its internal working machinery, namely (Putz, 2010c):

(i) considering the condition (4.538) as an invariant of the measurement theory since it assures the connection between the average over quantum fluctuation of the coordinate and the observed averaged coordinate;
(ii) specializing the quantum (average) relationship (4.541) for the condition given by Eq. (4.538);
(iii) obtaining the average of the second order coordinate (4.542);
(iv) combining the steps i) and ii) is computing the coordinate dispersion Δx as given by Eq. (4.539).

The present algorithm may be naturally supplemented with the analysis of the wave-particle duality. This is accomplished by means of considering further averages over the quantum fluctuations for the mathematical objects $\exp(-ikx)$ and $\exp(-k^2x^2)$ that are most suited to represent the waves and *particles*, due to their obvious shapes, respectively. Such computations of averages are best performed employing the Fourier k-transformation as resulted from the de Broglie packet, Eq. (4.537) with Eq. (4.537), equivalently rewritten successively as (Putz, 2009; Putz, 2010c):

$$\begin{aligned}
\langle f(x,k)\rangle_{a^2(x_0)} &= \frac{1}{\sqrt{2\pi a^2(x_0)}}\int_{-\infty}^{+\infty} dx f(x,k)\exp\left[-\frac{(x-x_0)^2}{2a^2(x_0)}\right] \\
&= \frac{1}{2\pi}\int_{-\infty}^{+\infty} dx f(x,k)\exp\left[-\frac{(x-x_0)^2}{2a^2(x_0)}\right]\int_{-\infty}^{+\infty} dk' \exp\left[-\frac{a^2(x_0)}{2}k'^2\right] \\
&= \frac{1}{2\pi}\int_{-\infty}^{+\infty} dx f(x,k)\exp\left[-\frac{(x-x_0)^2}{2a^2(x_0)}\right]\int_{-\infty}^{+\infty} dk \exp\left[\left(\frac{x-x_0}{\sqrt{2}a(x_0)} - i\frac{a(x_0)}{\sqrt{2}}k\right)^2\right] \\
&= \int_{-\infty}^{+\infty}\frac{dk}{2\pi}\int_{-\infty}^{+\infty} dx f(x,k)\exp(-ikx)\exp\left[ikx_0 - \frac{1}{2}a^2(x_0)k^2\right] \\
&= \int_{-\infty}^{+\infty}\frac{dk}{2\pi} f(k)\exp\left[ikx_0 - \frac{1}{2}a^2(x_0)k^2\right]
\end{aligned} \tag{4.548}$$

With the rule (4.548) one may describe the average behavior of the wave and particle, respectively as

$$\begin{aligned}
\left\langle \exp(-ikx) \right\rangle_{a^2(x_0)} &= \int_{-\infty}^{+\infty} \frac{dk}{2\pi} \exp\left[-ikx + ikx_0 - \frac{1}{2} a^2(x_0) k^2 \right] \\
&= \int_{-\infty}^{+\infty} \frac{dk}{2\pi} \exp\left[-ik(x - x_0) - \frac{1}{2} a^2(x_0) k^2 \right] \\
&= \exp\left[-\frac{(x - x_0)^2}{2a^2(x_0)} \right] \int_{-\infty}^{+\infty} \frac{dk}{2\pi} \exp\left\{ -\frac{a^2(x_0)}{2} \left[k + i\frac{x - x_0}{a^2(x_0)} \right]^2 \right\} \\
&= \frac{1}{2\pi} \exp\left[-\frac{(x - x_0)^2}{2a^2(x_0)} \right] \int_{-\infty}^{+\infty} dk' \exp\left\{ -\frac{a^2(x_0)}{2} k'^2 \right\} \\
&= \frac{1}{\sqrt{2\pi a^2(x_0)}} \exp\left[-\frac{(x - x_0)^2}{2a^2(x_0)} \right]
\end{aligned} \tag{4.549}$$

and

$$\begin{aligned}
&\left\langle \exp(-k^2 x^2) \right\rangle_{a^2(x_0)} = \int_{-\infty}^{+\infty} \frac{dk}{2\pi} \exp\left[-k^2 x^2 + ikx_0 - \frac{1}{2} a^2(x_0) k^2 \right] \\
&= \int_{-\infty}^{+\infty} \frac{dk}{2\pi} \exp\left[-k^2 \left(x^2 + \frac{a^2(x_0)}{2} \right) + ikx_0 \right] \\
&= \exp\left[-\frac{x_0^2}{4\left(x^2 + a^2(x_0)/2\right)} \right] \int_{-\infty}^{+\infty} \frac{dk}{2\pi} \times \exp\left\{ -\left(x^2 + \frac{a^2(x_0)}{2} \right) \left[k - i\frac{x_0}{2\left(x^2 + a^2(x_0)/2\right)} \right]^2 \right\} \\
&= \frac{1}{2\pi} \exp\left[-\frac{x_0^2}{4\left(x^2 + a^2(x_0)/2\right)} \right] \int_{-\infty}^{+\infty} dk' \exp\left\{ -\left(x^2 + \frac{a^2(x_0)}{2} \right) k'^2 \right\} \\
&= \frac{1}{\sqrt{2\pi\left[2x^2 + a^2(x_0) \right]}} \exp\left[-\frac{x_0^2}{2\left(2x^2 + a^2(x_0)\right)} \right]
\end{aligned} \tag{4.550}$$

It is worth observing that the practical rule (4.548) is indeed consistent since recovering in Eq. (4.549) the kernel of the Gaussian de Broglie wave-packet—for the *wave* behavior of a quantum object—as expected. As a consequence, the result (4.550) may be therefore considered as a viable analytical expression for characterizing the complementary *particle* nature of the quantum manifestation of an object.

Next, the ratio of Eqs. (4.549) and (4.550) is taken

$$\frac{Particle}{Wave} \equiv \frac{\left\langle \exp\left(-k^2x^2\right)\right\rangle_{a^2(x_0)}}{\left\langle \exp\left(-ikx\right)\right\rangle_{a^2(x_0)}} = \sqrt{\frac{a^2(x_0)}{2x^2+a^2(x_0)}}\exp\left[-\frac{x_0^2}{2\left(2x^2+a^2(x_0)\right)}+\frac{x^2-2xx_0+x_0^2}{2a^2(x_0)}\right] \quad (4.551)$$

giving the working tool in estimating the particle-to-wave content for a quantum object by considering various coordinate average information; this will be achieved by

(v) making the *formal* identity of the coordinate quantities in Equation (4.551) with the respective values as furnished by the steps i)-iii) of the above coordinate averages' algorithm

$$x_0 \leftrightarrow \left\langle x_0\right\rangle_{a^2(x_0)},\ x \leftrightarrow \left\langle x\right\rangle_{a^2(x_0)},\ x_0^2 \leftrightarrow \left\langle x_0^2\right\rangle_{a^2(x_0)},\ x^2 \leftrightarrow \left\langle x^2\right\rangle_{a^2(x_0)} \quad (4.552)$$

since they nevertheless emerge from quantum average operations (measurements).

Now we are ready for presenting the two possible scenarios for quantum evolutions along the associate HUR realization and the wave-particle behavior.

4.7.3.2 Observed Quantum Evolution

For the case of observed quantum evolution, the averaged observed position is considered in relation with the quantum fluctuation by the general relationship

$$\langle x \rangle_{a^2(x_0)} = \langle x_0 \rangle_{a^2(x_0)} = x_0 = na \, , \; n \in \Re \tag{4.553}$$

implying that the average of the second order of Feynman centroid looks like

$$\langle x_0^2 \rangle_{a^2(x_0)} = n^2 a^2 \tag{4.554}$$

When Eqs. (4.553) and (4.554) are considered into the identity (4.542), according with the step (iii) above, the actual average of the second order coordinate is obtained

$$\langle x^2 \rangle_{a^2(x_0)} = a^2 \left(1 + n^2\right) \tag{4.555}$$

Not surprisingly, when further combining relations (4.553) and (4.555) in computing the coordinate dispersion of Eq. (4.539), *i.e.*, fulfilling the step iv) above, one regains the value of Eq. (4.543) that recovers at its turn the standard HUR no matter how much the quantum fluctuation is modulated by the factor n. However, the P(article)/W(ave) ratio of Eq. (4.551) takes the form (Putz, 2010c)

$$\left(\frac{Particle}{Wave}\right)_{\substack{Observed\\Evolution}} = \frac{1}{\sqrt{3+2n^2}} \exp\left(\frac{3+n^2}{6+4n^2}\right) = \begin{cases} 0.952 \ldots n = 0 \\ 0.667 \ldots n = 1 \\ 0 \quad \ldots n \to \infty \end{cases} \tag{4.556}$$

showing that the wave-particle duality is indeed a reality that can be manifested in various particle-wave (complementary) proportions—yet never reaching the perfect equivalence (the ratio approaching unity). Moreover, because $(P/W)_{Obs} < 1$, it appears that the general behavior of a quantum object is merely manifested as wave when observed, from which arises the efficacy of spectroscopic methods in assessing the quantum properties of matter.

4.7.3.3 Free Quantum Evolution

Moving to the treatment of the *free quantum evolution*, the average of the first order coordinate is assumed as vanishing

$$\langle x \rangle_{a^2(x_0)} = \langle x_0 \rangle_{a^2(x_0)} = x_0 = 0 \tag{4.557}$$

since the quantum object, although existing, is not observed (see the spontaneous broken symmetry mechanism in the discussion Section 4.6.3.4 below).

The relation with quantum fluctuation may be nevertheless gained by the average of the second order of the Feynman centroid–considered under the form

$$\langle x_0^2 \rangle_{a^2(x_0)} = n^2 a^2 \tag{4.558}$$

Note that Eqs. (4.557) and (4.558) parallel the statistical behavior of error in measurements that being vanishing in the first case as mean deviation, is manifested in the second as squared deviation (dispersion), respectively.

Next, through recalling the referential Eq. (4.541)—the step (ii) in above algorithm—the average of the second order coordinate provides now the expression

$$\langle x^2 \rangle_{a^2(x_0)} = a^2 \left(1 - n^2\right) \tag{4.559}$$

The result (4.559) restrains the domain of the free evolution quantum fluctuation factor n to the realm $n \in [0,1]$. With Eqs. (4.557) and (4.559), the step (iii) in above algorithm, one finds the coordinate dispersion

$$\Delta x = a\sqrt{1 - n^2} \tag{4.560}$$

with the immediate consequence in adjusting the basic HUR as

$$\Delta x \Delta p \geq \frac{\hbar}{2}\sqrt{1 - n^2} \tag{4.561}$$

On the other hand, within conditions fixed by Eqs. (4.557)–(4.559) the P(article)/W(ave) index of Eq. (4.551) becomes (Putz, 2010c)

$$\left(\frac{Particle}{Wave}\right)_{\substack{Free\\Evolution}} = \frac{1}{\sqrt{3-2n^2}}\exp\left(\frac{3-3n^2}{6-4n^2}\right)$$
$$= \begin{cases} 0.952 \ldots n = 0 \\ 1 \quad \ldots n_\Omega = 0.54909 \\ 1.048 \ldots n = 0.87 \\ 1 \quad \ldots n_\alpha = 1 \\ \infty \quad \ldots n = \sqrt{3/2} = 1.22474 \end{cases} \tag{4.562}$$

Through characterizing the numerical results of Eq. (4.562), one firstly observes that they practically start from where the P/W function of Eq. (4.556) approaches its highest output. In other words, this tell us remarkable information according to which the *observed and free quantum evolutions are continuous realities, being smoothly accorded in the point of precise measurement* ($n = 0$). Another very interesting observation is that the P/W ratio symmetrically spans in Eq. (4.562) the existence domain either for wave P/W∈[0.952, 1) or particle P/W∈(1, 1.048] manifestations around their exact equivalence P/W = 1. However, the precise wave-particle equivalence is two-fold, namely in the socalled *omega* (Ω) and *alpha* (α) points of Eq. (4.562) characterized by the extended HUR versions of Eq. (4.561); written, respectively, as

$$(\Delta x \Delta p)_\Omega \geq 0.418\hbar \tag{4.563}$$

$$(\Delta x \Delta p)_\alpha \geq 0 \tag{4.564}$$

It is clear that whereas the omega case of Eq. (4.563) is characterized by the restrained quantum domain of ordinary HUR of Eq. (2.99), in which a quantum object's evolution may be grated, on the alpha point of Eq. (4.564) any quantum information is lost since no Planck constant exists there to drive the wave-particle quantum inter-conversion. It is this last case that may be eventually related with early cosmological stages when the quantum fields and particles are considered as absorbed in the universal gravity; nevertheless, this is just a hint for future possible use of the present extended-HUR phenomenology that may help in understanding the occurrence of the quantum information, entanglement, and the separation of the fields and particles towards the observed world.

4.7.3.4 Free vs. Observed Quantum Evolution

It is very instructive to present in a unitary manner the observed and free quantum evolution cases in the chart of Figure 4.6 by linking the HUR shapes of Eqs. (2.99) and (4.561) with the particle/wave ratios values of Eqs. (4.556) and (4.562), respectively. The P/W contribution spreads from the exclusively undulatory quantum manifestation (P/W = 0) in the *observed* domain of quantum evolution until the particle dominance (P/W > 1) in the *free* domain of quantum evolution.

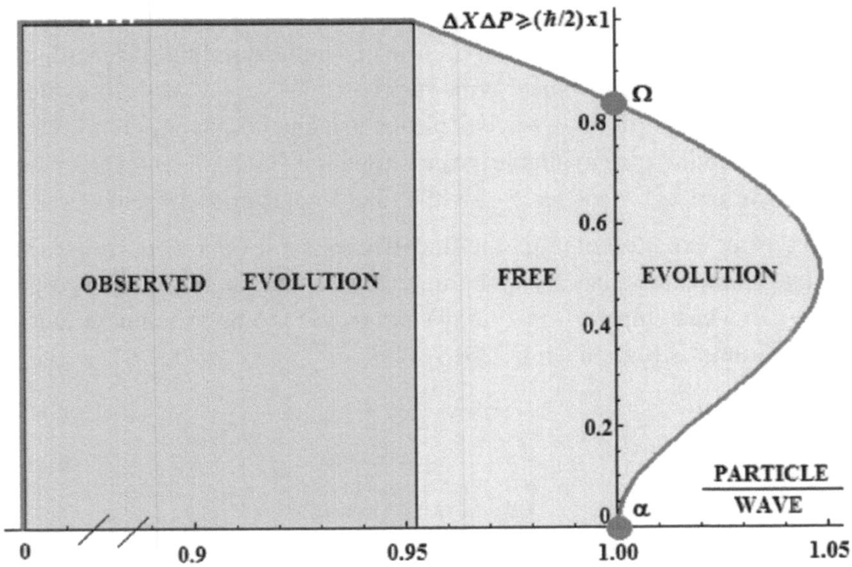

FIGURE 4.6 The chart of Heisenberg Uncertainty Relationship (HUR) appearance for observed and free quantum evolutions covering the complete scale of the particle to wave ratios as computed from the Eqs. (4.556) and (4.562), respectively; the points Ω and α correspond to wave-particle precise equivalence and to the special extended-HURs of Eqs. (4.563) and (4.564), respectively (Putz, 2010c).

Note that the possibility a quantum object is manifested *only* under particle behavior (*i.e.*, for P/W→∞) is forbidden; this is an important consequence of the present analytical discourse that is in agreement with the Copenhagen interpretation according which the quantum phenomena are merely manifested as undulatory (*viz.* Schrödinger equation) although some particle information may be contained but *never* in an exclusive manner (naturally, otherwise the Newtonian object would exist with no Planck constant and HUR relevance upon it).

However, the wave-particle *duality* matches perfectly and always with HUR in its standard (Schrödinger) formulation of Eq. (2.99); on the other side, the wave-particle exact equivalence (P/W = 1) may be acquired only in the free evolution regime that, in turn, it is driven by modified HUR as given by Eq. (4.563). In other words, it seems that any experiment or observation upon a quantum object or system would destroy the P/W balance specific for free quantum evolution towards the undulatory manifestation through measurement.

Yet, having the analytical expressions for both observed and free quantum evolutions may considerably refine our understanding of macro- and micro-universe. For instance, with various $(P/W)_{Observed}$, one can evaluate the appropriate particle-to-wave presence in a quantum complex for which experimental data are available: once knowing from a given measurement the quantities $\left\langle x_0^2 \right\rangle_{Exp}$ and $\left\langle x^2 \right\rangle_{Exp}$, with x_0 and x appropriately considered for each type of experiment (e.g., the statistical mean for classical records and the instantaneous values for quantum measurement of coordinate, respectively), one can employ Eqs. (4.554) and (4.555) to find the magnitude of the quantum fluctuation (Putz, 2010)

$$n = \sqrt{\frac{\left\langle x_0^2 \right\rangle_{Exp}}{\left| \left\langle x^2 \right\rangle_{Exp} - \left\langle x_0^2 \right\rangle_{Exp} \right|}} \tag{4.565}$$

that when replaced into Eq. (4.556) predicts the P/W ratio involved in that observation.

It is worth giving a working example for emphasizing the reliability of the present approach and to choose for this aim the fundamental Compton quantum experiment. In this case, the incoming photonic beam carries the wavelength λ_0 whilst the scattered one departs from that incident with the amount $\Delta\lambda = \lambda - \lambda_0$; such situation allows the immediate specialization of the quantum fluctuation magnitude (4.565) to its Compton form

$$n_{Compton} = \sqrt{\frac{\lambda_0^2}{(\lambda + \lambda_0)\Delta\lambda}} \tag{4.566}$$

Now we can interpret the various experimental situations encountered, employing the output of Eq. (4.566) to asses through Eq. (4.556) the wave-particle ratio degree present in specific measurements. For example, when the scattering is made on *free electrons*, then the higher and higher record for $\Delta\lambda$ implies the decrease of $n_{Compton}$ of Eq. (4.566) and consequently the increase of $(P/W)_{Compton}$ of Eq. (4.556); this is in accordance with the fact that *the scattered light on free electrons rises more and more its particle (photonic) behavior*. On the other side, when the scattering is made on *tight bonded electrons* (e.g., electrons in atoms of a material), the Compton wavelength departure is negligible, $\Delta\lambda \to 0$, leaving from Eq. (4.566) with the asymptotic higher quantum fluctuation magnitude $n_{Compton} \to \infty$ that corresponds at its turn with $(P/W)_{Compton} = 0$ in Eq. (4.556). This matches with the fact that this case corresponds with *complete wave manifestation of light that scatters bonded electrons*, resembling the (classical) interpretation according which the scattered bounded electron by a wave entering in resonance with it while oscillating with the same frequency. Therefore, the reliability of the present $(P/W)_{\text{Observed}}$ formalism was paradigmatically illustrated, easily applied to other quantum experiments, while giving the numerical P/W estimations once having particular data at hand. Equally valuable is the free evolution $(P/W)_{\text{Free}}$ ratio of Eq. (4.562) that may be employed for the wave-particle equivalency between the quantities (4.549) and (4.550) (Kleinert, 2004; Putz, 2009)

$$\left\langle \exp(-ikx) \right\rangle_{a^2(x_0)} \cong \left\langle \exp(-k^2x^2) \right\rangle_{a^2(x_0)} \tag{4.567}$$

with an important role in assessing the stability of matter, from atom to molecule. As an example, the justification of the Hydrogen stability was successfully proved through setting the ratio P/W = 1 in the omega point of function (4.562) or within its vicinity (Kleinert, 2004; Putz, 2009). Nevertheless, further applications of the $(P/W)_{Free}$ function (4.562) and of subsequent modified HUR may be explored also in modeling the various stages and parts of the Universe that cannot be directly observed, as well as when dealing with quantum hidden information in the sub-quantum or coherent states (Bohm & Vigier, 1954; Nielsen & Chuang, 2000).

On the other side, one would wish to further discuss the free quantum *vs.* observed quantum evolutions in terms of simple average of paths, *viz.* Eqs. (4.557) and (4.553), with practical examples, respectively. The best paradigm that can transform the first into the last one stands the *spontaneous symmetry breaking* (Goldstone, 1961) that has the role in turning the intrinsic zero ensemble averages of Eq. (4.557) to the finite observable quantum effects (and fluctuations) of Eq. (4.553). The best examples are the magnetization and the condensation phenomena: in the first case, due to the invariance under rotation of the Hamiltonian, the ensemble average of the total magnetic moment $\mathcal{M}$ is always zero, $<\mathcal{M}>= 0$, since $+\mathcal{M}$ and $-\mathcal{M}$ occur with the same probability (Anderson, 1984). In the case of condensation (for instance Bose-Einstein), the order parameter $\langle\psi\rangle$ that is obtained from averaging the bosonic fields on the canonical ensemble gives zero result in free (untouched) evolution, $\langle\psi\rangle = 0$, due to the inner annihilation nature of the bosonic field $\psi(x)$, beside the total Hamiltonian is global gauge invariant under the transformation $\psi(x) \to e^{i\theta}\psi(x), \forall\theta \in \Re$ that corresponds with the conservation of the total number of particles in the system (Huang, 1987).

However, either case is resolved within experiments by simple observation (e.g., the ferromagnets and the superfluid ^{4}He appear under natural conditions without special experimental conditions) through the so-called "Goldstone excitations" (spin waves and the phonons for ferromagnets and superfluids, respectively) that eventually turns (brakes) the microscopic (free evolution) Hamiltonian symmetry into the macroscopic (observed or directional evolution) symmetry. This mechanism of broken symmetry fits with the present free-to-observed quantum evolution picture since, when revealed, it involves a countless number of zero-energy (yet orthogonal) ground states, leading with the rising of the locally (Goldstone) excited state from one of the ground states that gradually changes over the space from the zero energy and infinity wavelength to some finite non-zero energy and long wavelength; such behavior parallels the turning of the condition of Eq. (4.557) into that of Eq. (4.553), where the exact Heisenberg principle is obeyed—however in different Particle/Wave ratios (depending on the phenomenon and experiment), see the above discussion and the Figure 4.6.

For advanced molecular physical chemistry, it is worth pointing out that the particle/wave ratio (P/W) of Eq. (4.551) may be used to re-shape the so-called *electronic localization function* (ELF) (Becke & Edgecombe, 1990; Silvi & Savin, 1994), which carries much information on the electronic probability to be manifested as wave or particle in chemical bonding, see Volume II of this five-volume set. As such, further identification of ELF with the quantity of P/W in the observed regime of Equation (4.556)

$$ELF_{P/W} = \left(\frac{Particle}{Wave}\right)_{\substack{Observed\\Evolution}} \leq 0.95 \tag{4.568}$$

tells us that, in accordance with the recent interpretation of ELF as error in electronic localization, the maximum prescribed error of localization of electrons in atoms and molecules is limited within the range [0,0.95] and can never be complete; *i.e.*, the electron is localizable at least as 5% from its particle contents. In other words, the present approach prescribes that any chemical bond contains at least 5% of particle nature of its pairing electrons, *i.e.*, the covalence is never complete while always coexisting with some ionicity!

This is a fundamental result of actual exact HUR treatment for chemical bonding. However, further application of the $ELF_{P/W}$ index (4.568) for explaining—for instance—the molecular aromaticity, see Volume III of this set, in terms of geometry of bonding and the amount of quantum fluctuation present, are in progress and will be in the future communicated.

Finally, for spectroscopic analysis, one could ask upon the corresponding time-energy uncertainty relationship (Busch, 2008) within the actual approach. Firstly, the correctness of such problem is conceptually guaranteed by the Heisenberg representation of a quantum evolution, where, for a cyclic vector of state (*viz.* the present periodical paths or orbits) and an unitary transformation U, the cyclic Hamiltonian H_U is accompanied by the time operator $t_U = -i\hbar\partial_\mu$ with the $\partial_\mu = d/d\mu(\varepsilon)$ relating the integrable measure $\mu(\varepsilon)$ as depending of the energetic spectra (ε) on the associate generalized Hilbert space (Ivanov, 2006). On the other side, quantitatively, the time-energy HUR faces with the practical problem in evaluating the general yield of the Hamiltonian variance

$$\Delta H = \sqrt{\langle H^2 \rangle - \langle H \rangle^2} \tag{4.569}$$

since containing the non-specified external potential dependency:

$$H = -\frac{\hbar^2}{2M}\partial_x + V(x) \tag{4.570}$$

Yet, the present periodic path approach may be eventually employed to assess the problem through reconsidering the width $a(x_0)$ of the de Broglie wave-function (4.537) as related with the averaged potential over the quantum fluctuations $\langle V(x) \rangle_{a^2(x_0)}$; a self-consistent equation is this way expected, while the final time-energy HUR may further depend on the ground or excited (Wigner) states considered, *i.e.*, within the inverse of the thermal energy limits $\beta \to \infty$ or $\beta \to 0$, respectively. Nevertheless, this remains a challenging subject that will be also approached in the near future.

4.7.3.5 Spectral-IQ Method

The wave-particle issue was in the "heart" of quantum mechanics, even in its very principles, Heisenberg one in particular (Putz, 2010c). Currently assumed as a complementarity reality, it was just recently quantified with the aim of the path integrals' quantum fluctuation factor (n) through considering the quantum averages for the Gaussian wave packet to the harmonic one for the particle and wave representations, respectively. The results were finite and apart of consistently explaining the atomic (and thus the matter) stability through particle-wave equivalency at the quantum level; they permit also a general formulation of the particle-to-wave ratio content for an observed event (4.556) as well as for the free quantum evolution (4.562).

One notes, for instance, that when quantum fluctuations asymptotically increase, the wave contents become infinite and cancel the particle observability, according to the Eq. (4.556).

Instead, even when a system hypothetically experiences zero quantum fluctuations, the wave nature of the system will be still slightly dominant over its particle side at both observed and free evolutions; see the upper branches of Eqs. (4.556) and (4.562). These extremes show that the wave nature of matter will never be fully transferred to particle contents and the mesosystems will never be fully characterized by pure particle (or mechanical) features.

However, it is apparent from Eq. (4.562) that free evolution of a stable system is merely associated with particle dominance, however, without being manifestly observable; in fact, such peculiar particle behavior of the free evolutions of stable matter confirms its inner quantum nature by quantifiable features.

Figure 4.7 depicts the main tendencies of the particle-to-wave ratios of a stable system in terms of its quantum fluctuation, in free or observed

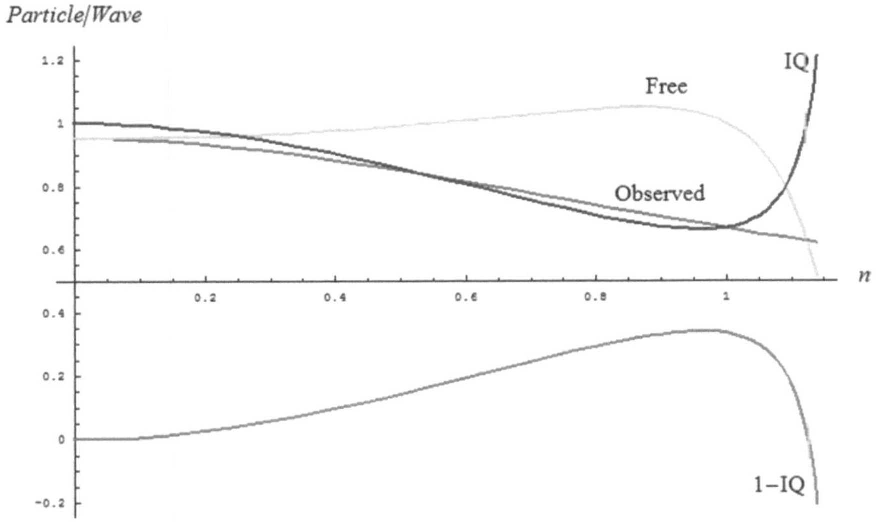

FIGURE 4.7 Depicted tendencies of the observed, free and inverse quantum (IQ) evolutions of the particle-to-wave ratio with respect to the quantum fluctuations (n) upon Eqs. (4.556), (4.562), and (4.571), respectively; the additional curve of residual inverse quantum index $RQ = 1 - IQ$ was added with the purpose of showing that free evolutions parallels RQ that is symmetrical with respect to the IQ factor (Putz & Putz, 2012).

conditions, alongside the present inverse quantum (IQ) index introduced as their competition.

Indeed, the inverse quantum index (4.571) showcases the manifestly inverse behavior respecting free quantum evolution while accompanying the observed evolution for the respective quantum fluctuations' range; therefore, it may constitute a suitable index for accounting the particle information degree in a general quantum evolution, from a free-to-observed one. Moreover, if one considers also the residual inverse quantum information $RQ = 1 - IQ$, one also gets a symmetrical tool with respect to IQ for treating the free evolution at the quantum level (Putz & Putz, 2012)

$$IQ_{Obs/Free} = \frac{\left(\frac{Particle}{Wave}\right)_{\substack{Observed\\Evolution}}}{\left(\frac{Particle}{Wave}\right)_{\substack{Free\\Evolution}}} = \sqrt{\frac{3-2n^2}{3+2n^2}}\exp\left(\frac{2n^2}{9-4n^4}\right) \tag{4.571}$$

Being the quantum fluctuation factor crucial for assessing the free and observed quantum behavior, it should be noted it may discriminate between these two quantum sides of motion, however, based solely on experimental measures of classical and quantum paths, since one considers their squared averages $\left\langle x_0^2 \right\rangle_{Exp}$ and $\left\langle x^2 \right\rangle_{Exp}$, respectively, as:

$$n \to n_{Obs} = \sqrt{\frac{\left\langle x_0^2 \right\rangle_{Exp}}{\left|\left\langle x^2 \right\rangle_{Exp} - \left\langle x_0^2 \right\rangle_{Exp}\right|}} \tag{4.572}$$

and

$$n \to n_{Free} = \sqrt{\frac{\left\langle x_0^2 \right\rangle_{Exp}}{\left\langle x^2 \right\rangle_{Exp} + \left\langle x_0^2 \right\rangle_{Exp}}} \tag{4.573}$$

It is obvious that for a given experimental set-up and records that the resulting observed evolution associates with higher quantum fluctuation than the corresponding free evolution, this feature being consistent with the (extended) Heisenberg uncertainty principle (Putz, 2010c).

However, when applied to spectroscopic data, they involve three classes of spectra information in terms of wave-numbers, namely:

- The maximum absorption line wave-number $\tilde{\upsilon}_0(A_{\max})$ that relates to the classical path, and the same for squared average measure in the inverse manner as:

$$\langle x_0 \rangle = \frac{1}{\tilde{\upsilon}_0(A_{\max})} \tag{4.574}$$

$$\langle x_0^2 \rangle = \frac{1}{\tilde{\upsilon}_0^2(A_{\max})} \tag{4.575}$$

- The left and right wave-numbers $\tilde{\upsilon}_L, \tilde{\upsilon}_R$ of the working absorption band, being arithmetically-to-geometrically averaged to get the average of quantum paths "inside" the band:

$$\langle x \rangle = \frac{\lambda_L + \lambda_R}{2} = \frac{1}{2}\left(\frac{1}{\tilde{\upsilon}_L} + \frac{1}{\tilde{\upsilon}_R}\right) \tag{4.576}$$

- The full width of a half maximum (FWHM) wave-number $\Delta\tilde{\upsilon}_{FWHM}$ of the concerned absorption band that is reciprocally associated with the dispersion of the quantum paths of vibrations within the band:

$$\Delta x = \frac{1}{\Delta\tilde{\upsilon}_{FWHM}} \tag{4.577}$$

Now, taken together, the quantum averaged path (4.576) and its dispersion (4.577) provide the average of the squared quantum paths, according to the general definition:

$$\langle x^2 \rangle = (\Delta x)^2 + \langle x \rangle^2 \tag{4.578}$$

Altogether, the classical and quantum paths' information of Eqs. (4.574)–(4.578) inversely correlate to the specific spectroscopic wave-numbers for a given absorption band and correlate the quantum fluctuations' indices of Eqs. (4.572) and (4.573) with the actual spectral-inverse quantum ones, respectively (Putz & Putz, 2012):

$$n_{Obs} = \frac{1}{\tilde{\upsilon}_0 \sqrt{\dfrac{1}{\Delta\tilde{\upsilon}_{FWHM}^2} + \dfrac{1}{4}\left(\dfrac{1}{\tilde{\upsilon}_L^2} + \dfrac{1}{\tilde{\upsilon}_R^2} + \dfrac{2}{\tilde{\upsilon}_L\tilde{\upsilon}_R}\right) - \dfrac{1}{\tilde{\upsilon}_0^2}}} \tag{4.579}$$

and

$$n_{Free} = \frac{1}{\tilde{\upsilon}_0 \sqrt{\dfrac{1}{\Delta\tilde{\upsilon}_{FWHM}^2} + \dfrac{1}{4}\left(\dfrac{1}{\tilde{\upsilon}_L^2} + \dfrac{1}{\tilde{\upsilon}_R^2} + \dfrac{2}{\tilde{\upsilon}_L\tilde{\upsilon}_R}\right) + \dfrac{1}{\tilde{\upsilon}_0^2}}} \tag{4.580}$$

They will be eventually used to compute the observed, free, inverse and residual inverse quantum indices to in depth characterizing of a given material for its porosity-to-free binding ordering through recognizing the particle *vs.* wave quantum tendency of the investigated state by spectroscopy in general and by absorption spectra in the present approach. Specific examples and analyses follow.

4.7.3.6 Spectral-IQ Results on Silica Sol-Gel-Based Mesosystems

Measurement of FT-IR absorption spectra (see the paradigmatic Figure 4.8) for samples under thermal treatment (Orcel et al., 1986; Neivandt et al., 1997; Paruchuri et al., 2005), e.g., same ionic liquid chain length, Cetyltrimethylammonium bromide (CTAB), respectively with DTAB or with their combination CTAB+DTAB, in different basic environment, are summarized in the Table 4.6, and are reported in Figures 4.9 and 4.10 for analysis at 60°C and 700°C, respectively.

The numerical Spectral-IQ results, as abstracted from Figures 4.9 and 4.10, are presented in Tables 4.7 and 4.8, for the particle-to-wave (P/W) ratio values in observed and free evolutions, Equations (4.556) and (4.562), as based on the quantum fluctuation factors of Eqs. (4.579) and (4.580),

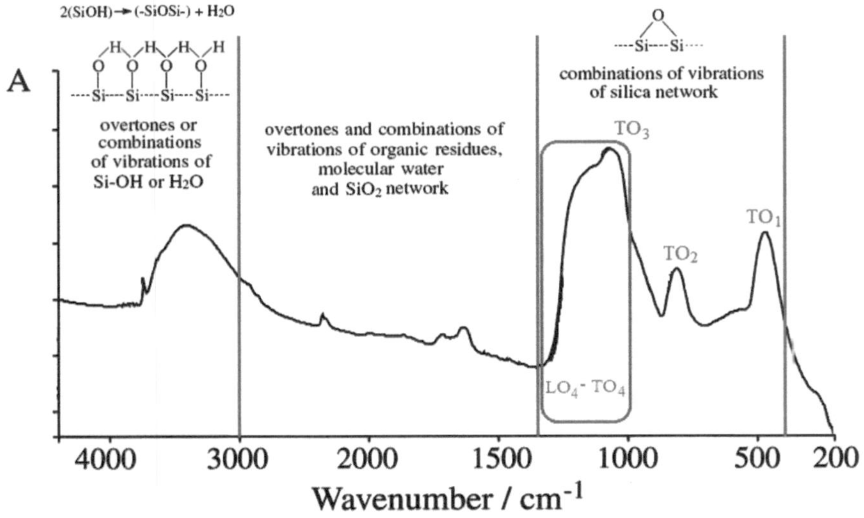

FIGURE 4.8 General pattern for wave-number domains of FTIR absorption spectra for silica sol-gel based materials, emphasizing the specific transversal optical (TO) main modes of rocking (TO_1), symmetric (TO_2) and antisymmetric (TO_3) vibrations of oxygen atoms in Si-O-Si bonds along the presently concerned disorder induced longitudinal-transverse vibrational mode (LO_4-TO_4) at the frontier of the silica network, along the remaining surface overtones and combinations of the network, residues and water vibrations, respectively. The marked LO_4-TO_4 band region is susceptible to wave-particle quantum "phase transition", thus regulating the physicochemical properties of meso-porosity and bonding at the network surface (Putz & Putz, 2012).

TABLE 4.6 Cases of the Ionic Liquid-Based Sol-Gel Synthesis Used in this Work (*)

Sample	Template	Base	Cosolvent	Solution
I	CTAB	NH_3		
II	CTAB	NaOH		
III	DTAB	NH_3	{Metoxy-ethanol}	{TEOS}
IV	DTAB	NaOH		
V	CTAB+DTAB	NH_3		
VI	CTAB+DTAB	NaOH		

(*)All chemicals were commercially available: Tetraethyl orthosilicate (TEOS), Metoxy-ethanol NH_4OH (25%), NaOH, CTAB (Cetyltrimethylammonium bromide), and DTAB (*n*-dodecyl trimethyl ammonium bromide) (Putz & Putz, 2012).

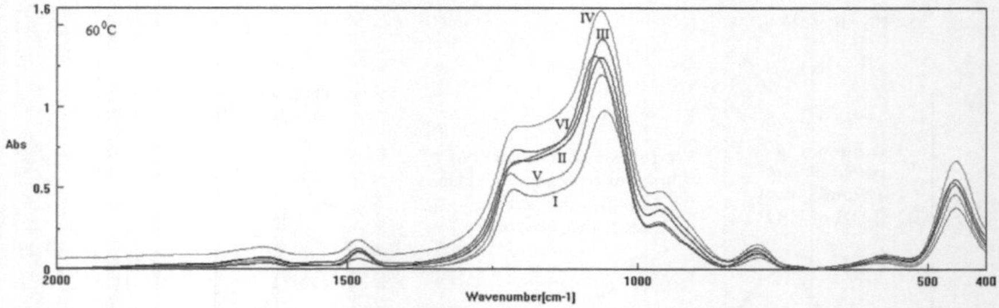

FIGURE 4.9 Absorption spectra of samples of Table 2 recorded at 60°C, with the TO_4 band of Figure 1 enhanced by the respective labeling (Putz & Putz, 2012).

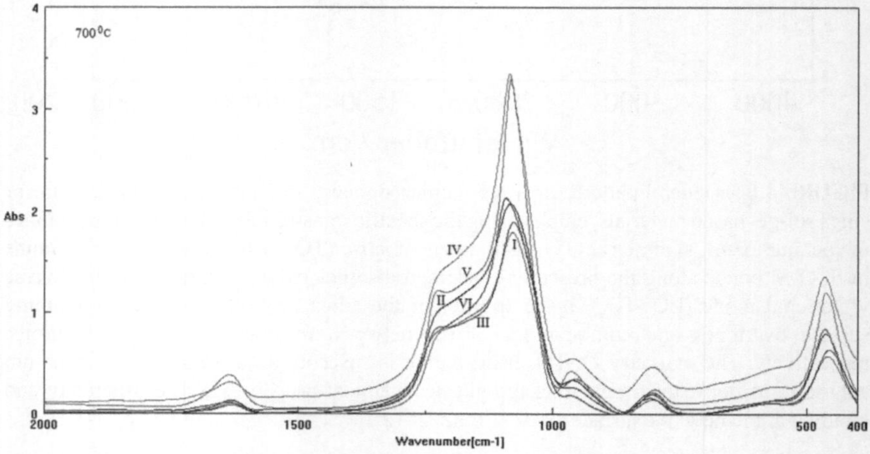

FIGURE 4.10 The same spectra records as in Figure 4.9, here for 700°C (Putz & Putz, 2012).

along the inverse quantum ratio of Eq. (4.571), respectively. Accordingly, one clearly observes the almost particle-to-wave equivalence throughout all samples, although the residual inverse quantum information $1-IQ$ makes the significant difference (in some cases, to adouble extent) in the wave- or free-binding content of samples; see for instance I-60 and V-60 with respect to II-60, IV-60 and VI-60 for samples at investigated at 60°C, and IV-700 *vs.* I-700, V-700 *vs.* III-700 and VI-700 *vs.* II-700 for samples investigated at 700°C, respectively.

However, in aiming to establish a hierarchy in binding potency, one should run on the residual IQ of the samples for identifying the decreased

TABLE 4.7 The Spectral-IQ Results, as Based on Eqs. (4.556), (4.562), and (4.571) with Quantum Fluctuation Factors (4.579) and (4.580) for the TO_4 Bands of Figure 4.9 ($\tilde{\upsilon}_L = 1299.787[cm^{-1}]$, $\tilde{\upsilon}_R = 999.910[cm^{-1}]$) at 60°C (Putz & Putz, 2012)

Sample	$\tilde{\upsilon}_0(A_{max})$	$\Delta\tilde{\upsilon}_{FWHM}$	n_{Obs}	$(P/W)_{\substack{Observed\\Evolution}}$	n_{Free}	$(P/W)_{\substack{Free\\Evolution}}$	IQ	$1-IQ$
I-60	1057.76	79.4905	0.0751762	0.94921	0.0747549	0.952777	0.996257	0.00374301
II-60	1063.55	107.4623	0.1011	0.947056	0.100083	0.95348	0.993266	0.00673417
III-60	1061.62	91.0061	0.0857609	0.948407	0.085137	0.95304	0.995138	0.00486168
IV-60	1063.55	112.1339	0.105501	0.946633	0.104346	0.953619	0.992674	0.00732553
V-60	1062.59	80.1571	0.0754605	0.94919	0.0750345	0.952783	0.996229	0.00377119
VI-60	1074.16	117.5866	0.109532	0.946227	0.108241	0.95375	0.992112	0.00788823

TABLE 4.8 The Same Type of Spectral-IQ Results as in Table 4.7, Here for the TO_4 Bands of Figure 4.10 ($\tilde{\upsilon}_L = 1299.787[cm^{-1}]$, $\tilde{\upsilon}_R = 999.910[cm^{-1}]$) at 700°C (Putz & Putz, 2012)

Sample	$\tilde{\upsilon}_0(A_{max})$	$\Delta\tilde{\upsilon}_{FWHM}$	n_{Obs}	$(P/W)_{\substack{Observed\\Evolution}}$	n_{Free}	$(P/W)_{\substack{Free\\Evolution}}$	IQ	$1-IQ$
I-700	1079.94	126.9705	0.117643	0.945365	0.116048	0.954029	0.990919	0.00908062
II-700	1083.8	156.5567	0.144573	0.942084	0.141643	0.955078	0.986395	0.0136049
III-700	1078.01	94.9767	0.0881346	0.948212	0.0874579	0.953104	0.994868	0.0051321
IV-700	1085.73	87.0955	0.0802383	0.948839	0.0797267	0.952899	0.99574	0.0042602
V-700	1083.8	70.1104	0.0647003	0.949903	0.0644312	0.952549	0.997223	0.00277721
VI-700	1092.48	101.4627	0.0929001	0.947807	0.0921086	0.953237	0.994304	0.00569644

potency of free bindings information. Accordingly, for 60°C, one notices from Table 4.7 the main set VI > IV > II followed by III > V > I, indicating two important features: both set contain all CTAB, DTAB and their combinations.

The used basic environment is the discriminating factor, here NaOH leaving with more free binding (and less porosity) potential for further interaction than NH_3, most probably due to the OH group ability (reactivity) to be further involved (and therefore blocked) in the Si surface through the related vibrations and overtones' combinations; see Figure 4.8.

Nevertheless, these binding potency series are apparently changing with the rising of the samples' temperature, as results in Table 4.8 provide for the 700°C case; however, agreement with 60°C is to be researched, while considering specific thermal analysis, as is exposed elsewhere.

On conceptual realm, when it relates to infrared spectroscopy (IR) investigations on sol-gel silica films (De et al., 1993; Al-Oweini & El-Rassy, 2009; Xue et al., 2007), a part of the consecrated transverse-optical vibrational modes, namely the rocking mode TO_1 (457–507 cm^{-1}) modeling the perpendicular motions of the bridging oxygen to the Si-O-Si plane, the symmetric mode TO_2 (810–820 cm^{-1}) modeling the stretching of oxygen atoms along the bisecting line of the Si-O-Si and the antisymmetric TO_3 (1070–1250 cm^{-1}) describing the motion in opposite distortion of the two neighboring SI-O bonds, there appears to be the so-called disorder-induced TO_4 modes (about 1200 cm^{-1}), interpreted as an increase in bonding strain with a longitudinal-transverse splitting recorded with lower wave numbers of LO (about 1170 cm^{-1}) with respect to TO.

However, in this last region within the band 1000–1300 cm^{-1} where the bonding on the surface should be better assigned to the ionic/covalent, porous/free binding or to the particle/wave quantum "phase transition" information, the importance of this assignment resides in the fact that the shown region characterizes the bulk-to-surface physico-chemical richest interaction, beyond which only the overtones and/or combination of vibrations of the network as a whole and of the organic residues and water are dominant; see Figure 4.8. Therefore, deeper understanding of the TO_4-LO_4 "phase transition" region at the frontier of the silica films by FTIR will give crucial information on the porosity of materials at the quantum-to-meso level in view of the hierarchical ordering of materials with higher potential for caring or hosting small molecules in/from organisms or the environment with direct consequences in pharmacology and ecotoxicology (Almeida et al., 1990; Primeau et al., 1997). Unfortunately, so far, the computational methods available for extracting from experimental spectra such information are missing and in favor of meso-to-macro analysis. Instead, this work makes the advancement of combining the observed data from FTIR spectra with the recent original method of modeling the wave/particle dual information by use of the spectroscopic assignment of the inverse quantum fluctuation factors (Putz, 2010c). This way, the present method fills the quantum-to-meso gap by the so-called *spectral-inverse quantum* (Spectral-IQ) algorithm, lending itself to being generalized and adapted for a wide type of spectra and gas-solid or sol-gel physicochemical interactions.

Actually, the present comparative analysis based on this IQ factor, as well as on its residual one, 1−IQ, showcases that, among a CTAB, DTAB and of their combination samples in various basic co-solvents, the simple CTAB+ammonia co-solvent provides the best porosity system for potentially carrying particles and effector interaction in various eco- and bio-environments; on the other extreme, the silica film obtained by the cosurfactant combination of CTAB+DTAB in NaOH basicity displays the highest free binding feature, thus being less specific and more associated with environmental hazard to be avoided.

The resulted method is, however, general, based on fundamental particle-to-wave dual quantum behavior; note that the present approach is based on the departure (then associated with the extended Heisenberg uncertainty) between the particle and psi-function description by the ratio between averages of Gaussian to stationary waves, further corresponding to the ratio between the real and the imaginary descriptions of the quantum objects, respectively (Putz, 2010c). Yet, considering the transformation of such a ratio to its nominator-denominator difference, the resulting "physical space" may be associated with the recently introduced inertons—a particle surrounded with its cloud of spatial excitations (Krasnoholovets, 2010), able to explain the photonic structure and the light-matter interaction in a deeper mechanistic (*i.e.*, deterministic) way; however, such a picture can be completely achieved when the scattered bonding (psi-function) states are also consistently described by their associated quantum particle—the recently introduced *bondon*, see the Volumes III and IV of the present five volume work—so that the inter-particle/ bosonic inerton-bondon interaction is finally modeling the obtained/observed spectra.

4.8 CONCLUSION

The main lessons to be kept for the further theoretical and practical investigations of the quantum chemistry by quantum mechanics that are presented in the present chapter pertain to the following:

- identifying the working path integral (Feynman) form for quantum fluctuation integral description for quantum history of an evolving system;
- employing the Feynman path integral towards recovering the Schrodinger equation so that revealing the generalized vision the path integral over all quantum histories has over the differential (segmented) observing of quantum evolution;
- writing the path integral for general quantum systems by Van Vleck-Pauli-Morette formulation;
- dealing with path integrals of harmonic oscillator, Bohr's atom and well potential for electronic paradigmatic motion in molecule, atom and solid, respectively, while recovering the eigen-energy quantifications by means of quantum principles applications (see Chapter 3);
- characterizing the many-electronic systems by density matrix formalism, at its turn related with path integral through the Bloch equation;
- understanding the quantum chemical behavior within the density matrix formalism derived from bra-ket Dirac formalization of quantum states, so appropriately introducing the exchange contribution, Slater representation and Hartree-Fock energy equation for N-electronic bonded systems;
- describing quantum chemical systems by orbital molecular picture: the Born-Oppenheimer of frozen nuclei and the Roothaan modeling of self-consistency in allied eigen-value of atoms-in-molecules orbitalic (and electronic density/population) problems;

- learning the role of semi-empirical approach of quantum systems: the various approximation levels of Fock matrix in exchange terms of eigen-energy, while recognizing the correlation behavior of the many-electronic systems complementing the mono-electronic orbitals occupancies;
- treating the quantum systems though the frozen electronic orbitals (Koopmans theorem) with the premiere role in assessing the chemical reactivity in general and to modeling it by introduced indices as electronegativity and chemical hardness directly related with the first and second total energy-to-total number of electrons derivations, respectively;
- solving the many-electronic observability problem by means of Density Functional Theory, so providing the existence Hohenberg-Kohn theorem of density-potential bijectivity along the energy density functional optimized form in terms of fictional Kohn-Sham mono-orbitals, yet with chemical reactivity value through the chemical potential (Lagrange multiplier) parameter assimilated with minus of the global electronegativity of the system;
- formulating the working forms for the popular density functionals of many-electronic energy, separately for kinetic, exchange and correlation contribution, as well as for the mixed exchanged-correlation combinations – custom for computational quantum chemistry;
- interpreting the observability in quantum chemistry by means of periodic yet fluctuating path, so employing the path integrals in providing the quantum amplitudes (equivalent with density matrix) and then to the partition function when closing the coordinates' ends in quantum orbits;
- connecting the Heisenberg uncertainty principle with averages on quantum paths' fluctuation, while advancing new modeling region of quantum evolution, especially the unknown free one beside that observe through measurement;
- developing new information-quantum (IQ) method in assignment the particle-to-wave regime for a given many-particle system under certain applied potential or environment (entrapment on chemical matrices, graphenes, and nanosystems etc.);
- finding applications for IQ method for nano- and meso-systems: here for silica sol-gel based methods in assessing the various bases entrapment on custom ionic liquids' templates by means of

wave-vs.-particle ratio of the free-vs.-observed quantum fluctuations as abstracted from FTIR spectra.

KEYWORDS

- **Bloch equation**
- **canonical density**
- **chemical hardness**
- **correlation energy**
- **density Functional Theory**
- **density matrix approach**
- **effective partition function**
- **electronic density matrices**
- **Energy density functionals**
- **exchange-correlation**
- **Feynman path integral**
- **free particle motion**
- **harmonic oscillator**
- **Hartree-Fock approximation**
- **Hohenberg-Kohn theorems**
- **Koopmans theorem**
- **mesosystems**
- **Molecular Orbitals**
- **observable quantum chemistry**
- **path integral for Bohr's atom**
- **periodic path integrals**
- **propagator equation**
- **quantum fluctuations**
- **quantum well**
- **Roothaan approach**
- **semi-empirical approximations**
- **Spectral-IQ method**

REFERENCES

AUTHOR'S MAIN REFERENCES

Putz, M. V. (2016a). *Quantum Nanochemistry. A Fully Integrated Approach: Vol. III. Quantum Molecules and Reactivity*. Apple Academic Press & CRC Press, Toronto-New Jersey, Canada-USA.

Putz, M. V. (2016b). *Quantum Nanochemistry. A Fully Integrated Approach: Vol. II. Quantum Atoms and Periodicity*. Apple Academic Press & CRC Press, Toronto-New Jersey, Canada-USA.

Putz, M. V. (2013). Koopmans' analysis of chemical hardness with spectral like resolution. *The Scientific World Journal*, 2013, 348415/14 pages (DOI, 10.1155/2013/348415).

Putz, A. M., Putz, M. V. (2012). Spectral inverse quantum (Spectral-IQ) method for modeling mesoporous systems. application on silica films by FTIR. *Int. J. Mol. Sci.* 13(12), 15925–15941 (DOI, 10.3390/ijms131215925).

Putz, M. V. (2011a). Electronegativity and chemical hardness: different patterns in quantum chemistry. *Curr. Phys. Chem.* 1(2), 111–139 (DOI, 10.2174/1877946811101020111).

Putz, M. V. (2011b). Conceptual density functional theory: from inhomogeneous electronic gas to Bose-Einstein condensat, In: Putz, M. V. (Ed.), *Chemical Information and Computational Challenges in 21st Century. A Celebration of 2011 International Year of Chemistry*, NOVA Science Publishers, Inc., New York, Chapter 1, pp. 1–60.

Putz, M. V., Putz, A.-M., Pitulice, L., Chiriac, V. (2010). On chemical hardness assessment of aromaticity for some organic compounds, *Int. J. Chem. Model.* 2(4), 343–354.

Putz, M. V. (2010a). On absolute aromaticity within electronegativity and chemical hardness reactivity pictures. *MATCH Commun. Math. Comput. Chem.* 64(2), 391–418.

Putz, M. V. (2010b). Compactness aromaticity of atoms in molecules. *International Journal of Molecular Sciences* 11(4), 1269–1310 (DOI, 10.3390/ijms11041269).

Putz, M. V. (2010c). On Heisenberg uncertainty relationship, its extension, and the quantum issue of wave-particle dualit. *Int. J. Mol. Sci.* 11(10), 4124–4139 (DOI, 10.3390/ijms11104124)

Putz, M. V. (2009). Path integrals for electronic densities, reactivity indices, and localization functions in quantum systems. *Int. J. Mol. Sci.* 10(11), 4816–4940 (DOI, 10.3390/ijms10114816).

Putz, M. V. (2008). Density functionals of chemical bonding. *Int. J. Mol. Sci.* 9(6), 1050–1095 (DOI, 10.3390/ijms9061050).

Putz, M. V., Chiriac, A. (2008). Quantum perspectives on the nature of the chemical bond. In: Putz, M. V. (Ed.), *Advances in Quantum Chemical Bonding Structures*. Transworld Research Network, Kerala, pp. 1–43.

Putz, M. V. (2007a). Unifying absolute and chemical electronegativity and hardness density functional formulations through the chemical action concept. In: Hoffman, E. O. (Ed.), *Progress in Quantum Chemistry Research*, Nova Science Publishers Inc., New York, pp. 59–121.

Putz, M. V. (2007b). Can quantum-mechanical description of chemical bond be considered complete? In: Kaisas, M. P. (Ed.), *Quantum Chemistry Research Trends*, Nova Science Publishers Inc., New York, Expert Commentary.

Putz, M. V., Russo, N., Sicilia, E. (2004). On the application of the HSAB principle through the use of improved computational schemes for chemical hardness evaluation. *J. Comp. Chem.* 25(7), 994–1003.

Putz, M. V. (2003). *Contributions within Density Functional Theory with Applications to Chemical Reactivity Theory and Electronegativity*, Dissertation.com, Parkland (Florida).

SPECIFIC REFERENCES

Abarbanel, H. D. I., Itzykson, C. (1969). Relativistic eikonal expansion. *Phys. Rev. Lett.* 23, 53–56.

Almeida, R. M., Guiton, T. A., Pantano, G. C. (1990). Characterization of silica gels by infrared reflection spectroscopy. *J. Non-Cryst. Solids* 121, 193–197.

Alonso, J. A., Girifalco, L. A. (1978). Nonlocal approximation to the exchange potential and kinetic energy of an inhomogeneous electron gas. *Phys. Rev. B* 17, 3735–3743.

Al-Oweini, R., El-Rassy, H. (2009). Synthesis and characterization by FTIR spectroscopy of silica aerogels prepared using several Si(OR)4 and R00Si(OR0)3 precursors. *J. Mol. Struct.* 919, 140–145.

Anderson, P. W. (1984). *Basic Notions of Condensed Matter Physics*. Benjamin-Cummings: Menlo Park (CA).

Anderson, W. P., Cundari, T. R., Zerner, M. C. (1991). An intermediate neglect of differential overlap model for second-row transition metal species. Int. J. Quantum Chem. 39, 31–45.

Anderson, W. P., Edwards, W. D., Zerner, M. C. (1986). Calculated spectra of hydrated ions of the first transition-metal series. *Inorg. Chem.* 25, 2728–2732.

Angeli, C. (1998). Physical interpretation of Koopmans' theorem: A criticism of the current didactic presentation. *J. Chem. Educ.* 75(11), 1494–1497.

Atkins, P. W., Friedman, R. S. (1997). *Molecular Quantum Mechanics*, Oxford University Press, New York.

Bacon, A. D., Zerner, M. C. (1979). An intermediate neglect of differential overlap theory for transition metal complexes: Fe, Co and Cu chlorides. *Theor. Chim. Acta* 53, 21–54.

Baird, N. C., Dewar, M. J. S. (1969). Ground states of σ-bonded molecules. IV. The MINDO method and its application to hydrocarbons. *J. Chem. Phys.* 50, 1262–1275.

Balàzs, N. (1967). Formation of stable molecules within the statistical theory of atoms. *Phys. Rev.* 156, 42–47.

Bamzai, A. S., Deb, B. M. (1981). The role of single-particle density in chemistry. *Rev. Mod. Phys.* 53, 95–126.

Bartlett, R. J., Grabowski, I., Hirata, S., Ivanov, S. (2004). The exchange-correlation potential in ab initio density functional theory. *J. Chem. Phys.* 122, 034104.

Bartlett, R. J., Lotrich, V. F., Schweigert, I. V. (2005). Ab initio DFT: The best of both worlds? *J. Chem. Phys.* 123, 062205.

Bartlett, R. J., Musial, M. (2007). Coupled-cluster theory in quantum chemistry. *Rev. Mod. Phys*. 79, 291–352.

Bartolotti, L. J. (1982). A new gradient expansion of the exchange energy to be used in density functional calculations on atoms. *J. Chem. Phys.* 76, 6057–6059.

Beaudry, C. M., Malerich, J. P., Trauner, D. (2005). Biosynthetic and biomimetic electrocyclizations. *Chem. Rev.* 105, 4757–4778.

Becke, A. D. (1986). Density functional calculations of molecular bond energies. *J. Chem. Phys.* 84, 4524–4529.

Becke, A. D. (1988). Density-functional exchange-energy approximation with correct asymptotic behavior. *Phys. Rev. A* 38, 3098–3100.

Becke, A. D., Edgecombe, K. E. (1990). A simple measure of electron localization in atomic and molecular systems. *J. Chem Phys.* 92, 5397–5403.

Berkowitz, M. (1986). exponential approximation for the density matrix and the Wigner's distribution. Ch*em Phys. Lett.* 129, 486–488.

Besley, N. A., Peach, M. J. G., Tozer, D. J. (2009). Time-dependent density functional theory calculations of near-edge X-ray absorption fine structure with short-range corrected functionals. *Phys. Chem. Chem. Phys.* 11, 10350–10358.

Bingham, R. C., Dewar, M. J. S., Lo, D. H. (1975a). Ground states of molecules. XXV. MINDO/3. Improved version of the MINDO semiempirical SCF-MO method. *J. Am. Chem. Soc.* 97, 1285–1293.

Bingham, R. C., Dewar, M. J. S., Lo, D. H. (1975b). Ground states of molecules. XXVI. MINDO/3 calculations for hydrocarbons. *J. Am. Chem. Soc.* 97, 1294–1301.

Bingham, R. C., Dewar, M. J. S., Lo, D. H. (1975c). Ground states of molecules. XXVII. MINDO/3 calculations for carbon, hydrogen, oxygen, and nitrogen species. *J. Am. Chem. Soc.* 97, 1302–1306.

Bingham, R. C., Dewar, M. J. S., Lo, D. H. (1975d). Ground states of molecules. XXVIII. MINDO/3 calculations for compounds containing carbon, hydrogen, fluorine, and chlorine. *J. Am. Chem. Soc.* 97, 1307–1311.

Blanchard, C. H. (1982). Density matrix and energy–time uncertainty. *Am. J. Phys.* 50, 642–645.

Bloch, F. (1932). Theorie des Austauschproblems und der Remanenzerscheinung der Ferromagnetika, *Z. Phys.* 74, 295–335.

Blum, K. (1981). *Density Matrix Theory and Applications*, Plenum Press, New York.

Bohm, D., Vigier, J. P. (1954). Model of the causal interpretation of quantum theory in terms of a fluid with irregular fluctuations. *Phys. Rev.* 96, 208–216.

Bokhan, D., Bartlett, R. J. (2006). Adiabatic ab initio time-dependent density-functional theory employing optimized-effective-potential many-body perturbation theory potentials. *Phys. Rev. A* 73, 022502.

Born, M., Oppenheimer, R. (1927). Zur Quantentheorie der Molekeln. *Ann. Physik (Leipzig)* 84, 457–484.

Boys, S. F. (1950). Electronic wavefunctions. I. A general method of calculation for stationary states of any molecular system. *Proc. Roy. Soc.* A200, 542–554.

Buckingham, A. D., Fowler, P. W., Galwas, P. A. (1987). Velocity-dependent property surfaces and the theory of vibrational circular dichroism. *Chem. Phys.* 112, 1–14.

Burke, K., Werschnik, J., Gross, E. K. U. (2005). Time-dependent density functional theory: past, present, and future. *J. Chem. Phys.* 123, 062206.

Busch, P. (2008). The time-energy uncertainty relation. *Lect. Notes Phys.* 734, 73–105.

Campbell, W. B., Finkler, P., Jones, C. E., Misheloff, M. N. (1975). Path-integral formulation of scattering theory. *Phys. Rev. D* 12, 2363–2369.

Capelle, K. (2006). A bird's-eye view of density-functional theory. (arXiv:cond-mat/0211443v5 [cond-mat.mtrl-sci]).
Cedillo, A., Robles, J., Gazquez, J. L. (1988). New nonlocal exchange-energy functional from a kinetic-energy-density Padé-approximant model. *Phys. Rev. A* 38, 1697–1701.
Chattaraj, P. K., Lee, H., Parr, R. G. (1991). Principle of maximum hardness. *J. Am. Chem. Soc.* 113, 1854–1855.
Chattaraj, P. K., Liu, G. H., Parr, R. G. (1995). The maximum hardness principle in the Gyftpoulos-Hatsopoulos three-level model for an atomic or molecular species and its positive and negative ions. *Chem. Phys. Lett.* 237, 171–176.
Chattaraj, P. K., Sarkar, U., Roy, D. R. (2007). Electronic structure principles and aromaticity. *J. Chem. Edu.* 84, 354–358.
Chen, J., Stott, M. J. (1991a.) V-representability for systems of a few fermions. *Phys. Rev. A* 44, 2809–2814.
Chen, J., Stott, M. J. (1991b). V-Representability for systems with low degeneracy. *Phys. Rev. A* 44, 2816–2821.
Chermette, H. (1999). Chemical reactivity indexes in density functional theory. *J. Comput. Chem.* 20, 29–154.
Christofferson, R. E. (1989). *Basic Principles and Techniques of Molecular Quantum Mechanics*, Springer, New York.
Ciesielski, A., Krygowski, T. M., Cyranski, M. K., Dobrowolski, M. A., Balaban, A. T. (2009). Are thermodynamic and kinetic stabilities correlated? A topological index of reactivity toward electrophiles used as a criterion of aromaticity of polycyclic benzenoid hydrocarbons. *J. Chem. Inf. Model. 49*, 369–376.
Clark, T. A. (1985). *Handbook of Computational Chemistry*, John Wiley and Sons, New York.
Clementi, E., Roetti, C. (1974). Roothaan-Hartree-Fock atomic wavefunctions: Basis functions and their coefficients for ground and certain excited states of neutral and ionized atoms, $Z \leq 54$. *At. Data Nucl. Data Tables* 14, 177–478.
Collins, J. B., Schleyer, P.v.R., Binkley, J. S., Pople, J. A. (1976). Self-consistent molecular orbital methods. XVII. Geometries and binding energies of second-row molecules. A comparison of three basis sets. *J. Chem. Phys.* 64, 5142–5152.
Corni, S., Cappelli, C., Del Zoppo, M., Tomasi, J. (2003). Prediction of solvent effects on vibrational absorption intensities and Raman activities in solution within the polarizable continuum model: a study on push−pull molecules. *J. Phys. Chem. A* 107(48), 10261–10271.
Corongiu, G. (2007). The Hartree-Fock-Heitler-London Method, III: Correlated diatomic hydrides. *J. Phys. Chem. A* 111, 5333–5342.
Cory, M. G., Stavrev, K. K., Zerner, M. C. (1997). An examination of the electronic structure and spectroscopy of high- and low-spin model ferredoxin via several SCF and CI techniques. *Int. J. Quant. Chem.* 63, 781–795.
Cuccoli, A., Macchi, A., Neumann, M., Tognetti, V., Vaia, R. (1992). Quantum thermodynamics of solids by means of an effective potential. *Phys. Rev. B* 45, 2088–2096.
Curtiss, L. A., Raghavachari, K., Redfern, P. C., Rassolov, V., Pople, J. A. Gaussian-3 (G3) theory for molecules containing first and second-row atoms. (1998). *J. Chem. Phys.* 109, 7764–7776.

Davis, L. P., Guidry, R. M., Williams, J. R., Dewar, M. J. S., Rzepa, H. S. (1981). MNDO calculations for compounds containing aluminum and boron. *J. Comp. Chem.* 2, 433–445.

de Broglie, L. (1987). Interpretation of quantum mechanics by the double solution theory, In: *Annales de la Fondation Louis de Broglie*, Fondation Louis de Broglie, Paris, France, Chapter 12, pp. 399–421.

De, G., Kundu, D., Karmakar, B., Ganguli, D. (1993). FTIR studies of gel to glass conversion in TEOS-fumed silica derived gels. *J. Non-Cryst. Solids* 155, 253–258.

De Proft, F., Geerlings, P. (2001). Conceptual and computational DFT in the study of aromaticity. *Chem. Rev.* 101, 1451–1464.

Del Bene, J., Jaffé, H. H. (1968a). Use of the CNDO method in spectroscopy. I. Benzene, pyridine, and the diazines. *J. Chem. Phys.* 48, 1807–1814.

Del Bene, J., Jaffé, H. H. (1968b). Use of the CNDO method in spectroscopy. II. Five-membered rings. *J. Chem. Phys.* 48, 4050–4056.

Del Bene, J., Jaffé, H. H. (1968c). Use of the CNDO method in spectroscopy. III. Monosubstituted benzenes and pyridines. *J. Chem. Phys.* 49, 1221–1229.

DePristo, A. E., Kress, J. D. (1987). Kinetic-energy functionals via Padé approximations. *Phys. Rev. A* 35, 438–441.

Derosa, P. A. (2009). A combined semiempirical-DFT study of oligomers within the finite-chain approximation, evolution from oligomers to polymers. *J Compu. Chem.* 30, 1220–1228.

Dewar, M. J. S., Dieter, K. M. (1986). Evaluation of AM1 calculated proton affinities and deprotonation enthalpies. *J. Am. Chem. Soc.* 108, 8075–8086.

Dewar, M. J. S., Hasselbach, E. (1970). Ground states of sigma-bonded molecules. IX. MINDO (modified intermediate neglect of differential overlap)/2 method. *J. Am. Chem. Soc.* 92, 590–598.

Dewar, M. J. S., Lo, D. H. (1972). Ground states of sigma-bonded molecules. XVII. Fluorine compounds. *J. Am. Chem. Soc.* 94, 5296–5303.

Dewar, M. J. S., Lo, D. H., Ramsden, C. A. (1975). Ground states of molecules. XXIX. MINDO/3 calculations of compounds containing third row elements. *J. Am. Chem. Soc.* 97, 1311–1318.

Dewar, M. J. S., McKee, M. L. (1977). Ground states of molecules. 41. MNDO results for molecules containing boron. *J. Am. Chem. Soc.* 99, 5231–5241.

Dewar, M. J. S., Rzepa, H. S. (1978). Ground states of molecules. 40. MNDO results for molecules containing fluorine. *J. Am. Chem. Soc.* 100, 58–67.

Dewar, M. J. S., Storch, D. M. (1985). Development and use of quantum molecular models. 75. Comparative tests of theoretical procedures for studying chemical reactions. *J. Am. Chem. Soc.* 107, 3898–3902.

Dewar, M. J. S., Thiel, W. (1977). Ground states of molecules. 38. The MNDO method. Approximations and parameters. *J. Am. Chem. Soc.* 99, 4899–4907.

Dewar, M. J. S., Zoebisch, E. G., Healy, E. F., Stewart, J. J. P. (1985). Development and use of quantum mechanical molecular models. 76. AM1: A new general purpose quantum mechanical molecular model. *J. Am. Chem. Soc.* 107, 3902–3909.

Dirac, P. A. M. (1929). Quantum mechanics of many-electron systems. *Proceedings of the Royal Society of London. Series A, Containing Papers of a Mathematical and Physical Character*, 123 (792), 714–733.

Dirac, P. A. M. (1933). The Lagrangian in quantum mechanics. *Phys. Z. Sowjetunion* 3, 64–72.

Dittrich, W., Reuter, M. (1994). *Classical and Quantum Dynamics: From Classical Paths to Path Integrals*, Springer, Berlin.

Duke, B. J., O'Leary, B. (1995). Non-Koopmans' molecules. *J. Chem. Educ.* 72(6), 501–504.

Ernzerhof, M. (1994). Density-functional theory as an example for the construction of stationarity principles. *Phys. Rev. A* 49, 76–79.

Fermi, E. (1927). Un metodo statistico per la determinazione di alcune priopriet à dell'atomo. *Rend. Accad. Naz. Lincei* 6, 602–607.

Feynman, R. P. (1948). Space-time approach to non-relativistic quantum mechanics. *Rev. Mod. Phys.* 20, 367–387.

Feynman, R. P. (1972). *Statistical Mechanics*, Benjamin, Reading (PA).

Feynman, R. P., Hibbs, A. R. (1965). *Quantum Mechanics and Path Integrals*, McGraw Hill, New York.

Feynman, R. P., Kleinert, H. (1986). Effective classical partition functions. *Phys. Rev. A* 34, 5080–5084.

Flores, J. A., Keller, J. (1992). Differential equations for the square root of the electronic density in symmetry-constrained density-functional theory. *Phys. Rev. A* 45, 6259–6262.

Fock, V. (1930). Näherungsmethode zur lösung des quantenmechanischen mehrkörperproblems. *Z. Physik* 61, 126–140.

Fung, S. Y., Duhamel, J., Chen, P. (2006). Solvent effect on the photophysical properties of the anticancer agent ellipticine. *J. Phys. Chem. A* 110 (40), 11446–11454.

Gaspar, R., Nagy, A. (1987). Local-density-functional approximation for exchange-correlation potential. Application of the self-consistent and statistical exchange-correlation parameters to the calculation of the electron binding. *Energies. Theor. Chim Acta* 72, 393–401.

Geerlings, P., De Proft, F., Langenaeker, W. (2003). Conceptual density functional theory. *Chem. Rev.* 103(5), 1793–1874.

Gerry, C. C., Singh, V. A. (1979). Feynman path-integral approach to the Aharonov-Bohm effect. *Phys. Rev. D* 20, 2550–2554.

Ghosh, S. K., Parr, R. G. (1986). Phase-space approach to the exchange energy functional of density-functional theory. *Phys. Rev. A* 34, 785–791.

Giachetti, R., Tognetti, V., Vaia, R. (1986). Quantum corrections to the thermodynamics of nonlinear systems. *Phys. Rev. B* 33, 7647–7658.

Glaesemann, K. R., Schmidt, M. W. (2010). On the ordering of orbital energies in high-spin ROHF. *J. Phys. Chem. A* 114, 8772–8777.

Goldstone, J. (1961). Field theories with «Superconductor» solutions. *Nuovo Cim.* 19, 154–164.

Greiner, W., Reinhardt, J. (1994). *Quantum Electrodynamics*, Springer, Berlin.

Gritsenko, O. V., Schipper, P. R. T., Baerends, E. J. (2000). Ensuring proper short-range and asymptotic behavior of the exchange-correlation kohn-sham potential by modeling with a statistical average of different orbital model potential *Int. J. Quantum Chem.* 76, 407–419.

Hall, M. J. W. (2001). Exact uncertainty relations. *Phys. Rev. A* 64, 052103.

Hartree, D. R. (1928a). The wave mechanics of an atom with a non-Coulomb central field. *Proc. Cam. Phil. Soc.* 24, 89–111.
Hartree, D. R. (1928b). The wave mechanics of an atom with a noncoulomb central field. Part, I. Theory and methods. Part II. Some results and discussions. *Proc. Cam. Phil. Soc.* 24, 111–132.
Hartree, D. R. (1957). *The Calculation of Atomic Structures*, John Wiley and Sons, New York.
Hehre, W. J., Stewart, R. F., Pople, J. A. Self-consistent molecular-orbital methods. I. Use of Gaussian expansions of Slater-type atomic orbitals. *J. Chem. Phys.* (1969). 51, 2657–2665.
Heitler, W., London, F. (1927). Wechselwirkung neutraler Atome und homöopolare Bindung nach der Quantenmechanik. *Z. Phys.* 44, 455–472.
Hickenboth, C. R., Moore, J. S., White, S. R., Sottos, N. R., Baudry, J., Wilson, S. R. (2007). Biasing reaction pathways with mechanical force. *Nature* 446, 423–427.
Hoffmann, R. (1963). An extended Hückel theory. I. Hydrocarbons. *J. Chem. Phys.* 39, 1397–1412.
Hoffmann, R., Woodward, R. B. (1968). The conservation of orbital symmetry. *Acc. Chem. Res.* 1, 17–22.
Hohenberg, P., Kohn, W. (1964). Inhomogeneous electron gas. *Phys. Rev.* 136:B864-B871.
Hohenberg, P., Kohn, W. (1964). Inhomogeneous Electronic Gas. *Phys. Rev.* 136, 864–871.
Huang, K. (1987). *Statistical Mechanics*, 2nd ed., John Wiley & Sons: New York, pp. 298–302.
Hückel, E. (1931). Quantentheoretische beiträge zum benzolproblem. *Z. Physik.* 71, 204–286 and 72, 310–337.
Hypercube (2002). Program Package, HyperChem 7.01; Hypercube, Inc.: Gainesville, FL, USA.
Infeld, L., Hull, T. E. (1951). The factorization method. *Rev. Mod. Phys.* 23, 21–68.
Isihara, A. (1980). *Statistical Physics*, Academic Press, New York.
Ivanov, A. L. (2006,) Energy-time uncertainty relations and time operators. *J. Math. Chem.* 43, 1–11.
Janak, J. F. (1978). Proof that $\partial E/\partial n_i = \varepsilon_i$ in density functional theory. *Phys. Rev. B* 18, 7165–7168.
Janke, W., Cheng, B. K. (1988). Statistical properties of a harmonic plus a delta-potential. *Phys. Lett. B* 129, 140–144.
Johnson, B. G., Gill, P. M. W., Pople, J. A. (1993). The performance of a family of density functional methods. *J. Chem. Phys.* 98, 5612–5627; Erratum: Johnson, B. G. J. (1994). *Chem. Phys.* 101, 9202.
Keller, J. (1986). On the formulation of the Hohenberg-Kohn-Sham theory. *Int. J. Quantum Chem.* 20, 767–768.
Kleinert, H. (1986). Effective potentials from effective classical potentials. *Phys. Lett. B* 181, 324–326.
Kleinert, H. (1989). Path collapse in Feynman formula. Stable path integral formula from local time reparametrization invariant amplitude. *Phys. Lett. B* 224, 313–318
Kleinert, H. (2004). *Path Integrals in Quantum Mechanics, Statistics, Polymer Physics, and Financial Markets* (3rd ed.), World Scientific, Singapore.

Klopman, G. (1964). A semiempirical treatment of molecular structures. II. Molecular terms and application to diatomic molecules. *J. Am. Chem. Soc.* 86, 4550–4557.

Kohn, W., Becke, A. D., Parr, R. G. (1996). Density functional theory of electronic structure. *J. Phys. Chem.* 100, 12974–12980.

Kohn, W., Sham, L. J. (1965). Self-consistent equations including exchange and correlation effects. *Phys. Rev.* 140, A1133–A1138.

Kohn, W., Sham, L. J. (1965). Self-consistent equations including exchange and correlation effects. *Phys. Rev.* 140, A1133–A1138.

Koopmans, T. (1934). Uber die zuordnung von wellen funktionen und eigenwerter zu den einzelnen elektronen eines atom. *Physica* 1, 104–113.

Krasnoholovets, V. (2010). Sub microscopic description of the diffraction phenomenon. *Nonlinear Opt. Quantum Opt.* 41, 273–286.

Kryachko, E. S., Ludeña, E. V. (1987). *Density Functional Theory in Quantum Chemistry*, Reidel, Dordrecht.

Kryachko, E. S., Ludena, E. V. (1991a). Formulation of N-and V-representable density-functional theory. I. Ground states. *Phys. Rev. A* 43, 2179–2192.

Kryachko, E. S., Ludena, E. V. (1991b). Formulation of N-and V-representable density-functional theory. II. Spin-dependent systems. *Phys. Rev. A* 43, 2194–2198.

Laidlaw, M. G. G., DeWitt-Morette, C. (1971). Feynman functional integrals for systems of indistinguishable particles. *Phys. Rev. D* 3, 1375–1378.

Laidlaw, M. G. G., DeWitt-Morette, C. (1971). Feynman functional integrals for systems of indistinguishable particles. *Phys. Rev. D* 3, 1375–1378.

Lam, K. C., Cruz, F. G., Burke, K. Viral (1998). Exchange-correlation energy density in Hooke's atom. *Int. J. Quantum Chem.* 69, 533–540.

Lee, C., Parr, R. G. (1987). Gaussian and other approximations to the first-order density matrix of electronic system, and the derivation of various local-density-functional-theories. *Phys. Rev. A* 35, 2377–2383.

Lee, C., Parr, R. G. (1990). Exchange-correlation functional for atoms and molecules. *Phys. Rev. A* 42, 193–199.

Lee, C., Yang, W., Parr, R. G. (1988). Development of the colle-salvetti correlation-energy formula into a functional of the electron density. *Phys. Rev. B* 37, 785–789.

Lee, C., Zhou, Z. (1991). Exchange-energy density functional: reparametrization of becke's formula and derivation of second-order gradient correction. *Phys. Rev. A* 44, 1536–1539.

Lee, H., Bartolotti, L. J. (1991). Exchange and exchange-correlation functionals based on the gradient correction of the electron gas. *Phys. Rev. A* 44, 1540–1542.

Lele, S. K. (1992). Compact finite difference schemes with spectral-like resolution. *J. Comput. Phys.* 103, 16–42.

Levy, M. (1991). Density-functional exchange correlation through coordinate scaling in adiabatic connection and correlation hole. *Phys. Rev. A* 43, 4637–4645.

Levy, M., Ernzerhof, M., Gorling, A. (1996). Exact local exchange potential from fock equations at vanishing coupling constant, and $\delta T_c/\delta n$ from wave-function calculations at full coupling constant. *Phys. Rev. A* 53, 3963–3973.

Levy, M., Gorling, A. (1996). Density-functional exchange identity from coordinate scaling. *Phys. Rev. A* 53, 3140–3150.

Levy, M., Perdew, J. The Constrained Search Formulation of Density Functional Theory, In: Dreizler, R. M., da Providencia, J. (eds.) (1985). *Density Functional Methods in Physics*, Plenum Press, New York, *NATO ASI Series B: Physics* 123, 11–31.

Liberman, D. A., Albritton, J. R., Wilson, B. G., Alley, W. E. (1994). Self-consistent-field calculations of atoms and ions using a modified local-density approximation. *Phys. Rev. A* 50, 171–176.

Lieb, E. H., Simon, B. (1977). The Thomas-Fermi theory of atoms, molecules and solids. *Adv. in Math.* 23, 22–116.

Liu, S., Nagy, A., Parr, R. G. (1999). Expansion of the density-functional energy components ec and tc in terms of moments of the electron density. *Phys. Rev. A* 59, 1131–1134.

Löwdin, P. O. (1955). Quantum theory of many-particle systems. I. Physical interpretations by means of density matrices, natural spin-orbitals, and convergence problems in the method of configurational interaction. *Phys. Rev.* 97, 1474–1489.

Löwdin, P.-O. (1993). Some remarks on the resemblance theorems associated with various orthonormalization procedures. *Int. J. Quantum Chem.* 48, 225–232.

Löwdin, P.-O. On the non-orthogonality problem connected with the use of atomic wave functions in the theory of molecules and crystals. J. Chem. Phys. (1950). 18, 365–376.

Manoli, S. D., Whitehead, M. A. (1988). Generalized-exchange local-spin-density-functional theory: calculation and results for non-self-interaction-corrected and self-interaction-corrected theories. *Phys. Rev. A* 38, 3187–3199.

March, N. H., Angel Rubio, A., Alonso, J. A. (1999). Lowest excitation energy in atoms in the adiabatic approximation related to the single-particle kinetic energy functional. *J. Phys. B: At. Mol. Opt. Phys.* 32, 2173–2179.

Messiah, A. (1961). *Quantum Mechanics* (Vols. 1 and 2), North-Holland: Amsterdam.

Moscardo, F., San-Fabian, E. (1991). Density-functional formalism and the two-body problem. *Phys. Rev. A* 44, 1549–1553.

Mulliken, R. S. (1934). A new electroaffinity scale: together with data on valence states and an ionization potential and electron affinities. *J. Chem. Phys.* 2, 782–793.

Murphy, D. R. (1981). Sixth-order term of the gradient expansion of the kinetic-energy density functional. *Phys. Rev. A* 24, 1682–1688.

Murrell, J. N., Harget, A. J. (1971). *Semi-empirical Self-consistent-field Molecular Orbital Theory of Molecules*, Wiley Interscience, New York.

Nagy, A. (1998). Kohn-Sham equations for multiplets. *Phys. Rev. A* 57, 1672–1677.

Nagy, A., Liu, S., Parr, R. G. (1999). Density-functional formulas for atomic electronic energy components in terms of moments of the electron density. *Phys. Rev. A* 59, 3349–3354.

Nandini, G., Sathyanarayana, D. N. (2003). Molecular conformation, vibrational spectra and solvent effect studies on glycyl-L-alanine zwitterion by ab initio method. *J. Mol. Struct. THEOCHEM* 638(1–3), 79–90.

Neal, H. L. (1998). Density functional theory of one-dimension two-particle systems. *Am. J. Phys.* 66, 512–516.

Neivandt, D. J., Gee, M. L., Tripp, C. P., Hair, M. L. (1997) Coadsorption of poly(styrenesulfonate) and cetyltrimethylammonium bromide on silica investigated by attenuated total reflection techniques. *Langmuir* 13, 2519–2526.

Nielsen, M., Chuang, I. (2000). *Quantum Computation and Quantum Information*, Cambridge University Press, Cambridge.

Ohlinger, W. S., Klunzinger, P. E., Deppmeier, B. J., Hehre, W. J. (2009). Efficient Calculation of Heats of Formation. *J. Phys. Chem. A* 113, 2165–2175.

Ohno, K. (1964). Some remarks on the Pariser-Parr-Pople method. *Theor. Chim. Acta* 2, 219–227.

Oleari, L., DiSipio, L., De Michelis, G. (1966). The evaluation of the one-center integrals in the semi-empirical molecular orbital theory. *Mol. Phys.* 10, 97–109.

Orcel, G., Phalippou, J., Hench, L. L. (1986). Structural changes of silica xerogels during low temperature dehydration. J. Non-Cryst. Solids 88, 114–130.

Ou-Yang, H., Levy, M. (1990). Nonuniform coordinate scaling requirements in density-functional theory. *Phys. Rev. A* 42, 155–159.

Ou-Yang, H., Levy, M. (1991). Theorem for functional derivatives in density-functional theory. *Phys. Rev. A* 4 4, 54–58.

Pariser, R., Parr, R. (1953). A semi-empirical theory of the electronic spectra and electronic structure of complex unsaturated molecules. I. & II. *J. Chem. Phys.* 21, 466–471; 767–776.

Park, J. L., Band, W., Yourgrau, W. (1980). Simultaneous measurement, phase-space distributions, and quantum state determination. *Ann. Phys.* 492, 189–199.

Parr, R. G., Craig, D. P., Ross, I. G. (1950). Molecular orbital calculations of the lower excited electronic levels of benzene, configuration interaction included. *J. Chem. Phys.* 18, 1561–1563.

Parr, R. G., Yang, W. (1989). *Density Functional Theory of Atoms and Molecules*, Oxford University Press, New York.

Paruchuri, V. K., Fa, K., Moudgil, B. M., Miller, J. D. (2005,) Adsorption density of spherical cetyltrimethylammonium bromide (CTAB) micelles at a silica/silicon surface. Appl. Spectrosc. 59, 668–672.

Pauli, W. (1940). The connection between spin and statistics. *Phys Rev.* 58, 716–722.

Peak, D., Inomata, A. (1969). Summation over Feynman histories in polar coordinates. *J. Math. Phys.* 10, 1422–1428.

Peak, D., Inomata, A. (1969). Summation over Feynman histories in polar coordinates. *J. Math. Phys.* 10, 1422–1428.

Pearson, R. G. (1997). *Chemical Hardness*. Wiley-VCH: Weinheim.

Perdew, J. P. (1986). Density-functional approximation for the correlation energy of the inhomogeneous electron gas. *Phys. Rev. B* 33, 8822–8824; with (1986). Erratum. *Phys. Rev. B* 34, 7406.

Perdew, J. P., Chevary, J. A., Vosko, S. H., Jackson, K. A., Pederson, M. R., Singh, D. J., Fiolhais, C. (1992). Atoms, molecules, solids, and surfaces: applications of the generalized gradient approximation for exchange and correlation. *Phys. Rev. B* 46, 6671–6687.

Perdew, J. P., Ernzerhof, M., Zupan, A., Burke, K. (1998). Nonlocality of the density functional for exchange and correlation: physical origins and chemical consequences. *J. Chem. Phys.* 108, 1522–1531.

Perdew, J. P., Yue, W. (1986). Accurate and simple density functional for the electronic exchange energy: generalized gradient approximation. *Phys. Rev. B* 33, 8800–8802.

Perdew, J. P., Zunger, A. (1981). Self-Interaction Correction to Density-Functional Approximations for Many-Electron System. *Phys. Rev. B* 23, 5048–5079.

Perdew, J. P. (1986). Density-functional approximation for the correlation energy of the inhomogeneous electron gas. *Phys. Rev. B* 33, 8822–8824.

Perdew, J. P., Burke, K., Ernzerhof, M. (1996). Generalized gradient approximation made simple. *Phys. Rev. Lett.* 77, 3865–3868.

Ploetner, J., Tozer, D. J., Dreuw, A. (2010). Dependence of excited state potential energy surfaces on the spatial overlap of the Kohn-Sham orbitals and the amount of nonlocal Hartree-Fock exchange in TDDFT. *J. Chem. Theor. Comput.* 6, 2315–2324.

Pople, J. A. (1953). Electron interaction in unsaturated hydrocarbons. *Trans. Faraday Soc.* 49, 1375–1385.

Pople, J. A., Beveridge, D. L., Dobosh, P. A. (1967). Approximate self-consistent molecular-orbital theory. V. Intermediate neglect of differential overlap. *J. Chem. Phys.* 47, 2026–2034.

Pople, J. A., Beveridge, D. V. (1970). *Approximate Molecular Orbital Theory,* McGraw-Hill, New York.

Pople, J. A., Head-Gordon, M., Raghavachari, K. (1987). Quadratic configuration interaction. A general technique for determining electron correlation energies. *J. Chem. Phys.* 87, 5968–35975.

Pople, J. A., Nesbet, R. K. (1954). Self-consistent orbitals for radicals. *J. Chem. Phys.* 22, 571–572.

Pople, J. A., Santry, D. P., Segal, G. A. (1965). Approximate self-consistent molecular orbital theory. I. Invariant procedures. *J. Chem. Phys.* 43:S129–S135.

Pople, J. A., Segal, G. A. (1965). Approximate self-consistent molecular orbital theory. II. Calculations with complete neglect of differential overlap. *J. Chem. Phys.* 43:S136–S151.

Pople, J. A., Segal, G. A. (1966). Approximate self-consistent molecular orbital theory. III. CNDO results for AB2 and AB3 systems. *J. Chem. Phys.* 44, 3289–3297.

Preuss, H. (1969). *Quantenchemie für Chemiker*, Verlag Chemie, Weinheim.

Primeau, N., Vautey, C., Langlet, M. (1997). The effect of thermal annealing on aerosol-gel deposited SiO2 films: A FTIR deconvolution study. *Thin Solid Films* 310, 47–56.

Purvis, G. D., Bartlett, R. J. (1982).. A full coupledcluster singles and doubles model: The inclusion of disconnected triples, *J. Chem. Phys.* 76, 1910–1919.

Rasolt, M., Geldart, D. J. W. (1986). Exchange and correlation energy in a nonuniform fermion fluid. *Phys. Rev. B* 34, 1325–1328.

Ridley, J. E., Zerner, M. C. (1976). Triplet states via intermediate neglect of differential overlap: benzene, pyridine and the diazines. *Theor. Chim. Acta* 42, 223–236.

Roos, B. O., Malmqvist, P.-Å. (2004). Relativistic quantum chemistry - the multiconfigurational approach. *Phys. Chem. Chem. Phys.* 6, 2919–2927.

Roos, B. O., Sadlej, A. J., Siegbahn, P. E. M. (1982). Complete-active-space self-consistent-field and contracted configuration-interaction study of the electron correlation in Ne, F-, Ne+, and, F. *Phys. Rev. A* 26, 1192–1199.

Roos, B. O., Taylor, P. R., Siegbahn, P. E. M. (1980). A complete active space SCF method (CASSCF) using a density matrix formulated super-CI approach. *Chem. Phys.* 48, 157–173.

Roothaan, C. C. J. (1951). New developments in molecular orbital theory. *Rev. Mod. Phys.* 23, 69–89.
Roothaan, C. C. J. (1951). New developments in molecular orbital theory. *Rev. Mod. Phys.* 23, 69–89.
Roothaan, C. C. J. (1958). Evaluation of molecular integrals by digital computer. *J. Chem. Phys.* 28, 982–983.

Roothaan, C. C. J. (1960). Self-consistent field theory for open shells of electronic systems. *Rev. Mod. Phys.* 32, 179–185.

Roothaan, C. C. J., Detrich, J. H. (1983). General quadratically convergent multiconfiguration self-consistent-field theory in terms of reduced matrix elements. *Phys. Rev. A* 27, 9–56.

Rubin, S. G., Khosla, P. K. (1977). Polynomial interpolation methods for viscous flow calculations. *J. Comp. Phys.* 24, 217–244.

Runge, E., Gross, E. K. U. (1984). Density-functional theory for time-dependent systems. *Phys. Rev. Lett.* 52, 997–1000.

Savin, A., Preuss, H., Stoll, H. *Non-Local Effects on Atomic and Molecular Correlation Energies Studies with a Gradient-Corrected Density Functional*, In: Erhahy, R., Smith, V. H. (Eds.) (1987). *Density Matrices and Density Functionals*, Reidel Publishing Company, pp. 457–465.

Savin, A., Stoll, H., Preuss, H. (1986). An application of correlation energy density functionals to atoms and molecules. *Theor Chim Acta* 70, 407–419.

Savin, A., Wedig, U., Preuss, H., Stoll, H. (1984). Molecular correlation energies obtained with a nonlocal density functional. *Phys. Rev. Lett.* 53, 2087–2089.

Schulman, L. S. (1981). *Techniques and Applications of Path Integration,* Wiley, New York.

Schulmann, L. S. (1968). A Path integral for spin. *Phys. Rev.* 176, 1558–1569.

Seidl, M., Perdew, J. P., Levy, M. (1999). strictly correlated electrons in density-functional theory. *Phys. Rev. A* 59, 51–54.

Senatore, G., March, N. H. (1994). Recent progress in the field of electron correlation. *Rev. Mod. Phys.* 66, 445–479.

Silvi, B., Savin, A. (1994). Classification of chemical bonds based on topological analysis of electron localization functions. Nature 371, 683–686.

Slater, J. C. (1929). Theory of complex spectra. *Phys. Rev.* 34, 1293–1322.

Slater, J. C. A (1951). Simplification of the Hartree-Fock method. *Phys. Rev.* 81, 385–390.

Slater, J. C., Johnson, K. H. (1972). Self-consistent-field Xα cluster method for polyatomic molecules and solids. *Phys. Rev. B* 5, 844–853.

Slater, J. I. (1960). *Quantum Theory of Atomic Structure*, McGraw-Hill Book Company, New York.

Snygg, J. (1982). The Heisenberg picture and the density operator. Am. J. Phys. 50, 906–909.

Stavrev, K. K., Zerner, M. C. (1995). On the Jahn–Teller effect on Mn^{2+} in zinc-blende ZnS crystal. *J. Chem. Phys.* 102, 34–39.

Stavrev, K. K., Zerner, M. C., Meyer, T. J. (1995). Outer-sphere charge-transfer effects on the spectroscopy of the [Ru(NH3)5(py)]2+ Complex. *J. Am. Chem. Soc.* 117, 8684–8685.

Stewart, J. J. P. (1990). MOPAC: A semiempirical molecular orbital program. *J. Comp. Aided Mol. Design* 4, 1–103.

Stewart, R. F. (1970). Small Gaussian expansions of Slater-type orbitals. *J. Chem. Phys.* 52, 431–439.
Stewart, J. J. P. (1989a). Optimization of parameters for semiempirical methods. I. Method, *J. Comput. Chem.* 10, 209–220.
Stewart, J. J. P. (1989b). Optimization of parameters for semiempirical methods. II Applications, *J. Comput. Chem.* 10, 221–264.
Szabo, A., Ostlund, N. S. (1996). *Modern Quantum Chemistry: Introduction to Advanced Electronic Structure Theory*, Dover, New York.
Taut, M. (1996). Generalized gradient correction for exchange: deduction from the oscillator model. *Phys. Rev. A* 53, 3143–3150.
Teller, E. (1962). On the stability of molecules in the Thomas-Fermi theory. *Rev. Mod. Phys.* 34, 627–631.
Thiel, W. (1988). Semiempirical methods: current status and perspectives. *Tetrahedron* 44, 7393–7408.
Thomas, L. H. (1927). The calculation of atomic fields. *Proc. Cambridge Phil. Soc.* 23, 542–548.
Thouless, D. J. (1972). *The Quantum Mechanics of Many-body Systems*, Academic Press, New York.
Tiwari, S., Mishra, P. C., Suhai, S. (2008). Solvent effect of aqueous media on properties of glycine: significance of specific and bulk solvent effects, and geometry optimization in aqueous media. *Int. J. Quantum Chem.* 108, 1004–1016.
Tozer, D. J., Handy, N. C. (1998). The development of new exchange-correlation functionals. *J. Chem. Phys.* 108, 2545–2555.
Vosko, S. J., Wilk, L., Nusair, M. (1980). Accurate spin-dependent electron liquid correlation energies for local spin density calculations: a critical analysis. *Can. J. Phys.* 58, 1200–1211.
Voth, G. A. (1991). Calculation of equilibrium averages with Feynman-Hibbs effective classical potentials and similar variational approximations. *Phys. Rev. A* 44, 5302–5305.
Wang, Y., Perdew, J. P. (1989). Spin scaling of the electron-gas correlation energy in the high-density limit. *Phys. Rev. B* 43, 8911–8916.
Wang, Y., Perdew, J. P., Chevary, J. A., Macdonald, L. D., Vosko, S. H. (1990). Exchange potentials in density-functional theory. *Phys. Rev. A* 41, 78–85.
Wiegel, F. W. (1986). *Introduction to Path-Integral Methods in Physics and Polymer Science*, World Scientific, Singapore.
Wiener, N. (1923). Differential space. *J. Math. Phys.* 2, 131–174.
Wilson, L. C., Levy, M. (1990). Nonlocal wigner-like correlation-energy density functional through coordinate scaling. *Phys. Rev. B* 41, 12930–12932.
Wilson, L. C., Levy, M. (1990). Nonlocal wigner-like correlation-energy density functional through coordinate scaling. *Phys. Rev. B* 41, 12930–12932.
Woodward, R. B., Hoffmann, R. (1965). Stereochemistry of electrocyclic reactions. *J. Am. Chem. Soc.* 87, 395–397.
Xue, W., He, H., Zhu, J., Yuan, P. (2007). FTIR investigation of CTAB–Al–montmorillonite complexes. *Spectrochim. Acta Part A* 67, 1030–1036.
Zhao, Q., Morrison, R. C., Parr, R. G. (1994). From electron densities to kohn-sham kinetic energies, orbital energies, exchange-correlation potentials, and exchange-correlation energies. *Phys. Rev. A* 50, 2138–2142.

Zhao, Q., Parr, R. G. (1992). Local exchange-correlation functional: numerical test for atoms and ions. *Phys. Rev. A* 46:R5320-R5323.

FURTHER READINGS

Cramer, C. J. (2002). *Essentials of Computational Chemistry*, John Wiley & Sons, Chichester.
Dreizler, R. M., Gross, E. K. U. (1990). Density Functional Theory, Springer Verlag, Heidelberg.
Fiolhais, C., Nogueira, F., Marques, M., Eds. (2003). *A Primer in Density Functional Theory*, Springer-Verlag, Berlin.
Hartree, D. R. (1957). *The Calculation of Atomic Structures*, Wiley & Sons, New York.
Jensen, F. (2007). *Introduction to Computational Chemistry*, John Wiley and Sons, Chichester.
Koch, W., Holthausen, M. C. (2002). A Chemist's Guide to Density Functional Theory, Wiley-VCH, Weinheim, 2nd edition.
Kohanoff, J. (2006). *Electronic Structure Calculations for Solids and Molecules: Theory and Computational Methods*, Cambridge University Press, Cambridge (UK).
March, N. H. (1991). *Electron Density Theory of Many-Electron Systems*, Academic Press, New York.
Nesbet, R. K. (2002). *Variational Principles and Methods in Theoretical Physics and Chemistry*, Cambridge University Press, New York.
Richard, M. M. (2004). *Electronic Structure: Basic Theory and Practical Methods*, Cambridge University Press, Cambridge (UK).
Rzewuski, J. (1969). *Field Theory*, Hafner, New York.
Sholl, D., Steckel, J. A. (2009). *Density Functional Theory: A Practical Introduction*, Wiley-Interscience, New Jersey.
Slater, J. C. (1963). *Theory of Molecules and Solids, Vol. 1, Electronic Structure of Molecules*, McGraw-Hill, New York.
Streitwieser, A. (1961). *Molecular Orbital Theory for Organic Chemists*, Wiley, New York.

APPENDICES

CONTENTS

This section is organized as a short course of advanced mathematics by elementary methods for physical-chemists. Therefore, the topics while being not comprehensive are unitarily presented although not necessary in an order paralleling their use in the textbook.

A.1 USEFUL POWER SERIES

A.1.1 GENERALIZED BINOMIAL SERIES

A useful application dealing with power series expansion regards the calculation of special limits, otherwise difficult to solve. As an example one can immediately evaluate:

$$\lim_{x\to 0}\frac{\sin x}{x}=\lim_{x\to 0}\left[\frac{1}{x}\left(\frac{x}{1!}-\frac{x^3}{3!}+\frac{x^5}{5!}-\frac{x^7}{7!}+...\right)\right]=\lim_{x\to 0}\left(1-\frac{x^2}{3!}+\frac{x^4}{5!}-\frac{x^6}{7!}+...\right)=1$$

The (Newton) binomial series expansion generally looks like

$$(a+b)^n=\sum_{k=0}^{\infty}\binom{n}{k}a^{n-k}b^k$$

where the number

$$\binom{n}{k}=\frac{n!}{k!(n-k)!}$$

represents the number of ways a collection of k-objects may be realized from a set of n-objects without regarding order. It nevertheless satisfies the fundamental identities:

$$\binom{n}{k}=\binom{n}{n-k}=(-1)^k\binom{k-n-1}{k}$$

as one may immediately check with the above definition formula. Yet, this formula may be regard also as holding in general, i.e., also for negative powers of n, in which case it is called as generalized or negative (Newton) binomial series and bears the successive forms abstracted from the original one through replacing $n\to -n$ along employing the appropriate coefficient replacements:

$$(a+b)^{-n}=\sum_{k=0}^{\infty}\binom{-n}{k}a^{-n-k}b^k$$

$$=\sum_{k=0}^{\infty}(-1)^k\binom{k+n-1}{k}a^{-n-k}b^k$$

Worth specializing this formula for the case when

$$a=1,\ b=-x$$

so that we get the working formula:

$$\frac{1}{(1-x)^n}=\sum_{k=0}^{\infty}(-1)^k\binom{k+n-1}{k}(-x)^k=\sum_{k=0}^{\infty}\binom{k+n-1}{k}x^k$$

that immediately recovers the geometric series for $n=1$:

$$\frac{1}{1-x}=\sum_{k=0}^{\infty}\binom{k}{k}x^k=\sum_{k=0}^{\infty}x^k$$

A.1.2 THE POISSON (COMB FUNCTION) FORMULA FOR SERIES

The Poisson formula permits the transformation of the series $\sum_{m=-\infty}^{+\infty} f(m)$ into another one for which the evaluation is (most of the times) easier. One will start from the consideration of the "comb" function:

$$S(q)=\sum_{m=-\infty}^{+\infty}\delta(q-m)$$

seen as series of the delta (Dirac) functions, with the representation in the Figure A.1.1.

From the above definition and/or from Figure A.1.1 there can be observed the periodicity property of the "comb" function:

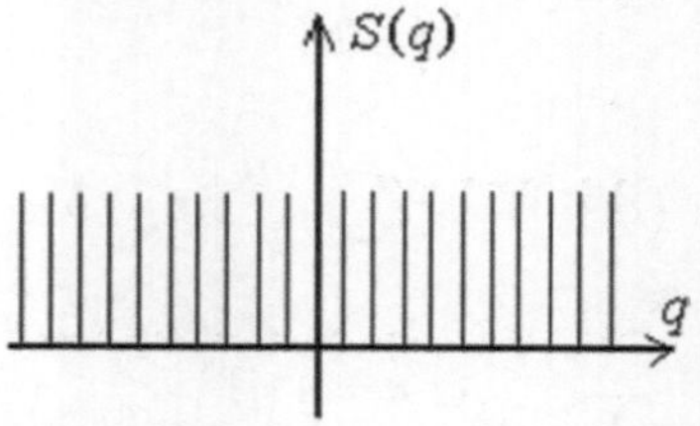

FIGURE A.1.1 The representation of the "comb" function.

$$S(q+n) = S(q);\ n \in Z$$

that allows to develop the original function in Fourier series:

$$S(q) = \sum_{n=-\infty}^{+\infty} S_n \exp(-i2\pi qn)$$

In this series the coefficient of the Fourier expansion has the expression:

$$S_n \equiv \int_{-1/2}^{+1/2} dqS(q)\exp(i2\pi nq) = 1$$

where, for getting the last identity the above expression for $S(q)$ as well as the properties of the delta Dirac function have been used. With the result back in the exponential series one finds also the identity:

$$\sum_{m=-\infty}^{+\infty} \delta(q-m) = \sum_{n=-\infty}^{+\infty} \exp(-i2\pi\, qn)$$

Moreover, when the terms of the last relation are multiplied with the integral $\int_{-\infty}^{+\infty} dqf(q)$ and the properties of the delta Dirac function counts again, the final Poisson formula for series shapes as:

$$\sum_{m=-\infty}^{+\infty} f(m) = \sum_{n=-\infty}^{+\infty} \left[\int_{-\infty}^{+\infty} dqf(q)\exp(-i2\pi\, qn) \right]$$

The final expression seems, at the first sight, like a complicated version of the initial series (the left hand side), whereas in the concrete applications it proves to be a very useful transformation of the initial series into a convergent one.

A.2 EULER'S BETA AND GAMMA FUNCTIONS: APPLICATIONS ON POISSON INTEGRALS AND STIRLING'S APPROXIMATION

One starts from the integral identity (through parts) for the function Gamma-Euler function:

$$
\begin{aligned}
\Gamma(a)_{\substack{a>0\\ a\in\Re}} &= \int_0^\infty x^{a-1}e^{-x}dx \\
&= -x^{a-1}e^{-x}\Big|_0^\infty + (a-1)\int_0^\infty x^{a-1}e^{-x}dx \\
&= (a-1)\Gamma(a-1)
\end{aligned}
$$

with which one can be formed the following recursive identities

$$\Gamma(1) = \int_0^\infty e^{-x}dx = -e^{-x}\Big|_0^\infty = 1$$

$$\Gamma(2) = (2-1)\Gamma(1) = 1$$

$$\Gamma(3) = (3-1)\Gamma(2) = 2$$

$$\Gamma(n) = (n-1)!$$

$$\Gamma(n+1) = n!$$

Moreover, the Gamma Euler function allows the next product formation

$$
\begin{aligned}
\Gamma(a)\Gamma(b) &= \int_0^\infty x^{a-1}e^{-x}dx\int_0^\infty x^{b-1}e^{-y}dy \\
&\overset{y=u-x}{=} \int_0^\infty e^{-u}du\int_0^u e^{x}(u-x)^{b-1}x^{a-1}e^{-x}dx \\
&= \int_0^\infty e^{-u}du\int_0^u (u-x)^{b-1}x^{a-1}dx \\
&\overset{x=ut}{=} \int_0^\infty e^{-u}du\int_0^1 u^{b-1}(1-t)^{b-1}u^{a-1}t^{a-1}udt \\
&= \underbrace{\int_0^\infty u^{a+b-1}e^{-u}du}_{\Gamma(a+b)}\underbrace{\int_0^1 t^{a-1}(1-t)^{b-1}dt}_{\equiv B(a,b)}
\end{aligned}
$$

where the Euler's Beta function had been introduced

$$\beta \text{ or } B(a,b) = \int_0^1 t^{a-1}(1-t)^{b-1}\,dt$$

thus having also the equivalent expression in terms of Euler`s Gamma functions

$$B(a,b) = \frac{\Gamma(a)\Gamma(b)}{\Gamma(a+b)}$$

The last relationship seems to be particularly useful for the computation of the Gamma function with semi-integer arguments. This process involves several steps. Firstly, worth remarking the identity

$$B(a,1-a)_{0<a<1} = \frac{\Gamma(a)\Gamma(1-a)}{\Gamma(1)} = \Gamma(a)\Gamma(1-a)$$

On the other side the Euler's Beta integral representation has the successive transformations

$$\begin{aligned}
B(a,1-a)_{0<a<1} &= \int_0^1 t^{a-1}(1-t)^{-a}\,dt \\
&\overset{t=\frac{x}{1+x}}{=} \int_0^\infty \left(\frac{x}{1+x}\right)^{a-1}\left(\frac{1}{1+x}\right)^{-a}\frac{1}{(1+x)^2}\,dt \\
&= \int_0^\infty \frac{x^{a-1}}{1+x}\,dx \\
&\overset{x=e^y}{=} \int_{-\infty}^\infty \frac{e^y e^{ay-y}}{e^y+1}\,dy \\
&= \int_{-\infty}^\infty \frac{e^{az}}{e^z+1}\,dz
\end{aligned}$$

Recognizing in the last integral a complex integral, it can be solved by identification of poles through the complex equation

$$\left.\begin{aligned} e^z+1=0 \\ z=i\alpha \end{aligned}\right\} \Rightarrow \cos\alpha + i\sin\alpha = -1 \Rightarrow \alpha=(2k+1)\pi \Rightarrow z_k=(2k+1)\pi i$$

From it, there follows that the poles are on the complex axis of the complex plane (and not found in origin) which allows the choice of a contour integration in the positive part of the complex plane (closed counterclockwise) so that directly apply the residue theorem for the last form of the above Beta Euler integral

$$B(a,1-a)_{0<a<1} = \int_{-\infty}^{\infty} \frac{e^{az}}{e^z+1} dz = 2\pi i \sum_{z_k} \mathbf{Res}_{\mathrm{k}} B(a,1-a)$$

According to the residue theorem, if the integral has the form $h(z) = f(z)/g(z)$, with $f(z)$ a regular function, and $g(z)$ has simple poles at the points $z = z_k$, then the sum of the residues of the function $h(z)$ at poles z_k, is written as

$$\sum_{z_k} \mathbf{Res}_{\mathbf{k}} h(z) = \sum_{z_k} \frac{f(z)}{\partial_z g(z)}\bigg|_{z=z_k}$$

With this, the last form of the Beta Euler function becomes

$$\begin{aligned}
B(a,1-a) &= 2\pi i \sum_{k\geq 0} \frac{e^{az}}{\partial_z\left(e^z+1\right)}\bigg|_{z=(2k+1)i\pi} \\
&= 2\pi i \sum_{k\geq 0} \frac{e^{a(2k+1)i\pi}}{e^{(2k+1)i\pi}} = -2\pi i \sum_{k\geq 0} e^{a(2k+1)i\pi} \\
&= -2\pi i e^{ia\pi}\left(1+e^{i2a\pi}+e^{i4a\pi}+\ldots\right) \\
&= -2\pi i e^{ia\pi} \frac{1}{1-e^{i2a\pi}} \\
&= \pi \frac{2i}{e^{ia\pi}-e^{-ia\pi}} \\
&= \pi \sin\left(a\pi\right)
\end{aligned}$$

Corroborating this result with the relation with the Gamma Euler functions the new identity is formed, namely

$$\Gamma(a)\Gamma(1-a) = \pi \sin\left(a\pi\right)$$

wherefrom one may evaluate the Gamma Euler functions with semi-integer argument

$$\text{for } a = \frac{1}{2} : \left[\Gamma\left(\frac{1}{2}\right)\right]^2 = \pi \Rightarrow \Gamma\left(\frac{1}{2}\right) = \sqrt{\pi}$$

but also the Poisson integrals directly as

$$\sqrt{\pi} = \Gamma\left(\frac{1}{2}\right) = \int_0^\infty x^{-\frac{1}{2}} e^{-x} dx$$

$$\overset{x=\alpha t^2}{=} 2\alpha \int_0^\infty \alpha^{-\frac{1}{2}} t^{-1} e^{-\alpha t^2} t dt = 2\alpha^{1/2} \int_0^\infty e^{-\alpha t^2} dt$$

$$\Rightarrow \int_0^\infty e^{-\alpha t^2} dt = \frac{1}{2}\sqrt{\frac{\pi}{\alpha}}$$

as well as the allied integral with doubled interval of integration

$$\int_{-\infty}^{\infty} e^{-\alpha t^2} dt = \sqrt{\frac{\pi}{\alpha}}$$

More generally, one can establish *the general expression for Poisson integrals*

$$I(\alpha)_{\substack{m=0,1,2,...\\ n=1,2,3,...}} = \int_{-\infty}^{+\infty} x^m \exp\left(-\alpha x^n\right) dx = \begin{cases} 0 ... n = 1,3,5,... \\ 2\int_0^\infty x^m \exp\left(-\alpha x^n\right) dx ... n = 2,4,6,... \end{cases}$$

with special values for the integrals with $n = 2,4,6,...$

$$I(\alpha)_{\substack{m=0,1,2,...\\ n=2,4,6...}} = \int_0^\infty x^m \exp\left(-\alpha x^n\right) dx$$

$$\overset{y=\alpha x^n}{=} \int_0^\infty \left(\frac{y}{\alpha}\right)^{\frac{m}{n}} \exp(-y) \frac{1}{\alpha n} \left(\frac{y}{\alpha}\right)^{\frac{1-n}{n}} dy$$

$$= \int_0^\infty \left(\frac{y}{\alpha}\right)^{\frac{m+1}{n}-1} \exp(-y) \frac{1}{\alpha n} dy$$

$$= \frac{1}{n\alpha^{\frac{m+1}{n}}} \underbrace{\int_0^\infty y^{\frac{m+1}{n}-1} \exp(-y) dy}_{\Gamma\left(\frac{m+1}{n}\right)}$$

$$\Rightarrow I(\alpha)_{\substack{m=0,1,2,...\\ n=2,4,6...}} = \frac{1}{n\alpha^{\frac{m+1}{n}}}\Gamma\left(\frac{m+1}{n}\right)$$

This result generates an important particularization for $n = 2$ case, depending on the even or odd order of the m parameter

$$I(\alpha)_{\substack{m=0,1,2,..\\ n=2}} = \begin{cases} I(\alpha)_{\substack{2m+1\\ n=2}} = \dfrac{m!}{2\alpha^{m+1}} \\ I(\alpha)_{\substack{2m\\ n=2}} = \dfrac{(2m-1)!!}{2^{m+1}}\left(\dfrac{\pi}{\alpha^{2m+1}}\right)^{1/2} \end{cases}$$

where

$$(2m-1)!! = 1 \cdot 3 \cdot 5 ... (2m-1)$$

allowing frequent specializations in Poisson classic formula for integrals, as:

$$I_0 = \int_0^\infty e^{-\alpha x^2} dx = \frac{1}{2}\left(\frac{\pi}{\alpha}\right)^{1/2}$$

$$I_1 = \int_0^\infty x e^{-\alpha x^2} dx = \frac{1}{2\alpha}$$

$$I_2 = \int_0^\infty x^2 e^{-\alpha x^2} dx = \frac{1}{4\alpha}\left(\frac{\pi}{\alpha}\right)^{1/2} = -\frac{dI_0}{d\alpha}$$

$$I_3 = \int_0^\infty x^3 e^{-\alpha x^2} dx = \frac{1}{2\alpha^2} = -\frac{dI_1}{d\alpha}$$

$$I_4 = \int_0^\infty x^4 e^{-\alpha x^2} dx = \frac{3}{4\alpha^2}\left(\frac{\pi}{\alpha}\right)^{1/2} = -\frac{dI_2}{d\alpha}$$

...

Nevertheless, the Stirling`s approximation formula can be also formulated, for large numbers, using the Euler Gamma functions under the form

$$\Gamma(n+1)_{\substack{n>>1\\(n-1\cong n)}} = \int_0^\infty x^n e^{-x} dx = \int_0^\infty e^{n\ln x - x} dx$$

to be solved by the saddle point method (*saddle point approximation*), or by the stationary phase: this implies the developing the function under the integral, around its maximum

$$0 = \partial_x (n \ln x - x)_{x_{\max}} = \frac{n}{x_{\max}} - 1 \Rightarrow x_{\max} = n$$

for instance, until the second order of the Taylor series

$$\underbrace{n\ln x - x}_{f(x)} \cong \underbrace{n\ln n - n}_{f(x_{\max}=n)} + \underbrace{f'(x)\big|_{x=x_{\max}}}_{0}(x-n) + \frac{1}{2}\underbrace{f''(x)\big|_{x=x_{\max}}}_{-1/n}(x-n)^2$$

$$= n\ln n - n - \frac{(x-n)^2}{2n}$$

this way allowing the expansion of the genuine integral also over the negative infinite domain (assuring the completeness, since the presence of the quadratic contribution of the second order Taylor approximation). All in all one has successively

$$n! = \Gamma(n+1)_{\substack{n>>1\\(n-1\cong n)}} = \int_{-\infty}^{\infty} \exp\left[n\ln n - n - \frac{(x-n)^2}{2n}\right] dx$$

$$= e^{n\ln n - n} \int_{-\infty}^{\infty} e^{-\frac{(x-n)^2}{2n}} dx \overset{y=x-n}{=} e^{n\ln n - n} \underbrace{\int_{-\infty}^{\infty} e^{-\frac{y^2}{2n}} dy}_{\substack{\sqrt{2n\pi}\\(Poisson)}}$$

$$= \sqrt{2n\pi}\exp(n\ln n - n)$$

Finally, through logarithmation and the neglection of the numerical terms for $n>>1$, the Stirling's approximation formula is found as

$$\ln n! \cong n\ln n - n + \frac{1}{2}\ln n + \underbrace{\frac{1}{2}\ln(2\ \ )}_{\textbf{neglected}}$$

$$\cong \underbrace{\left(n+\frac{1}{2}\right)}_{\cong n}\ln n - n$$

$$\Rightarrow \ln n! \cong n\ln n - n$$

This relation is customary for determining the probability distribution from the Boltzmann, Fermi-Dirac and Bose-Einstein statistics.

A.3 ELEMENTARY CALCULATION OF RIEMANN-ZETA SERIES: APPLICATION ON STATISTICAL INTEGRALS

Having to solve the statistical integral, such as those, which intervene in the Stefan-Boltzmann law

$$\int_0^\infty \frac{x^3}{\exp(x)-1}dx = ?$$

one recognizes it belongs to a larger family of integrals, namely

$$J_{n+1} = \int_0^\infty \frac{x^n}{\exp(x)-1}dx,\ n \in \mathbf{N}$$

To resolve them, the equivalent sequence is formed

$$\begin{aligned} J &= \int_0^\infty \frac{x^n e^{-x}}{1-e^{-x}}dx \\ &= \int_0^\infty x^n e^{-x}\sum_{k=0}^\infty e^{-kx}dx \\ &= \int_0^\infty x^n e^{-x}\left(1+e^{-x}+e^{-2x}+e^{-3x}+\ldots\right)dx \\ &= \int_0^\infty x^n\left(e^{-x}+e^{-2x}+e^{-3x}+\ldots\right)dx \\ &= \int_0^\infty x^n\sum_{k=1}^\infty e^{-kx}dx \end{aligned}$$

$$\underset{x^n=y^n/k^n}{\overset{y=kx}{\underset{dx=dy/k}{=}}} \int_0^\infty \frac{y^n}{k^n} \sum_{k=1}^\infty e^{-y} \frac{dy}{k}$$

$$= \sum_{k=1}^\infty \frac{1}{k^{n+1}} \int_0^\infty y^n e^{-y} dy$$

$$= \zeta(n+1)\Gamma(n+1)$$

where the Euler's Gamma special function was recognized

$$\Gamma(n+1) = \int_0^\infty y^n e^{-y} dy$$

Allowing to identify the power series form for the zeta-Riemann series

$$\zeta(n+1) = \sum_{k=1}^\infty \frac{1}{k^{n+1}}$$

In other words, the concerned integrals can be calculate by the products

$$J_n = \int_0^\infty \frac{x^{n-1}}{\exp(x)-1} dx = \zeta(n)\Gamma(n) = \zeta(n)(n-1)!$$

where, in order to express a complete analytical result, the Riemann series remain to be assessed in various orders. In the first instance, one notes that with the aid of the integral criterion (since the ratio criterion does not appear as conclusive) there can be assure the convergence of the Riemann series for $n>1$ order through the series' coefficient

$$\int^\infty a(k)dk = \int^\infty \frac{dk}{k^n} = -\frac{1}{n-1}\frac{1}{k^{n-1}}\Bigg|^\infty \underset{n>1}{\to} 0$$

Then, while passing to the actually assessments, the $\zeta(n)$ series will be employed for the case of $n = 2l, l \in \mathbf{N}^*$. However, in what follows, the McLaurin series expansion will be employed

$$f(x) \cong \sum_{k=0}^\infty \frac{1}{n!}\left(\frac{d^k f(x)}{dx^k}\right)_{x=0} x^k$$

for the logarithm function to have

$$\ln(1-x) = -\left(\frac{1}{1-x}\right)_{x=0} x - \frac{1}{2!}\left[\frac{1}{(1-x)^2}\right]_{x=0} x^2 - \frac{1}{3!}\left[\frac{2}{(1-x)^3}\right]_{x=0} x^3 - \dots$$

$$\ln(1-x) = -\sum_{k=1}^{\infty} \frac{x^k}{k}$$

Then, with the complex variable specialization (analytical continuation)

$$x \to e^{ix}$$

one may write the equivalent identities

$$\sum_{k=1}^{\infty} \frac{e^{ikx}}{k} = -\ln\left(1-e^{ix}\right)$$

$$\Leftrightarrow \sum_{k=1}^{\infty} \frac{\cos(kx) + i\sin(kx)}{k} = \mathrm{Re}\left[-\ln\left(1-e^{ix}\right)\right] + i\,\mathrm{Im}\left[-\ln\left(1-e^{ix}\right)\right]$$

from where the selection of the imaginary part leaves with trigonometric series resumation

$$\sum_{k=1}^{\infty} \frac{\sin(kx)}{k} = -\mathrm{Im}\left[\ln\left(1-(\cos x + i\sin x)\right)\right]$$

Further processing of this relation will be performed by taking into consideration the complex variable form

$$z = |z|e^{i\varphi} = \mathrm{Re}\, z + i\,\mathrm{Im}\, z$$

with its logarithm formed as

$$\ln z = \ln|z| + i\phi$$

from where the imaginary part looks like

$$\mathrm{Im}[\ln z] = \phi = \arctan\left(\frac{\mathrm{Im}\, z}{\mathrm{Re}\, z}\right)$$

Under these conditions, the last trigonometric series is successively evaluated

$$\sum_{k=1}^{\infty}\frac{\sin(kx)}{k}=-\arctan\left(\frac{-\sin x}{1-\cos x}\right)=\arctan\left(\frac{2\sin\frac{x}{2}\cos\frac{x}{2}}{2\sin^2\frac{x}{2}}\right)=\arctan\left(\cot\frac{x}{2}\right)$$

Moreover, with the r.h.s. of the last expression may be rewritten, through the recourse of the trigonometric identity

$$\tan^{-1}x+\cot^{-1}x=\frac{\pi}{2}$$

on which one applies the replacement

$$x\rightarrow\cot y$$

$$\tan^{-1}(\cot y)+\cot^{-1}(\cot y)=\frac{\pi}{2}$$

to result in the identity

$$\arctan(\cot y)=\frac{\pi}{2}-y$$

so that providing with the series limit

$$\sum_{k=1}^{\infty}\frac{\sin(kx)}{k}=\frac{\pi-x}{2}$$

While integrated the last relation yields the companion series

$$\sum_{k=1}^{\infty}\frac{\cos(kx)}{k^2}=\frac{x^2-2\pi x}{4}+C$$

With the new series notation

$$f(x)=\sum_{k=1}^{\infty}\frac{\cos(kx)}{k^2}$$

one can recognize the constant of integration as representing the Riemann zeta series of second order

$$C=f(0)=\sum_{k=1}^{\infty}\frac{1}{k^2}=\zeta(2)$$

Consequently, finding the Riemann zeta series of second degree is the equivalent with identify the constant of integration from de cosine series above; for this, the classical method of determining the constants is applied, considering particular conditions of the series involved. Especially if there are considered the particular expansions

$$f(0)=\frac{1}{1^2}+\frac{1}{2^2}+\frac{1}{3^2}+\frac{1}{4^2}+...$$

$$f(\pi)=-\frac{1}{1^2}+\frac{1}{2^2}-\frac{1}{3^2}+\frac{1}{4^2}-...$$

one can form the equation

$$\frac{f(0)+f(\pi)}{2}=\frac{1}{2^2}+\frac{1}{4^2}+...=\frac{1}{2^2}f(0)$$

wherefrom resulting the relation

$$f(\pi)=-\frac{f(0)}{2}$$

which back-transposed in the terms of the values of these functions as the above series with constants of integration, generates the simple equation (and solution)

$$\frac{\pi^2-2\pi^2}{4}+C=-\frac{1}{2}C\Rightarrow C=\frac{\pi^2}{6}$$

This result immediately associates with analytical finding of the (limit of the) series Riemann zeta of second order

$$\zeta(2)=\sum_{k=1}^{\infty}\frac{1}{k^2}=\frac{\pi^2}{6}$$

This value is useful for calculate the statistical integral

$$J_2=\int_0^{\infty}\frac{x}{\exp(x)-1}dx=\zeta(2)\Gamma(2)=\zeta(2)(2-1)!=\frac{\pi^2}{6}$$

For the evaluation of the integrals of higher order, one will continue with the integration of the series, under the complete cast now complete

$$\sum_{k=1}^{\infty}\frac{\cos(kx)}{k^2}=\frac{x^2-2\pi x}{4}+\frac{\pi^2}{6}$$

Thus, a new first integration generates the new series

$$\sum_{k=1}^{\infty}\frac{\sin(kx)}{k^3}=\frac{x^3-3\pi x^2+2\pi^2 x}{12}+C^*$$

With the new constant of integration

$$C^*=\sum_{k=1}^{\infty}\frac{\sin(kx)}{k^2}\bigg|_{x=0}=0$$

allowing for further integration with further new constant

$$\sum_{k=1}^{\infty}\frac{\cos(kx)}{k^4}=\frac{-x^4+4\pi x^3-4\pi^2 x^2}{48}+C^{\#}$$

The last relation, with the current constant of integration is processed as in the previous case of the Riemann-zeta series second order, now introducing for the fourth order the new function

$$g(x)=\sum_{k=1}^{\infty}\frac{\cos(kx)}{k^4}$$

with which it can be recognized through identity with Riemann-zeta function of fourth order, leaving with the last constant of integration

$$C^{\#}=g(0)=\sum_{k=1}^{\infty}\frac{1}{k^4}=\zeta(4)$$

Thus, by considering the expansions

$$g(0)=\frac{1}{1^4}+\frac{1}{2^4}+\frac{1}{3^4}+\frac{1}{4^4}+\ldots$$

$$g(\pi)=-\frac{1}{1^4}+\frac{1}{2^4}-\frac{1}{3^4}+\frac{1}{4^4}-\ldots$$

and of their linear combination

$$\frac{g(0)+g(\pi)}{2}=\frac{1}{2^4}+\frac{1}{4^4}+\ldots=\frac{1}{2^4}g(0)$$

one has the simple solution as the relationship

$$g(\pi) = -\frac{7}{8} g(0)$$

to be rewritten under the equation for the associated constant of their series limits, with the elementary solution

$$\frac{-\pi^4 + 4\pi^4 - 4\pi^4}{48} + C^{\#} = -\frac{7}{8} C^{\#} \Rightarrow C^{\#} = \frac{\pi^4}{90}$$

Eventually yielding the expression of the Riemann-zeta fourth order series

$$\zeta(4) = \sum_{k=1}^{\infty} \frac{1}{k^4} = \frac{\pi^4}{90}$$

and of the related statistical integral

$$J_4 = \int_0^{\infty} \frac{x^3}{\exp(x) - 1} dx = \zeta(4)\Gamma(4) = \zeta(4)(4-1)! = \frac{\pi^4}{15}$$

specific for the Stefan-Boltzmann law.

A.4 LAGRANGE INTERPOLATION AND NUMERICAL INTEGRATION: APPLICATION ON ERROR FUNCTION

Being given an n-dimensional set of discrete points $\{x_i\}_{i=\overline{0,n}}$ and of the observed (or measured) values on them $\{y_i = f(x_i)\}_{i=\overline{0,n}}$, one is interested in establishing the analytical function $f(x)$ that fits with all given data. In these respect, the Lagrange interpolation method provides the solution of this problem by the associate interpolation polynomial

$$f(x) \to L_n(x) = \sum_{k=0}^{n} p_k(x) y_k = p_0(x) y_0 + p_1(x) y_1 + ... + p_n(x) y_n$$

where $p_k(x)$ stands as parametric coefficients of expansion. The whole Lagrange n-th degree polynomial has to satisfy the constraints on which aim was introduced, namely:

$$L_n(x_i) = y_i \ , i = \overline{0,n}$$

that in the unfold writing reads

$$\begin{cases} p_0(x_0)y_0 + p_1(x_0)y_1 + \ldots + p_n(x_0)y_n = y_0 \\ p_0(x_1)y_0 + p_1(x_1)y_1 + \ldots + p_n(x_1)y_n = y_1 \\ \vdots \\ p_0(x_n)y_0 + p_1(x_n)y_1 + \ldots + p_n(x_n)y_n = y_n \end{cases}$$

leaving with the general condition for that this system is fulfilled:

$$p_k(x_i) = \delta_{ki} = \begin{cases} 1, \ i = k \\ 0, \ i \neq k \end{cases}$$

from where there is obvious that the general polynomial equation

$$p_k(x_i) = 0 \ , \forall x_i$$

has n-roots, i.e., for all $i \neq k = \underbrace{0,1,\ldots,n}_{n+1 \ indices}$ specializations. Therefore, the general p_k polynomial looks like:

$$p_k(x) = A_k\left(x - x_0\right)\left(x - x_1\right)\ldots\left(x - x_{k-1}\right)\left(x - x_{k+1}\right)\ldots\left(x - x_n\right)$$

where the constant A_k follows from the remaining polynomial constraint:

$$p_k(x_k) = 1 \Rightarrow A_k = \frac{1}{\left(x_k - x_0\right)\left(x_k - x_1\right)\ldots\left(x_k - x_{k-1}\right)\left(x_k - x_{k+1}\right)\ldots\left(x_k - x_n\right)}$$

providing the closed form of the *Lagrange interpolation* polynomial

$$L_n(x) = \sum_{k=0}^{n} \frac{\left(x - x_0\right)\left(x - x_1\right)\ldots\left(x - x_{k-1}\right)\left(x - x_{k+1}\right)\ldots\left(x - x_n\right)}{\left(x_k - x_0\right)\left(x_k - x_1\right)\ldots\left(x_k - x_{k-1}\right)\left(x_k - x_{k+1}\right)\ldots\left(x_k - x_n\right)} y_k$$

with the only restriction that the data points to be different:

$$x_0 \neq x_1 \neq x_2 \neq \ldots \neq x_n$$

In next we will consider the case where the data points are chosen as a net with equidistant nodes, with h-the net step:

$$x_{i+1} = x_i + h, \quad i = \overline{0,n}$$

from where follows the useful relationships:

$$x_k = x_0 + kh;$$

$$x_i - x_j = (i-j)h;$$

$$x - x_k = x - (x_0 + kh) = (x - x_0) - kh = (q-k)h$$

where we have introduced the notation:

$$q = \frac{x - x_0}{h} = \begin{cases} 0, & x = x_0 \\ \vdots \\ n, & x = x_0 \end{cases}$$

In this framework the Lagrange polynomial rewrites as:

$$L_n(q) = \sum_{k=0}^{n} \frac{(q-0)h(q-1)h...(q-(k-1))h(q-(k+1))h...(q-n)h}{(k-0)h(k-1)h...(k-(k-1))h(k-(k+1))h...(k-n)} \underbrace{\left[\frac{q-k}{q-k}\right]}_{additional} y_k$$

$$= \sum_{k=0}^{n} (-1)^{n-k} \frac{q(q-1)...(q-n)}{k!(n-k)!(q-k)} \underbrace{\left[\frac{n!}{n!}\right]}_{added} y_k$$

$$= (-1)^n \frac{q(q-1)...(q-n)}{n!} \sum_{k=0}^{n} (-1)^k \binom{n}{k} \frac{y_k}{(q-k)}$$

With the help of this form the numerical integrations for smooth functions may be undertaken by the approximation:

$$I = \int_a^b f(x)dx \cong \int_{a=x_0}^{b=x_n} L_n(x)dx = \frac{b-a}{n} \int_0^n L_n(q)dq$$

noting that for

$$dx = hdq$$

we have

$$x = a : q = \frac{a-a}{h} = 0$$

$$x = b : q = \frac{b-a}{h} = \frac{x_n - x_0}{h} = \frac{x_0 + nh - x_0}{h} = n$$

With the above equidistant Lagrange formula, the integral approximation explicitly casts in the so-called *Newton-Côtes numerical integration formula*:

$$I = \frac{b-a}{n}\sum_{k=0}^{n}\left[(-1)^{n-k}\binom{n}{k}\frac{1}{n!}\int_0^n \frac{q(q-1)...(q-n)}{(q-k)}dq\right]y_k = (b-a)\sum_{k=0}^{n} H_k y_k$$

with the coefficients:

$$H_k = \frac{(-1)^{n-k}}{n}\binom{n}{k}\frac{1}{n!}\int_0^n \frac{q(q-1)...(q-n)}{(q-k)}dq$$

and the rest (error of approximation):

$$R_{NC}\left(\int f\right) = \frac{\sup\limits_{q\in[0,n]}\left|f^{(n+1)}(q)\right|}{(n+1)!}\int_0^n q(q-1)...(q-n)dq$$

where

$$f(q) = hf(x) = hf(x_0 + hq)$$

Now, the application of this formula may be undertaken in various ways.

One is the so-called trapezoidal approximation and sees the equal division of the interval a, b as:

$$a = x_0 < x_1 < ... < x_{i-1} < x_i < ... < x_n = b,$$

$$h = x_i - x_{i-1} = \frac{b-a}{n}$$

in order to apply the *linear interpolation*, i.e., the above formula for each two consecutive pairs of points of whose prototype stand the first two:

$$I_T^{0\to1} = \int_{x_0}^{x_1} f(x)dx = h\sum_{k=0,1} H_{k,n=1}y_k = h\left(H_{0,n=1}y_0 + H_{1,n=1}y_1\right)$$
$$= h\left[\left(-\int_0^1 (q-1)dq\right)y_0 + \left(\int_0^1 qdq\right)y_1\right] = \frac{h}{2}(y_0 + y_1)$$

while in general will be:

$$I_T^{i-1\to i} = \int_{x_{i-1}}^{x_i} f(x)dx = \frac{h}{2}(y_{i-1} + y_i)$$

With this the overall trapezoidal formula sums for all such intervals (and integrals) to the final result:

$$I_T^{0\to n} = \int_a^b f(x)dx = \sum_{i=1}^{n}\int_{x_{i-1}}^{x_i} f(x)dx = \sum_{i=1}^{n} I_T^{i-1\to i}$$
$$= \frac{h}{2}(y_0 + y_1) + \frac{h}{2}(y_1 + y_2) + \ldots + \frac{h}{2}(y_{n-1} + y_n)$$
$$= h\left(\frac{y_0 + y_n}{2} + \sum_{i=1}^{n-1} y_i\right)$$

or rewritten in more familiar terms:

$$\int_a^b f(x)dx \underset{TRAPEZOID}{\approx} \frac{b-a}{n}\left[\frac{f(a = x_0) + f(b = x_n)}{2} + \sum_{i=1}^{n-1} f(x_i)\right]$$

Note that this formula provides quite good approximation if the function to be integrated is smoothly enough and the number of intervals is enough large as well. However, the error in the trapezoid approximation account for the fact that the interpolation is based on a single interval (so with n=1 I the above definition) and therefore reads for a single interval as:

$$R_T^{i-1\to i}(\int f) = \frac{\sup\limits_{q_i\in[i-1,i]}\left|f^{(2)}(q_i)\right|}{2!}\int_0^1 q(q-1)dq = -\frac{\sup\limits_{q_i\in[i-1,i]}\left|f^{(2)}(q_i)\right|}{12}$$

with the second order derivation being evaluated like:

$$f^{(2)}(q_i) = \frac{d^2}{dq_i^2}\left[hf\left(x_0 + hq_i\right)\right] = h^3 f^{(2)}\underbrace{(x_0 + hq_i)}_{x_i}$$

to provide the total error of the approximation

$$R_T^{0\to n}(\int f) = \sum_{i=1}^{n} R_T^{i-1\to i}(\int f) = -\frac{h^3}{12}\sum_{i=1}^{n}\sup_{\xi_i\in[x_{i-1},x_i]}\left|f^{(2)}(\xi_i)\right|$$

that, since recognizing we always can write

$$\exists\xi\in[a,b]\Big|\ \frac{1}{n}\sum_{i=1}^{n}\left|f^{(2)}(\xi_i)\right| = f^{(2)}(\xi)$$

such that to conclude with the error result:

$$\left|R_T^{[a,b]}(\int f)\right| \le n\frac{h^3}{12}\sup_{\xi\in[a,b]}\left|f^{(2)}(\xi)\right| = \frac{(b-a)^3}{12n^2}\sup_{\xi\in[a,b]}\left|f^{(2)}(\xi)\right|$$

telling us that the precision is as higher as the number of intervals n increases with the square power.

The second way of employing the Newton-Côtes numerical integration formula is to consider it with *quadratic interpolations* or with three points approximation; nevertheless, for that the interval $[a,b]$ is now more grainy divided, namely in $2n$ equal sub-intervals:

$$a = x_0 < x_1 < x_2 < \ldots < x_{2i} < x_{2i+1} < x_{2i+2} < \ldots < x_{2n} = b\,,$$

$$h = x_{j+1} - x_j = \frac{b-a}{2n},\ \ j = 0,1,\ldots,2n-1$$

With this approach the so-called *Simpson formula* is delivered; however, as before, firstly the "template" interval formula is considered, that now reads as

$$
\begin{aligned}
I_T^{0\to 2} &= \int_{x_0}^{x_2} f(x)dx = 2h\sum_{k=0,1,2} H_{k,n=2}y_k \\
&= 2h\left(H_{0,n=2}y_0 + H_{1,n=2}y_1 + H_{2,n=2}y_2\right) \\
&= 2h\begin{bmatrix} \left(\frac{1}{4}\int_0^2 (q-1)(q-2)dq\right)y_0 + \left(-\frac{1}{2}\int_0^2 q(q-2)dq\right)y_1 \\ +\left(\frac{1}{4}\int_0^2 q(q-1)dq\right)y_2 \end{bmatrix} \\
&= 2h\left(\frac{1}{6}y_0 + \frac{2}{3}y_1 + \frac{1}{6}y_2\right) = \frac{h}{3}\left(y_0 + 4y_1 + y_2\right)
\end{aligned}
$$

with the direct generalization

$$
I_S^{2i\to 2i+2} = \int_{x_{2i}}^{x_{2i+2}} f(x)dx = \frac{h}{3}\left(y_{2i} + 4y_{2i+1} + y_{2i+2}\right)
$$

and the final summed up approximation:

$$
\begin{aligned}
I_S^{0\to 2n} &= \int_a^b f(x)dx = \sum_{i=0}^{n-1}\int_{x_{2i}}^{x_{2i+2}} f(x)dx = \sum_{i=0}^{n-1} I_S^{2i\to 2i+2} \\
&= \frac{h}{3}\left(y_0 + 4y_1 + y_2\right) + \frac{h}{3}\left(y_2 + 4y_3 + y_4\right) + \ldots + \frac{h}{3}\left(y_{2n-2} + 4y_{2n-1} + y_{2n}\right) \\
&= \frac{h}{3}\left[y_0 + y_{2n} + 4\left(y_1 + y_3 + \ldots + y_{2n-1}\right) + 2\left(y_2 + y_4 + \ldots + y_{2n-2}\right)\right]
\end{aligned}
$$

allowing the general workable formulation:

$$
\int_a^b f(x)dx \underset{SIMPSON}{\approx} \frac{b-a}{6n}\begin{bmatrix} f(a = x_0) + f(b = x_{2n}) \\ +4\sum_{i=0}^{n-1} f\left(x_{2n+1}\right) + 2\sum_{i=1}^{n-1} f\left(x_{2n}\right) \end{bmatrix}
$$

The remaining issue regarding the appreciation of the error assumed by this model has to be here approached in different manner than the general recipe prescribes, since vanishing of the integral

$$\int_0^2 q(q-1)(q-2)dq = 0$$

Yet, the error is established on individual three-points interval through evaluating the difference:

$$R_S^{2i\to 2i+2} = \int_{x_{2i}}^{x_{2i+2}} ydx - \frac{h}{3}\left(y_{2i} + 4y_{2i+1} + y_{2i+2}\right)$$

equivalently rewritten as:

$$R_S^{j-h\to j+h}(h) = \int_{x_j-h}^{x_j+h} ydx - \frac{h}{3}\left[y\left(x_j-h\right)+4y\left(x_j\right)+y\left(x_j+h\right)\right],\quad j=2i+1$$

Now, this expression is successively differentiated respecting the h-step until a closely form is obtained, namely:

$$\begin{aligned}
&\partial_h R_S^{j-h\to j+h}(h)\\
&= y\left(x_j+h\right)+y\left(x_j-h\right)\\
&\quad -\frac{1}{3}\begin{bmatrix} y\left(x_j-h\right)+4y\left(x_j\right)\\ +y\left(x_j+h\right)\end{bmatrix} - \frac{h}{3}\left[-\partial_h y\left(x_j-h\right)+\partial_h y\left(x_j+h\right)\right]\\
&= \frac{2}{3}\begin{bmatrix} y\left(x_j+h\right)\\ +y\left(x_j-h\right)\end{bmatrix} - \frac{4}{3}y\left(x_j\right) - \frac{h}{3}\begin{bmatrix} -\partial_h y\left(x_j-h\right)\\ +\partial_h y\left(x_j+h\right)\end{bmatrix},\\
&\partial_h^2 R_S^{j-h\to j+h}(h)\\
&= \frac{1}{3}\left[-\partial_h y\left(x_j-h\right)+\partial_h y\left(x_j+h\right)\right] - \frac{h}{3}\left[\partial_h^2 y\left(x_j-h\right)+\partial_h^2 y\left(x_j+h\right)\right],\\
&\partial_h^3 R_S^{j-h\to j+h}(h)\\
&= -\frac{h}{3}\underbrace{\left[\partial_h^3 y\left(x_j+h\right)-\partial_h^3 y\left(x_j-h\right)\right]}_{\left[\left(x_j+h\right)-\left(x_j-h\right)\right]\partial_h^4 y\left(\xi_j\right)} = -\frac{2h^2}{3}\partial_h^4 y\left(\xi_j\right)\xi_j, \in\left(x_j-h, x_j+h\right)
\end{aligned}$$

Next, noting that

$$R_S^{j-h\to j+h}(h)=0,\quad \partial_h R_S^{j-h\to j+h}(h)=0,\quad \partial_h^2 R_S^{j-h\to j+h}(h)=0$$

one may undergo the reverse way in doing successive integration respecting h; so we have

$$\partial_h^2 R_S^{j-h\to j+h}(h) = \underbrace{\partial_h^2 R_S^{j-h\to j+h}(0)}_{0} + \int_0^h \partial_t^3 R_S^{j-h\to j+h}(t)dt = -\frac{2h^3}{9}\partial_h^4 y(\xi_j),$$

$$\partial_h R_S^{j-h\to j+h}(h) = \underbrace{\partial_h R_S^{j-h\to j+h}(0)}_{0} + \int_0^h \partial_t^2 R_S^{j-h\to j+h}(t)dt = -\frac{h^4}{18}\partial_h^4 y(\xi_j),$$

$$R_S^{j-h\to j+h}(h) = \underbrace{R_S^{j-h\to j+h}(0)}_{0} + \int_0^h \partial_t R_S^{j-h\to j+h}(t)dt = -\frac{h^5}{90}\partial_h^4 y(\xi_j)$$

With this, the error on the whole interval $[a,b]$ will be written summing the last estimation while returning to the initial indicial notations:

$$R_S^{[a,b]}(\int f) = -\frac{h^5}{90}\sum_{i=0}^{n-1}\partial_h^4 y(\xi_{2i}), \quad \xi_{2i} \in (x_{2i}, x_{2i+2})$$

that can be further resumed on the ground that for continuous 4-th time derivable function on $[a,b]$,

$$\exists \xi \in [a,b] \| \ \frac{1}{n}\sum_{i=0}^{n-1}\left|y^{(4)}(\xi_{2i})\right| = y^{(4)}(\xi)$$

to became

$$R_S^{[a,b]}(\int f) = -\frac{h^5}{90}ny^{(4)}(\xi) = -\frac{n}{90}\frac{(b-a)^5}{(2n)^5}y^{(4)}(\xi) = -\frac{(b-a)^5}{2880n^4}y^{(4)}(\xi)$$

so that

$$\left|R_S^{[a,b]}(\int f)\right| \le \frac{(b-a)^5}{180(2n)^4}\sup_{\xi\in[a,b]}\left|f^{(4)}(\xi)\right|$$

with the clear message that for the same intermediary points, the Simpson method lower the error considerably, in fact as much as with fourth power of the number of intervals $2n$ considered.

At this point one standard application regards the computation out of the error function:

$$\operatorname{erf}(x) = \frac{2}{\sqrt{\pi}} \int_0^x \exp\left(-t^2\right) dt$$

its value $\operatorname{erf}(1) = ?$ There is immediate that recognizing the function to integrate as

$$y = \exp\left(-t^2\right)$$

and by considering the interval $[0,1]$ divided in ten equidistant sub-intervals

$$a = 0 < 0.1 < 0.2 < \ldots < 0.9 < 1 = b$$

there is more than enough for a valuable estimation of the integral

$$\operatorname{erf}(1) = \frac{2}{\sqrt{\pi}} \int_0^1 \exp\left(-t^2\right) dt$$

Actually, one may easily found that

$$\sup_{\xi \in [0,1]} \left| y^{(2)}(\xi) \right|^{\xi=1} = 0.735759, \quad \sup_{\xi \in [0,1]} \left| y^{(4)}(\xi) \right|^{\xi=0} = 12$$

so that the respective errors in evaluating this integral with Trapezoid or Simpson formulas give:

$$\left| R_T^{[0,1]} \left(\operatorname{erf}(1) \right) \right| \leq 6.13132 \times 10^{-4}, \quad \left| R_S^{[0,1]} \left(\operatorname{erf}(1) \right) \right| \leq 6.66667 \times 10^{-6}$$

For immediate application of the Trapezoid and Simpson numerical integral formulas, one firstly calculates the function values in the considered points (nodes) of the integration interval, with the results:

$$\begin{gathered}
\{x_0 = 0 : f(x_0) = 1\}; \{x_1 = 0.1 : f(x_1) = 0.99005\}; \\
\{x_2 = 0.2 : f(x_2) = 0.960789\}; \{x_3 = 0.3 : f(x_3) = 0.913931\}; \\
\{x_4 = 0.4 : f(x_4) = 0.852144\}; \{x_5 = 0.5 : f(x_5) = 0.778801\}; \\
\{x_6 = 0.6 : f(x_6) = 0.697676\}; \{x_7 = 0.7 : f(x_7) = 0.612626\}; \\
\{x_8 = 0.8 : f(x_8) = 0.527292\}; \{x_9 = 0.9 : f(x_9) = 0.444858\}; \\
\{x_{10} = 1 : f(x_{10}) = 0.367879\}
\end{gathered}$$

providing the respective results for the integral:

$$\int_0^1 \exp\left(-t^2\right)dt \underset{TRAPEZOID}{\cong} 0.746211; \int_0^1 \exp\left(-t^2\right)dt \underset{SIMPSON}{\cong} 0.746825$$

and finally for the searched error function:

$$\operatorname{erf}(1) \underset{TRAPEZOID}{\cong} 0.842009; \operatorname{erf}(1) \underset{SIMPSON}{\cong} 0.842702$$

In the same manner other values of the error function may be provided; some results are presented in the Table A.4.1.

TABLE A.4.1 Common Values for the Error Function, $erf(x)$ Approximated by the Simpson Numerical Integration Method

x	**0.05**	**0.10**	**0.15**	**0.20**	**0.25**	**0.30**	**0.35**	**0.40**	**0.45**	**0.5**
$\operatorname{erf}(x)$	0.056	0.113	0.168	0.223	0.276	0.329	0.379	0.428	0.476	0.521
x	**0.55**	**0.60**	**0.65**	**0.70**	**0.75**	**0.80**	**0.85**	**0.90**	**0.95**	**1**
$\operatorname{erf}(x)$	0.563	0.604	0.642	0.678	0.701	0.742	0.771	0.797	0.821	0.843

A.5 GALILEAN SPACE-TIME TRANSFORMATIONS: THE MICHELSON EXPERIMENT

The classical (Galilei, Newton) approach of the mechanics of material bodies is based on several fundamental principles of time, space, mass, event, dynamics. They are best explained in the treatise *Philosophiae Naturalis Principia Mathematica (Mathematical Principles of Natural Philosophy)* by the following assertions (classical postulates, CP):

CP1: The time is absolute (it is an invariant): "*the absolute, true and mathematical time, of itself, and from its own nature, equably flows, without regard to anything external, and by another name is called duration*". The consequence of this postulate: the instantaneous propagation of light signals, which leads to absolute simultaneity, i.e., the simultaneous events in a system remain simultaneous in any other system, according to the absolute length which is conserved.

CP2: The space is absolute (it is an invariant) "*The absolute space, in its own nature, without regard to anything external, always remains*

similar and immovable". The consequences of this postulate: preserving the shape of a body between inertial reference systems (which is moving towards each another with constant speed), including angles, surfaces, and volumes.

CP3: The mass of a body is absolute (it is an invariant): it is constant to a solid observer, the body in question, as to the others observers, all well, found in inertial reference system (IRS) from them. As the result, the laws of (Newtonian) dynamics are conserved between inertial reference systems.

CP4: The occurrence of an event is unique, in any inertial reference system, in which it may be noticed. As a result: the transformation laws of space and time are linear (allow for unique solution).

Further, we will apply these principles/postulates for analytical determination of space-time transformations, based on the construction in Figure A.5.1.

From the last classical principle (CP4), for the M body in the Figure A.5.1 we have:

$$\begin{cases} x = f(x',t') = \alpha x' + \beta t' \\ y = y' \\ z = z' \\ t = g(x',t') = \gamma x' + \delta t' \end{cases}$$

Further, the constants $\alpha, \beta, \gamma, \delta$, will be determinated, applying **CP1-4** principles, in order to define the classical (also-called Galilei) transformations.

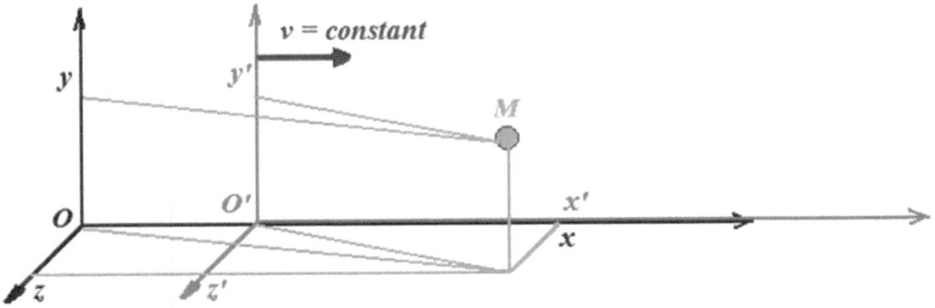

FIGURE A.5.1 Coordinates' representation of an object (mobile) M in two inertial reference systems, Oxyz and O'x'y'z', being the last in uniform rectilinear motion with constant velocity (v) from the first (Putz, 2010).

We will consider only the motion along the direction Ox//Ox', with immediate generalization; thus, we will analyze just the first and the last relation from the system above; by differentiating them, there is firstly obtained:

$$\begin{cases} dx = \alpha dx' + \beta dt' \\ dt = \gamma dx' + \delta dt' \end{cases}$$

then by making their ratio one has

$$v = \text{constant} = \frac{dx}{dt} = \frac{\alpha dx' + \beta dt'}{\gamma dx' + \delta dt'} = \frac{\alpha \frac{dx'}{dt'} + \beta}{\gamma \frac{dx'}{dt'} + \delta} = \frac{\alpha v' + \beta}{\gamma v' + \delta}$$

The last relationship actually gives the definition of the inertial reference system: if the M body moves within a reference system $O'x'$, at its turn moving at a constant speed (linear and uniform), v, respecting another reference system (Ox), then that body is moving with a constant speed (rectilinear and uniform) v' respecting the system $O'x'$, and can be considered as the origin ($M \equiv O''$) of a new reference system, so-called inertial. As an important consequence, it says that inertial systems are equivalent, in the classical sense.

Returning to the determination of the constants of actual movement, i.e., the parameters $\alpha, \beta, \gamma, \delta$, since the speed does not explicitly appear in the above relation of space-time transformations, we will consider it as a general default dependence

$$\begin{cases} x = \alpha(v)x' + \beta(v)t' \\ t = \gamma(v)x' + \delta(v)t' \end{cases}$$

which we will specialized for $M \equiv O' \Rightarrow M\ moves\ with\ v-velocity$, fulfilling the transformations

$$\begin{cases} x = \beta(v)t' \\ t = \delta(v)t' \end{cases} \Rightarrow \begin{cases} dx = \beta(v)dt' \\ dt = \delta(v)dt' \end{cases} \Rightarrow v = \frac{dx}{dt} = \frac{\beta(v)}{\delta(v)} \Rightarrow \beta(v) = v\delta(v)$$

The reference systems Ox and $O'x'$, being equivalent, the above f transformation relations should remain valid also for the reverse writing, when considering the velocity inversion

$$v \to -v \Rightarrow \beta(-v) = -v\delta(-v)$$

We have to remember these relationships and also continue to use the *time as invariant* of classical motion, as explained in above **CP1** principle. Thus, the relationship

$$t = t'$$

can be equivalently considered in each inertial systems in question, obtaining respectively

$$t = \gamma(v)x' + \delta(v)t' = t' \Rightarrow \begin{cases} \delta(v) = 1 \\ \gamma(v) = 0 \end{cases}$$

$$t' = \gamma(-v)x + \delta(-v)t = t \Rightarrow \begin{cases} \delta(-v) = 1 \\ \gamma(-v) = 0 \end{cases}$$

The latest results, corroborated with the previous ones, allow the rewriting of the space-time transformation relations as follow

$$Observer\ \ in\ O: \begin{cases} x = \alpha(v)x' + vt' \\ t = t' \end{cases}$$

$$Observer\ \ in\ \ O': \begin{cases} x' = \alpha(-v)x - vt \\ t' = t \end{cases}$$

By mutual replacing of spatial transformations between the (equivalent) inertial referential systems, one successively obtains

$$\begin{aligned} x &= \alpha(v)x' + vt' = \alpha(v)\left[\alpha(-v)x - vt\right] + vt \\ &= \alpha(v)\alpha(-v)x - \alpha(v)vt + vt; \end{aligned}$$

$$x = \alpha(v)\alpha(-v)x - vt\left[1 - \alpha(v)\right] \Rightarrow \begin{cases} \alpha(v)\alpha(-v) = 1 \\ 1 - \alpha(v) = 0 \end{cases} \Rightarrow \begin{cases} \alpha(-v) = 1 \\ \alpha(v) = 1 \end{cases}$$

which allows the complete writing of Galilean transformations

$$in\ O:\begin{cases} x = x' + vt' \\ y = y' \\ z = z' \\ t = t' \end{cases}, \quad in\ O':\begin{cases} x' = x - vt \\ y' = y \\ z' = z \\ t' = t \end{cases}$$

Remarkably, if passing from the coordinates to velocities, and then to accelerations, we obtain

$$\dot{x} = \dot{x}' + v \Rightarrow \ddot{x} = \ddot{x}' \Rightarrow m\ddot{x} = m\ddot{x}' \Leftrightarrow F = F'$$

where, by applying the postulate of the invariance mass **CP3**, on the multiplication of the body mass in question, we get the conservation of the Newton's 2nd principle, revealed by equalizing the measured forces in the two inertial reference systems. From here, there results that the entire Newtonian dynamics is conserved under the Galilean transformations between inertial reference systems (and for the observers attached to them).

The crisis of this approach of the Universe, came one with the analytical explanation of the electromagnetic field equations (Maxwell theory) that does not seem to remain invariant between inertial reference systems (IRS), such as is the Newtonian dynamics of material bodies. Without going into the electrodynamics' details, we can still notice that the Galilean transformations have an inconsistency during the identical transformation of temporal derivatives, *per se* admitted as being rightful done in above temporal derivative, when it was concluded the equivalence of Newtonian dynamics in all IRS. To highlight this limitation, we rewrite the Galilei transformations (reverse) into the so-called general covariance

$$\begin{cases} x'_\alpha = x_\alpha - v_\alpha t_\alpha \\ t'_\alpha = t_\alpha \end{cases}$$

with indices of Cartesian directions $\alpha = 1(Ox), 2(Oy), 3(Oz)$. When considering the delta- Kronecker tensor definition

$$\delta_{\alpha\beta} = \begin{cases} 1 \ldots \alpha = \beta \\ 0 \ldots \alpha \neq \beta \end{cases}$$

there can be confirmed the invariance nature of the spatial derivate

$$\partial_x = \frac{\partial}{\partial x_\alpha} = \frac{\partial x'_\beta}{\partial x_\alpha}\frac{\partial}{\partial x'_\beta} = \frac{\partial\left(x_\beta - v_\beta t_\beta\right)}{\partial x_\alpha}\frac{\partial}{\partial x'_\beta}$$

$$= \frac{\partial x_\beta}{\partial x_\alpha}\frac{\partial}{\partial x'_\beta} = \delta_{\alpha\beta}\frac{\partial}{\partial x'_\beta} = \frac{\partial}{\partial x'_\alpha} = \partial'_x$$

but not for the temporal one since the succession of equivalences

$$\partial_t = \frac{\partial}{\partial t_\alpha} = \frac{\partial t'_\beta}{\partial t_\alpha}\frac{\partial}{\partial t'_\beta} - \frac{\partial x'_\beta}{\partial t_\alpha}\frac{\partial}{\partial x'_\beta}$$

$$= \underbrace{\frac{\partial t_\beta}{\partial t_\alpha}}_{\delta_{\alpha\beta}}\frac{\partial}{\partial t'_\beta} - \frac{\partial\left(x_\beta - v_\beta t_\beta\right)}{\partial t_\alpha}\frac{\partial}{\partial x'_\beta}$$

$$= \delta_{\alpha\beta}\frac{\partial}{\partial t'_\beta} + v_\beta\frac{\partial t_\beta}{\partial t_\alpha}\frac{\partial}{\partial x'_\beta} = \delta_{\alpha\beta}\frac{\partial}{\partial t'_\beta} + v_\beta\delta_{\alpha\beta}\frac{\partial}{\partial x'_\beta}$$

$$= \frac{\partial}{\partial t'_\alpha} + v_\alpha\frac{\partial}{\partial x'_\alpha} = \partial'_t + v\partial'_x$$

This "weakness" of Galilean transformations, which are not invariant to the temporal derivative transformations, will be further remedied by introducing the so-called quadric-vector, in order to combine the temporal evolution with the spatial one (within the so-called Minkowski Universe- see below).

Also, we will highlight the "crisis" of Galilean transformations from the perspective of the consequences of Maxwell's theory of the propagation of light, regarding the nature of speed of light; from the electromagnetic field theory one considers the propagation of light as having finite (non-infinite) velocity: actually from the wave equation ensues the necessity of a finite speed of propagation, denoted by "*c*", for both the electric (**E**) and magnetic (**B**) fields

$$\left(\nabla^2 - \frac{1}{c^2}\frac{\partial^2}{\partial t^2}\right)\mathbf{E} = 0, \quad \left(\nabla^2 - \frac{1}{c^2}\frac{\partial^2}{\partial t^2}\right)\mathbf{B} = 0$$

Otherwise, the electromagnetic wave propagation itself would be compromised, with the all arsenal of the experimental undisputed facts!

This hypothesis was checked by Michelson, in a famous experiment, described in Figure A.5.2 and also with a detail in Figure A.5.3. Note that Michelson, by the arrangement from Figure A.5.2, wanted to demonstrate that the value of the speed of light does not affect the times in which the initial beam from the source S (after crossing two equal optical paths) reaches the detector at the same time:

$$t_1 = t_2$$

by the hypothesis the universal ether's existence (considered mobile, although with material, continuous and elastic medium properties, and existing everywhere in the Universe) supports this motion, respectively the propagation of information.

Excluding the common paths of light, from the S (source) to O(1,2) (semi-reflective mirrors) to the detector, the times on different parts O-A-O and O-B-O will be calculated taking into account the parallel propagation and, respectively, the transversal one between the light front and the speed

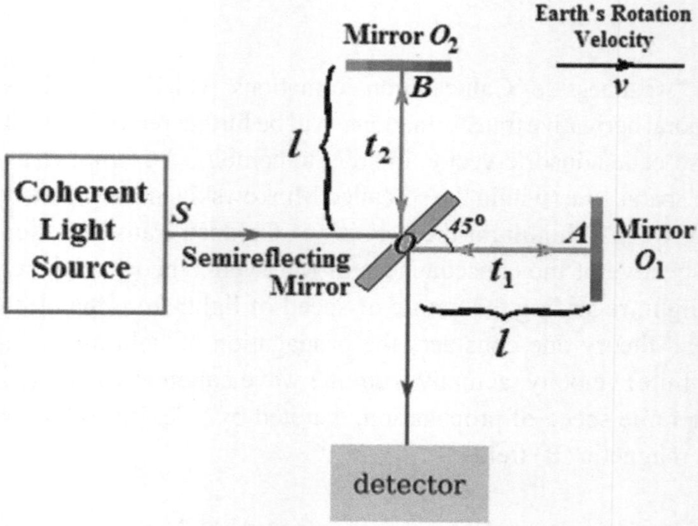

FIGURE A.5.2 Michelson experiment scheme (1881) to demonstrate the existence of the universal ether; adapted from (HyperPhysics, 2010; Putz, 2010).

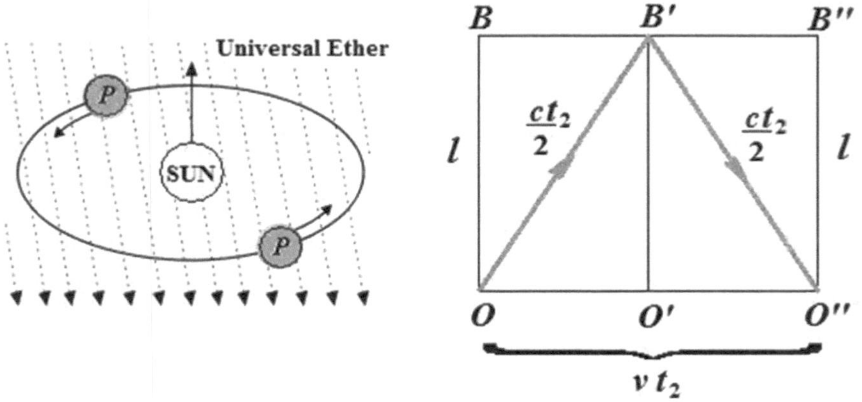

FIGURE A.5.3 Details for Michelson experiment from Figure A.5.2: left - the hypothesis of the existence of the universal ether, as supporting the propagation of light and the interaction at a distance; right: the path of light front (molded on the universal ether) reflecting by the mirror O2 from Figure A.5.2, taking into account the rotation of the Earth at a constant speed *v*; adapted from (HyperPhysics, 2010; Putz, 2010).

of the Earth's rotation, which is considered constant *v*. Then, on the direction parallel to the Earth's rotation, we have the time to go through

$$t_1 = \frac{l}{\underbrace{c+v}_{O-A}} + \frac{l}{\underbrace{c-v}_{A-O}} = \frac{2lc}{c^2 - v^2}$$

For the second path of the light, on the transversal direction front respecting the direction of Earth's rotation, as is considered in the Figure A.5.3 (right side), we write the identity

$$c^2 t_2^2 = 4l^2 + v^2 t_2^2$$

wherefrom immediately results the second searched time

$$t_2 = \frac{2l}{\sqrt{c^2 - v^2}}$$

The ratio of the two durations has the expression

$$\frac{t_1}{t_2} = \frac{c}{\sqrt{c^2 - v^2}}$$

which becomes *unity* in three instances:

1. The Earth is immobile ($v = 0$), a non-realistic situation;
2. The speed of light is infinite ($c \to \infty$), which would mean that also $t_1 = t_2 = 0$, in agreement with the instantaneous propagation at a distance, provided by the Newtonian description of Nature, but in disagreement with the electromagnetic description of light with finite speed, according to Maxwell equations (above) - so unlikely hypothesis too;
3. The speed of light is finite, but for the parallel propagation of light with the Earth's rotation speed the shrinkage of distance traveled occurs with the so-called Lorentz-Fitzgerald factor (longitudinal contraction factor)

$$\sqrt{1-\frac{v^2}{c^2}}$$

so that the real time to go through on the route O-A-O actually looks like

$$t_1 = \underbrace{\frac{l\sqrt{1-v^2/c^2}}{c+v}}_{O-A} + \underbrace{\frac{l\sqrt{1-v^2/c^2}}{c-v}}_{A-O} = \frac{2lc\sqrt{1-v^2/c^2}}{c^2-v^2} = \frac{2l}{\sqrt{c^2-v^2}}$$

hence automatically resulting in the report $t_1 / t_2 = 1$, without any additional assumptions concerning the numerical value of the speed of light or of the speed of the Earth!

The introduction of the longitudinal contraction factor in present exposition raises a conceptual problem demanding its clarification in a consistent manner. This will be analyzed as follows!

A.6 LORENTZIAN SPACE-TIME TRANSFORMATIONS: THE MINKOWSKI UNIVERSE

The necessity of reconsideration of Galilei transformations was above exposed, with the opportunity the non-invariance time derivation of these transformations was highlighted. In addition, it was introduced the idea that light propagates at finite speed (by Maxwell – in theoretical way, in 1873, through the publication of his great *Treatise on Electricity and Magnetism*, as a “response” to the experiment carried out by Michelson, in 1881, and not only). Under these conditions, Einstein (in 1905) reformulates the classical postulates of Newtonian mechanics, which will be called special postulates (SP) of space-time transformation of physical laws; there are two postulates: the first refers to the necessity that the laws of Nature have the same forms between IRS, while the second refers to the invariance of light speed for the observations from various IRS (inertial reference systems). In Einstein’s terms, we have the postulates/principles of special relativity formulated as:

SP1: “The laws by which the laws of physical system are changed do not depend on the choice of the coordinate system at which those changes are reported, from the set of reference systems in uniform translation of each other.”

SP2: “Each ray of light moves in the coordinate system <<at rest>> with the definite velocity *c*, independent of whether it is emitted by a body at rest or a body in motion.”

With these principles, we will reload the Galilean transformation from the point in which we can write the general relation

$$x' = \alpha x + \beta t$$

that we particularize (as in the Galilean case of movement) for $M \equiv O' \Rightarrow M \text{ has the velocity } v$ and thus $x = vt$, resulting in

$$0 = \alpha vt + \beta t \Rightarrow \beta = -\alpha v$$

with consequence in the rewriting of the reciprocal transformations of the space as

$$x' = \alpha\left(x - vt\right)$$

$$x = \alpha\left(x' + vt'\right)$$

thus already customaring of the IRS equivalence asserted by SP1. The last two relationships can be combined so that we can successively write the time in the system O'

$$t' = \frac{x - \alpha x'}{\alpha v} = \frac{x - \alpha^2 (x - vt)}{\alpha v} = \alpha t - \frac{\alpha^2 - 1}{\alpha v} x$$

Now, the second postulate of special relativity, written at the level of the invariance of speed of light, in any IRS is enforced

$$c = \frac{x}{t} = \frac{x'}{t'}$$

to allow the appropriate replacement in (above presented) special (reciprocal) transformations in expressing t' in two equivalent ways by the system of equations

$$\begin{cases} ct' = \alpha (c - v) t \\ t' = \alpha t - \dfrac{\alpha^2 - 1}{\alpha v} ct \end{cases}$$

This system is solved (for instance, by forming the ratio of component equations) for the search parameter

$$\alpha = \frac{1}{\sqrt{1 - \frac{v^2}{c^2}}}$$

With this, there can be immediately written the coordinate relativistic transformation to be

$$x' = \frac{x - vt}{\sqrt{1 - \frac{v^2}{c^2}}}$$

but also for the time "coordinate" transformation

$$t' = \frac{t - \frac{v}{c^2} x}{\sqrt{1 - \frac{v^2}{c^2}}}$$

relations which are acknowledged as the *Lorentz transformations*.

Next, let's explore the consequences of these new relations of space and time relativistic transformations. If we rewrite (as above) the Lorentz transformation in a covariant form

$$\begin{cases} x'_\alpha = \alpha\left(x_\alpha - v_\alpha t_\alpha\right) \\ t'_\alpha = \alpha\left(t_\alpha - v x_\alpha / c^2\right) \end{cases}$$

we can condense them into a space-time continuum (to so-called Minkowski Universe) offered by the generalized Lorentz transformations in covariant form (in 4D)

$$x'^i = \Lambda^i{}_j x^j$$

with the differential form

$$dx'^i = \Lambda^i{}_j dx^j$$

$\Lambda^i{}_j$ is written with the indices $i, j = 0, \{a, b, c\}$ thus resulting in the 4D information embodying 4x4 matrices which mix the time-space information by indices using Latin letters:

- when $i, j = 0 \rightarrow$ temporal coordinate
- when $i, j = a(Ox), b(Oy), c(Oz) \rightarrow$ spatial coordinate

The condensed covariant/anti-covariant writing is called as Grosmann-Einstein type and roughly shows all the natural information from a spatial-temporal transformation in the generalized sense (special relativistic). Let's see what advantages we may have from this writing. First of all, one observes how from a "single one shot" the invariance problem from spatial-temporal derivatives (specific for Galilei transformations) is immediately solved

$$\partial'_i = \frac{\partial}{\partial x'^i} = \frac{\partial x^j}{\partial x'^i}\frac{\partial}{\partial x^j} = \Lambda_i{}^j \partial_j$$

showcasting the same transformation law, as coordinates itself. There can be noted that both the anticovariant indexing rule was considered for the denominator coordinates, in order to obtain a unitary writing (formula), very useful for related analysis, see also in the following.

Accordingly, the so-called spatial-temporal interval (quadratic) Minkowski can be formed,

$$ds^2 = g_{kp} dx^k dx^p$$

written like a generalization of the Euler metric, denoted by the tensor g_{ij} (called as the space-time Lorentz metric). The point is that the principles **CP1** and **CP2** must be now combined, through this spatial-temporal common metric, in such way the invariance of Minkowski's interval to be produced, according the above Lorenz transformations. Further, we are going to explore this compatibility. Firstly, let's notice that the Minkovski invariance

$$ds'^2 = g_{ij} dx'^i dx'^j = g_{ij} \Lambda^i{}_k \Lambda^j{}_p dx^k dx^p = g_{kp} dx^k dx^p = ds^2$$

implies the existence of Minkowski equality

$$g_{ij} \Lambda^i{}_k \Lambda^j{}_p = g_{kp}$$

leaving with the matrix transcript

$$\left(\Lambda^T\right)(g)(\Lambda) = (g)$$

with the consequence that

$$\left[\det(\Lambda)\right]^2 = 1 \Rightarrow \det(\Lambda) = \pm 1$$

This relationships, being the consequences of the invariance of the Minkowski interval, let's check whether such construct is in agreement with Lorentz transformations. Thereby, we firstly consider the quadrivector form in the Minkowski Universe, in the covariant variants (indices above) and anticovariant (indices bottom), respectively

$$x^i = \left(ct, x^1, x^2, x^3\right),\ x_i = \left(ct, -x_1, -x_2, -x_3\right)$$

and the same for the differential (infinitesimal) distance quadrivectors

$$dx^i = \left(dx^0, dx^\alpha\right),\ dx_i = \left(dx_0, -dx_\alpha\right)$$

Next, the pseudo-Euclidean metric $(+,-,-,-)$ is considered just to prevent the Galilei transformation non-invariance at the generalized derivative operations in the continuous Minkowski spatial-temporal Universe that is having the metric

$$g_{ij} = \begin{pmatrix} g_{00} & 0 & 0 & 0 \\ 0 & g_{11} & 0 & 0 \\ 0 & 0 & g_{22} & 0 \\ 0 & 0 & 0 & g_{33} \end{pmatrix} = \begin{pmatrix} +1 & 0 & 0 & 0 \\ 0 & -1 & 0 & 0 \\ 0 & 0 & -1 & 0 \\ 0 & 0 & 0 & -1 \end{pmatrix}$$

Under these conditions the quadratic Minkowski interval merely can be written

$$ds^2 = c^2dt^2 - \left(dx^1\right)^2 - \left(dx^2\right)^2 - \left(dx^3\right)^2 = c^2dt^2 - d\mathbf{r}^2$$

while the square length of the positional quadrivetors and also of the partial derivatives, also unfold as

$$x^i x_i = c^2t^2 - \mathbf{r}^2, \; \partial_i \partial^i = \partial_0^2 - \vec{\partial}^2$$

We can check now the Lorentz transformations, based on these rational assumptions. One employs the following particular cases

$$\begin{cases} dx'^i = \Lambda^i{}_0 c dt \\ dt' = \Lambda^0{}_0 dt \end{cases}$$

whose ratio generates the equality

$$-v^i = \frac{dx'^i}{dt'} = c\frac{\Lambda^i{}_0}{\Lambda^0{}_0} \Rightarrow \Lambda^i{}_0 = -\frac{1}{c}v^i \Lambda^0{}_0$$

The last relationship has to be completed with the temporal component of the Minkovski 4D tensor through the appropriate particularization of Minkowski equality

$$g_{00} = g_{ij}\Lambda^i{}_0 \Lambda^j{}_0 \Leftrightarrow 1 = \left(\Lambda^0{}_0\right)^2 - \sum_i \left(\Lambda^i{}_0\right)^2$$

rewritten under the equivalent equation

$$1=\left(\Lambda^0{}_0\right)^2-\frac{1}{c^2}\left(\Lambda^0{}_0\right)^2\underbrace{\sum_i\left(v^i\right)^2}_{v^2}$$

with the immediate solution

$$\Lambda^0{}_0=\frac{1}{\sqrt{1-v^2/c^2}}$$

which implies the immediate knowledge of the remaining components

$$\Lambda^i{}_0=\Lambda^0{}_i=-\frac{1}{c}v^i\Lambda^0{}_0=\frac{-v^i/c}{\sqrt{1-v^2/c^2}}$$

Further, we will analyze the situation from Figure A.5.1, for which the 3D vector of velocity has the components $\mathbf{v}=(v,0,0)$. Although non-unique in general, there is convenient to choose the sub-matrix $\Lambda^\alpha{}_\beta$, $\alpha,\beta=\{1,2,3\}$, in such way that together with the above (time-time and space-time) components to generate Lorentz transformations as phenomenologically deduced; this form can be

$$\Lambda^\alpha{}_\beta=\delta_{\alpha\beta}+\frac{v_\alpha v_\beta}{v^2}\left(\frac{1}{\sqrt{1-v^2/c^2}}-1\right)$$

for which one can verify the space transformation relativistic law

$$\begin{aligned}x'^1&=x'=\Lambda^1{}_jx^j\\&=\Lambda^1{}_0x^0+\sum_\alpha\Lambda^1{}_\alpha x^\alpha\\&=\frac{-v^1/c}{\sqrt{1-v^2/c^2}}ct+\sum_{\alpha=1}^{3}\left[\delta_{1\alpha}+\frac{v_1v_\alpha}{v^2}\left(\frac{1}{\sqrt{1-v^2/c^2}}-1\right)\right]x^\alpha\end{aligned}$$

$$= \frac{-vt}{\sqrt{1-v^2/c^2}} + x + \frac{x\left(1-\sqrt{1-v^2/c^2}\right)}{\sqrt{1-v^2/c^2}}$$
$$= \frac{x-vt}{\sqrt{1-v^2/c^2}}$$

along that of the relativistic time transformation

$$x'^0 = ct' = \Lambda^0{}_j x^j$$
$$= \Lambda^0{}_0 x^0 + \sum_\alpha \Lambda^0{}_\alpha x^\alpha$$
$$= \frac{ct}{\sqrt{1-v^2/c^2}} + \underbrace{\Lambda^0{}_1}_{\neq 0\left(v^1=v\right)} x^1 + \underbrace{\Lambda^0{}_2}_{=0\left(v^2=0\right)} x^2 + \underbrace{\Lambda^0{}_3}_{=0\left(v^3=0\right)} x^3$$
$$= \frac{ct}{\sqrt{1-v^2/c^2}} + \frac{-v/c}{\sqrt{1-v^2/c^2}} x$$
$$\Rightarrow t' = \frac{t-vx/c^2}{\sqrt{1-v^2/c^2}}$$

both resulting in agreement with previous Lorentz determinations.

Returning to the phenomenological approach, beyond the matrix and tensor formalism just sketched, several fundamental conclusions can be expressed, seen as consequences of the postulates of special relativity (Einstein), namely:

- Minkowski space-time Universe is a reality, just like the quadrivectors; in Minkowski's terms, expressed at a conference in Cologne on September 21 (1908). "*The views on space and time that I want to develop for you were born on the experimental-physical ground. In this lays their strength. Their tendency is to give radical demonstrations. From now on, the space for itself should completely disappear in shadow, considering only the existence of an association of the two.* "
- The interval (square distance) Minkowski is a size invariant at the 4D transformations, actually bringing together the separate

Galilean-Newtonian postulates of space and time, in a single absolute space-time continuum. Therefore, this universal value is also-called the *Universe line*

$$ds^2 = c^2dt^2 - dl^2$$

The *universe line* has different forms of achievement that limits the spatial and temporal knowledge, fixed by the Observer placed into the origin of time-space ($ds^2 = 0$), with such information confined inside the cone of light perceiving between past and future events, Figures A.6.1 and A.6.2.

The consequences of spatial-temporal knowledge, derived from an event location on the Universe line, realizes the manifested relative (relativity) knowledge. The general rule is that for two IRS, with an event measured in both of them, the so-called proper/self system is the system where the event was produced, while the system in which it is measured/received by the transmitted signal is denoted as the observer system. With them, the panel of possibilities for proper system vs. observer system for space and time, can be composed, see Table A.6.1.

From the Table A.6.1, the fundamental ideas related to the relative variation of the distance and duration are deduced; remarkably, while two separate space events in their IRS are measured in another IRS, mobile to

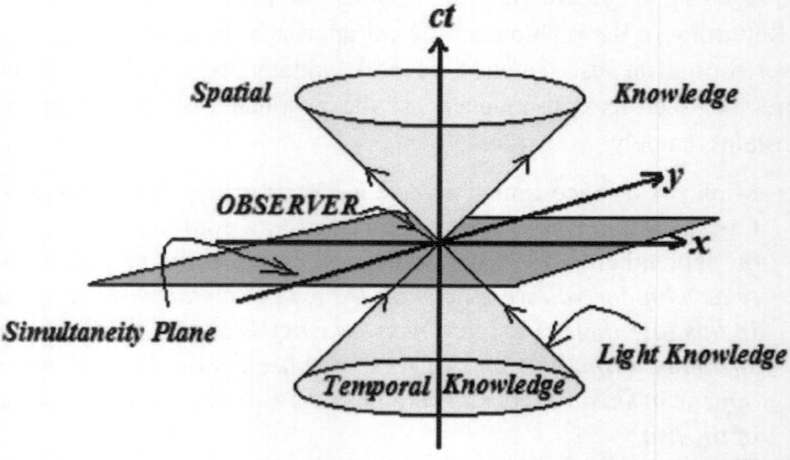

FIGURE A.6.1 Light cone representation and types of knowledge related to the Minkowski Universe; adapted from (HyperPhysics, 2010; Putz, 2010).

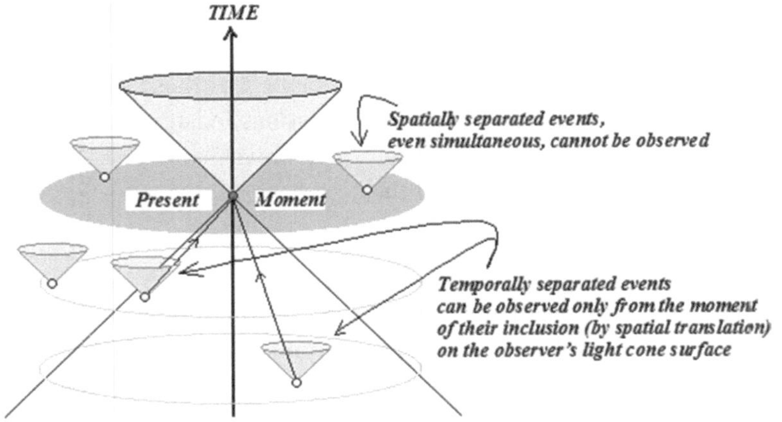

FIGURE A.6.2 The interpretation of knowledge of separate events, in space or time, based on light cone applied on the observer (located at the origin of the present moment); adapted from (HyperPhysics, 2010; Putz, 2010).

TABLE A.6.1 The Relative Knowledge of the Distance and the Duration in Special Relativity (Based on Lorentz Transformations) in Minkovski's Universe among the inertial referential systems (IRS) events (Putz, 2010)

Measured size	*IRS Event*	*Event characterization*	*Equation for working*	*Relative reception*
DISTANCE	*Events in IRS-O':* x'_1, x'_2	*Simultaneously measured in IRS-O:* $t_1 = t_2$	$x' = \frac{x - vt}{\sqrt{1 - v^2 / c^2}}$	$\underbrace{\Delta x'}_{self-IRS} = \frac{\Delta x}{\sqrt{1 - v^2 / c^2}}$
	Events in IRS-O: x_1, x_2	*Simultaneously measured in IRS -O':* $t'_1 = t'_2$	$x = \frac{x' + vt'}{\sqrt{1 - v^2 / c^2}}$	$\underbrace{\Delta x}_{self-IRS} = \frac{\Delta x'}{\sqrt{1 - v^2 / c^2}}$
DURATION	*Events in IRS-O':* t'_1, t'_2	*Collocated in IRS-O':* $x'_1 = x'_2$	$t = \frac{t' + vx'/c^2}{\sqrt{1 - v^2 / c^2}}$	$\underbrace{\Delta t'}_{self-IRS} = \Delta t \sqrt{1 - v^2 / c^2}$
	Events in IRS-O: t_1, t_2	*Collocated in IRS-O:* $x_1 = x_2$	$t' = \frac{t - vx / c^2}{\sqrt{1 - v^2 / c^2}}$	$\underbrace{\Delta t}_{self-IRS} = \Delta t' \sqrt{1 - v^2 / c^2}$

the proper, they cannot be simultaneously observed in their own systems (to each other), while temporal events from IRS can be collocated by reception in the same system. The representation of these ideas is shown in Figure A.6.2 in the light cone representation, located on an observer (proper IRS) at the present moment in Minkowski universe. Moreover, from the Table A.6.1 there is noted that:

- The length of an object, measured in the direction of mobile system movement is maximum in the proper system of the object; here worth to recall that the effect of contraction, manifested in the observer system (not the proper one) is in agreement with the Lorentz-FitzGerald longitudinal contraction, previously related with the temporal paradox likely to solve the optical path in Michelson experiment (see the Section A.5).
- The duration for the development of a process is minimal in its proper system, in front of which the phenomenon (the event) is at rest.

At the end of this section, there is worth to be mentioning the relativist verses of Romanian Literature in two ways: those related of the relative length were admirably synthesized by Tudor Arghezi (1980-1967) in the poetry *The separation (Despărțirea,* in Romanian*)* (of the volume *Suitable Words*):

... When I left, a pendulum was slowly bitten in the fog,
So scanty that hour passed over the time...

Instead, the lyrics about space and its traveling by the finite, delayed light (propagated with the speed of light) are brilliantly embodied by the dammed poet Mihai Eminescu (1850–1889) in the mythical poem *Toward the star/ Ad Astra (La Steaua,* in Romanian) (Romanian Voice, 2012), written in 1886 (!) long before the formulation of postulates of special relativity theory by Einstein, in 1905:

Behold, that star that's shining
up there, so far away;
Her light has traveled on eonsin
reaching our eyes today.

Perhaps it died while on its way
through infinite dark spaces.
Yet only now does its light stray
to shine upon your faces.

The icon of the fading star
slow rises in the sky.
She was entire, while we didn't see her fly.
Despite what we observe today, she's dying.

A.7 DYNAMICAL CONSEQUENCES OF SPECIAL RELATIVITY: EINSTEIN'S MASS AND ENERGY

There is commonly accepted that the dynamic properties of a system may be elegantly determined as based on the associated Lagrangian (L); at its turn, the Lagrangian of a system determines the so-called action of the system defined by the integral

$$S = \int_a^b L dt$$

For the relativistic systems, even in the absence of applied forces, the time evolution is intrinsically accounted by the inner universe line linking the temporal transformation respecting the proper time (with the considered innertially system)

$$ds = c \underbrace{dt\sqrt{1 - v^2 / c^2}}_{PROPER\ TIME}$$

so that the physical action takes the "observed" form

$$S = -\alpha c \int_a^b \sqrt{1 - v^2 / c^2}\, dt$$

The minus sign is added for illustrating the loss of action in the environment by time passing, yet, even without having the minus sign explicitly here it will be reached in the final expression of the α constant to be determined upon the following procedure. Firstly one recognizes from the last relation the working relativistic Lagrangian

$$L_{relativist} = -\alpha c\sqrt{1-\frac{v^2}{c^2}}$$

Under the absence of the external forces (equivalently with free motion anyway) the above Lagrangian has to recover in the non-relativistic limit $v << c$ the classical Lagrangian limit (the kinetic energy in fact) of the free particle

$$L_{classic} = \frac{m_0 v^2}{2}$$

Accordingly the relativistic Lagrangian is Taylor expanded in the first order of the moving velocity respecting the velocity of light and eventually equated with its classical version

$$L_{relativist} = -\alpha c\left(1-\frac{v^2}{c^2}\right)^{1/2} \underset{v<<c}{\cong} -\alpha c\left(1-\frac{1}{2}\frac{v^2}{c^2}\right)$$

$$= \underbrace{-\alpha c}_{\substack{CONSTANT\\POTENTIAL}} + \frac{\alpha}{2}\frac{v^2}{c} \cong \frac{\alpha}{2}\frac{v^2}{c} \rightarrow \frac{m_0 v^2}{2} = L_{classic}$$

To obtain the searched factorization constant

$$\alpha = m_0 c$$

and the relativistic Lagrangian

$$L_{relativist} = -m_0 c^2\sqrt{1-\frac{v^2}{c^2}}$$

With the Lagrangian form known one can unfold the relativistic dynamics according with the analytical mechanics principles; actually, for the momentum one uses the form of the conjugated canonical momentum, widely checked in the classical framework with the analytical Lagrangian formulation

$$p_{classic} = \frac{\partial L_{classic}}{\partial v} = \frac{\partial}{\partial v}\left[\frac{m_0 v^2}{2} - V(q,t)\right] = m_0 v$$

to provide now at the relativistic level the successive forms

$$p_{relativist} = \frac{\partial L_{relativist}}{\partial v}$$

$$= -m_0 c^2 \frac{\partial}{\partial v}\left(1 - \frac{v^2}{c^2}\right)^{1/2} = -m_0 c^2 \frac{1}{2}\left(1 - \frac{v^2}{c^2}\right)^{-1/2} \frac{1}{c^2}(-1)2v$$

$$= \frac{m_0 v}{\sqrt{1 - \frac{v^2}{c^2}}} = m_d v$$

From where one identifies the celebrated dynamic relativistic mass

$$m_d = \frac{m_0}{\sqrt{1 - \frac{v^2}{c^2}}}$$

Worth quoting on the experimental reliability of this fundamental expression of Nature by the cathodic tube of Nacken (Figure A.7.1) who in 1935 adjusted the electromagnetic fields by collecting-deflecting bobbins such that to produce the deflecting trajectories of electrons (the cathodic rays) such that upon their unification they are accelerated with various velocities up to v/c=0.7! The observed dynamical masses were calculated from the ratio of the applied fields with the results listed in Table A.7.1 with an excellent agreement between theoretical and experimental data – this way validating the basic dynamical consequences of special relativity theory.

Going now to the total energy calculation for the relativistic free system, one employs the Hamiltonian relationship with Lagrangian up on the general form (see also Section 2.2), with the velocity $\dot{q} = v$

$$H(t,q,p) = p\dot{q} - L(t,q,\dot{q})$$

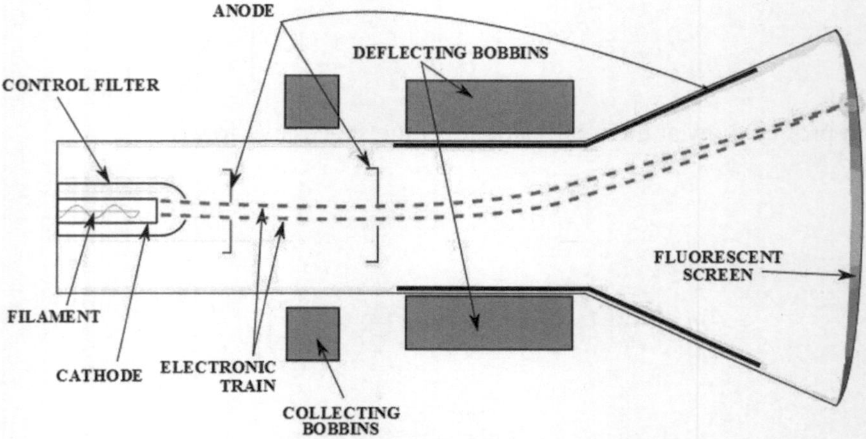

FIGURE A.7.1 Schematic representation of a cathodic tube for obtaining the accelerated electrons as cathodic rays: adapted after (Wikimedia, 2010; Putz, 2010).

TABLE A.7.1 Measured vs. Calculated Values for the Electronic Dynamic Masses Recorded for the Cathodic Rays of Figure A.7.1 (Nacken, 1935; Putz, 2010)

v/c	$(m/m_0)_{OBS}$	$[(m/m_0)_{CAL}]/[(m/m_0)_{OBS}]$
0.630	1.175	0.988
0.643	1.240	0.988
0.695	1.395	0.988

As previous was the case with momentum analytical definition, we firstly check the classical reliability of this Hamiltonian-Lagrangian connection: under a general potential action one has

$$E_{classic} = p_{classic}\dot{q} - L_{classic} = (m_0\dot{q})\dot{q} - L_{classic}$$
$$= m_0 v^2 - \left[\frac{m_0 v^2}{2} - V(q,t)\right] = \frac{m_0 v^2}{2} + V(q,t)$$

Confirming the total energy expressions as the conventional sum between kinetic and potential energy components. Having the Hamiltonian-Lagrangian expression proven as reliable in classical side its relativistic extension naturally follows in the same way with relativistic Lagrangian as

$$E_{relativistic} = p_{relativistic}\dot{q} - L_{relativistic} = (m_d v)v - L_{relativistic}$$
$$= \frac{m_0 v^2}{\sqrt{1-\frac{v^2}{c^2}}} - \left(-m_0 c^2\sqrt{1-\frac{v^2}{c^2}}\right) = \frac{m_0 v^2 + m_0 c^2\left(1 - v^2/c^2\right)}{\sqrt{1-\frac{v^2}{c^2}}} = \frac{m_0 c^2}{\sqrt{1-\frac{v^2}{c^2}}} = m_d c^2$$

The result is consecrated as the famous Einstein relation for the total relativistic energy of a system relating its dynamical mass. Worth noting that to the same results one arrives when takes the force way first, namely employing the force-momentum relationship according with the 2nd Newton law yet with the relativistic momentum

$$F_{relativist} = \frac{dp_{relativist}}{dt} = \frac{d}{dt}\frac{m_0 v}{\left(1 - v^2/c^2\right)^{1/2}} = \frac{m_0 \dot{v}}{\left(1 - v^2/c^2\right)^{3/2}}$$

then involved in the mechanical work (energy) produced/consumed under the finite time $\Delta t = t_2 - t_1$ on the finite distance $\Delta q = q_2 - q_1$ with the form

$$\Delta E = \int_{x_1}^{x_2} F_{relativist}\, dx = \int_{t_1}^{t_2} \frac{m_0}{\left(1 - v^2/c^2\right)^{3/2}}\frac{dv}{dt} v dt = \int_{v_1}^{v_2} \frac{m_0 v}{\left(1 - v^2/c^2\right)^{3/2}} dv$$
$$= \left[\frac{m_0 c^2}{\sqrt{1 - v^2/c^2}}\right]_1^2 = \left[m_d c^2\right]_1^2 = \left(m_{d2} - m_{d1}\right)c^2 = c^2 \Delta m$$

where the elementary integral transformation was considered

$$\int \frac{XdX}{\left(1 - aX^2\right)^{3/2}} = \frac{1}{2a}\int \frac{d\left(aX^2\right)}{\left(1 - aX^2\right)^{3/2}} \overset{aX^2=Y}{=} \frac{1}{2a}\int \frac{dY}{\left(1-Y\right)^{3/2}} \overset{1-Y=Z}{=} -\frac{1}{2a}\int \frac{dZ}{Z^{3/2}}$$
$$= \frac{1}{aZ^{1/2}}$$

The result represents the Einstein relation for variation of the total energy as being proportionally with the dynamical mass variation, while having the square of the light velocity as the proportionality constant. Together one records the symmetry in momentum and energy relativistic formulations when they both are considered under the relativistic system

$$\begin{cases} p = \dfrac{m_0 v}{\sqrt{1 - v^2 / c^2}} \\ E = \dfrac{m_0 c^2}{\sqrt{1 - v^2 / c^2}} \end{cases}$$

Which solved for the particle/system's velocity generated the energy-momentum relativistic relationship, very useful in nuclear and theory of fields and quantum particles

$$E = \sqrt{m_0^2 c^4 + p^2 c^2}$$

When about electron, there is most interesting considering the equation formed between the rest energy of an electron with its produced electrostatic potential resulting in the so-called electronic "classical" radii

$$m_e c^2 = \frac{1}{4\pi\varepsilon_0} \frac{e^2}{r_0} \Rightarrow r_0 = \frac{1}{4\pi\varepsilon_0} \frac{e^2}{m_e c^2} = 2.8 \cdot 10^{-15} [m]$$

while suggesting the electronic dimension or of its spatial limit till which it can be confined; being the electronic classical radii in the real of nuclei radii appears as reasonably the Fermi theory of beta (electronic) emission of the nuclei, essentially in the proton-nucleon transformations as well as in the nuclear scatterings and nucleo-synthesis theory.

Remarkably, as Hans Bethe was remarking on the growing days of quantum-relativistic mechanics (pioneering by Einstein and Enrico Fermi) a tiny particle may release an enormous quantity of energy due to the multiplication of its small mass with the huge value obtained from squaring the light velocity. This appears even clearer when one consider the energy-momentum relationship and expands it under the non-relativistic limit ($pc << m_0 c^2 \Leftrightarrow v << c$)

$$E = \left(m_0^2 c^4 + p^2 c^2\right)^{1/2} = m_0 c^2 \left(1 + \frac{p^2 c^2}{m_0^2 c^4}\right)^{1/2} = m_0 c^2 \left(1 + \frac{p^2 c^2}{2 m_0^2 c^4} + \ldots\right)$$

$$\cong m_0c^2 + \frac{p^2}{2m_0} = m_0c^2 + \frac{m_0v^2}{2}$$

yet obtaining that the rest energy dominates by far the kinetic term, although not visible/observable until the disintegration/fusion or scattering of it takes place in nuclear or elementary particle realm. This becomes even more apparent with the Einstein example

1 gram ...releases cca. 9×10^{13} Joules ... 56.17×10^{31} eV (!!)

through the inter-conversion relation

$$1\ Joule = 6.24150974 \times 10^{18}\ electrons'volts\ (eV)$$

Giving an ideas of the disintegration power (think the electron in the Hydrogen atom has only 13.6 eV!) This way there is natural considering the relativistic expressions for mass, momentum and energy when dealing with nuclear reactions in general and when modeling the stellar combustion in special.

A.8 FUNDAMENTAL CONSTANTS AND CONVERSION FACTORS

A.8.1 ENERGY EQUIVALENTS

TABLE A.8.1 Energetic Conversion Factors Between Conventional Atomic Units (a.u. or Hartree Energy E_h), Electron-Volts (eV) and Chemical (Kcal/Mol) and Physical (Joule) Energies

	a.u. (E_h)	eV	Kcal/Mol	J
a.u. (E_h)	1	27.21138386(68)	627.71	4.35974394 (4.397)$\times 10^{-18}$
eV	3.67493254 (4.430)$\times 10^{-2}$	1	23.069	1.602176487 (40)$\times 10^{-19}$
Kcal/ Mol	1.5931$\times 10^{-3}$	4.3348$\times 10^{-2}$	1	6.947700141$\times 10^{-21}$
J	2.29371269 (4.386)$\times 10^{17}$	6.24150965 (4.391)$\times 10^{18}$	1.439325215$\times 10^{20}$	1

A.8.2 FREQUENCY EQUIVALENTS

TABLE A.8.2 Common Frequency Inter-Conversion Factors Within Conventional Energetic Framework; $E = mc^2 = hc/\nu = h\nu = k_B T$; $E_h = 2R_\infty hc = \alpha^2 m_0 c^2$; $1\ u = m_u = (1/12)\ m(^{12}C) = 10^{-3}\ kg\ mol^{-1}/N_A$

	Hz
J	$(1\ J)/h = 1.509\ 190\ 450(4.413) \times 10^{33}$ Hz
kg	$(1\ kg)c^2/h = 1.356\ 392\ 733(68) \times 10^{50}$ Hz
m^{-1}	$(1\ m^{-1})c = 299\ 792\ 458$ Hz
K	$(1\ K)k_B/h = 2.083\ 6644(36) \times 10^{10}$ Hz
1eV	$(1\ eV)/h = 2.417\ 989\ 454(4.374) \times 10^{14}$ Hz
a.u. (E_h)	$(1\ Eh)/h = 6.579\ 683\ 920\ 722(44) \times 10^{15}$ Hz
Kcal/Mol	$(1\ Kcal)/h = 6.318\ 678\ 576(4.413) \times 10^{36}$ Hz
u	$(1\ u)c^2/h = 2.252\ 342\ 7369(4.407) \times 10^{23}$ Hz

A.8.3 PHYSICAL-CHEMICAL CONSTANTS

TABLE A.8.3 Custom Physical Quantities with Their Common Symbols, Measured Values (Relative Errors in Parenthesis) and Units Within the International System Framework

Quantity	***Symbol***	***Value***	***Unit***
Planck constant	h	$6.626\ 068\ 96(4.408) \times 10^{-34}$	J s
		$4.135\ 667\ 33(4.385) \times 10^{-15}$	eV s
	$\hbar$	$1.054\ 571\ 628(53) \times 10^{-34}$	J s
		$6.582\ 118\ 99(4.391) \times 10^{-16}$	eV s
speed of light in vacuum	c, c_0	299 792 458	$m\ s^{-1}$
magnetic constant	μ_0	$4\pi \times 10^{-7} = 12.566\ 370\ 614... \times 10^{-7}$	$N\ A^{-2}$
electric constant	$\varepsilon_0 = \frac{1}{\mu_0 c^2}$	$8.854\ 187\ 817... \times 10^{-12}$	$F\ m^{-1}$
electron mass	m_0	$9.109\ 382\ 15(45) \times 10^{-31}$	kg
proton mass	m_p	$1.672\ 621\ 637(4.421) \times 10^{-27}$	kg

TABLE A.8.3 Continued

Quantity	*Symbol*	*Value*	*Unit*
alpha particle mass		$6.644\ 656\ 20(4.408) \times 10^{-27}$	kg
proton-electronmass ratio	m_p / m_0	1836.152 672 47(4.418)	
elementary charge	e	$1.602\ 176\ 487(40) \times 10^{-19}$	C
fine-structure constant	$\alpha = \frac{e^2}{4\pi\varepsilon_0 \hbar c} = \frac{e_0^2}{\hbar c}$	$7.297\ 352\ 5376(50) \times 10^{-3}$	
inverse fine-structure constant	$1/\alpha$	137.035 999 679(4.432)	
Rydberg constant	$R_\infty = \frac{\alpha^2 m_e c}{2h}$	10 973 731.568 527(4.411)	m^{-1}
Bohr radius	$a_0 = \frac{\alpha}{4\pi R_\infty}$	$0.529\ 177\ 208\ 59(36) \times 10^{-10}$	m
Compton wavelength	λ_C	$2.426\ 310\ 2175(4.408) \times 10^{-12}$	m
	$\bar{\lambda}_C = \frac{\lambda_C}{2\pi} = \alpha a_0 = \frac{\alpha^2}{4\pi R_\infty}$	$386.159\ 264\ 59(53) \times 10^{-15}$	m
Boltzmann constant	k_B	$1.380\ 6504(4.399) \times 10^{-23}$ $8.617\ 343(4.390) \times 10^{-5}$	$J\ K^{-1}$ $eV\ K^{-1}$

REFERENCES

AUTHOR'S MAIN REFERENCE

Putz, M. V. (2010). *Environmental Physics and the Universe* (in Romanian), The West University of Timişoara Publishing House, Timişoara.

SPECIFIC REFERENCES

Hassani, S. (1999). *Mathematical Physics: A Modern Introduction to Its Foundations*, Springer-Verlag, Berlin.

HyperPhysics (2010). http://hyperphysics.phy-astr.gsu.edu/Hbase/hframe.html

Nacken, M. (1935). Messungen der massenveränderlichkeit des elektrons an schnellen kathodenstrahlen, *Annalen der Physik*, 415(4), 313–329.

Romanian Voice (2012). http://www.romanianvoice.com/

Wikimedia (2010). Cathode ray tube: http://commons.wikimedia.org/wiki/File:Cathode_ray_tube_diagram-en.svg

FURTHER READINGS

Arfken, G. B., Weber, H. J. (1995). *Mathematical Methods for Physicists* (4th ed.), Academic Press, San Diego.

Arnold, V. I., Vogtmann, K., Weinstein, A. (1997). *Mathematical Methods of Classical Mechanics* (2nd ed.), Springer-Verlag, New York.

Boas, M. L. (2006). *Mathematical Methods in the Physical Sciences* (3rd ed.), John Wiley & Sons, Hoboken.

Courant, R., Hilbert, D. (1989). *Methods of Mathematical Physics*, Interscience Publishers, New York.

Glimm, J., Jaffe, A. (1987). *Quantum Physics: A Functional Integral Point of View* (2nd ed.), Springer-Verlag, New York.

Haag, R. (1996). *Local Quantum Physics: Fields, Particles, Algebras* (2nd ed.), Springer-Verlag, Berlin & New York.

Hawking, S. W., Ellis, G. F. R. (1973). *The Large Scale Structure of Space-Time*, Cambridge University Press, Cambridge.

Kato, T. (1995). *Perturbation theory for linear operators* (2nd repr. 1980 edition), Springer-Verlag, Berlin.

Kusse, B. R. (2006). *Mathematical Physics: Applied Mathematics for Scientists and Engineers'* (2nd ed.), Wiley-VCH, Weinheim.

Margenau, H., Murphy, G. M. (1976). *The Mathematics of Physics and Chemistry* (2nd ed.), R. E. Krieger Pub. Co., Huntington.

Mathews, J., Walker, R. L. (1970). *Mathematical Methods of Physics* (2nd ed.), W. A. Benjamin, New York.

Morse, P. McCord, Feshbach, H. (1999). *Methods of Theoretical Physics* (repr. of the original 1953 edition), McGraw Hill, Boston.

INDEX

F

N

O

P

Q

R

Y

Z